Electrical Installation and Inspection

Based on the 2002
NATIONAL ELECTRICAL CODE®

Electrical Installation and Inspection

CHARLES M. TROUT

DELMAR
THOMSON LEARNING

Australia   Canada   Mexico   Singapore   Spain   United Kingdom   United States

Electrical Installation and Inspection
by Charles M. Trout

Business Unit Director: Alar Elken	**Executive Marketing Manager:** Maura Theriault	**Project Editor:** Ruth Fisher
Executive Editor: Sandy Clark	**Channel Manager:** Fair Huntoon	**Art/Design Coordinator:** Rachel Baker
Acquisitions Editor: Mark Huth	**Marketing Coordinator:** Brian McGrath	**Cover Design:** Charles Cummings, Advertising
Editorial Assistant: Dawn Daugherty	**Executive Production Manager:** Mary Ellen Black	
Development Editor: Jennifer A. Thompson	**Production Coordinator:** Toni Hansen	

COPYRIGHT © 2002 by Delmar, a division of Thomson Learning, Inc.
Thomson Learning™ is a trademark used herein under license.

Printed in the United States of America
1 2 3 4 5 XX 05 04 03 02 01

For more information contact Delmar,
3 Columbia Circle, PO Box 15015,
Albany, NY 12212-5015.

Or find us on the World Wide Web at http://www.delmar.com

ALL RIGHTS RESERVED. No part of this work covered by the copyright hereon may be reproduced or used in any form or by any means—graphic, electronic, or mechanical, including photocopying, recording, taping, Web distribution or information storage and retrieval systems—without written permission of the publisher.

For permission to use material from this text or product, contact us by
Tel   (800) 730-2214
Fax (800) 730-2215
www.thomsonrights.com

Library of Congress Cataloging-in-Publication Data

Trout, Charles M.
    Electrical installation and inspection / Charles M. Trout.
      p. cm.
    Includes index.
    ISBN 0-7668-2058-0 (alk. paper)—ISBN 0-7668-2059-9 (alk. paper)
    1. Electric apparatus and appliances—Installation. 2. Electric wiring, Interior. 3. Electric wiring—Inspection. I. Title.
    TK452 .T76 2002
    621.319'24—dc21   2001042256

NOTICE TO THE READER

Publisher does not warrant or guarantee any of the products described herein or perform any independent analysis in connection with any of the product information contained herein. Publisher does not assume, and expressly disclaims, any obligation to obtain and include information other than that provided to it by the manufacturer.

The reader is expressly warned to consider and adopt all safety precautions that might be indicated by the activities herein and to avoid all potential hazards. By following the instructions contained herein, the reader willingly assumes all risks in connection with such instructions.

The Publisher makes no representation or warranties of any kind, including but not limited to, the warranties of fitness for particular purpose or merchantability, nor are any such representations implied with respect to the material set forth herein, and the publisher takes no responsibility with respect to such material. The publisher shall not be liable for any special, consequential, or exemplary damages resulting, in whole or part, from the readers' use of, or reliance upon, this material.

CONTENTS

Preface xiii
Acknowledgments xvii

CHAPTER 1 Basic Electricity 1
Key Terms 1
Introduction 2
Getting Started 2
Basic Electrical Circuits 3
Review Questions 18
Frequently Asked Questions 21

CHAPTER 2 Residential Branch-Circuit, Feeder, and Service Calculations 23
Key Terms 23
Introduction 24
Residential Electrical Branch-Circuits 25
Residential Feeder and Service Calculations 34
Review Questions 40
Frequently Asked Questions 43

CHAPTER 3 Residential Outlets 45
Key Terms 45
Introduction 46
Residential Receptacle Outlets Required 46
Residential Lighting Outlets Required 59
Review Questions 64
Frequently Asked Questions 67

CHAPTER 4 Grounding 69
Key Terms 69
Introduction 70
Fundamentals of Grounding and Bonding 70
Practical Grounding 76
Review Questions 86
Frequently Asked Questions 89

CHAPTER 5 Electrical Inspection—Rough-in 93
Key Terms 93
Introduction 93
Inspection Procedures—Rough-in 94

Inspection Procedures—Service and Grounding ... 109
Review Questions ... 112
Frequently Asked Questions ... 115

CHAPTER 6 Circular Raceway Systems ... 117
Key Terms ... 117
Introduction ... 118
Metal Raceways ... 118
Nonmetallic Raceways ... 126
Review Questions ... 129
Frequently Asked Questions ... 133

CHAPTER 7 Appliances ... 135
Key Terms ... 135
Introduction ... 135
General ... 136
Individual Appliances ... 136
Review Questions ... 143
Frequently Asked Questions ... 146

CHAPTER 8 Optional Load Calculations ... 149
Key Terms ... 149
Introduction ... 149
General ... 149
Dwelling Unit ... 150
Existing Dwelling Unit ... 152
Multifamily Dwelling ... 153
Two Dwelling Units ... 155
Schools ... 155
New Restaurants ... 157
Review Questions ... 159
Frequently Asked Questions ... 160

CHAPTER 9 Transformers ... 163
Key Terms ... 163
Introduction ... 164
General ... 164
Protection by Primary Overcurrent Device ... 170
Primary and Secondary Protection ... 172
Review Questions ... 179
Frequently Asked Questions ... 182

CHAPTER 10 Motors and Controllers ... 183
Key Terms ... 183
Introduction ... 183
Six Steps to a Motor Installation ... 184
Motor Controllers ... 189
Review Questions ... 190
Frequently Asked Questions ... 193

CHAPTER 11 Commercial Load Calculations ... 197
Key Terms ... 197
Introduction ... 197
General ... 198
Store Building ... 198
Office Building ... 201
Marinas and Boatyards ... 203
Recreational Vehicle Parks ... 205
Farms ... 206
Motels ... 209
Review Questions ... 211
Frequently Asked Questions ... 213

CHAPTER 12 Overcurrent Protection ... 215
Key Terms ... 215
Introduction ... 215
General ... 215
Purpose of Overcurrent Devices ... 216
Selection of Overcurrent Devices ... 218
Summary ... 226
Review Questions ... 226
Frequently Asked Questions ... 229

CHAPTER 13 Hazardous (Classified) Locations ... 233
Key Terms ... 233
Introduction ... 233
General ... 234
Class I Locations ... 234
Class II Locations ... 239
Class III Locations ... 239
Intrinsically Safe Systems ... 241
Zone 0, 1 and 2 Locations ... 242

Summary	242
Review Questions	246
Frequently Asked Questions:	247

CHAPTER 14 Swimming Pools — 249

Key Terms	249
Introduction	250
General	250
Permanently Installed Pools	252
Grounding and Bonding Swimming Pools	256
Review Questions	261
Frequently Asked Questions	265

CHAPTER 15 Emergency Systems — 267

Key Terms	267
Introduction	268
Emergency Systems	268
Legally Required Standby Systems	273
Interconnected Electric Power Production Sources	273
Summary	276
Review Questions	277
Frequently Asked Questions	279

CHAPTER 16 Carnivals, Circuses, Fairs, and Similar Events Temporary Installations — 281

Key Terms	281
Introduction	282
General	282
Temporary Installations	288
Summary	289
Review Questions	290
Frequently Asked Questions	292

CHAPTER 17 Health-Care Facilities — 293

Key Terms	293
Introduction	294
General	294
Essential Electrical Systems	299
Summary	302
Review Questions	302
Frequently Asked Questions	305

CHAPTER 18 Mobile Homes, Manufactured Homes, and Mobile Home Parks ... 309

Key Terms ... 309
Introduction ... 310
General ... 310
Summary ... 326
Review Questions ... 327
Frequently Asked Questions ... 330

CHAPTER 19 Recreational Vehicles, Recreational Vehicle Parks, and Park Trailers ... 333

Key Terms ... 333
Introduction ... 334
General ... 334
Summary ... 341
Review Questions ... 343
Frequently Asked Questions ... 345

CHAPTER 20 Floating Buildings, Marinas, and Boatyards ... 347

Part I—Floating Buildings ... 347

Key Terms ... 347
Introduction ... 347
General ... 347
Summary ... 350

Part II—Marinas and Boatyards ... 351

Key Terms ... 351
Introduction ... 351
General ... 351
Summary ... 355
Review Questions ... 355
Frequently Asked Questions ... 358

CHAPTER 21 Electric Signs ... 361

Key Terms ... 361
Introduction ... 361
General ... 362
Summary ... 368
Review Questions ... 369
Frequently Asked Questions ... 371

CHAPTER 22 Office Furnishings 375
Key Terms 375
Introduction 375
General 375
Summary 378
Review Questions 378
Frequently Asked Questions 381

CHAPTER 23 Cranes and Hoists 383
Key Terms 383
Introduction 383
General 383
Summary 389
Review Questions 392
Frequently Asked Questions 393

CHAPTER 24 Elevators, Dumbwaiters, Escalators, Moving Walks, Wheelchair Lifts, and Stairway Chair Lifts 397
Key Terms 397
Introduction 398
General 398
Summary 409
Review Questions 411
Frequently Asked Questions 414

CHAPTER 25 Electric Vehicle Charging System 415
Key Terms 415
Introduction 415
General 416
Summary 421
Review Questions 422
Frequently Asked Questions: 424

CHAPTER 26 Information Technology Rooms 425
Key Terms 425
Introduction 425
General 425
Summary 428
Review Questions 429
Frequently Asked Questions 430

CHAPTER 27 Fire Pumps ... 433
Key Terms ... 433
Introduction ... 433
General ... 433
Summary ... 440
Review Questions ... 441
Frequently Asked Questions ... 442

CHAPTER 28 Class 1, Class 2, Class 3 Circuits ... 445
Key Terms ... 445
Introduction ... 446
General ... 446
Summary ... 454
Review Questions ... 456
Frequently Asked Questions ... 457

Appendices ... 459
Appendix A ... 459
Appendix B ... 459

Code Index ... 461

Subject Index ... 467

PREFACE

INTENDED USE AND LEVEL

This book, *Electrical Installation and Inspection*, illustrates to the reader the basic requirements for installing electrical work in residential, commercial or industrial buildings or structures and tells how to improve on the basic requirements in order to attain an installation that will be adequate for the present needs of the user and to provide for future expansion.

As a prerequisite for easily understanding the information contained in the text, Chapter 1 concentrates on basic electricity, electrical formulas, and the assembly of component parts into electrical circuits. This chapter gives the student the necessary knowledge concerning electrical theory to use this book to its full advantage. Chapter 1 also serves as a review of basic electricity for the student who has previous experience in the electrical field and desires to upgrade his knowledge of electrical installations and inspections.

It is recognized there are many hazards involving shock or fire that are present when electrical installations are being made and a comprehensive knowledge of the requirements of the *National Electrical Code®* is necessary to minimize these hazards. The information in each chapter of this book is referenced to the appropriate section of the *National Electrical Code.*®

SUBJECT AND APPROACH

Many electrical installations are made based solely on the *National Electrical Code®* requirements, with far too much effort spent trying to "beat the Code" by doing less than what is considered the minimum requirements for a safe and efficient installation. The end user has a right to an installation that is free from hazard and which, if properly maintained, will last the life of the building or structure.

Electrical installations based solely on the *National Electrical Code®* are considered to be essentially free from hazard, but are not "*necessarily efficient, convenient, or adequate for good service or future expansion of electrical use.*" This book demonstrates to the reader that electrical work can be installed incorporating the requirements of the *National Electrical Code,*® without being limited to these requirements.

There is a practice that is prevalent in the electrical installation field today, called "value engineering," in which the electrical installer can reduce an adequate comprehensive design to one that satisfies the requirements of the *National Electrical Code,*® but generally results in a design that does not provide adequate or reasonable provisions for future equipment changes or future increases in the use of electricity.

There are many books on the market today that dwell on the minimum requirements as outlined in the *National Electrical Code®* and fail to alert the reader that these are "minimum" requirements and they may not meet the users' needs. This book demonstrates that while maintaining the provisions of the *NEC,*® good design will provide for the electrical service the end user is entitled to.

Inspection procedures will be described to acquaint the student with the manner in which electrical inspections are completed. Code enforcement personnel, the Authority Having Jurisdiction (AHJ), make their inspections based on the ordinances adopted by the municipality that employs them. For the most part, the ordinances are patterned after the requirements of the

National Electrical Code.® However, there are also many local requirements, in the form of ordinances, developed by municipalities to cover individual problems that particular municipality has concerns about. Before starting any electrical installation the installer must carefully review the ordinances governing electrical installations in that area.

FEATURES

- **Key Terms**—As in any technical book, there will be terminology that may be difficult for the student to understand. Immediately following the explanation of the objectives of each chapter there is a list of key terms used in that chapter with accompanying definitions.

- **Review Questions**—Following each chapter are review questions covering the material presented in that chapter. While the examination procedures are dictated by the individual instructor, these "open book" questions will help demonstrate the student's ability to use this book and the *National Electrical Code*® effectively. Each question relating to a Code requirement should have the applicable Code reference identified by the student.

- **Frequently Asked Questions**—Following each chapter is a "Frequently Asked Questions" section. A review of these questions will help the students in any problem areas they have encountered.

NEW FOR THE 2002 CODE

All code references in this book refer to the 2002 Edition of the *National Electrical Code.*® The major change in this edition is the renumbering of all Chapter 3 Articles with the exception of *Articles 300* and *310*. *Articles 380* and *384* have been moved into Chapter 4 since they reference equipment rather than wiring methods and materials. Other significant changes are the moving of all references to receptacles, cord connectors and attachment caps from *Article 410* to a new *Article 406* and a change in *Article 334* to permit the use of Type NM cable, with certain restrictions, in buildings over three floors in height.

Please note that this textbook was completed after all the normal steps in the NFPA 70 review cycle—Proposals to Code-Making Panels, review by Technical Correlating Committee, Report on Proposals, Comments to Code-Making Panels, review by Technical Correlating Committee, Report on Comments, NFPA Annual Meeting, and ANSI Standards Council—and before the actual publication of the 2002 edition of the *NEC.*® Every effort has been made to be technically correct, but there is always the possibility of typographical errors or appeals made to the NFPA Board of Directors after the normal review cycle that could change the appearance or substance of the Code. If changes do occur after the printing of this book, they will be included in the Instructor's Guide and will be incorporated into the textbook upon its next printing.

Please note also that the Code has a standard method to introduce changes between review cycles, called "Tentative Interim Amendment," or TIA. These TIAs and typographical errors can be downloaded from the NFPA website, www.nfpa.org, to make your copy of the Code current.

Metrics (SI) and the *NEC*®

▶The United States is the last major country in the world not using the metric system as the primary system. We have been very comfortable using English (United States Customary) values. But this is changing. Manufacturers are now showing both inch-pound and metric dimensions in their catalogs. Plans and specifications for governmental new construction and renovation projects started after January 1, 1994 have been using the metric system. You may not feel comfortable with metrics, but metrics are here to stay. You might just as well get familiar with the metric system.

The *NEC®* and other NFPA standards are becoming international standards. All measurements in the 2002 *NEC®* are shown with metrics first, followed by the inch-pound value in parentheses. For example, 600 mm (24 in.).

In *Electrical Installation and Inspection*, ease in understanding is of utmost importance. Therefore, inch-pound values are shown first, followed by metric values in parentheses. For example, 24 in. (600 mm).

A *soft metric conversion* describes a product already designed and manufactured to the inch-pound system, with its dimensions converted to metric dimensions. The product does not change in size.

A *hard metric measurement* describes a product designed to SI metric dimensions. No conversion from inch-pound measurement units is involved. A *hard conversion* describes an existing product redesigned into a new size.

In the 2002 edition of the *NEC®*, existing inch-pound dimensions did not change. Metric conversions were made, then rounded off. Where rounding off would create a safety hazard, the metric conversions are mathematically identical.

For example, if a dimension is required to be six ft, it is shown in the *NEC®* as 1.8 m (6 ft). Note that the 6 ft remains the same, and the metric value of 1.83 m has been rounded off to 1.8 m. This edition of *Electrical Installation and Inspection* reflects these rounded-off changes, except that the inch-pound measurement is shown first, i.e., 6 ft (1.8 m).

Trade Sizes

A unique situation exists. Strange as it may seem, what electricians have been referring to for years has not been correct!

Raceway sizes have always been an approximation. For example, there has never been a ½-in. raceway! Measurements taken from the *NEC®* for a few types of raceways are shown in the Table below.

TRADE SIZE	INSIDE DIAMETER (I.D.)
½ Electrical Metallic Tubing	0.622 in.
½ Electrical Nonmetallic Tubing	0.560 in.
½ Flexible Metal Conduit	0.635 in.
½ Rigid Metal Conduit	0.632 in.
½ Intermediate Metal Conduit	0.660 in.

You can readily see that the cross-sectional areas, critical when determining conductor fill, are different. It makes sense to refer to conduit, raceway, and tubing sizes as *trade sizes*. The *NEC®* in *90.9(C)(1)* states that "where the actual measured size of a product is not the same as the nominal size, trade size designators shall be used rather than dimensions. Trade practices shall be followed in all cases." This edition of *Electrical Installation and Inspection* uses the term *trade size* when referring to conduits, raceways, and tubing. For example, past editions may have referred to a ½-in. EMT. In this edition, it is referred to as trade size ½ EMT.

The *NEC®* also uses the term *metric designator*. A ½-in. EMT is shown as *metric designator 16 (½)*. A 1-in. EMT is shown as *metric designator 27 (1)*. The numbers 16 and 27 are the *metric designator* values. The (½) and (1) are the *trade sizes*. The metric designator is the raceway's inside diameter—in rounded off millimeters (mm). The following table shows some of the more common sizes of conduit, raceways, and tubing. A complete table is found in the *NEC,® Table 300.1(C)*. Because of possible confusion, this text uses only the term *trade size* when referring to conduit and raceway sizes.

METRIC DESIGNATOR AND TRADE SIZE	
Metric Designator	**Trade Size**
12	⅜
16	½
21	¾
27	1
35	1¼
41	1½
53	2
63	2½
78	3

Conduit knockouts in boxes do not measure up to what we call them. Here are some examples.

TRADE SIZE KNOCKOUT	ACTUAL MEASUREMENT
½	⅞ in.
¾	1³⁄₃₂ in.
1	1⅜ in.

Outlet boxes and device boxes use their nominal measurement as their *trade size*. For example, a 4 in. × 4 in. × 1½ in. does not have an internal cubic-inch volume of 4 in. × 4 in. × 1½ in. = 24 in.³ *Table 314.16(A)* shows this size box as having 21 cubic-in. volume. This table shows *trade sizes* in two columns—millimeters and inches.

In this text, a square outlet box is referred to as trade size 4 × 4 × 1½. Similarly, a single-gang device box would be referred to as a trade size 3 × 2 × 3 box.

Trade sizes for construction material will not change. A 2 × 4 is really a *name*, not an actual dimension. A 2 × 4 will still be referred to as a 2 × 4. This is its *trade size*.

In this text, measurements directly related to the *NEC®* are given in both inch-pound and metric units. In many instances, only the inch-pound units are shown. This is particularly true for the examples of raceway, box fill, and load calculations for square foot areas.

Because the *NEC®* rounded off most metric conversion values, a computation using metrics results in a different answer when compared to the same computation done using inch-pounds. For example, load calculations for a residence are based on 3 volt-amperes per square foot or 33 volt-amperes per square meter.

For a 40 ft × 50 ft dwelling: 3 VA × 40 ft × 50 ft = 6000 volt-amperes.

In metrics, using the rounded off values in the *NEC®*: 33 VA × 12 m × 15 m = 5940 volt-amperes.

The difference is small, but nevertheless, there is a difference.

To show calculations in both units throughout this text would be very difficult to understand and would take up too much space. Calculations in either metrics or inch-pounds are in compliance with the *NEC,® 90.9(D)*. In *90.9(C)(3)* we find that metric units are not required if the industry practice is to use inch-pound units. ◄

It is interesting to note that the examples in *Chapter 9* of the *NEC®* use inch-pound units, not metrics.

Please note that when comparing computations made by both the English and Metric systems, slight differences will occur due to the conversion method used. These differences are not significant, and computations for both systems are therefore valid.

SUPPLEMENTS

- **Instructor's Guide**—Includes all the solutions to questions contained in the book, as well as a set of transparency masters covering all 28 chapters.
- **E. resource**—A CD-ROM containing a Computerized Test Bank and PowerPoint presentations that outline important electrical concepts in each chapter.

ABOUT THE AUTHOR

Charles "Charlie" Trout began his electrical career in 1942 as an electrician helper for the Pennsylvania Railroad. In 1952, he became a member of Local Union 134 IBEW in Chicago, Illinois, and began work as a journeyman electrician for an electrical contractor.

In 1963, Charlie joined the International Association of Electrical Inspectors as an associate member and now has almost 40 years of continuous membership in the IAEI. He was President and CEO of Main Electric Company of Chicago for twenty years and holds an Electrical Contractors License and a Supervising Electricians License in the City of Chicago.

Charlie is well known for his classes and seminars for the Supervising Electricians Examination for the City of Chicago. His class workbook, *Basic Electrical Inspection Techniques—City of Chicago Electrical Code*, was widely used as a Chicago Code reference book. He also taught classes in "Code Enforcement—Electrical" at Harper College in Palatine, Illinois, for many years using his class workbook, *Basic Electrical Inspection Techniques—National Electrical Code,®* as the study guide.

In recognition of his widely acclaimed "Code Breakfast" for the Illinois Chapter of the IAEI, this regularly scheduled semi-annual event has been named the "Charlie Trout Code Breakfast" in his honor.

Charlie has been a member of the National Electrical Contractors Association (NECA) since 1980 and has been a member of its Codes and Standards Committee since 1989. He has represented NECA on *National Electrical Code®* Making Panels No. 6 and No. 18 and is currently Chairman of Code-Making Panel No. 12, a position he has held since 1993.

ACKNOWLEDGEMENTS

I would like to thank my wife, Carole, for her untiring help, patience, and encouragement while I labored timelessly to complete this book.

I would also like to thank my friend, Dewain Belote, from St. Petersburg, Florida, electrical code consultant and instructor for the Independent Electrical Contractors, for his valuable assistance in the preparation of this book.

In addition I offer my sincere thanks for all the valuable assistance and direction given to me by the Delmar staff.

The Author and Publisher also wish to thank the following reviewers for their contribution to the development of this book:

DeWain Belote Pinellas Technical Education Ctr St. Petersburg, FL	David Gehlauf Tri-County Vocational School Glouster, OH	Fred Johnson Champlain Valley Tech Plattsburg, NY
Orville Lake Augusta Technical Institute Evans, GA	Ronald Murray New England Technical Institute New Britain, CT	Paul Owens San Juan College Aztec, NM
Michael Pederson NW Iowa CC Sheldon, IA	Vohn Peeler Rowan Cabarrus CC Salisbury, NC	Rodney Stanley Morehead State University Morehead, KY
Vern Willson Tucson Electric Power Co. Tucson, AZ	Timmy Young Carl D. Perkins Job Corps Ctr Prestonburg, KY	

Applicable tables and section references are reprinted with permission from NFPA 70-2002, *National Electrical Code,*® copyright 2002. National Fire Protection Association, Quincy, MA 02269. This reprinted material is not the complete and official position of the NFPA on the referenced subject, which is represented only by the standard in its entirety.

CHAPTER 1

Basic Electricity

OBJECTIVES

After completing this chapter, the student should understand:
- basic electrical formulas.
- basic components of an electrical circuit.
- the difference between series and parallel circuits.
- complete circuit requirements.
- neutral conductor.
- volt, ampere, watt, ohm, and resistor.
- magnetic field.

KEY TERMS

Ammeter: An electrical instrument used to measure the flow of electric current

Ampere: Unit of measurement of electrical current

Bus Bar: A conductor in the form of a bar or rod

Circuit Breaker: A device designed to open and close a circuit by non-automatic means and to open a circuit automatically at a predetermined overcurrent

Electromotive Force: Potential difference between the terminals of a source of electrical energy expressed in volts

Fuse: A device containing a conductor or link that melts when excessive current flows through an electric circuit, the melting causes it to open, thereby protecting the circuit

Ground-fault: Unintentional contact between an energized conductor and a grounded surface

Inductive Reactance: Opposition to the flow of electrical current produced by the collapsing of the magnetic field surrounding a conductor

Intensity of Current: Magnitude of current flow expressed in amperes

Magnetic Field: A field that is set up around a conductor whenever current passes through the conductor

Negative: Electric charge that has an excess of electrons

Ohm: Unit of measurement of electrical resistance

Overcurrent Protective Device: A device designed to open a circuit automatically at a predetermined overcurrent

Power: The rate at which work is done

Power Factor: The ratio between true power and apparent power

Positive: Electric charge with a deficiency of electrons

Resistance: Opposition to the flow of current in an electrical circuit

Short-Circuit: Unintentional contact between two energized conductors of opposite polarity, short of the load

Transformer Action: The transfer of energy in an ac circuit from one winding to another by means of electromagnetic induction

Volt: Unit of measurement of electromotive force

Voltmeter: An electrical instrument used to measure voltage

Watt: Unit of measurement of electrical power

INTRODUCTION

This chapter will cover basic electrical circuits and components and how to assemble them into practical combinations for use in any electrical installation we may undertake. Each student must make a careful study of these basic circuits as a prerequisite for a complete understanding of the contents of this book.

GETTING STARTED

What is electricity? It is the term used for the electron theory. The electron theory states that all matter is made of electricity. All matter is made up from molecules and all molecules are made up of atoms. Atoms are composed of electrons that are particles of negatively charged electricity, held together by a positively charged nucleus. Electricity occurs when some of the electrons are moved. The electrons can only be moved by some force or pressure that pushes the electrons from atom to atom. This action can only occur in a *closed loop*. The movement of the electrons is called an electric current. Figure 1-1 shows a display of nature's electrical power.

Figure 1-1 A display of nature's electricity, *Courtesy of* Gary Nelson

Now, all we need to get started is this copy of *Electrical Installation and Inspection*, a *NEC®* book, and a calculator. From our Key Terms we should understand the following terminology and be ready to start our electrical calculations. First, instead of *V* for **volts**, and *A* for **amperes**, the electrical industry has for use in formulas adopted the use of *E* for volts and *I* for amperes. *E* stands for **electromotive force** measured in volts. The use of *I* for current comes from **intensity of current** measured in amperes. The **power** (P) of a circuit is expressed in **watts**. Watts and power are used interchangeably. Therefore, electrical power is equal to voltage times amperes, this can be expressed as W = E × I, or P = E × I.

Electrical formulas use the algebraic practice of placing two letters side by side when their equivalents are to be multiplied. An example of this is *VA*, which in electrical work is volt-amperes. Volt-amperes are voltage multiplied by amperes. We just learned (in the previous paragraph) that voltage × amperes is equal to watts. When do we use E × I and when do we use V × A? E × I is used for work in formulas, and V × A is used in alternating current load calculations. When solving these formulas, there are generally two known factors, and we use our formula to find an unknown quantity. In our study of direct current circuits, we will use E and I, and when we study alternating current, we will further explain the use of V and A.

When we turn on an electric appliance or device, we obtain some desired effect. A lightbulb lights, a flashlight shines a beam, or an electric heater begins to radiate heat (Figure 1-2).

What happens when we turn the switch on? By closing the switch, we establish what is known as a completed electrical circuit. The completed electrical circuit forms the closed loop previously mentioned. To establish an electrical circuit, we must have a voltage source to push or force the electricity through the circuit, along or by means of a wire or conductor. The electricity we move through the conductor is called current and is measured in amperes. When the voltage pushes the current through the conductor and through the filament of a lightbulb, the filament heats up to incandescence and gives off light. A filament is made of a special material, usually tungsten, which is of very high **resistance** to the flow of current and will heat up when current is forced through it. Remember, a voltage source has

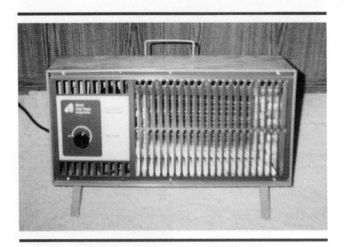

Figure 1-2 Electric heater used as supplemental heat in a storage room

two ends or terminals. A circuit is not complete until there is a closed loop from one end of the source and back through the source. The relationship between voltage, current, and resistance will be explained in the following material.

BASIC ELECTRICAL CIRCUITS

Basic electrical circuits include direct current series circuits, direct current parallel circuits, and alternating current circuits

Direct Current Series Circuit

An electric dc series circuit is a completed electrical conducting path consisting of all, or some, of the components shown in Figure 1-4. It includes the path from the source, through the load, and returning through the source. Now let us look at what we need to make an electrical circuit. Electrical circuits have five basic components:

1. Source of power, which is usually a battery, generator, or transformer. The power you are getting to your building comes from a utility company transformer. The flashlight you use has a battery or a group of batteries for its power. Emergency power is obtained from generators, which operate when your source of power from the utility company is interrupted (Figure 1-3). Generators and transformers will be covered in more detail when we study alternating current circuits. For now, we will use a battery as our power source.

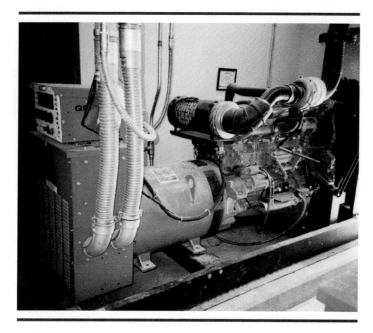

Figure 1-3 Emergency generator installation, *Courtesy of Caddo Parish 9.1-1*

2. Conductor or path to carry the current flow. This is generally a wire or cable, but in some installations, it is a copper or aluminum **bus bar**.

3. Load, which is a luminaire (lighting fixture), a motor, or an appliance such as a toaster or an electric range. The load is what the power feeds. Another term for the load is *utilization equipment*.

4. Control, which is necessary to turn the load on and off. This may be a switch, or perhaps a thermostat to turn off a furnace.

5. **Overcurrent protective device.** This is the **fuse** or **circuit breaker**, which protects the circuit from **short-circuits** or **ground-faults**. Do not worry now about shorts or grounds, we will cover them in detail later.

Figure 1-5 illustrates a circuit with all of these components. A battery is used as a source of power. The symbol for the battery is drawn with alternate short and long lines to depict the plates inside the battery. The short lines indicate **negative** plates, and the longer lines indicate **positive** plates. It takes two opposite plates to make one battery cell. Each cell of a lead-acid battery is calculated at 2 volts. A lantern battery has three cells and is rated at 6 volts. A battery is a direct current (dc) source that does not rise and fall as we will learn ac current does. It has an initial buildup to its peak voltage, and it remains constant until the circuit opens and the voltage falls to the "0" axis line. An initial **magnetic field** does build up as the voltage rises to its peak and remains constant until the circuit is opened.

Closing the switch (control) (see Figure 1-5) will allow current to flow on the path (conductor) from the positive side of the battery (source), through the lamp (load), and return to the negative side of the battery. Opening the switch (control) interrupts the path (conductor), or circuit, and turns off the lamp.

In Figure 1-5, we are using a lightbulb for our load. In the electrical trade, a lightbulb is called a lamp and the socket you place the lamp into is a lampholder. So the floor or table lamps you have at home are really lampholders, with lamps for light, and lampshades to direct the light (Figure 1-6).

In Figure 1-5, we used a battery, which is a dc source. Remember, dc is a current that always flows in the same direction. As previously stated, a magnetic field does build up as the voltage rises to its peak.

All of the dc series circuits have one important feature: All of the current that leaves the positive

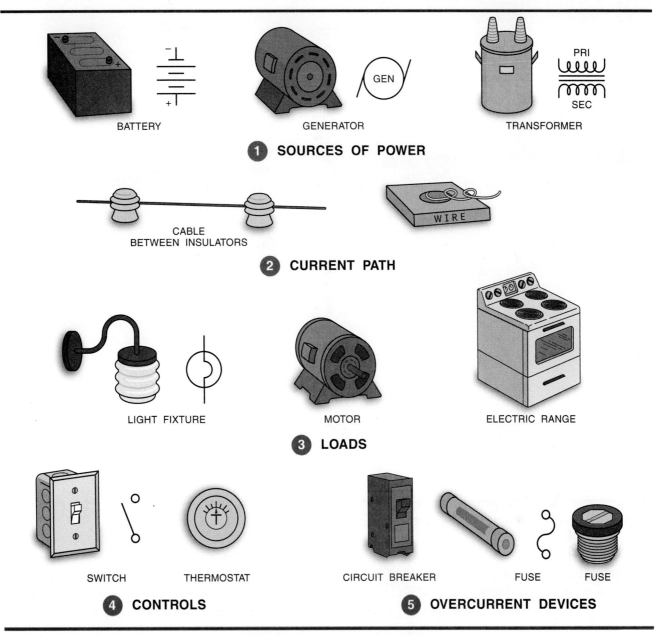

Figure 1-4 Circuit components

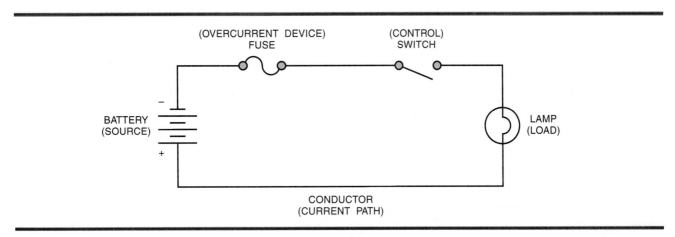

Figure 1-5 Circuit with components

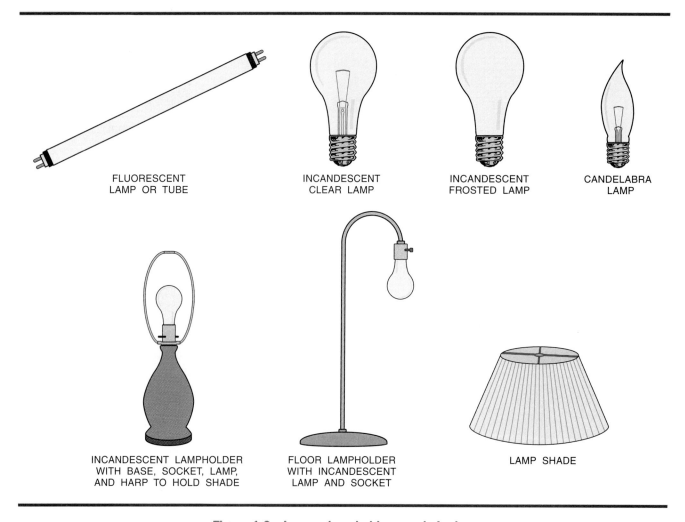

Figure 1-6 Lamps, lampholders, and shades

terminal of the battery or other dc source of power must flow through every part of the circuit to reach the negative terminal. The current has no alternate paths available to it to return to the source. The current is the same in all parts of a dc series circuit. You can have more than one load in series, but the same amount of current must flow through each load. If any of the loads are opened, then the entire circuit is open. The string of lights (see Figure 1-7) is a good illustration of this. If the lamps (bulbs) are in a series, it means the filaments in the lamps are in a series, and if one lamp burns out, the whole string of lights goes out. Remember when Christmas tree lights would all go out when one lamp was burned out? We would then have the job of finding the burned-out bulb. We will learn in our study of parallel circuits how this problem was solved. The Christmas tree lights in our homes are used in an alternating circuit, but the principle is the same.

We often use a resistor to represent the load in a circuit (Figure 1-8). To calculate the voltage, current, and resistance relationship in a dc series circuit, we use what is called Ohm's Law. This law is named after George Ohm, a German scientist, who discovered the relationship between voltage, current, and resistance. Ohm's Law states: "The current in a dc series circuit is directly proportional to the voltage and inversely proportional to the resistance." The term **ohm** is used as a unit of measurement of electrical resistance. Written as an equation, Ohm's Law is as follows:

$$E = I \times R \quad \text{or} \quad I = \frac{E}{R} \quad \text{or} \quad R = \frac{E}{I}$$

Directly proportional to the voltage means that when the voltage is doubled, the current is doubled. When the voltage is reduced by one-half, the current is reduced by one-half. Inversely proportional to the

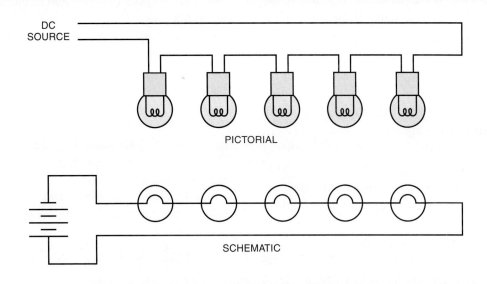

Figure 1-7 String of lights with filaments in a series

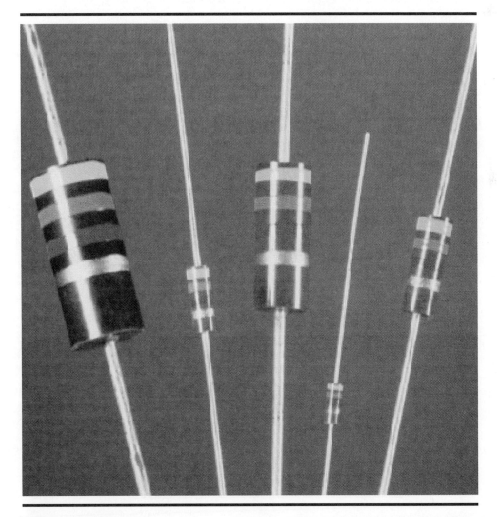

Figure 1-8 Resistors. *Courtesy of* Allen Bradley Co., Inc., a Rockwell International Company

resistance means that if the resistance is doubled, the current is reduced by one-half. If the resistance is reduced by one-half, the current is doubled. The diagrams (see Figure 1-9) demonstrate the use of Ohm's Law equations. In Figure 1-9, the symbol for an **ammeter**, a circle with a capital A, is shown with a question mark signifying that the current is an unknown quantity in the equation.

Although the three equations representing Ohm's Law are quite simple, they are the most important of all electrical equations and must be thoroughly understood for a complete understanding of electrical theory.

As shown in Figure 1-10, more than one load can be connected in a dc series circuit. To solve the Ohm's Law equations, it is necessary to total the loads in the series circuit into one equivalent load. This is simple when resistances are in a series since it is only necessary to add the resistances together to gain one total equivalent. Figure 1-10 shows resistances in series and demonstrates the use of Ohm's Law.

Figure 1-10 consists of three resistors in series having values of 5 ohms, 10 ohms, and 15 ohms. To find the amount of current flow, we must first find the total resistance.

$$R_t = R_1 + R_2 + R_3$$
$$R_t = 5 + 10 + 15$$
$$R_t = 30 \text{ ohms}$$

Using the total resistance and the known applied voltage, we can calculate the current.

$$I = \frac{E}{R} \quad \text{or} \quad I = \frac{120}{30} \quad \text{or} \quad I = 4 \text{ amperes}$$

In any series circuit, the sum of the voltage drops across the loads (resistors) is equal to the applied voltage. This can be proven in the circuit mentioned previously. We have already determined

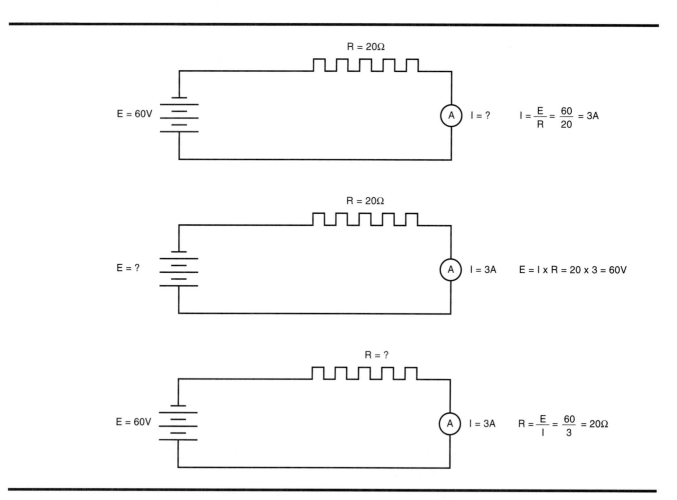

Figure 1-9 Solving Ohm's Law equations

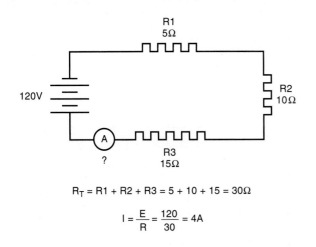

Figure 1-10 Ohm's Law with resistance in series

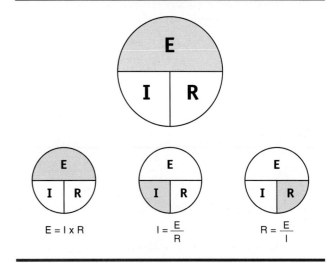

Figure 1-11 Ohm's Law circle (pie)

that the current in that circuit is 4 amperes, and since we know that the current is the same in all parts of a series, by using Ohm's Law, we can multiply the value of each resistor by 4 amperes.

$$R1 = 5 \times 4 = 20$$
$$R2 = 10 \times 4 = 40$$
$$R3 = 15 \times 4 = 60$$
$$120 \text{ volts}$$

In summary, the rules for dc series circuits are as follows:

1. The current is the same in all parts of a series circuit.
2. The sum of the voltage drops across the individual resistance is equal to the source voltage.
3. The voltage drop across each resistor is proportional to the size of the resistor.
4. The total resistance in a series circuit is the sum of all the resistances in the circuit.

A handy way to remember the Ohm's Law equations is by using the Ohm's Law Circle or the Pie as it is sometimes called (see Figure 1-11). By covering the letter of the value you are looking for, the remaining letters indicate which equation to use. For example, by covering E we have IR left, which of course is $I \times R$. If we cover I we have E over R left, and if we cover R, we have E over I left.

Another aid to finding the correct electrical equation to use is the summary of basic formulas circle. This circle (see Figure 1-12) consists of the

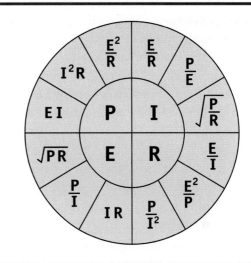

Figure 1-12 Summary of basic formulas circle

four basic quantities: P for Power, I for Current, E for Voltage, and R for Resistance. P for power is many times replaced with W for watts. Beside each quantity are three spaces, each with two of the other basic quantities.

Direct Current Parallel Circuits

Many electrical circuits have more than one path to follow. Figure 1-13 is an example of this type of circuit. The same principles applied to series circuits can be used to find quantities in parallel circuits. A parallel circuit can be defined as one having more than one current path to follow from a single voltage source.

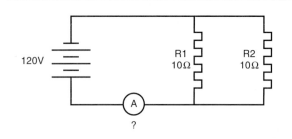

Figure 1-13 Parallel circuit with two loads

Remember the string of lights in the dc series circuit? In Figure 1-14, those same lights are shown in a parallel configuration with one lamp burned out. The rest of the lamps stay lit. An open part of the circuit will not necessarily stop current from flowing in other parts of the circuit.

We learned that in a series circuit the voltage divides proportionately across the loads in the circuit. In a parallel circuit, the applied voltage is present across all the loads in the circuit.

In a series circuit, the current is the same in all parts of the circuit. In a parallel circuit, the current at the source is equal to the sum of all the currents in each part of the circuit. The current in a parallel circuit divides in relation to the value of the resistors (loads) in all paths of the circuit. For a given voltage, the current varies inversely with the resistance. Ohm's law still applies. Figure 1-15 shows voltage and current comparisons in parallel circuits.

Use Ohm's Law as follows: with a source voltage of 120 volts, a **voltmeter** connected across R1 would read 120 V, and a voltmeter connected across R2 would read 120 V. In the current comparison

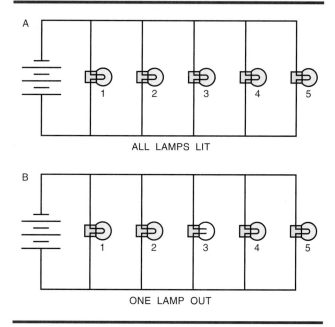

Figure 1-14 String of lamps in parallel with lamp B3 filament open

using the same values, the current flow in resistor R1 would be 3 amperes. The current flow in resistor R2 would be 2 amperes. This is shown by:

$$I = \frac{E}{R} \quad \text{or} \quad I = \frac{120}{40} \quad \text{or} \quad I = 3$$

$$I = \frac{E}{R} \quad \text{or} \quad I = \frac{120}{60} \quad \text{or} \quad I = 2$$

The R1 current plus the R2 current would total 5 amperes at the source.

Another way to prove this is by finding the equivalent values of the resistors and using Ohm's Law. In a series circuit, we totaled the values of

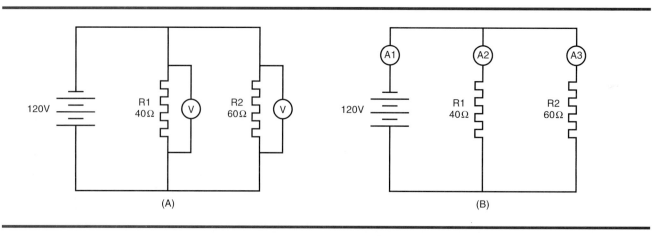

Figure 1-15 (A) voltage, (B) current—voltage and current comparisons

the resistors by adding them. In a parallel circuit, finding the equivalent resistance of resistors in parallel is done by taking the *reciprocal of the sum of the reciprocals*. Sounds difficult, but follow the text carefully:

$$R_{eq} = \frac{1}{\frac{1}{R1}+\frac{1}{R2}} \text{ or } \frac{1}{\frac{1}{40}+\frac{1}{60}} \text{ or } \frac{1}{\frac{3}{120}+\frac{2}{120}}$$

$$\text{or } \frac{1}{\frac{5}{120}} \text{ or } \frac{120}{5} = 24 \text{ ohms}$$

(LCD)   (Invert)

$$I = \frac{E}{R} \text{ or } I = \frac{120}{24} \text{ or } I = 5 \text{ amperes}$$

The third step required finding the lowest common denominator (LCD) for the values of 40 and 60, which is 120. In the fourth step, we inverted and dropped the number 1, which is not necessary to hold since any number over 1 results in that whole number. Therefore, in a parallel circuit, the current at the source would be the total of the currents across all the resistances of the loads in the circuit.

A quick method of finding the equivalent resistance of resistances in parallel is using a calculator.

Let us use the same values we previously used.

Step #1: Clear your calculator and make sure it is not in a memory mode.

Step #2: Press 1; press ÷; press 40; press M+; your calculator reads 0.025.

Step #3: Press 1; press ÷; press 60; press M+; your calculator reads 0.01666.

Step #4: Press 1; press ÷; press MR; press =; your calculator reads 24.

A few quick tips on finding equivalent resistances for resistors in parallel:

1. The equivalent resistance of two parallel resistors of equal value is one-half the value of one resistor.
2. The equivalent resistance of resistors in parallel is always less than the resistance of the smallest resistor.
3. The equivalent resistance of any number of equal value resistors can be found by dividing the resistance of one resistor by the number of resistors.

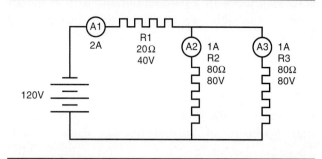

Figure 1-16 Series/parallel circuit

There are electrical circuits that have some parts wired in series and some parts wired in parallel (Figure 1-16). These circuits may look very complicated, but there is a simple solution to solving problems concerning these circuits. Just reduce all the parallel resistor values to an equivalent value and add it to the resistors in series.

In Figure 1-16, we have two 80 ohm resistors in parallel that are in series with a 20 ohm resistor. We can find the equivalent resistance of the two 80 ohm resistors that are in parallel by using a calculator, using the reciprocal of the sum method, or we can use the helpful hints that tell us that the equivalent resistance of two resistors of equal value is one-half the value of one resistor. Therefore, the equivalent resistance of the two 80 ohm resistors is 40 ohms.

For practice and to prove our theory, take out your calculators: one divided by 80, press M+, your calculator should read .0125; one divided by 80, press M+, your calculator should read .0125; one divided by MR, press equal, your calculator should read 40. We add the 40 ohms to the 20 ohms to arrive at a total of 60 ohms in the series/parallel circuit.

The rest is even easier. Using Ohm's Law, we divide the source voltage of 120 volts by the total resistance of 60 ohms and find a total current flow of 2 amperes in our circuit. But how is this current divided up through all three resistors? Let us take them individually. Again using Ohm's Law, we must first find the voltage across each resistor. We learned earlier in this chapter that the same amount of current flow must return to the source as leaves the source. We also learned that when more than one load is in a series in a dc circuit, the voltage divides according to the resistance. The voltage used by each load in a series circuit is called the IR drop, or the voltage is equal to I × R. The voltage is the same across all loads in a parallel circuit. To find the IR drop across R1, we multiply the current of 2 amperes times the resistance of 20 ohms with a result of 40 volts. The remaining 80 volts (120 − 40 = 80) is now impressed across both of the resistors in parallel. Dividing the 80 volts across each parallel resistor by the 80 ohm value of the resistor, we find 1-ampere of current through each resistor. These two 1-ampere values combine to return to the source as 2 amperes, the same current that we show leaving the source.

Alternating Current Circuit

An alternating current is one that is reversed at regular intervals. An alternating current rises above a *0* axis line until it reaches its peak value and then descends back down to the axis line completing one-half cycle. It then descends below the axis line until it reaches its peak and then rises back to the axis line completing a full cycle. The frequency of an alternating current is dependent upon how many times this full cycle is completed in one second. Each full cycle is 360° (Figure 1-17).

The voltage of the alternating circuit rises and falls with the current in an exact line if the power factor is *0*.

What is **power factor**? Power factor is the ratio between true power and apparent power. Apparent power is obtained when power factor is disregarded. Power factor is almost always less than unity, or less than 1, therefore, true power is almost always a percentage of apparent power.

This can be shown in the formula:

$$\text{Power factor} = \frac{\text{True Power (Watts)}}{\text{Apparent Power (VA)}}$$

For example, if the voltage in a circuit was measured with a voltmeter at 120 volts and the current was measured with an ammeter at 10 amperes, then using our formula we would have 1200 VA of apparent power. But if the wattmeter on this circuit reads 1000 watts, the power factor would be:

$$\text{Power Factor} = \frac{1000 \text{ (Watts)}}{1200 \text{ (VA)}} = .83$$

One reason for power factor loss in ac circuits is that the magnetic field that forms around the conductor is constantly building up and collapsing with the frequency of the alternating current (Figure 1-18). When the magnetic fields cut across the conductor, they produce a current in the conductor that is opposite to the applied current and tend to make the current lag the voltage. This opposing current is called **inductive reactance**. The symbol for inductance is *L*, and for inductive reactance is X_L, which is called X sub L.

Another reason for power factor loss in ac circuits is the effect of capacitance. This will produce an opposition to a change in voltage and tends to make the current lead the voltage. This opposition to a change in voltage is called capacitive reactance. The symbol for capacitance is *C*, and for capacitive reactance is X_c, which is called X sub C.

A convenient way to remember these leading or lagging currents is the phrase *ELI the ICE man*. The

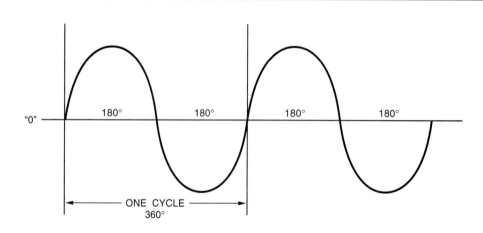

Figure 1-17 Graph showing alternating current sine wave through two cycles

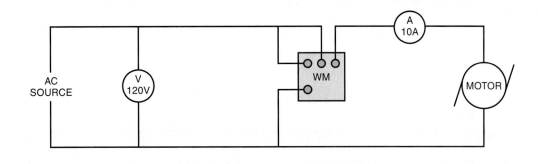

Figure 1-18 Power factor loss in an alternating circuit

word ELI has E for voltage, L for inductance, and I for current. The current (I) is after, or lagging, the voltage (E) in inductive (L) circuits. The word ICE has I for current, C for capacitance, and E for voltage. The current (I) is before, or leading, the voltage (E) in capacitive (C) circuits.

From the previous text, it is seen that if you have an inductive circuit (voltage leading current), by introducing a capacitance in the circuit (current leading voltage), you can accomplish power factor correction by having these opposing forces cancel each other out. See Appendix A for power factor.

If the power factor is less than unity, the current either leads or lags the voltage depending upon whether the circuit is capacitive or inductive. Anything *less than 0* power factor results in power being wasted by currents opposing each other. The watthour meter records only true power and the higher value of apparent power, although used, is not recorded. For this reason, utility companies require power factor correction to a set value so that they are compensated for the real power they supply.

Figure 1-19, (A) shows voltage and current sine waves in phase with each other. Both voltage and current begin and finish the cycle together. (B) shows current lagging the voltage which indicates an (ELI) inductive circuit.

In Figure 1-20, a transformer is used as the power source. The symbol for a transformer is two coils separated by two lines signifying an iron core. Generally, the primary coil is omitted and just the secondary coil is shown on drawings where the secondary circuit is all that we are concerned with. When an alternating current is supplied to the primary coil of a transformer, the rising and falling magnetic field cuts across the secondary winding as well as the primary winding. This occurs because both windings are wound on top of one another on the iron core. The difference in the ratio of turns on each winding establishes the voltage ratio.

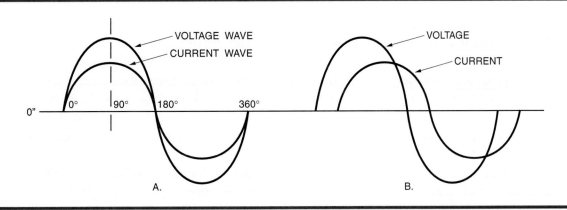

Figure 1-19 Sine wave showing: (A) Voltage and current in phase (B) Current lagging voltage

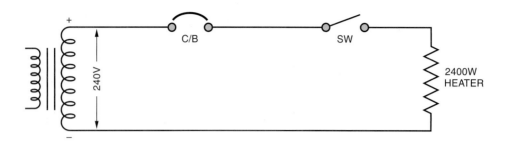

Figure 1-20 Circuit with transformer source, circuit breaker overcurrent protection, switch control, and heater load

A transformer with 200 turns in its primary coil and 100 turns in its secondary coil has a two-to-one ratio. If the primary voltage is 480 volts, the secondary voltage is 240 volts. This is called a stepdown transformer. The voltage and turns ratio are directly proportional. This inductive action is sometimes referred to as **transformer action**. The electrical power produced on the secondary side of the transformer is electrically isolated from the primary electrical power. The current is inversely proportional to the voltage. When the voltage is increased, the current is decreased. When the voltage is decreased, the current is increased. See Appendix B for transformer basics.

In Figure 1-20, we have marked the transformer with a negative end and a positive end of the secondary winding. When doing circuit analysis using alternating current, it is necessary to assume a particular instant of time to designate a negative and a positive end of the secondary winding. The negative and positive ends of the transformer secondary windings alternate with the frequency of the alternating current, requiring us to pick a particular moment of the alternations.

Assuming a secondary voltage of 240 volts and a load of 2400 watts, when the switch is closed, current flows from the positive end of the transformer secondary through the closed switch and through the 2400-watt heater load and back through the transformer. Current also flows through the circuit breaker overcurrent device. The overcurrent device is installed to protect the conductors of the circuit. It is not sized to protect the heater (load). Equipment protection is handled differently and will be discussed in detail in later chapters.

This is an ac series circuit and values or quantities are found in much the same manner as we did in dc series circuits. This circuit is a resistive circuit because the only resistance in the circuit is the resistive heating coils. There is some resistance in the conductor, but for our study purposes we will not consider it.

As we learned earlier in "Getting Started" in this Chapter, we can use a power formula to determine quantities in ac circuits. This formula is much like the Ohm's Law equation except that our known quantity for the load is rated in watts rather than in ohms. The equation we will use here is $W = E \times I$.

Where: W – is the load in watts
                    (P is sometimes used instead of W)
              E – is the source voltage
              I – is the current flow in amperes

This formula can be transposed into the following variations:

$$E = \frac{W}{R} \quad \text{or} \quad I = \frac{W}{E}$$

Using this formula to determine the current flow in Figure 1-20:

$$I = \frac{W}{E} \quad \text{or} \quad I = \frac{2400}{240} = 10 \text{ amperes}$$

The circuit shown in Figure 1-20 is a simple 2-wire circuit. The following circuit shown in Figure 1-21 is a 3-wire circuit used extensively in electrical wiring for construction. This circuit is known as the Edison 3-wire System. In the early days of electrical distribution it was common practice for the utility company (Edison) to provide a 2-wire service entrance to buildings. If large loads were present, a third wire was added to provide 240-volt power to the building. These 3-wire systems were generally referred to as *Edison 3-wire Systems*.

The term *Edison 3-wire System* dates back to Thomas Edison's introduction of his concept that such a supply system could be produced by connecting two 120-volt power systems in series in such a way that their polarities would cause the voltages to be additive.

The most common way to accomplish this is by use of a center-tapped secondary winding of a transformer. The secondary winding is actually two 120-volt windings in series. See Figure 1-21.

The load should be distributed as equally as possible across the two windings. A perfectly balanced system would result in no current flow in the neutral conductor.

The secondary of the transformer is center tapped to provide a return path that is common to each of the phase conductors. The return path is called the common or neutral conductor. Closing switch #1 allows current to flow from the positive end of the transformer secondary winding, through the 2400-watt heater, and return through the common conductor to the center tap on the transformer.

Using our formula:

$$I = \frac{W}{E} \quad \text{or} \quad I = \frac{2400}{120} = 20 \text{ amperes}$$

Closing switch #1 and #2 allows current to flow from the positive end of the transformer secondary winding, through both 2400-watt heaters, and return to the negative end of the secondary winding.

Using our formula:

$$I = \frac{W}{E} \quad \text{or} \quad I = \frac{4800}{240} = 20 \text{ amperes}$$

No current flows in the common or neutral conductor when the same size loads are connected across the two phase conductors. The common or neutral conductor carries only the unbalanced portion of the load. Since there is no unbalance (the loads are the same for each phase conductor), the common conductor is in a neutral state. Remember, the current flows in the same direction in both parts of the transformer secondary windings. Current cannot flow in both directions, simultaneously, in a conductor. The currents will cancel each other out in direct proportion.

If the current flowing through the two loads would attempt to use the common conductor formed by our center tapping the transformer, there could be no current flow in the common or neutral conductor. The equal currents would be bucking each other and canceling each other out. Our circuit is from the positive end of the transformer winding, through closed switch 1, and through both of the 2400-watt heaters, through closed switch 2, and back through the secondary winding. Figure 1-22 shows the canceling effect of opposite current flow.

In Figure 1-23, we will use the same circuit, but we will introduce the principle of grounding. Grounding will be covered in much more detail in another chapter. Grounding is a safety practice and only the aspect of intentional grounding will be covered here.

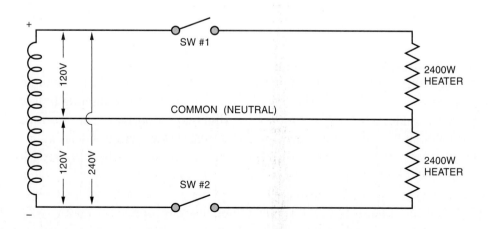

Figure 1-21 Edison 3-wire System

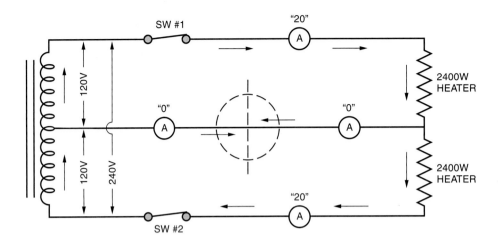

Figure 1-22 Three-wire circuit showing canceling effect of opposite current flow

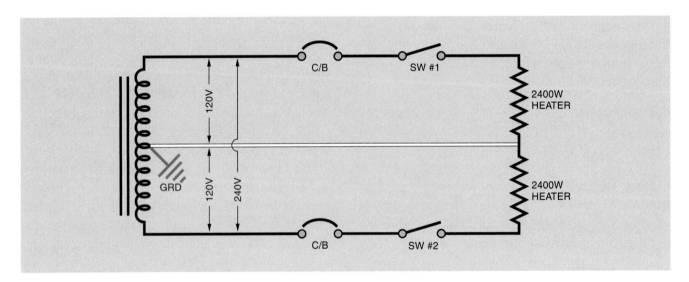

Figure 1-23 Three-wire circuit showing intentional ground

In Figure 1-23, we have now intentionally grounded the center tap of the transformer. The neutral or common conductor, if present, is always the conductor that is grounded. Grounded means connected to earth by some conducting body that serves in the place of earth. Grounding is generally accomplished by means of a conductor connected between the electrical system and a water pipe system, or a rod driven into the earth. Grounding is indicated by the ground symbol shown in Figure 1-23.

The ac circuits we have looked at have been series circuits. Figure 1-24 shows a 3-wire circuit with parallel loads.

The circuit shown is typical of many circuits used in electrical construction. It consists of two phase conductors and a common (neutral) conductor feeding luminaires (lighting fixtures). The luminaire (lighting fixture) loads are shown as 240 watts each. We have two circuits sharing a common neutral return conductor. There are six 240-watt luminaires (fixtures) on each circuit. The circuits are rated at 15 amperes each because the rating of the circuit is dependent upon the rating of the overcurrent devices. These overcurrent devices are 15-ampere circuit breakers.

When switch #1 is closed, current flows from

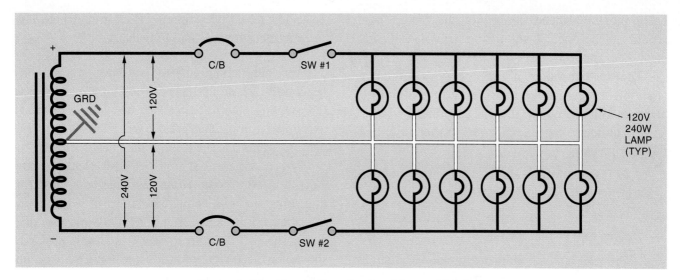

Figure 1-24 Three-wire circuit with parallel loads

the transformer through each of the loads, and returns through the common conductor (neutral) back through the transformer. The current flow in that circuit is found by totaling the loads in watts and dividing by the voltage.

$$I = \frac{W}{E} \quad \text{or} \quad I = \frac{1440}{120} = 12 \text{ amperes}$$

Or we could take the loads individually and total them.

$$I = \frac{W}{E} \quad \text{or} \quad I = \frac{240}{120} = 2 \text{ amperes} \times 6 = 12 \text{ amperes}$$

In Figure 1-24, if both switch #1 and switch #2 were closed, current would flow from the transformer through all of the loads and back through switch #2 to the transformer. No current would flow through the common (neutral) conductor. Remember, current will not flow through a conductor in both directions. The currents bucking each other would cancel out proportionately. The source voltage would now be 240 volts. Current flow can be found by taking all of the loads (12 × 240 = 2880) and dividing by the source voltage.

$$I = \frac{W}{E} \quad \text{or} \quad I = \frac{2800}{240} = 12 \text{ amperes}$$

If one luminaire (light fixture) load should burn out and open that part of the circuit as shown in Figure 1-25, the common (neutral) conductor would

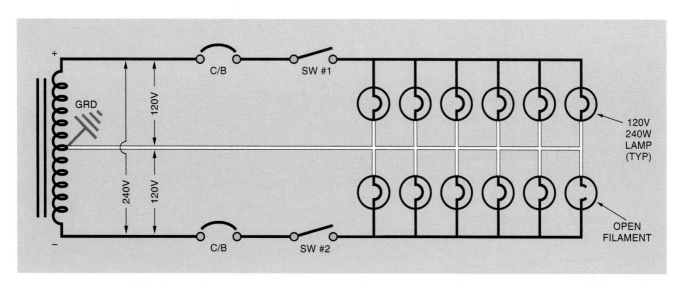

Figure 1-25 Three-wire circuit with parallel loads and one lamp filament open

carry the unbalanced portion of the load. There would not be equal loads in the neutral canceling each other out.

The current value of the one open luminaire (lighting fixture) would cause a current flow of 2 amperes in the common (neutral) conductor.

In summary, the rules for an alternating circuit are as follows:

1. The current at the source is equal to the total of the currents in each of the loads.
2. The voltage across each load in the circuit is equal to the source voltage.
3. The total opposition in an alternating circuit is equal to the sum of the resistance, capacitive reactance, and inductive reactance in the circuit.
4. The total opposition is called impedance. The symbol for the total impedance is Z.

The alternating current we have studied has been single-phase current, as shown by a sine wave representing current flow rising and falling over a *0* axis line. In future chapters, we will learn about three-phase currents and their use in the construction field.

This completes our study of basic electrical circuits, and after completing the question review for this unit, we will begin our study of Chapter 2, "Residential Branch-Circuit, Feeder, and Service Calculations." Be sure to work out each problem shown in the review questions so there will be no question about your complete understanding of "Basic Electricity."

REVIEW QUESTIONS

MULTIPLE CHOICE

1. The current flow in a circuit with 120 volts at the source and a load of 1200 watts is:
   - ____ A. 10 amperes.
   - ____ B. 20 amperes.
   - ____ C. 5 amperes.
   - ____ D. 1.73 amperes.

2. The current flow in a circuit with a voltage of 240 volts and a resistive load of 30 ohms is:
   - ____ A. 3 amperes.
   - ____ B. 6 amperes.
   - ____ C. 8 amperes.
   - ____ D. 5 amperes.

3. The voltage to the neutral conductor (ground) in a 3-wire 120/240-volt circuit is:
   - ____ A. 120 volts.
   - ____ B. 208 volts.
   - ____ C. 173 volts.
   - ____ D. 240 volts.

4. The current flow in the neutral conductor of a balanced 3-wire circuit of 120/240 volts with a 1200-watt load on each phase is:
   - ____ A. 10 amperes.
   - ____ B. 20 amperes.
   - ____ C. 0 amperes.
   - ____ D. 12 amperes.

5. The current flow in a 2-wire, 120-volt circuit is determined by the following formula:
   ___ A. I = E × W
   ___ B. I = W / E
   ___ C. I = E / W
   ___ D. I = E × 1.73

6. The unit of measurement of electrical resistance is:
   ___ A. ampere.
   ___ B. volt.
   ___ C. ohm.
   ___ D. watt.

7. An ampere is a unit of measurement of:
   ___ A. electrical power.
   ___ B. electrical current.
   ___ C. electrical resistance.
   ___ D. electrical inductance.

8. Power factor is the ratio between:
   ___ A. voltage and amperage.
   ___ B. voltage and resistance.
   ___ C. amperage and resistance.
   ___ D. apparent power and true power.

9. In ac circuits, if the current leads the voltage, the circuit is:
   ___ A. capacitive.
   ___ B. inductive.
   ___ C. resistive.
   ___ D. direct current.

10. The neutral conductor is always the:
    ___ A. largest conductor.
    ___ B. smallest conductor.
    ___ C. intentionally grounded conductor.
    ___ D. bus bar.

TRUE OR FALSE

1. The *ampacity* of a conductor is its voltage to ground?
   ___ A. True
   ___ B. False

2. A circuit breaker is an overcurrent device?
   ___ A. True
   ___ B. False

3. The neutral conductor is the conductor that is intentionally grounded?
   ___ A. True
   ___ B. False

4. Ohm's Law can be proven by the formula E = I × R.
   ___ A. True
   ___ B. False

5. The common conductor is the only conductor that can become neutral.
   ____ A. True
   ____ B. False
6. A battery is a source of direct current.
   ____ A. True
   ____ B. False
7. In a capacitive circuit, the current leads the voltage.
   ____ A. True
   ____ B. False
8. In a series circuit, the sum of the voltage drops across the loads is equal to the applied voltage.
   ____ A. True
   ____ B. False
9. The current is the same in all parts of a series circuit.
   ____ A. True
   ____ B. False
10. Power factor is the ratio between true power and apparent power.
    ____ A. True
    ____ B. False

FILL IN THE BLANKS

1. Referring to the figure below, enter the voltage where indicated by each letter.

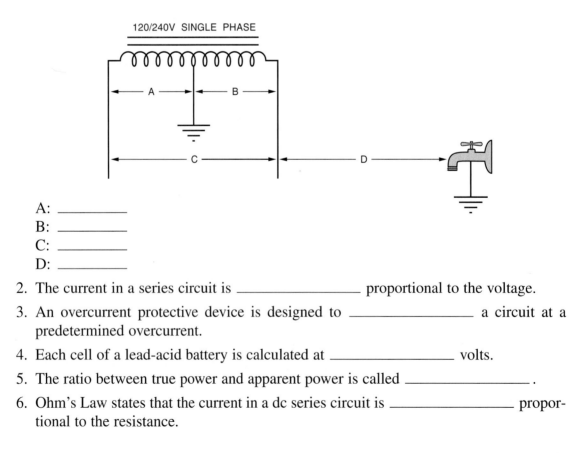

   A: _____
   B: _____
   C: _____
   D: _____

2. The current in a series circuit is _____ proportional to the voltage.
3. An overcurrent protective device is designed to _____ a circuit at a predetermined overcurrent.
4. Each cell of a lead-acid battery is calculated at _____ volts.
5. The ratio between true power and apparent power is called _____.
6. Ohm's Law states that the current in a dc series circuit is _____ proportional to the resistance.

7. The equivalent resistance of two parallel resistors of equal value is _____ the value of one resistor.

8. The equivalent resistance of two resistors in parallel is always _____ than the resistance of the smallest resistor.

9. The current value at the source of a circuit is always equal to the _____ of the currents in each of the loads in the circuit.

10. In a series circuit, the current is the _____ in all parts of the circuit.

11. Finding the equivalent resistance of resistors in parallel is done by the _____ of the sum of the _____ method.

FREQUENTLY ASKED QUESTIONS

Question 1: Why do the voltmeter and ammeter read apparent power and the wattmeter read true power in the same circuit?

Answer: The voltmeter and the ammeter are reading the circuit separately or individually. The wattmeter is reading both the current and the voltage together and this picks up the lagging or leading characteristic between the current and voltage of that circuit resulting in a true power reading on the wattmeter.

Question 2: What causes the inductance in a circuit that results in the voltage leading the current?

Answer: The alternating current is the basic reason. The rising and collapsing magnetic fields cutting across a conductor is a result of alternating current. Circuit components such as motors or ballasts in fluorescent fixtures have windings that produce a lot of inductance when the magnetic fields cut across them.

Question 3: How can some municipalities change the requirements for the installation of electrical work from those in the *NEC*®?

Answer: The *NEC*® is made available for Adoption by Reference by public authorities. It is referenced in ordinances or similar instruments. It can be adopted in whole or in part. Changes, additions, or deletions must be noted separately.

Question 4: What is meant by the "impedance of a circuit"?

Answer: Impedance is the name given to the total opposition to the flow of alternating current in that circuit. It is the combined opposition resulting from resistance, inductive reactance, and capacitive reactance. It is measured in ohms designated by the symbol Z.

Question 5: What is meant by capacitance?

Answer: Capacitance is the ability of a circuit to store energy. When two conductors are separated by an insulating material, they have the ability to store energy. An arrangement of two conducting materials separated by an insulating material is called a capacitor. When a capacitor is connected in a dc circuit, current will flow until the capacitor is charged. The capacitor will remain charged until discharged by connecting the two ends of the capacitor to a resistor (load). When a capacitor is charged, a voltage is present between the two plates of the capacitor. When a capacitor is connected to an ac circuit, the alternating current continues to flow. The current is first in one direction, charging the capacitor, and then in the other or opposite direction, discharging the capacitor.

In an ac circuit, the conductors or wires, separated by air, form a capacitor. When this capacitor is charged and discharged by the changing current, the discharge current is in opposition to the flow of the alternating current. This opposition is called capacitive reactance. The effect of capacitance in an ac circuit is that it tends to make the current lead the voltage.

Question 6: How can you define one end of a transformer as being negative and the other end positive?

Answer: Current could not flow unless there was opposite polarity. In alternating current circuits, the polarities change in conjunction with the frequency. When doing calculations, it is necessary to pick a moment in time and label the transformer ends as being a certain polarity at that moment in time.

Question 7: What is meant by a magnetic field?

Answer: When current flows through a conductor, a magnetic flux or field is set up around the conductor. If the current is changed as in alternating current, the flux collapses around the conductor. A phenomenon occurs whereby a current is induced in the conductor. This induced current is in opposition to the current flow, which caused the magnetic field. This opposition to the flow of current is called inductive reactance.

Question 8: What is meant by transformer action?

Answer: Transformer action is an induction process. A transformer has a primary winding and a secondary winding. The windings are generally wound on an iron core and are wound adjacent to each other, or one on top of the other. When alternating current flows through the primary winding, a magnetic flux is set up around that winding. This flux also encompasses the secondary winding. When the alternating current cycles with the applied frequency, the magnetic field rises and collapses over the secondary winding thereby inducing a voltage in the secondary winding. This induced voltage is directly proportional to the voltage in the primary. The ratio between the primary and secondary voltage are directly proportional to the number of turns in the windings.

Question 9: What is the difference between ohm and impedance?

Answer: They are both units of measurement of electrical opposition to current flow. An ohm is a unit that can be used to designate the amount of electrical resistance in a dc circuit. An ohm is also used to designate the total opposition in an ac circuit that consists of inductive reactance, capacitive reactance, and resistance. The total opposition is called impedance and is measured in ohms.

Question 10: What do they mean by short-circuit?

Answer: Basically, it is a circuit without a load. The circuit conductors are touching each other without some type of load to limit the flow of current. Imagine taking the two leads from a battery and touching them together. The current flow would be limited only by the capacity of the battery. It is called a short-circuit because the circuit is short of the load.

CHAPTER 2

Residential Branch-Circuit, Feeder, and Service Calculations

OBJECTIVES

After completing this chapter, the student should understand:

- the meaning of a branch-circuit.
- how the branch-circuit rating is determined.
- a multiwire branch-circuit.
- how to do load calculations for:
  - lighting branch-circuits.
  - small appliance branch-circuits.
  - laundry branch-circuits.
  - furnace branch-circuits.
  - electric range circuits.
- a bathroom branch-circuit.
- a ground-fault circuit-interrupter.

KEY TERMS

Ampacity: The current in amperes that a conductor can carry continuously under the conditions of use without exceeding its temperature rating

Branch-Circuit: The circuit conductors between the final overcurrent device protecting the circuit and the outlet(s)

Branch-Circuit, Multiwire: A branch-circuit that consists of two or more ungrounded conductors that have a potential difference between them, and a grounded conductor that has an equal potential difference between it and each ungrounded conductor of the circuit and that is connected to the neutral or grounded conductor of the circuit

Feeder: Circuit conductors between the service equipment, the source of a separately derived system, or other power supply source, and the final branch-circuit overcurrent device

Ground-fault Circuit-interrupter: A device that is intended to open a circuit within an established period of time when it detects a current to ground that exceeds a predetermined value that is less than that required to operate the overcurrent protective device of the circuit

Outlet Box: A box installed in the wiring system to house the connection to a utilization device or equipment

Receptacle: A contact device installed at the outlet for the connection of an attachment plug

Service: The conductors and equipment for delivering electric energy from the serving utility to the wiring system of the premises being served

Utilization Device: A device for connection to utilization equipment, such as a receptacle

Utilization Equipment: Equipment that utilizes or uses electrical energy for lighting, heating, or similar purposes

INTRODUCTION

Unlike circuiting in commercial or industrial occupancies where the circuits are generally installed to meet the individual needs of the occupant, the circuits in dwelling units are applicable to all residences. For example, in a factory installation, receptacle outlets are not required every 12 ft as in some rooms of dwellings, but are required only where equipment will be placed for the industrial process. Experience has indicated where residential circuits should be required to meet the level of safety demanded by the *NEC.* The advances in modern technology have brought new appliances into everyday use in residential occupancies. Kitchen appliances, video technology, computers, printers, heaters, and bathroom appliances are commonplace in residences today (Figure 2-1).

Recognizing the need for circuiting to meet these new demands, the NFPA, through their code-making process, has established minimum requirements to maintain the level of safety consistent with the purpose of the *NEC.*

Designers should plan their installations beyond the scope of the *NEC* to meet requirements for future expansion of the electrical systems, as well as ensuring satisfactory performance for the needs of today.

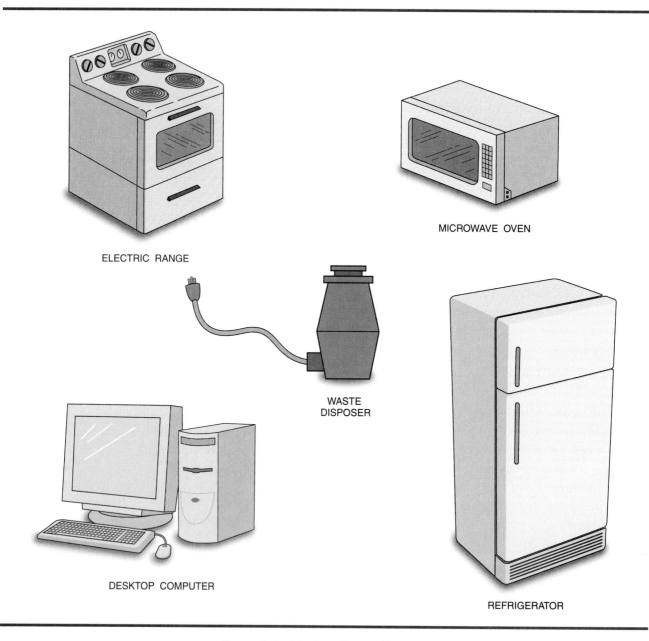

Figure 2-1 Modern day appliances

RESIDENTIAL ELECTRICAL BRANCH-CIRCUITS

Article 210 of the *NEC®* covers **branch-circuits**. Branch-circuits are the circuit conductors between the final overcurrent device protecting the circuit and the outlet. This simply means that when two circuit conductors, usually a black or some other color wire and a white wire, leave the panel where the colored wire is connected to a fuse or circuit breaker and the white wire is connected to the neutral bar, and are run to an **outlet box** for connection to a **receptacle** or some other **utilization device**, they constitute a branch-circuit, (Figure 2-2).

The rating of a branch-circuit is determined by the rating of the overcurrent device. If a 20-ampere circuit breaker is used, it is called a 20-ampere branch-circuit. If a 20-ampere fuse is used, it is a 20-ampere branch-circuit. If a 15-ampere overcurrent device is used, it is a 15-ampere branch-circuit. Unless specifically permitted otherwise, in other parts of the Code, the **ampacity** of the circuit conductors must be of a value that will be protected by the branch-circuit rating. The ampacity of a wire or conductor is the current in amperes that it can carry continuously without damage to the conductor insulation.

Branch-circuits shall be permitted as **multiwire circuits**. A multiwire circuit shall be permitted to be considered as multiple circuits. This means that the overcurrent device protecting the multiwire circuit can be a single device, such as a two- or three-pole circuit breaker, or it can be two or three individual devices, such as two or three single-pole circuit breakers, or two or three fuses (Figure 2-3).

The important thing to remember is that whether you are using two or three overcurrent devices, there will be one return or neutral conductor shared by all of the phase conductors. Figure 2-4 illustrates a permitted multiwire circuit, and one that is *not* permitted because the phase conductors do not have a difference of potential between them. The same phase conductor is being used for both circuits.

Remember, in order to qualify as a multiwire circuit, the phase conductors must have a difference

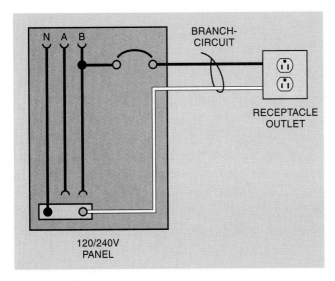

Figure 2-2 Branch-circuit from final overcurrent device to outlet

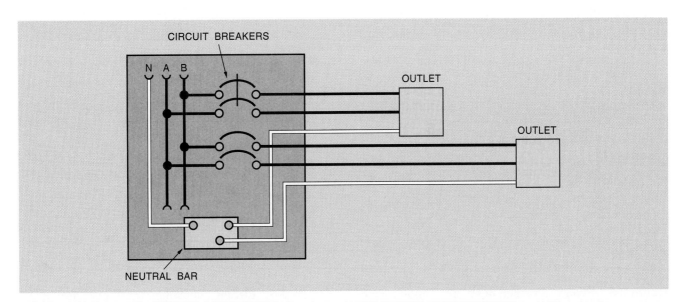

Figure 2-3 Multiwire circuits using a 2-pole circuit breaker or two 1-pole circuit breakers

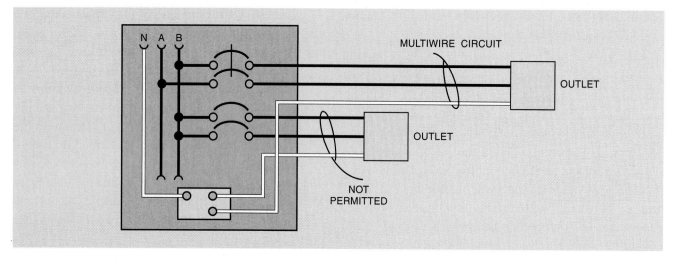

Figure 2-4 Multiwire circuit with opposite phase conductors and a not permitted circuit using same phase conductors

of potential between each other and the common (neutral) conductor.

What do we mean when we say a difference in potential, or a potential difference between two conductors? We are talking about a voltage potential. This potential can only exist when there is a phase displacement between the voltages present on the conductors. Remember our transformer secondary in Chapter 1, "Alternating Current Circuit"? The two ends of the transformer corresponded with a 180° phase displacement. The center tap we provided corresponded to a 90° displacement with either of the two ends of the transformer (Figure 2-5).

As shown in Figure 2-6, if we were to take a 3-wire circuit using two phase conductors from the same end of the transformer and a neutral from the center tap of the transformer, we would not have a multiwire circuit.

In order to share a neutral conductor to carry the return current back to the transformer source, we must use phase conductors with a difference of potential between each other and the neutral conductor (Figure 2-7).

If we do not have a difference of potential between the phase conductors, all of the current in the circuits must use the neutral to return to the source. Unless the neutral was sized to carry the load of two circuits, it would be overloaded (see Figure

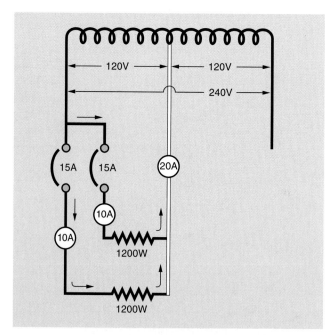

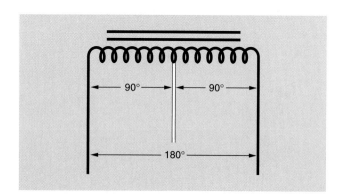

Figure 2-5 Phase displacement on single-phase transformer secondary

Figure 2-6 Circuit showing incorrect wiring of multiwire circuit and overloaded neutral conductor

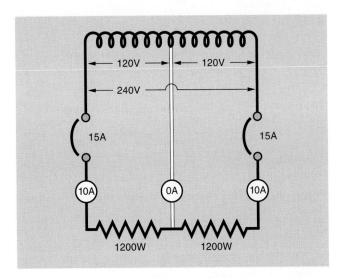

Figure 2-7 Three-wire circuit using phase conductors with a difference in potential

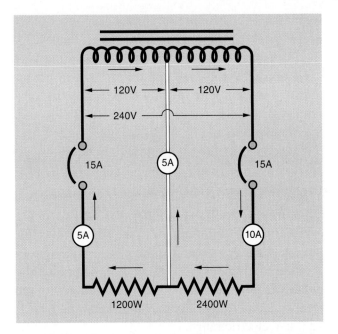

Figure 2-8 Multiwire circuit with unequal loads showing unbalanced current flow in the neutral conductor

2-4 and Figure 2-6), and the neutral conductor would heat up, causing the insulation to deteriorate.

In a multiwire circuit, the neutral conductor will carry the maximum unbalance of the load on the circuits. If you have two circuits with equal loads, there is no unbalance, and the neutral conductor does not carry any current, it is neutral. If one load is turned off, the neutral carries the entire load of the other circuit. If the loads are unequal, the neutral conductor will carry the unbalanced portion between the two circuits (see Figure 2-8).

Now with our knowledge of branch-circuits, let us study in a step-by-step manner, the required circuits and load calculations for a one-family dwelling. First, what is a one-family dwelling? Figure 2-9 shows a one-family dwelling under construction. According to the *NEC*, a one-family dwelling is a building that consists solely of one dwelling unit, and a dwelling unit is one or more

Figure 2-9 Typical one-family dwelling under construction

rooms for the use of one or more persons as a housekeeping unit with space for eating, living, and sleeping, and permanent provisions for cooking and sanitation. You can have a dwelling unit in a one-family dwelling, a two-family dwelling, or a multi-family dwelling. For our study, we will use a one-family dwelling.

Circuits for a One-Family Dwelling

Article 220 of the *NEC,® Branch-Circuit, Feeder and Service Calculations*, contains provisions for the electrical circuits required in a one-family dwelling. Using *Article 220*, we will determine the circuits required for a one-family dwelling with the following criteria: (1) floor area—1500 square ft (2) 12.0 kW electric range.

General Lighting Load. *NEC® 220.3(A)* requires a minimum per sq ft value be taken in accordance with *Table 220.3(A)* (Figure 2-10). For a dwelling occupancy, a unit load per sq ft of 3 volt-amperes is required. To find this information, go down the left hand column of *Table 220.3(A)*, titled *Type of Occupancy*, until you reach "Dwelling Units." Then go across to the right hand column, titled "Unit Load Per Square Foot (Volt-Amperes)," where you will find 3 as the required unit load. 1500 sq ft @ 3 volt-amperes = 4500 VA. Note the small *a* used after "Dwelling Units." It refers you to Note *a* at the bottom of the Table. This Note refers you to *220.3(B)(10)*.

NEC® 220.3(B)(10), Dwelling Occupancies, tells us that *In one-family, two-family, and multi-family dwellings and in the guest rooms of hotels and motels, the outlets specified in (1), (2), and (3) are included in the general lighting calculations of 220.3(A).* No additional load calculations shall be required for these outlets. The outlets included are:

1. Under (A), all general use receptacle outlets of 20-ampere rating or less, and this includes the receptacles connected to the circuits in *210.11(C)(3)* which reads: *At least one 20-ampere branch-circuit shall be provided to supply the bathroom receptacle outlets. Such circuits shall have no other outlets.* General use receptacles are the receptacles in the living room, bedrooms, hallways, family rooms, and the like. They do not include receptacles on the small appliance branch-circuits or laundry

Table 220.3(A) General Lighting Loads by Occupancy

Type of Occupancy	Volt-Amperes per Square Meter	Volt-Amperes per Square Foot
Armories and auditoriums	11	1
Banks	39[b]	3½[b]
Barber shops and beauty parlors	33	3
Churches	11	1
Clubs	22	2
Court rooms	22	2
Dwelling units[a]	33	3
Garages — commercial (storage)	6	½
Hospitals	22	2
Hotels and motels, including apartment houses without provision for cooking by tenants[a]	22	2
Industrial commercial (loft) buildings	22	2
Lodge rooms	17	1½
Office buildings	39	3½[b]
Restaurants	22	2
Schools	33	3
Stores	33	3
Warehouses (storage)	3	¼
In any of the preceding occupancies except one-family dwellings and individual dwelling units of two-family and multifamily dwellings:		
Assembly halls and auditoriums	11	1
Halls, corridors, closets, stairways	6	½
Storage spaces	3	¼

[a]See 220.3(B)(10).
[b]In addition, a unit load of 11 volt-amperes/m² or 1 volt-ampere/ft² shall be included for general-purpose receptacle outlets where the actual number of general-purpose receptacle outlets is unknown.

Figure 2-10 *NEC® 2002 Table 220.3(A)*. (Reprinted with permission from NFPA 70-2002)

circuits, but they do include the bathroom receptacles (Figure 2-11).

2. Under (B), the receptacle outlets specified in *210.52(E)* and *(G)* are excluded. For a one-family dwelling and each unit of a two-family dwelling that is at grade level, at least one receptacle outlet accessible at grade level, and not more than 6½ ft (2.0 m) above grade, shall be installed in front and back of the dwelling (Figure 2-12). See *210.8(A)(3)* for **ground-fault circuit-interrupter** (GFCI) protection. For a one-family dwelling, at least one receptacle outlet, in addition to any provided for the laundry equipment, shall be installed in each basement, each attached garage, and each detached garage with electric power. See *210.8(A)(2)* and *(A)(5)* for GFCI protection requirements.

3. Under (C), the lighting outlets specified in *210.70(A)* and *(B)* are located in every habitable room, bathrooms, hallways, stairways, attached and detached garages, attics, utility rooms, and basements.

All of the previously mentioned receptacle outlets and lighting outlets are included in the 3 VA unit load per sq ft, as shown in *Table 220.3(A)*, for a dwelling type occupancy.

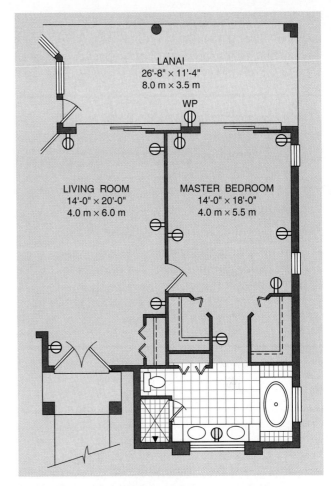

Figure 2-11 Partial floor plan showing general use receptacles in bath, living room, master bedroom, and lanai

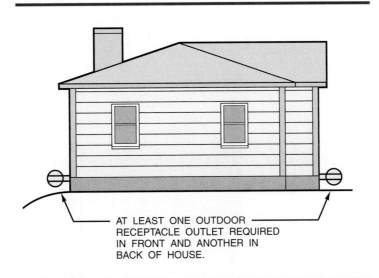

Figure 2-12 At least one outdoor receptacle installed in front and back of the dwelling

30 CHAPTER 2 Residential Branch-Circuit, Feeder, and Service Calculations

NEC® 210.8 provides the requirements for GFCI protection for personnel (Figure 2-13). A GFCI is a device that is intended to open a circuit within an established period of time when it detects a current to ground that exceeds a predetermined value that is less than that required to operate the overcurrent

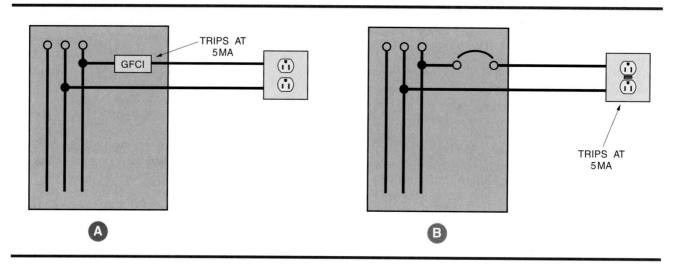

Figure 2-13 Ground-fault circuit-interrupter protection (A) circuit breaker (B) receptacle

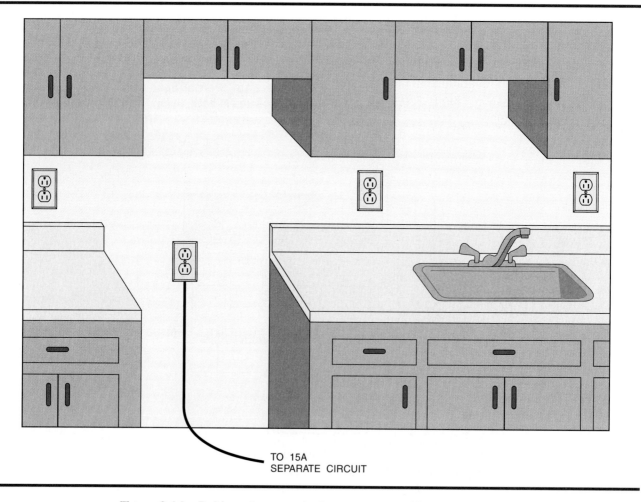

Figure 2-14 Refrigerator receptacle on separate 15-ampere circuit

protective device of the circuit. This means that in a 15-ampere rated circuit, one that has a 15-ampere circuit breaker protecting it, a ground-fault would not open the circuit until current in excess of 15 amperes has passed through the fault. If a Class A ground-fault circuit breaker were installed, the circuit breaker would open in ¼ of a cycle at a predetermined level of approximately 5 milliamperes of current flow. This would protect any personnel who might be exposed to the ground-fault condition. GFCIs are explained in detail in Chapter 4, "Grounding."

We have determined by use of *Table 220.3(A)* that our one-family dwelling requires 3 volt-amperes per sq ft of floor area for computation of the general lighting load. The floor area of our one-family dwelling is 1500 sq ft × 3 VA = 4500 VA general lighting load.

Lighting Branch-Circuits Required. To determine the lighting branch-circuits required, we will use the power formula learned in Chapter 1 showing that I = W/E.

$$I = \frac{4500 \text{ VA}}{120 \text{ volts}} = 37.5 \text{ amperes;}$$

$$\frac{37.5 \text{ A}}{15 \text{ A (per circuit)}} = 3, 15 \text{ A circuits;}$$

$$\text{or} \quad \frac{37.5 \text{ A}}{20 \text{ A (per circuit)}} = 2, 20 \text{ A circuits}$$

Generally, in residential wiring, 15-ampere circuits are used for wiring the general lighting loads as shown in *220.3(B)(10)*.

Small-Appliance Load. *NEC® 210.11(C)(1)* requires two or more 20-ampere small-appliance branch-circuits. *NEC® 220.16(A)* requires that each small-appliance branch-circuits be computed at 1500 volt-amperes.

Small-appliance branch-circuits required by *210.11(C)(1)* feed all the receptacle outlets covered by *210.52(B)*. These receptacle outlets are those serving the kitchen, pantry, breakfast room, dining room, or similar areas of a dwelling unit. There is an exception for the refrigeration receptacle that is permitted to be supplied by an individual branch-circuit rated 15 amperes or greater (Figure 2-14). Another

Table 220.11 Lighting Load Demand Factors

Type of Occupancy	Portion of Lighting Load to Which Demand Factor Applies (Volt-Amperes)	Demand Factor (Percent)
Dwelling units	First 3000 or less at From 3001 to 120,000 at Remainder over 120,000 at	100 35 25
Hospitals*	First 50,000 or less at Remainder over 50,000 at	40 20
Hotels and motels, including apartment houses without provision for cooking by tenants*	First 20,000 or less at From 20,001 to 100,000 at Remainder over 100,000 at	50 40 30
Warehouses (storage)	First 12,500 or less at Remainder over 12,500 at	100 50
All others	Total volt-amperes	100

*The demand factors of this table shall not apply to the computed load of feeders or services supplying areas in hospitals, hotels, and motels where the entire lighting is likely to be used at one time, as in operating rooms, ballrooms, or dining rooms.

Figure 2-15 *NEC® 2002 Table 220.11.* (Reprinted with permission from NFPA 70-2002)

exception for *210.52(B), Exception No. 2*, permits the refrigeration circuit to be excluded from the load calculations.

The small-appliance branch-circuit loads are permitted to be included with the general lighting load, and subjected to the demand factors provided in *Table 220.11* (Figure 2-15).

Laundry Branch-Circuit. *NEC® 210.11(C)(2)* requires at least one 20-ampere circuit to supply the laundry receptacle outlet(s). *NEC® 220.16(B)* requires that the laundry circuits be computed at 1500 volt-amperes each. The laundry load is permitted to be included with the general lighting load and is subjected to the demand factors provided in *Table 220.11* (see Figure 2-16).

Furnace Branch-Circuit. *NEC® 422.12* requires central heating equipment be supplied by an individual branch-circuit. An individual branch-circuit is one that supplies only one piece of utilization equipment. Therefore, no other equipment can be on the furnace circuit. Auxiliary equipment, such as a pump, valve, humidifier, or

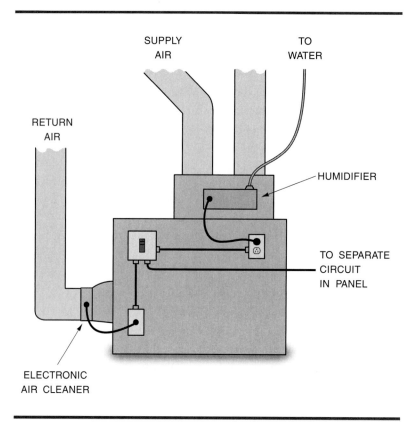

Figure 2-16 Gas-fired furnace with humidifier and electronic air cleaner on individual branch-circuit

electrostatic air cleaner directly associated with the heating equipment, may be connected to the same branch-circuit (Figure 2-16). There is no load requirement shown in *422.12*, but *220.4* requires that for circuits supplying loads consisting of motor-operated utilization equipment that is fastened in place and that has a motor larger than ⅛ horsepower in combination with other loads, the total computed load shall be based on 125% of the largest motor load, plus the sum of the other loads.

Note that the requirement in *422.12* applies only to central heating equipment. Central heating equipment would most likely be a gas-fired or oil-fired furnace or boiler. Baseboard heaters, heating cables or panels, and the like are not covered by this requirement.

Bathroom Branch-Circuits. *210.11(C)(3)* requires at least one 20-ampere branch-circuit to be provided to supply the bathroom receptacle outlet(s). This circuit may supply more than one bathroom, and this circuit shall have no other outlet(s) (Figure 2-17).

If the 20-ampere circuit supplies a single bathroom, outlets for other equipment within the same bathroom can be supplied from this circuit. If other equipment is supplied from the 20-ampere branch-circuit, then *210.23(A)* restricts the total rating of **utilization equipment** fastened in place, other than luminaires (lighting fixtures), to not exceed more than 50% of the branch-circuit ampere rating where cord- and plug-connected utilization equipment not fastened in place is also supplied (Figure 2-18).

Electric Range. Other than a large central air-conditioning unit or an outlet for electric vehicle charging equipment, the electric range is the largest load you will find in a residential occupancy. Electric ranges, electric cooking tops, and electric ovens are installed in a variety of combinations. Experience has shown that these units are rarely, if ever, utilized to the full extent of their capacity. When all four electric burners are on, one of them is thermostatically controlled and cycles with the thermostat operation. If the oven is on at the same time, it is also thermostatically controlled and cycles with

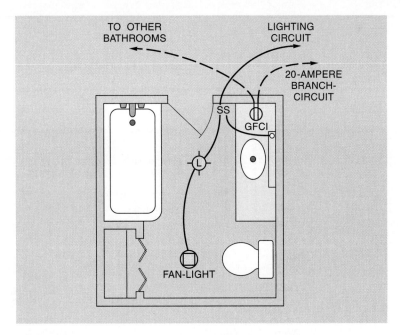

Figure 2-17 Minimum of one 20-ampere branch-circuit to supply bathroom receptacles. Other outlets not permitted on bathroom receptacle circuit

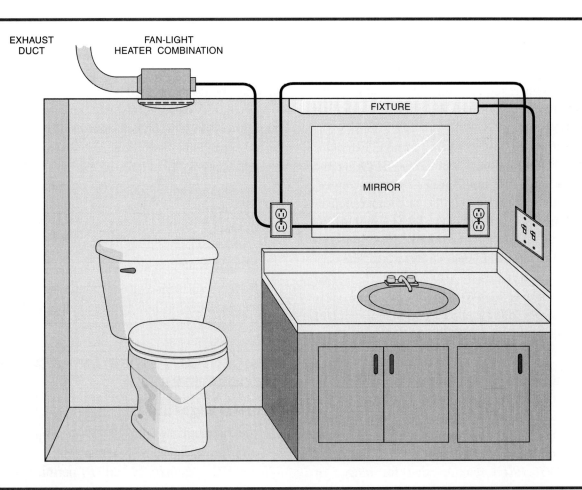

Figure 2-18 Bathroom circuit supplying receptacles and a fan-light-heater in a single bathroom

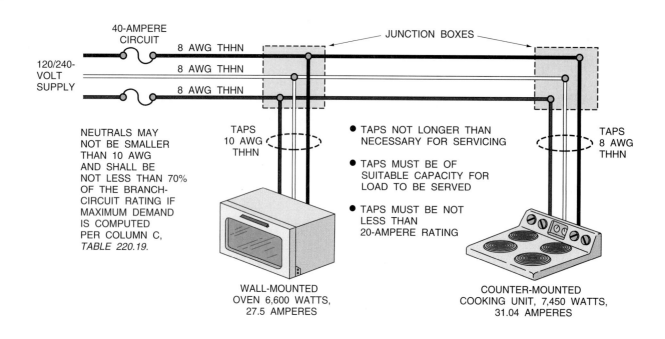

Figure 2-19 Electric cooking top and wall-mounted oven

the thermostat setting. Settings are frequently changed, and the result is that a 12.0 kW range probably will never use 12.0 kW while in operation (Figure 2-19).

The *NEC®* has established, through extensive testing and observation, *Table 220.19* showing the demand loads for household electric ranges, counter-mounted cooking units, and other household cooking appliances (Figure 2-20). Column C is used for determining the maximum demand in kilowatts. The term kilowatt means 1000 watts. The decimal symbol kilo stands for 1000. Column C is used for ranges not over 12 kW rating.

NEC® Table 220.19 allows a demand factor of 8.0 kW (8000 VA) for a single range.

$$\frac{8000 \text{ VA}}{240 \text{ V}} = 33.3 \text{ amperes}$$

NEC® Table 310.16 shows that 8 AWG THW CU (copper) conductors would be the required size for a 33.3 ampere load (Figure 2-21).

NEC® 210.19(C) requires that for ranges of 8¾ kW or more rating, the minimum branch-circuit rating shall be 40 amperes. *Exception No. 2* allows for a reduction in the size of the neutral conductor to not less than 70% of the branch-circuit rating and shall not be less than a 10 AWG.

From the calculations we have made to this point, the following branch-circuits will be required for this one-family dwelling:

General Lighting — 3-1P15A
Small Appliance — 2-1P20A
Laundry   — 1-1P20A
Furnace   — 1-1P20A
Bathroom   — 1-1P20A
Electric Range — 1-2P40A

Now we will make our calculations to determine the required size of the service conductors.

RESIDENTIAL FEEDER AND SERVICE CALCULATIONS

General Lighting

The general lighting load is subject to the demand factors of *220.11* which refers us to *Table 220.11*. *NEC® 220.16(A)* and *(B)* permit the small-appliance and laundry loads to be included with the general lighting load and to be subjected to the demand factors provided in *220.11*.

Table 220.19 Demand Loads for Household Electric Ranges, Wall-Mounted Ovens, Counter-Mounted Cooking Units, and Other Household Cooking Appliances over 1¾ kW Rating (Column C to be used in all cases except as otherwise permitted in Note 3.)

Number of Appliances	Demand Factor (Percent) (See Notes)		Column C Maximum Demand (kW) (See Notes) (Not over 12 kW Rating)
	Column A (Less than 3½ kW Rating)	Column B (3½ kW to 8¾ kW Rating)	
1	80	80	8
2	75	65	11
3	70	55	14
4	66	50	17
5	62	45	20
6	59	43	21
7	56	40	23
8	53	36	23
9	51	35	24
10	49	34	25
11	47	32	26
12	45	32	27
13	43	32	28
14	41	32	29
15	40	32	30
16	39	28	31
17	38	28	32
18	37	28	33
19	36	28	34
20	25	28	35
21	34	26	36
22	33	26	37
23	32	26	38
24	31	26	39
25	30	26	40
26–30	30	24	15 kW + 1 kW for each range
31–40	30	22	
41–50	30	20	25 kW + ¾ kW for each range
51–60	30	18	
61 and over	30	16	

1. Over 12 kW through 27 kW ranges all of same rating. For ranges individually rated more than 12 kW but not more than 27 kW, the maximum demand in Column C shall be increased 5 percent for each additional kilowatt of rating or major fraction thereof by which the rating of individual ranges exceeds 12 kW.
2. Over 8¾ kW through 27 kW ranges of unequal ratings. For ranges individually rated more than 8¾ kW and of different ratings, but none exceeding 27 kW, an average value of rating shall be computed by adding together the ratings of all ranges to obtain the total connected load (using 12 kW for any range rated less than 12 kW) and dividing by the total number of ranges. Then the maximum demand in Column C shall be increased 5 percent for each kilowatt or major fraction thereof by which this average value exceeds 12 kW.
3. Over 1¾ kW through 8¾ kW. In lieu of the method provided in Column C, it shall be permissible to add the nameplate ratings of all household cooking appliances rated more than 1¾ kW but not more than 8¾ kW and multiply the sum by the demand factors specified in Column A or B for the given number of appliances. Where the rating of cooking appliances falls under both Column A and Column B, the demand factors for each column shall be applied to the appliances for that column, and the results added together.
4. Branch-Circuit Load. It shall be permissible to compute the branch-circuit load for one range in accordance with Table 220.19. The branch-circuit load for one wall-mounted oven or one counter-mounted cooking unit shall be the nameplate rating of the appliance. The branch-circuit load for a counter-mounted cooking unit and not more than two wall-mounted ovens, all supplied from a single branch circuit and located in the same room, shall be computed by adding the nameplate rating of the individual appliances and treating this total as equivalent to one range.
5. This table also applies to household cooking appliances rated over 1¾ kW and used in instructional programs.

Figure 2-20 *NEC® 2002 Table 220.19.* (Reprinted with permission from NFPA 70-2002)

General Lighting	4500 volt-amperes
Small Appliance	3000 volt-amperes
Laundry	1500 volt-amperes
	9000 volt-amperes

Table 220.11 Demand Factors:

1st 3000 VA @ 100%	3000 volt-amperes
6000 VA @ 35%	2100 volt-amperes
Net Computed Load:	5100 volt-amperes
Range Load:	8000 volt-amperes
Net Computed Load:	13,100 volt-amperes

Net Computed Load divided by applied voltage: ($I = W/E$)

$$\frac{13{,}100}{240} = 54.58 \text{ amperes}$$

Demand factors are variances given because study and experience has shown that the actual demand load or usage end up being far less than the computed load. Almost every building has a watthour meter that registers the amount of electricity that is used. Studies of these readings have shown that demand factor diversities are safe to use. Demand factors are not used to save money, but rather to not waste money.

A good example of this would be a bedroom in a dwelling unit. If the measurements of this bedroom were 12 ft × 12 ft, or 144 sq ft, we would use *Table 220.3(A)* and multiply this measurement by 3 volt-amperes. The computed load for this room would be 432 volt-amperes. Rarely, if ever, would you find a bedroom using 400 watts of power. If by some chance you were using 400 watts in this location, the probability is that you would not be using any power in other rooms for which you had also provided 3 volt-amperes per sq ft.

Neutral Feeder Conductor Computation

Net Computed Load:	5100 VA
Range 8000 × .70 (*220.19*)	5600 VA
	10,700 VA

$$\frac{10{,}700 \text{ VA}}{240 \text{ V}} = 44.58 \text{ amperes}$$

Table 310.16 Allowable Ampacities of Insulated Conductors Rated 0 Through 2000 Volts, 60°C Through 90°C (140°F Through 194°F), Not More Than Three Current-Carrying Conductors in Raceway, Cable, or Earth (Directly Buried), Based on Ambient Temperature of 30°C (86°F)

	Temperature Rating of Conductor (See Table 310.13.)						
	60°C (140°F)	75°C (167°F)	90°C (194°F)	60°C (140°F)	75°C (167°F)	90°C (194°F)	
Size AWG or kcmil	Types TW, UF	Types RHW, THHW, THW, THWN, XHHW, USE, ZW	Types TBS, SA, SIS, FEP, FEPB, MI, RHH, RHW-2, THHN, THHW, THW-2, THWN-2, USE-2, XHH, XHHW, XHHW-2, ZW-2	Types TW, UF	Types RHW, THHW, THW, THWN, XHHW, USE	Types TBS, SA, SIS, THHN, THHW, THW-2, THWN-2, RHH, RHW-2, USE-2, XHH, XHHW, XHHW-2, ZW-2	Size AWG or kcmil
	COPPER			ALUMINUM OR COPPER-CLAD ALUMINUM			
18	—	—	14	—	—	—	—
16	—	—	18	—	—	—	—
14*	20	20	25	—	—	—	—
12*	25	25	30	20	20	25	12*
10*	30	35	40	25	30	35	10*
8	40	50	55	30	40	45	8
6	55	65	75	40	50	60	6
4	70	85	95	55	65	75	4
3	85	100	110	65	75	85	3
2	95	115	130	75	90	100	2
1	110	130	150	85	100	115	1
1/0	125	150	170	100	120	135	1/0
2/0	145	175	195	115	135	150	2/0
3/0	165	200	225	130	155	175	3/0
4/0	195	230	260	150	180	205	4/0
250	215	255	290	170	205	230	250
300	240	285	320	190	230	255	300
350	260	310	350	210	250	280	350
400	280	335	380	225	270	305	400
500	320	380	430	260	310	350	500
600	355	420	475	285	340	385	600
700	385	460	520	310	375	420	700
750	400	475	535	320	385	435	750
800	410	490	555	330	395	450	800
900	435	520	585	355	425	480	900
1000	455	545	615	375	445	500	1000
1250	495	590	665	405	485	545	1250
1500	520	625	705	435	520	585	1500
1750	545	650	735	455	545	615	1750
2000	560	665	750	470	560	630	2000

CORRECTION FACTORS

Ambient Temp. (°C)	For ambient temperatures other than 30°C (86°F), multiply the allowable ampacities shown above by the appropriate factor shown below.						Ambient Temp. (°F)
21–25	1.08	1.05	1.04	1.08	1.05	1.04	70–77
26–30	1.00	1.00	1.00	1.00	1.00	1.00	78–86
31–35	0.91	0.94	0.96	0.91	0.94	0.96	87–95
36–40	0.82	0.88	0.91	0.82	0.88	0.91	96–104
41–45	0.71	0.82	0.87	0.71	0.82	0.87	105–113
46–50	0.58	0.75	0.82	0.58	0.75	0.82	114–122
51–55	0.41	0.67	0.76	0.41	0.67	0.76	123–131
56–60	—	0.58	0.71	—	0.58	0.71	132–140
61–70	—	0.33	0.58	—	0.33	0.58	141–158
71–80	—	—	0.41	—	—	0.41	159–176

* See 240.4(D).

Figure 2-21 *NEC® 2002 Table 310.16.* (Reprinted with permission from NFPA 70-2002)

When calculating the load on the neutral conductor of an electric range, *220.22* permits the maximum unbalanced load to be considered as 70% of the load on the ungrounded conductors as determined by *Table 220.19*.

Service Conductor Size

Ungrounded Conductors
54.58 amperes   *Table 310.16*
6 AWG THW copper

Grounded (Neutral) Conductor
44.48 amperes   *Table 310.16*
8 AWG THW copper

These sizes would be correct for dwelling units in other than one-family dwellings. However, *230.79(C)* requires that for a one-family dwelling, the **service** disconnecting means shall have a rating of not less than 100 amperes, 3-wire. NEC® *230.42(B)* requires that ungrounded conductors shall have an ampacity of not less than the minimum rating of the disconnecting means specified in *230.79(C)*.

Using *Table 310.16*, for 100 amperes, a 3 AWG THW copper conductor may be used, and for the grounded (neutral conductor), we may still use the 8 AWG THW copper conductor since that has been sized using *220.22* for the neutral **feeder** conductor.

There is another diversity we may take by using *310.15(B)(6)*. For a one-family dwelling, the conductors as listed in *Table 310.15(B)(6)* may be used for 120/240 volt, 3-wire, single-phase service-entrance conductors. Using *Table 310.15(B)(6)*, we may use 4 AWG THW copper conductors for the ungrounded conductors and 8 AWG THW copper for the grounded conductor.

A comparison of the two sizes of conductors that may be used is shown in the following tables. Both copper and aluminum sizes are shown in *Table 310.16* or *Table 310.15(B)(6)*.

Using *Table 310.16*
2–3 AWG THW Copper   2–1 AWG THW Aluminum
1–8 AWG THW Copper   1–4 AWG THW Aluminum

Or using *Table 310.15(B)(6)*
2–4 AWG THW Copper   2–2 AWG THW Aluminum
1–8 AWG THW Copper   1–6 AWG THW Aluminum

The dwelling unit we have just used for our computations assumed very basic conditions. Let's do a calculation on a more involved unit.

Floor Area: 2400 sq ft
Appliances: 12.0 kW Electric Range
               4.5 kW Dryer
               1.2 kW Dishwasher
               4.5 kW Water Heater
               ⅓ hp Disposal
               20.0 kW Electric Heat

Calculation:

General Lighting, *Table 220.3(A)*
2400 sq ft × 3 VA/sq ft   =   7200 VA

Small Appliance, *220.16(A)*
1500 VA × 2 VA/sq ft   =   3000 VA

Laundry, *220.16(B)*
1500 VA × 1 VA/sq ft   =   1500 VA

Total General Lighting and
Small Appliance and Laundry   11,700 VA

Remember, we are permitted to include small appliances and laundry with the general lighting and be subjected to the demand factors of *220.11*.

Demand Factors (*Table 220.11*)
3000 VA @ 100%   =   3000 VA
8700 VA @ 35%   =   3045 VA

Net General Lighting and
Small Appliance Load   6045 VA

12.0 kW Electric Range
(*Table 220.19*)   8000 VA
4.5 kW Dryer (*220.18*)   5000 VA
1.2 kW Dishwasher   1200 VA
4.5 kW Water Heater   4500 VA
⅓ horsepower Disposal
(7.2 A × 115 V × 1.25)   1035 VA
20.0 kW Electric Heat   20,000 VA

Total Calculated Load: 45,780 VA

$$\frac{W\,(VA)}{V} = \frac{45{,}780}{240} = 190.75 \text{ amperes}$$

Neutral (grounded) feeder conductor computation:

Net General Lighting and
Small Appliance Load: 6045 VA

Electric Range (*220.22*)
8000 VA × .70   5600 VA

Electric Dryer (*220.22*)
5000 VA × .70   3500 VA

Dishwasher   1200 VA
Disposal (7.2 A × 115 V)   828 VA

Total Computed Neutral (Grounded)
Conductor Load   17,173 VA

$$\frac{W\ (VA)}{V} = \frac{17{,}173}{240} = 71.55 \text{ amperes}$$

Conductor Size Required:
Using *Table 310.16*

(Ungrounded)   190.75 A
  2–3/0 AWG THW CU or
  2–250 kcmil AWG THW AL

Neutral (grounded) conductor   71.55 A
  1–4 AWG THW CU or
  1–3 AWG THW AL

Using *Table 310.15(B)(6)*

(Ungrounded)   190.75 A
  2–2/0 AWG THW CU or
  2–4/0 AWG THW AL

Neutral (grounded) conductor   71.55 A
  1–4 AWG THW CU or
  1–3 AWG THW AL

We have added a few items in this computation we did not discuss previously, so let us review what we have done in this calculation.

The general lighting load was computed exactly as we did for the first one-family dwelling unit. Again, the small-appliance and laundry loads were permitted to be included with the general lighting and were subjected to the demand factors of *220.11*. This gave us our Net Computed General Lighting and Small-Appliance Load.

To the Net Computed General Lighting and Small-Appliance Load we add the 12.0 kW electric range. *NEC® 220.19* allows us to use the demand loads of *Table 220.19*. Based on one range in the left-hand column, we go across to Column C (not over 12.0 kW rating) and find the maximum demand required is 8.0 kW.

Now we must add the other appliances. First add the 4.5 kW dryer. *NEC® 220.18* requires that we use 5000 watts (volt-amperes) or the nameplate rating, whichever is greater. The nameplate rating on the dryer specified is 4500 watts (4.5 kW) so we use 5000 watts (volt-amperes).

Next we add the dishwasher at its nameplate rating of 1200 watts (volt-amperes) according to *220.3(B)(1)*.

Now we add the hot water heater at its nameplate rating of 4.5 kW (4500 volt-amperes) in compliance with *220.3(B)(1)*.

Next we add the ⅓ horsepower disposal. We must convert the ⅓ horsepower to volt-amperes. Here we are making reference to *Sections* and *Tables* that will be covered in depth in future chapters. For now, we will use these references as directed. Referring to *Table 430.148, Full Load Currents in Amperes, Single Phase Alternating Current Motors*, we find the ⅓ horsepower motor at 115 volts, which is the nameplate rating of our disposal unit, has a full load current of 7.2 amperes. Now we look at *220.3(C), Motor Loads*, that directs us to *430.22* and find we must use 125% of the full load current of the motor. Using our power formula W = I × E, we multiply 7.2 A × 115 V = 828 VA × 1.25 = 1035 VA.

Finally, we add our electric heating load of 20.0 kW (20,000 VA). *NEC® 220.15* requires that fixed electric space-heating loads shall be computed at 100% of the total connected load.

The last step was dividing the total computed load in volt-amperes by the applied voltage to obtain our load in amperes (I = W/E).

Looking at *220.17*, we find that if we had four or more appliances fastened in place, we could take a demand factor of 75%. These appliances cannot include electric ranges, clothes dryers, space-heating equipment, or air-conditioning equipment. We have only two eligible appliances, the disposal and the water heater. This section becomes very useful when computing multifamily dwellings.

NEC® 220.21, Noncoincident Loads, allows us to use either the electric-heating load or the air-conditioning load, whichever is greater. We did not specify an air-conditioning load since the electric-heating load would be the larger of the two.

The calculation of the neutral (grounded) feeder load is based only on those loads that, if unbalanced, would require a return conductor back to the source. These loads are the 120-volt loads which always require a neutral, and the 240-volt loads which may require a neutral, such as the electric range. We are required by *220.22* to provide for a load of 70% of the electric range load and of the electric dryer load for the neutral feeder. The central electric-heating load and the electric water heater are 240-volt loads that do *not* require a neutral. The general lighting load that includes the small-appliance and laundry loads is always a part of the neutral feeder computation.

Now let us take a look at another way to compute the ampere load on the service conductors.

	Amperes		
	Line A	Line B	Neutral (Grounded)
General Lighting 6045 ÷ 240	25.2	25.2	25.2
Electric Range 8000 ÷ 240	33.3	33.3	(70%) 23.3
Electric Dryer 5000 ÷ 240	20.8	20.8	(70%) 14.6
Dishwasher 1200 ÷ 120	10.0		10.0
Water Heater 4500 ÷ 240	18.7	18.7	
Disposal 1035 ÷ 120		9.0	
Electric Heat 20,000 ÷ 240	83.3	83.3	
	191.3	190.3	73.1

In the computation above, Line A and Line B represent Phase A and Phase B in the circuit breaker or fuse panel. The 70% computation on the electric range and the electric dryer are in accordance with *220.22*. In real life, there is probably no current flow in the neutral (grounded) conductor of the range circuit since the ranges today do not use any 120-volt circuitry for the heating elements. Any 120-volt power required is accomplished through the use of a small transformer located in the control panel of the range.

No neutral current was allowed for the disposal because it would balance out with the dishwasher. If the disposal and the dishwasher were on a 3-wire multiwire circuit, which is the logical way to circuit these two appliances, there would be little or no current in the neutral (grounded) conductor (Figure 2-22).

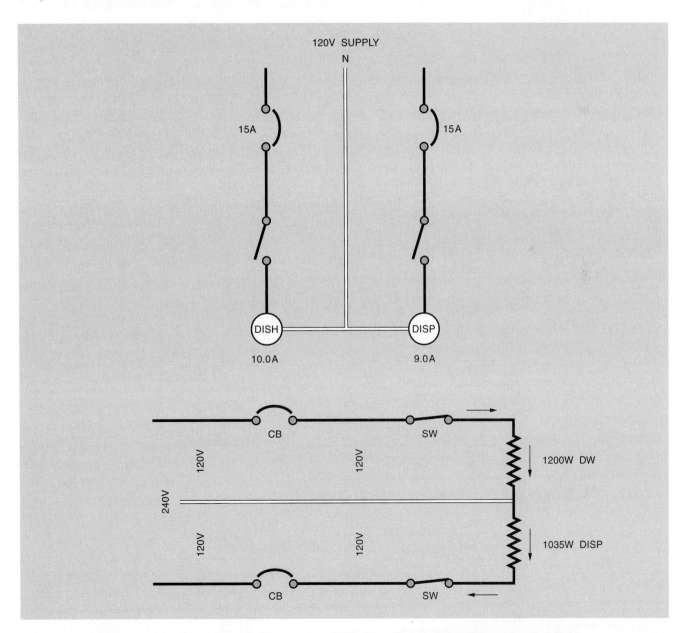

Figure 2-22   Three-wire multiwire circuit feed for disposal and dishwasher in kitchen

Referring to Figure 2-22, if both loads were energized, the total load would be 2235 VA ÷ 240 V = 9 A. There would be little or no current flow in the neutral (grounded) conductor. If the dishwasher were de-energized, the current flow through the disposal and the neutral (grounded) conductor would be 9 amperes. If the disposal were de-energized, the current flow through the dishwasher and the neutral (grounded) would be 9 amperes. In either instance, the current flow in the neutral (grounded) conductor would never exceed 9 amperes. Therefore, in our chart, we only show neutral current for one of the two appliances.

Remember, there is no current flow in the neutral conductor of a perfectly balanced system. The neutral conductor is then in a neutral state or condition and there is no current flow.

In the previously mentioned system, if everything was turned on and drawing full current, there would be approximately one ampere of current flow in the neutral (grounded) conductor. In its most unbalanced condition, there would be approximately 73.1 amperes flowing in the neutral (grounded) conductor.

In the previous computations, we have been using the 75° portion of *Table 310.16*. In order to properly use these 75° conductors, we must be sure that the terminals on the equipment they will be attached to are rated for 75°C. Circuit breaker and panelboard terminals rated at 75° will be marked as suitable for 75°. This information may be found in *110.14* and will be discussed further in a later chapter.

REVIEW QUESTIONS

MULTIPLE CHOICE

1. The general lighting load for each sq ft of floor area in a dwelling unit shall not be less than:
   - ____ A. 5 volt-amperes.
   - ____ B. 3 volt-amperes.
   - ____ C. 1500 watts.
   - ____ D. 8 volt-amperes.

2. The rating of the small-appliance branch-circuits shall be:
   - ____ A. 1500 volt-amperes.
   - ____ B. 20 amperes.
   - ____ C. 15 amperes.
   - ____ D. 1500 watts.

3. The current flow in the neutral conductor of a balanced 3-wire circuit of 120/240 volts with a 1200-watt load is:
   - ____ A. 20 amperes.
   - ____ B. 10 amperes.
   - ____ C. 0 amperes.
   - ____ D. 5 amperes.

4. The demand factor for a single household 12.0 kW range is:
   - ____ A. 12.0 kW.
   - ____ B. 14.0 kW.
   - ____ C. 8.0 kW.
   - ____ D. 10.0 kW.

CHAPTER 2 Residential Branch-Circuit, Feeder, and Service Calculations

5. The demand factor for the first 3000 volt-amperes of the general lighting load is:
   ____ A. 25%.
   ____ B. 35%.
   ____ C. 50%.
   ____ D. 100%.

6. The load on the small-appliance branch-circuit may be included with the general lighting load and subjected to demand factors not to exceed:
   ____ A. 50%.
   ____ B. 1500va.
   ____ C. 35%.
   ____ D. None of the Above

7. The ampacity of a 3 AWG THW copper conductor is:
   ____ A. 100 amperes.
   ____ B. 110 amperes.
   ____ C. 115 amperes.
   ____ D. 173 amperes.

8. The minimum ampacity of the ungrounded conductors serving a one-family dwelling shall be:
   ____ A. 120/240 volt.
   ____ B. 60 amperes.
   ____ C. 100 amperes.
   ____ D. 240 amperes.

9. Central heating equipment must be connected to a _____ branch-circuit.
   ____ A. individual
   ____ B. multiwire
   ____ C. appliance
   ____ D. general purpose

10. Demand factor for four or more qualifying appliances is:
    ____ A. 50%.
    ____ B. 60%.
    ____ C. 75%.
    ____ D. 25%.

TRUE OR FALSE

1. The ampacity of a conductor is its voltage to ground.
   ____ A. True
   ____ B. False
   Code reference _____ .

2. A circuit breaker is an overcurrent device.
   ____ A. True
   ____ B. False
   Code reference _____ .

3. The neutral conductor is never grounded.
   ____ A. True
   ____ B. False
   Code reference _____ .

4. A one-family dwelling is found in multifamily buildings.
   ____ A. True
   ____ B. False
   Code reference _____ .

5. The general lighting unit load per sq ft for a dwelling unit is 6 VA.
   ____ A. True
   ____ B. False
   Code reference _____ .

6. A GFCI is an overcurrent protective device.
   ____ A. True
   ____ B. False
   Code reference _____ .

7. The small appliance load is permitted to be subjected to the demand factors of *220.11*.
   ____ A. True
   ____ B. False
   Code reference _____ .

8. The calculated load on the neutral conductor of an electric range is considered to be 70% of the load on the ungrounded conductors as determined by *Table 220.19*.
   ____ A. True
   ____ B. False
   Code reference _____ .

9. Laundry receptacles require a 20-ampere rated circuit, but no load requirement is included in the load calculations.
   ____ A. True
   ____ B. False
   Code reference _____ .

10. A unit load of 1500 volt-amperes is required for the refrigeration circuit.
    ____ A. True
    ____ B. False
    Code reference _____ .

FILL IN THE BLANKS

1. It shall be acceptable to apply demand factors for range loads in accordance with _____ , including Note 4.

2. Fixed electric space-heating loads shall be computed at _____ of the total connected load.

3. The load for household electric dryers shall be _____ or the nameplate rating, whichever is greater.

4. In each dwelling unit, the load shall be computed at _____ for each 2-wire small-appliance branch-circuit.

5. In addition to the number of branch-circuits required by other parts of this section, at least one _____ shall be provided to supply the bathroom receptacle outlet(s).

6. The rating of a branch-circuit is determined by the _____ of the _____ device.

7. A service is the conductors and equipment for delivering electric energy from the _____ to the wiring system of the premises being wired.

8. A branch-circuit is the circuit conductor between the final _____ device protecting the circuit and the outlet(s).

9. The neutral conductor is _____ by all of the phase conductors in a multi-wire circuit.

10. In a multiwire circuit, the neutral conductor will carry the maximum _____ of the load on the circuits.

11. A building that consists solely of one dwelling unit is a _____ .

12. A _____ is a device that is intended to open a circuit within an established period of time when it detects a current to ground that exceeds a predetermined value that is less than that required to operate the overcurrent protective device of the circuit.

13. The laundry load is permitted to be included with the _____ subjected to the demand factors of *220.11*.

14. The bathroom 20-ampere branch-circuit may supply more than one bathroom, and this circuit _____ outlet(s).

15. *NEC® 220.21*, Non-Coincident Loads, allows us to use either the electric heating load or the air-conditioning load _____ .

16. The general lighting load that includes the small appliance and _____ is always a part of the neutral feeder computation.

17. For a dwelling occupancy, a unit load per sq ft of _____ is required.

FREQUENTLY ASKED QUESTIONS

Question 1: Why do they call the white wire a neutral when it is not part of a multiwire circuit?

Answer: It is incorrect to call the grounded conductor a neutral when it is not used in a multiwire circuit. It cannot be neutral to one wire. It must be neutral to two or more wires. The white wire is connected to the neutral bar and common practice has caused this wire to be mistakenly called "neutral."

Question 2: Why do we call this neutral wire the grounded conductor?

Answer: The neutral wire is always the conductor that is intentionally grounded. See *250.26* of the *NEC*.

Question 3: What does it mean when they say the neutral conductor is shared?

Answer: When a neutral conductor is used as a return conductor in a circuit that consists of two or more ungrounded conductors that have a potential difference between them and the neutral conductor (a multiwire circuit), the neutral conductor is being shared or used by each of the ungrounded conductors.

Question 4: Does the *NEC®* require color-coding for the ungrounded conductors of a circuit?

Answer: *NEC® 210.4(D)* contains the requirements for identification of ungrounded conductors. This identification is only required where more than one voltage system exists in a building. Each ungrounded conductor of a multiwire branch-circuit, where accessible, shall be identified by phase and system. Identification shall be permitted to be by color-coding, marking tape, tagging, or other approved means.

Question 5: Is the neutral wire always the white wire?

Answer: No, and the white wire is not always the neutral wire. In a 2-wire branch-circuit, the white wire is the return conductor to the source. It is never neutral. To be neutral it must have a relationship to more than one wire. It cannot be neutral to one wire.

Question 6: Is the neutral wire always a grounded wire?

Answer: No. As indicated previously, it is always the wire that is intentionally grounded, but we have systems that are ungrounded. These systems also have multiwire circuits that have conductors that become neutral and are not grounded. Chapter 4 on "Grounding" will cover this in more detail.

Question 7: Why do we use the term "wire" sometimes and "conductor" at other times?

Answer: The term conductor is more correct. The conductors of systems and circuits are not always a wire. In Chapter 1, we learned that a conductor may be a wire, or a busbar, or a printed circuit, or the use of some other conductive material.

Question 8: Are service-entrance conductors considered feeders?

Answer: No. Service-entrance conductor requirements are given in *Article 230* and are defined in *Article 100* as conductors between the service equipment and the point of connection to the serving utility at the service point. Feeder requirements are given in *Article 215* and are defined as the conductors between the service equipment and the final branch-circuit overcurrent device.

Question 9: What does nominal voltage mean?

Answer: It means *in name only* according to Webster. In electrical work when we say the system voltage will be 120/240 volts *nominal*, we mean 120/240 volts subject to the variations caused by resistance, voltage drop, overloading of utility facilities, and any reactance not compensated for.

Question 10: When computing loads, how do we handle fractions of an ampere?

Answer: Computations resulting in fractions of an ampere smaller than 0.5 shall be permitted to be dropped.

CHAPTER 3

Residential Outlets

OBJECTIVES

After completing this chapter, the student should understand:
- how to define a dwelling unit.
- receptacle outlets required in a dwelling unit.
- lighting outlets required in a dwelling unit.
- GFCI protection.
- receptacle outlet spacing requirements.
- small-appliance receptacle outlet requirements.
- laundry receptacle outlet requirements.
- general use receptacles.

KEY TERMS

Attached Garage: A garage for automotive vehicles that is a physical part of the residential structure and can be entered without leaving the residential structure

Dedicated Space: A space designated for the use of a defined appliance or other equipment

Detached Garage: A garage for automotive vehicles which is separate from the residential structure and can only be accessed by leaving the residential structure

Dwelling, Multifamily: A building that contains three or more dwelling units

Dwelling, One-family: A building that consists solely of one dwelling unit

Dwelling, Two-family: A building that consists solely of two dwelling units

Dwelling Unit: One (or more) rooms for the use of one (or more) persons as a housekeeping unit with space for eating, living, and sleeping, and permanent provisions for cooking and sanitation

Habitable Room: A room in a residential structure designed for human occupancy with facilities for human existence such as, a living room, dining room, kitchen, bathroom, bedroom, and similar rooms

Lighting Outlet: An outlet intended for the direct connection of a lampholder, a luminaire (lighting fixture), or a pendant cord terminating in a lampholder

Outlet: A point on a wiring system at which current is taken to supply utilization equipment

Receptacle: A contact device installed at the outlet for connection to an attachment plug. A single receptacle is a single contact device with no other contact device on the same yoke. A multiple receptacle is two or more contact devices on the same yoke

Receptacle Outlet: An outlet where one or more receptacles are installed

INTRODUCTION

The necessary **outlets** for a dwelling unit are those that will satisfy the users' needs. There are, however, requirements for dwelling unit outlets that must be followed in order to attain the safety that the *NEC®* demands.

RESIDENTIAL RECEPTACLE OUTLETS REQUIRED

General Provisions

The *NEC®* requirements for receptacle outlets are found in *210.52*. The **receptacle outlets** required by this section are in addition to any **receptacle** that is a part of a luminaire (lighting fixture) or appliance, located within cabinets, or located more than 5½ ft (1.7 m) above the floor (Figure 3-1).

Receptacles that are factory installed in separate assemblies, as part of permanently installed electric baseboard heaters, are permitted as the required receptacle outlets in the wall space they occupy. These receptacles are allowed because some baseboard heater instructions will *not* permit the heaters to be installed below receptacle outlets. The temperature on the top of baseboard heaters is great enough to deteriorate the insulation on cords that would be plugged into the receptacles installed above the baseboard heaters. Remember, according to *210.52*, these receptacle outlets are not permitted to be connected to the heating circuits of the baseboard heaters (Figure 3-2).

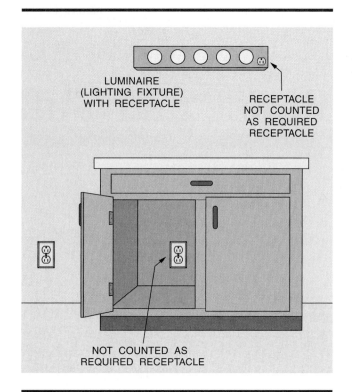

Figure 3-1 Receptacle outlets in cabinets, part of a luminaire (lighting fixture), or located more than 5½ ft (1.7 m) above the floor are not counted as required receptacles

Receptacle outlets are required in every kitchen, family room, dining room, living room, parlor, library, den, sunroom, bedroom, recreation room, or similar rooms of **dwelling units**. (Figure 3-3 shows various types of receptacles.)

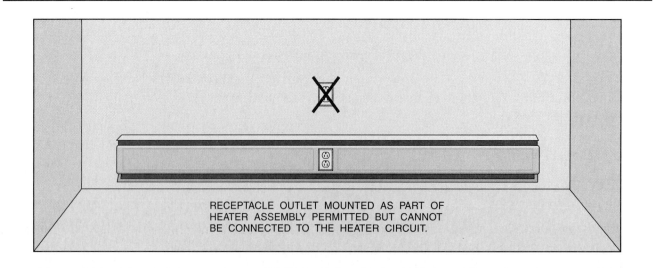

Figure 3-2 Receptacle outlets are not permitted above baseboard heaters, factory-installed receptacle outlets are permitted

CHAPTER 3   Residential Outlets   47

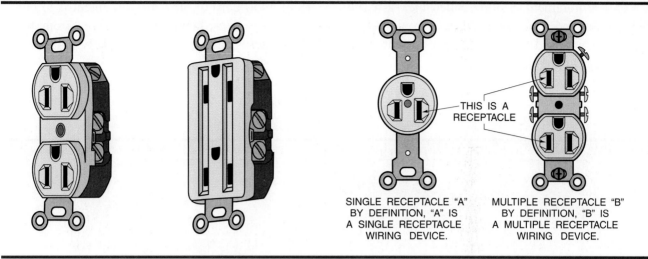

Figure 3-3   Receptacle types

Receptacle outlets must be installed so that no point along the floor line, in any wall space, is more than 6 ft (1.8 m), measured horizontally, from a receptacle outlet in that space. A wall space is any space 2 ft (600 mm) or more in width, measured around corners, and unbroken by doorways, fireplaces, and similar openings. Wall spaces include the space occupied by fixed panels in exterior walls (excluding sliding panels) and the space afforded by fixed room dividers such as freestanding bar-type counters or railings. Placing receptacle outlets in this manner will allow a lamp or portable appliance to be placed anywhere in the room without the use of extension cords (Figure 3-4).

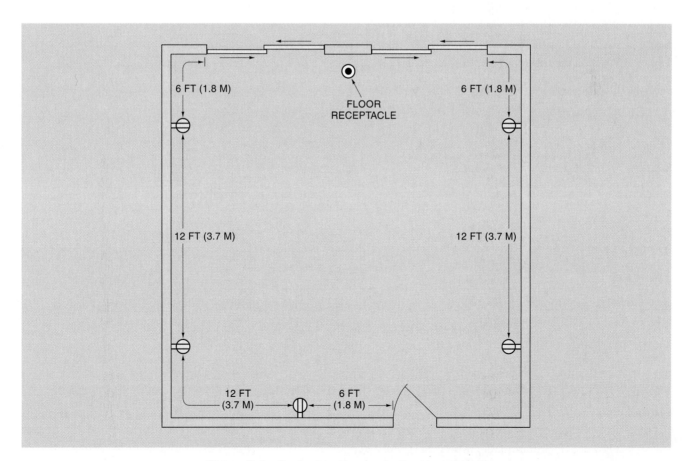

Figure 3-4   Typical wall receptacle arrangement

This requirement is often called the 12-ft rule because the end result is a receptacle outlet every 12 ft. Bear in mind that there is much more in the spacing requirements than a 12 ft rule.

Receptacle outlets in floors shall not be counted as part of the required outlets unless located within 18 in. (450 mm) of the wall. The intent is to eliminate cords from lamps or appliances from being placed across areas where it is likely that persons will be walking (Figure 3-5).

A receptacle outlet shall be installed wherever flexible cords with attachment plugs are used.

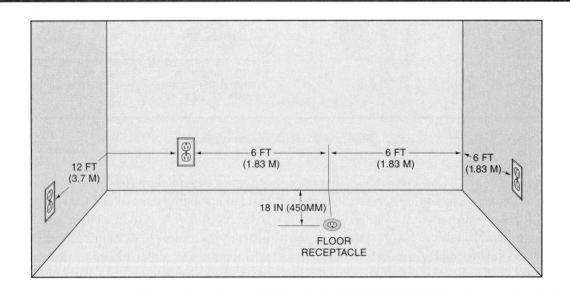

Figure 3-5 Floor receptacles must be within 18 in. (450 mm) of the wall to be counted as required receptacle

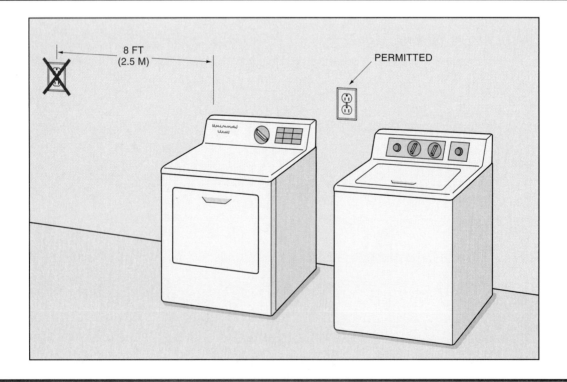

Figure 3-6 Receptacle outlets for laundry equipment within 6 ft (1.8 m) of appliance location

Flexible cords are not permitted to be permanently connected in a dwelling unit. Flexible cords shall not be substituted for the fixed wiring of a building.

Receptacle outlets installed in a dwelling unit for specific appliances, such as laundry equipment, shall be installed within 6 ft (1.8 m) of the intended location of the appliance. This will help ensure that the flexible cord furnished with the appliance will reach the receptacle outlet without the use of extension cords (Figure 3-6).

Small Appliance Receptacles. In the kitchen, pantry, breakfast room, dining room, or similar area of a dwelling unit, the two or more small-appliance branch-circuits required by *210.11(C)(1)* shall serve all of the receptacle outlets required. The refrigeration equipment receptacle is permitted to be supplied from an individual branch-circuit rated 15 amperes or greater. This provision in *210.52(B)(1), Exception No. 2* allows the refrigeration receptacle to be taken from the small-appliance branch-circuit and put on an individual branch-circuit. Remember, an individual branch-circuit is defined as a branch-circuit that supplies only one utilization equipment.

Wall Counter Spaces. The spacing of small-appliance receptacles is shown in *210.52(A)(1)*, but there are special requirements for specific locations. A receptacle outlet must be installed at each wall counter that is 12 in. (300 mm) or wider. Receptacle outlets must be installed so that no point along the wall line is more than 24 in. (600 mm) measured horizontally, from a receptacle outlet in that space. Receptacle outlets shall not be installed more than 18 in. (450 mm) above the countertop (Figure 3-7).

Receptacle outlets installed to serve countertop surfaces are required to have GFCI protection for personnel. See *210.8(A)(5)*.

The wall space located behind the kitchen sink need not be counted when installing kitchen wall receptacles. Figure 3-8 illustrates the placement of wall receptacles at the kitchen sink.

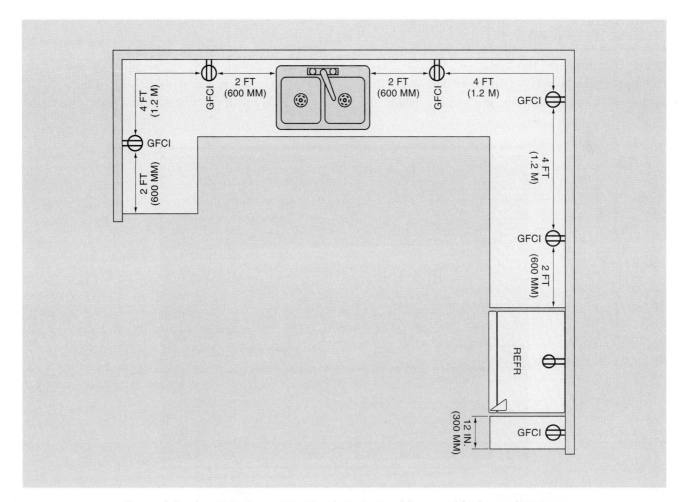

Figure 3-7 Countertop receptacle placement with ground-fault requirements

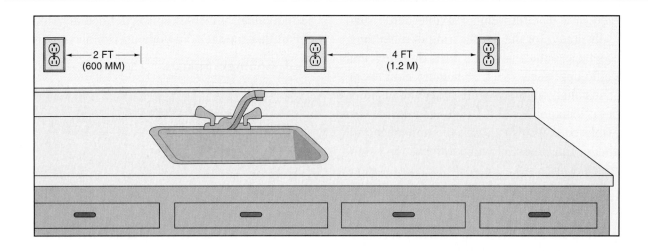

Figure 3-8 Countertop wall receptacle spacing at sink

Island Counter Spaces. At least one receptacle outlet must be installed at each island counter space that has a long dimension of 24 in. (600 mm), or greater, and a short dimension of 12 in. (300 mm), or greater (Figure 3-9).

Receptacle outlets shall not be installed in a face-up position in an island countertop. Where the island countertop is flat across its entire surface, receptacle outlets may be installed not more than 12 in. (300 mm) below the countertop, where the countertop does not extend more than 6 in. (150 mm) beyond its support base (Figure 3-10).

Receptacle outlets installed to serve countertop surfaces are required to have GFCI protection for personnel. See *210.8(A)(6)*.

Peninsular Counter Spaces. At least one receptacle outlet must be installed at each peninsular counter space with a long dimension of 24 in. (600 mm), or greater, and a short dimension

Figure 3-9 Receptacle outlet at island counter

CHAPTER 3 Residential Outlets 51

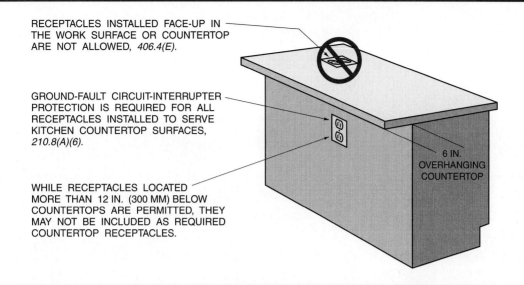

Figure 3-10 Island countertop receptacle placement

Figure 3-11 Peninsular receptacle in residence

of 12 in. (300 mm), or greater (Figure 3-11). A peninsular countertop is measured from the connecting edge.

Receptacle outlets installed to serve countertop surfaces are required to have GFCI protection for personnel. See *210.8(A)(6)* (Figure 3-12).

Receptacle outlets shall not be installed in a face-up position in a peninsular countertop. When the peninsular countertop is flat across its entire surface, receptacle outlets may be installed not more than 12 in. (300 mm) below the countertop, where the countertop does not extend 6 in. (150 mm) beyond its support base (Figure 3-13). The required receptacle outlet may be installed in a cabinet, if available, above the countertop, but not more than 18 in. (450 mm) above the countertop.

52 CHAPTER 3 Residential Outlets

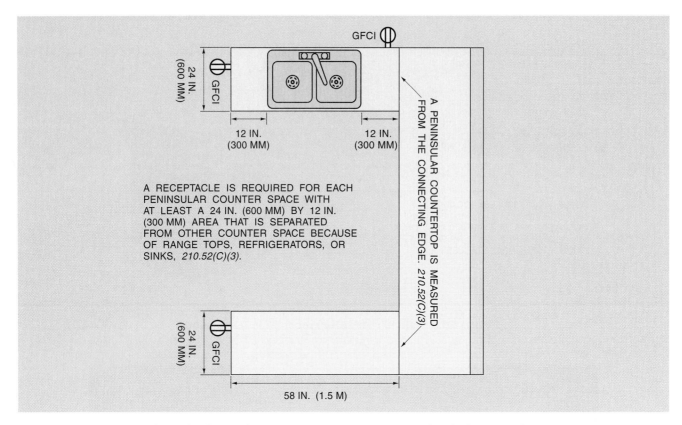

Figure 3-12 Peninsular receptacle placement with GFCI protection

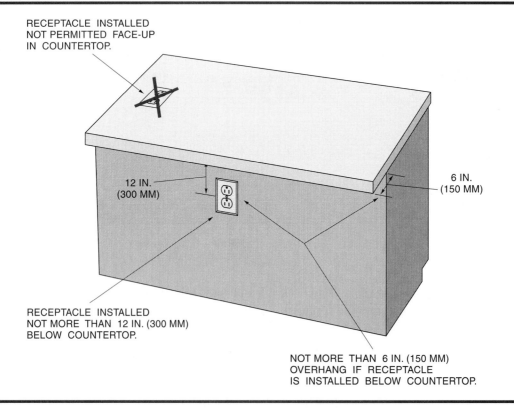

Figure 3-13 Island receptacle placement

Figure 3-14 Bathroom receptacle with GFCI protection

Bathrooms. At least one wall receptacle outlet shall be installed in dwelling unit bathrooms within 36 in. (900 mm) of the outside edge of each basin (Figure 3-14). The receptacle outlet must be installed on a wall that is adjacent to the basin.

Receptacle outlets must *not* be installed in a face-up position in a countertop or work surface in a bathroom basin location.

Receptacle outlets installed in bathrooms are required to have GFCI protection for personnel. See *210.8(A)(1)* (Figure 3-15).

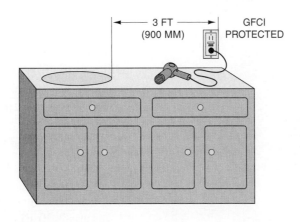

Figure 3-15 Bathroom receptacle within 36 in. (900 mm) of sink

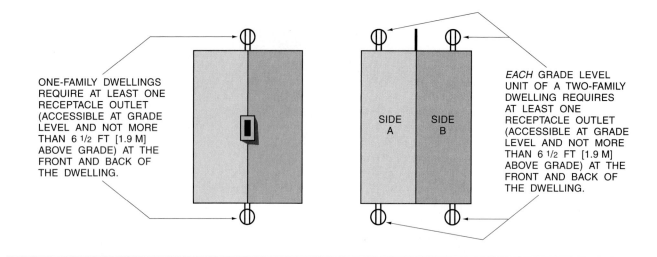

Figure 3-16 Required outdoor receptacles

Outdoor Receptacle Outlets. For a **one-family dwelling** and each unit of a **two-family dwelling** that is at grade level, at least one receptacle outlet accessible at grade level, and not more than 6½ ft (2.0 m) above grade, shall be installed at the front and the back of the dwelling (Figure 3-16).

Outdoor receptacle outlets must have GFCI protection for personnel in accordance with *210.8(A)(3)* (Figure 3-17).

The only exception to this requirement is outdoor receptacle outlets that are supplied by a dedicated branch-circuit for electric snow-melting or deicing equipment and are not readily accessible (Figure 3-18). The important thing is whether the outdoor receptacle outlet is readily accessible and could be used for some other purpose than its intended use.

Laundry Areas. The laundry areas of dwelling units require at least one receptacle outlet to serve this area (Figure 3-19).

There are some exceptions to this requirement; a dwelling unit in a **multifamily dwelling** where laundry facilities are provided for the use of all occupants, a laundry receptacle outlet is not required. In other than a one-family dwelling where laundry facilities are not permitted, a laundry receptacle outlet shall not be required.

The receptacles installed in the laundry area are permitted to serve other purposes than the washer and dryer. A receptacle may be installed for an ironing facility. These receptacle outlets may all be installed on the laundry circuit required by *210.11(C)(2)* (Figure 3-20).

Figure 3-20 illustrates a receptacle outlet for a gas dryer. If an electric dryer were used, a 240-volt circuit would be required for this outlet as shown by the line.

Figure 3-17 Residential outdoor receptacle ground-fault protected label

CHAPTER 3 Residential Outlets 55

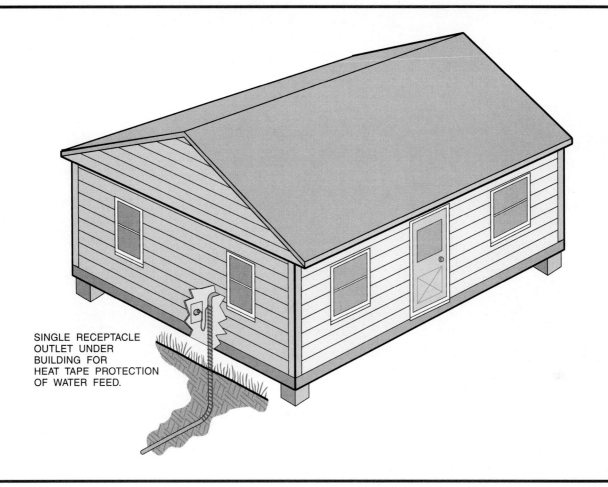

Figure 3-18 Outdoor receptacle outlet for deicing or snow-melting. Does not require ground-fault protection if not readily accessible

Figure 3-19 Laundry area receptacle

CHAPTER 3 Residential Outlets

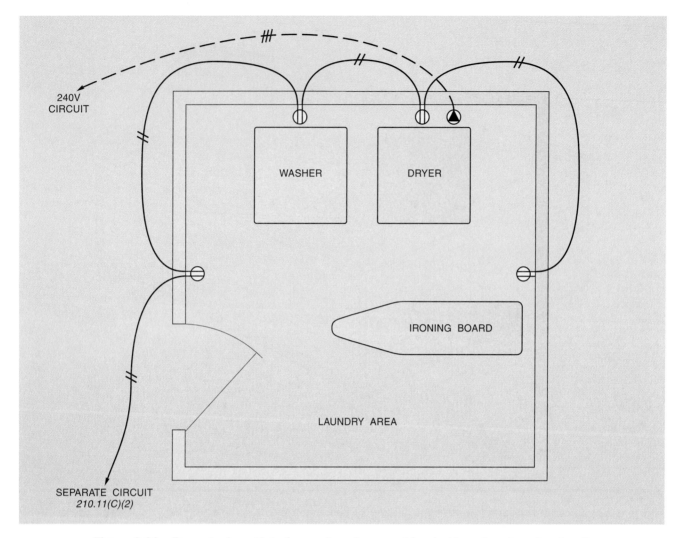

Figure 3-20   Receptacle outlets for washer, dryer, and ironing board on laundry circuit

Garages. For a one-family dwelling, at least one receptacle outlet shall be installed in an **attached garage** (Figure 3-21).

At least one receptacle outlet must be installed in each **detached garage** with electric power. If a detached garage does not have any electric power run to it, then, of course, no receptacle outlet is required. All receptacles installed in garages, either attached or detached, must have GFCI protection for personnel. See *210.8(A)(2)* (Figure 3-22).

There are two exceptions to this requirement. *First*, receptacle outlets that are not readily accessible do not need GFCI protection. An example of this is the receptacle outlet used for the garage door opener (Figure 3-23).

This receptacle outlet is not considered readily accessible because a ladder is required to reach this outlet. *Second*, receptacle outlets installed in

Figure 3-21   Receptacle in attached garage

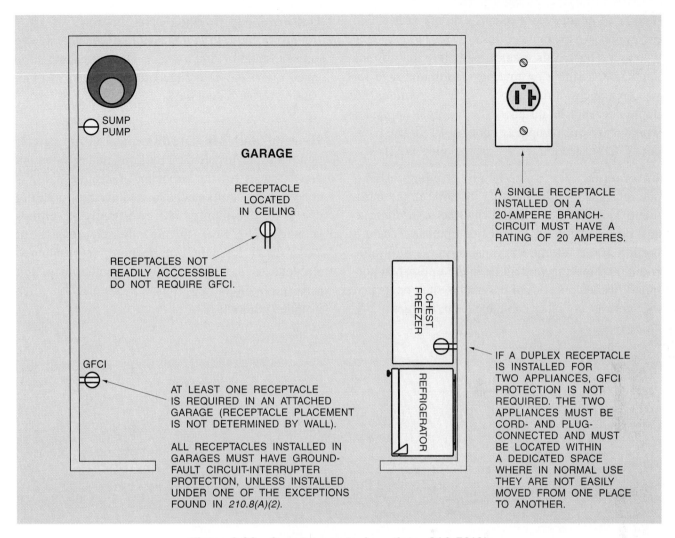

Figure 3-22 Garage receptacle outlets, *210.52(G)*

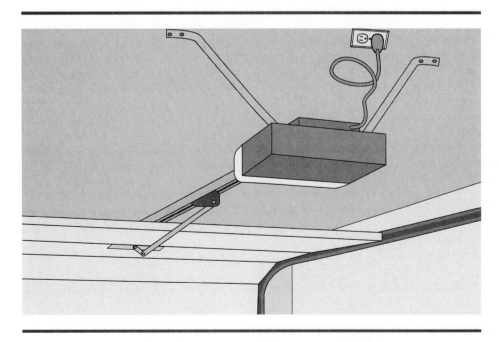

Figure 3-23 Garage door opener receptacle not readily accessible

58 CHAPTER 3 Residential Outlets

dedicated spaces for appliances do not need GFCI protection. An example of this is a freezer located in a garage. This requirement is explicit in that for one appliance, a single receptacle is permitted, and for two appliances, a duplex receptacle is permitted. A duplex receptacle installed for a single appliance would leave one receptacle outlet open for use without GFCI protection for the user (Figure 3-22).

Basements. For a one-family dwelling, at least one receptacle outlet must be installed in the basement. This receptacle outlet shall be *in addition* to any provided for the laundry equipment. Where a portion of the basement is finished into a **habitable room**, this receptacle shall be in the unfinished portion. Finished portions of basements shall be treated the same as any other habitable room (Figure 3-24).

Hallways. In dwelling units, hallways of 10 ft (3.0 m) or more in length shall have one receptacle outlet installed (Figure 3-25). This measurement is taken down the centerline of the hallway, without passing through doorways.

Heating and Air Conditioning. A receptacle outlet must be installed on rooftops, and in attics and crawl spaces for the servicing of heating and air-conditioning equipment. This requirement does not apply to one-family and two-family dwellings (Figure 3-26). Keep in mind the requirements of *210.8(A)(3)* for GFCI protection for personnel. Receptacles for heating and air-conditioning unit servicing are required to have this protection when installed outdoors.

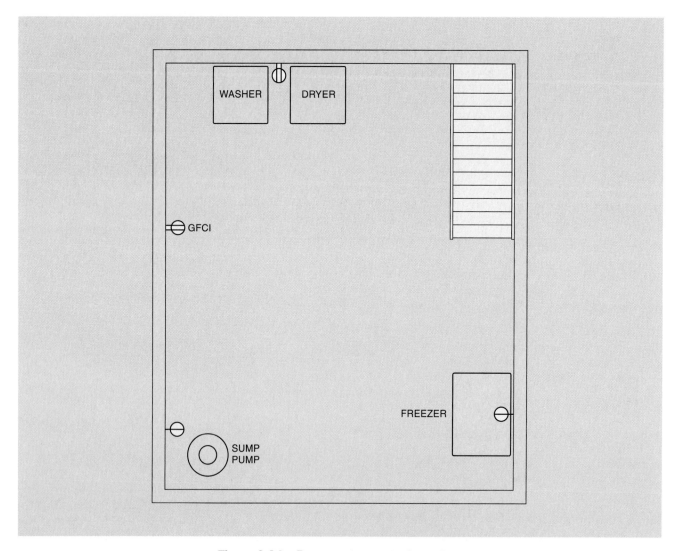

Figure 3-24 Basement receptacle outlets

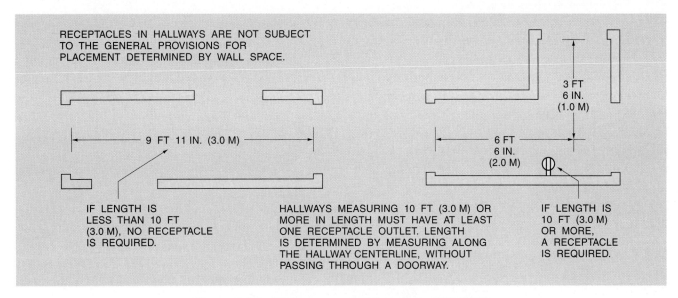

Figure 3-25   Hallway receptacle outlet, *210.52(H)*

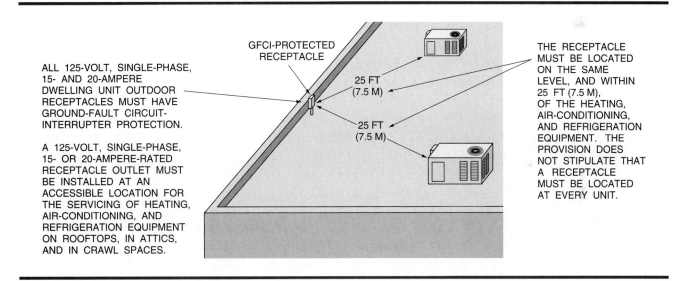

Figure 3-26   Receptacle outlet for servicing HVAC equipment on roof tops, *210.63*

RESIDENTIAL LIGHTING OUTLETS REQUIRED

In dwelling units, in accordance with *210.70(A)(1)*, **lighting outlets** must be installed in every habitable room. The lighting outlets must be wall switch-controlled (Figure 3-27). In rooms other than kitchens and bathrooms, one or more receptacles controlled by a wall switch may be used in lieu of lighting outlets. Occupancy sensors shall be permitted to be used in addition to the wall switch, or if they are located at the switch and are equipped with a manual override that will allow the sensor to act as a switch.

In accordance with *210.70(A)(2)*, wall switch-controlled lighting outlets are required in hallways, stairways, garages, detached garages with electric power, and on the exterior side of outdoor entrances or exits that have grade level access (Figure 3-28). Remember, grade level access can mean a stairway from grade level to the entrance.

60 CHAPTER 3 Residential Outlets

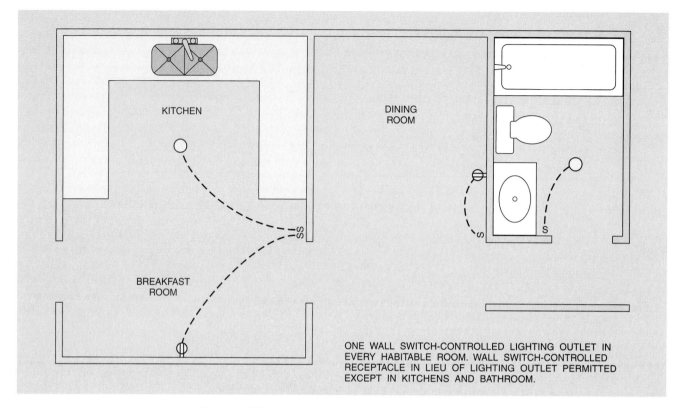

Figure 3-27 Lighting outlets required, *210.70(A)*

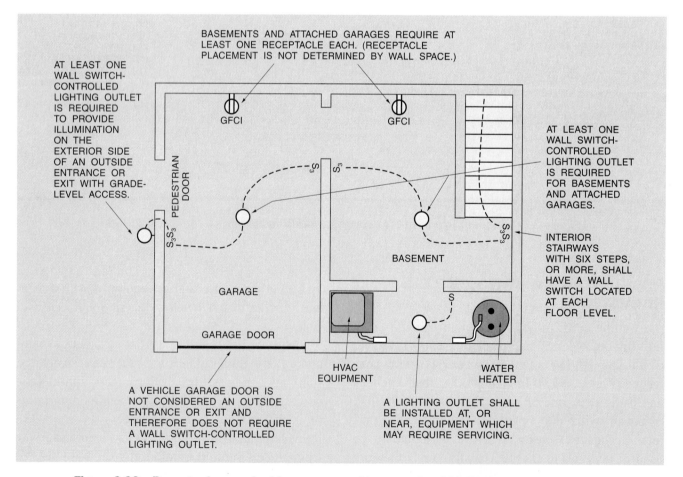

Figure 3-28 Receptacles required in garages and basements, *210.52(G)* and *210.70(A)(2)*

Where lighting outlets are installed in interior stairways, *210.70(A)(2)(c)* requires that a wall switch must be installed at each level to control the lighting where the difference between floor levels is six steps or more. To accomplish this requirement, 3-way and 4-way switches must be used to control the lighting outlets (Figure 3-29).

The following diagrams will illustrate the proper wiring of 3-way and 4-way switches. Because all the wiring in dwelling units are required to be grounded, we will use the terms grounded and ungrounded to identify the conductors. Remember, the grounded conductor must never be switched.

In wiring 3-way switches, the single (common) screw terminal of one switch is connected to the source. The other two screw terminals are connected to the two identical screw terminals of the second switch. These are known as the travelers. The common screw on the second switch is connected to the luminaire (lighting fixture). The grounded conductor is connected directly to the load. In practical wiring, there will always be three wires between switches. Note that only the ungrounded conductor is switched.

When it is desired to control a luminaire (lighting fixture) from more than two locations, then 4-way switches must be added to the circuit. You can add as many 4-way switches as locations are desired. The circuit wiring is the same as for the 3-way switches, except the travelers are broken through the 4-way switches (Figure 3-30).

Closet lights are not required to be installed by the *NEC,®* but there are specific requirements for closet lights if they are installed. Closet lights are addressed in *410.8*. This section is very explicit in defining a storage space in a closet and should be studied very carefully. Luminaires (fixtures) in closets are limited to two types of luminaires (fixtures). Surface-mounted or recessed incandescent luminaires (fixtures) with completely enclosed lamps and

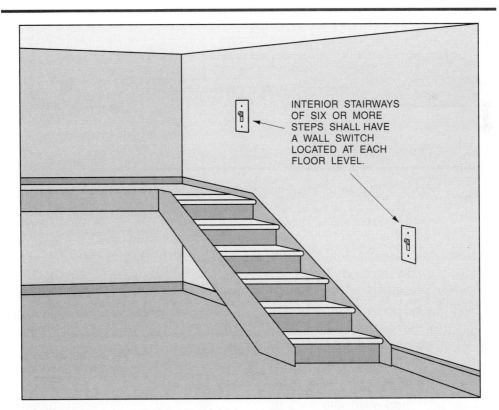

A MINIMUM OF ONE LIGHTING OUTLET IS REQUIRED IN EACH HALLWAY AND STAIRWAY. THE LIGHT MUST BE WALL-SWITCH CONTROLLED, EXCEPT WHERE CONTROLLED BY REMOTE, CENTRAL, OR AUTOMATIC MEANS. A RECEPTACLE CONTROLLED BY A WALL SWITCH IS NOT PERMITTED.

Figure 3-29 Stairway lighting switch-controlled *210.70(A)(2)(c)*

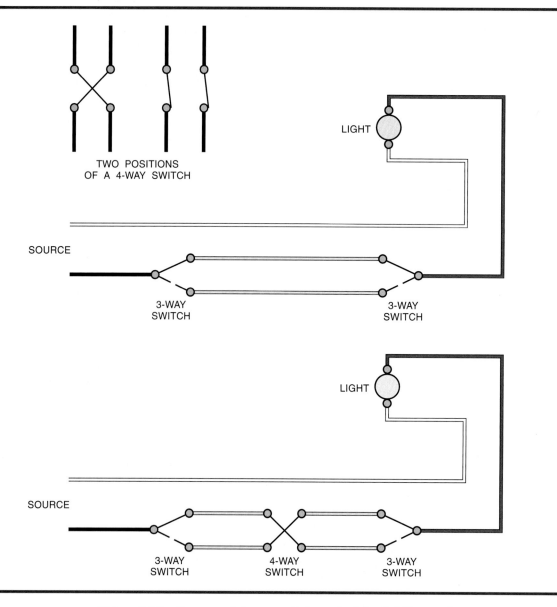

Figure 3-30 3-way and 4-way switch control for lighting outlets

surface-mounted or recessed fluorescent luminaires (fixtures) are permitted. Incandescent luminaires (fixtures) with open or partially enclosed lamps and pendant fixtures are not permitted.

Surface-mounted incandescent luminaires (fixtures) can be installed on the wall above the door or on the ceiling, provided there is a minimum of 12 in. (300 mm) clearance between the luminaire (fixture) and the nearest point of storage. This means that a closet with a 12 in. (300 mm) shelf and a 12 in. (300 mm) clearance would have to be 24 in. (600 mm) plus the luminaire (fixture) dimension in depth to use this type of luminaire (fixture). A standard closet is usually 24 in. (600 mm) deep. This would appear to rule out the use of surface-mounted incandescent luminaires (fixtures) in standard closets (Figure 3-31).

Surface-mounted fluorescent luminaires (fixtures) can be installed on the wall above the door or on the ceiling, provided there is a minimum clearance of 6 in. (150 mm) between the luminaire (fixture) and the nearest point of storage. This type of luminaire (fixture) would not appear to be allowed in a standard closet with a depth of 24 in. (600 mm) (Figure 3-31).

Recessed incandescent luminaires (fixtures) with a completely enclosed lamp or a recessed fluorescent luminaire (fixture) may be installed in the wall or in the ceiling, provided there is a minimum clearance of 6 in. (150 mm) between the luminaire (fixture) and the nearest point of storage (Figure 3-31).

REVIEW QUESTIONS

MULTIPLE CHOICE

1. Receptacles required by *210.52* cannot be installed more than _____ above the floor.
   ____ A. 5½ ft (1.7 mm)
   ____ B. 6 ft (1.8 mm)
   ____ C. 12 in. (300 mm)
   ____ D. 24 in. (600 mm)

2. Receptacle outlets shall not be installed more than _____ above the kitchen countertop.
   ____ A. 24 in. (600 mm)
   ____ B. 12 in. (300 mm)
   ____ C. 18 in. (450 mm)
   ____ D. 6 in. (150 mm)

3. Wall switch-controlled lighting outlets must be installed in which of these areas?
   ____ A. garages
   ____ B. hallways
   ____ C. stairways
   ____ D. all of the above

4. Rooftop receptacle outlets must be installed for air-conditioning servicing on which of the following dwellings?
   ____ A. one-family
   ____ B. two-family
   ____ C. multifamily
   ____ D. none of the above

5. At least one bathroom wall receptacle shall be installed not more than _____ from the outside edge of each basin.
   ____ A. 12 in. (300 mm)
   ____ B. 18 in. (450 mm)
   ____ C. 36 in. (900 mm)
   ____ D. 24 in. (600 mm)

6. The required outdoor receptacles in the front and rear of a one-family dwelling must not be more than _____ above grade.
   ____ A. 6 ft (1.8 m)
   ____ B. 6½ ft (2.0 m)
   ____ C. 36 in. (900 mm)
   ____ D. 24 in. (600 mm)

7. In dwelling units, hallways of _____ or more in length shall have at least one receptacle outlet installed.
   ____ A. 10 ft (3.0 m)
   ____ B. 6 ft (1.8 m)
   ____ C. 20 ft (6.0 m)
   ____ D. 15 ft (4.5 m)

CHAPTER 3 Residential Outlets

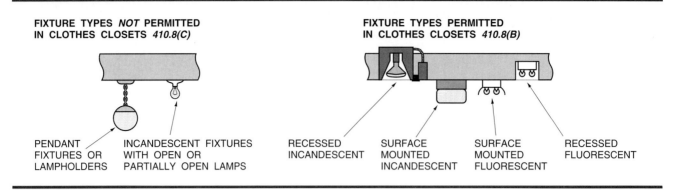

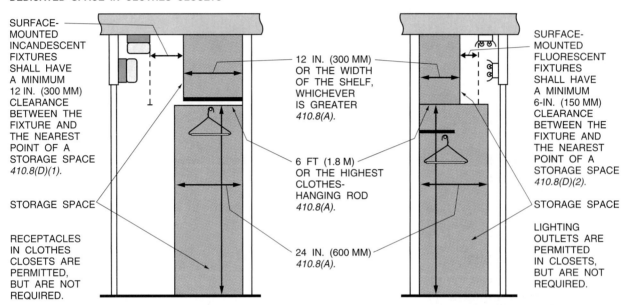

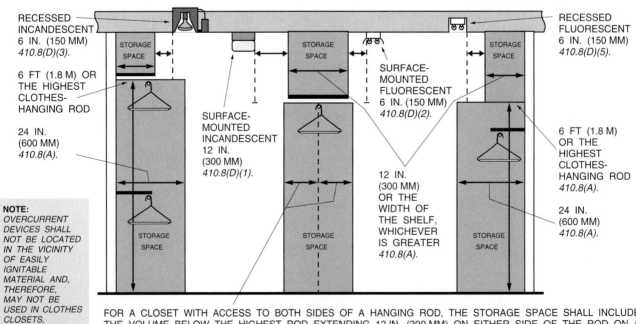

Figure 3-31 Clothes closets

8. In dwelling units, on interior stairways, a wall switch must be installed at each level to control the lighting where the difference between floor levels is _____ steps or more.
   _____ A. three
   _____ B. four
   _____ C. six
   _____ D. eight

9. The two conductors run between 3-way switch terminals are known as the:
   _____ A. switch legs.
   _____ B. grounds.
   _____ C. travelers.
   _____ D. none of the above

10. The requirements for GFCI protection in dwelling units is found in _____ .
    _____ A. *NEC® 210.8*
    _____ B. *NEC® 220.3*
    _____ C. *NEC® 210.11*
    _____ D. *NEC® 220.16*

TRUE OR FALSE

1. Wall space occupied by sliding panels in exterior walls should not be included in determining receptacle outlets required.
   _____ A. True
   _____ B. False
   Code reference _____ .

2. No point in any wall space should be more than 12 ft (3.7 m) from a receptacle outlet in that wall space.
   _____ A. True
   _____ B. False
   Code reference _____ .

3. Outdoor receptacles may be supplied by the small-appliance branch-circuit.
   _____ A. True
   _____ B. False
   Code reference _____ .

4. Receptacle outlets installed in the kitchen to serve countertop surfaces shall be supplied by not less than two small-appliance branch-circuits.
   _____ A. True
   _____ B. False
   Code reference _____ .

5. Countertop receptacle outlets shall be installed so that no point along the wall line is more than 24 in. (600 mm) from a receptacle outlet in that space.
   _____ A. True
   _____ B. False
   Code reference _____ .

66 CHAPTER 3 Residential Outlets

6. Factory-installed receptacle outlets in baseboard heaters may be connected to the heater circuits.
   ____ A. True
   ____ B. False
   Code reference _____ .

7. Receptacle outlets in floors may not be counted as part of the required outlets in a dwelling unit unless located within 18 in. (450 mm) of the wall.
   ____ A. True
   ____ B. False
   Code reference _____ .

8. A receptacle outlet installed in a dwelling unit for the washer must be installed within 3 ft of the washing machine.
   ____ A. True
   ____ B. False
   Code reference _____ .

9. Receptacle outlets may be installed face-up in island countertop surfaces.
   ____ A. True
   ____ B. False
   Code reference _____ .

10. The required bathroom receptacle outlet must be installed on a wall that is adjacent to the basin.
    ____ A. True
    ____ B. False
    Code reference _____ .

FILL IN THE BLANKS

1. An outlet is _____ at which current is taken to supply utilization equipment.

2. Receptacle outlets in floors _____ as part of the required outlets unless located _____ .

3. A receptacle outlet shall be installed _____ with attachment plugs are used.

4. The _____ are permitted to be supplied from a branch-circuit rated _____ or greater.

5. At least one receptacle outlet shall be installed _____ with electric power.

6. In dwelling units, _____ must be installed in every habitable room. These _____ must be controlled by a wall switch.

7. Surface-mounted incandescent luminaires (fixtures) can be _____ or on the ceiling, provided there is a minimum of 12 in. (300 mm) clearance between the luminaire (fixture) and the _____ storage.

8. Surface-mounted fluorescent luminaires (fixtures) can be _____ or on the ceiling in closets, provided there is a minimum clearance of 6 in. (150 mm) between the luminaire (fixture) and the _____ storage.

9. In dwelling units, hallways of _____ or more in length shall have one receptacle outlet installed.
10. Wall spaces include _____ in exterior walls (excluding sliding panels).

FREQUENTLY ASKED QUESTIONS

Question 1: Why does the *NEC®* use the metric system of measure as well as the inch-pound units or English system as it is sometimes called?

Answer: The *NEC®* has since the 1990 edition used both of these systems of measurement with the inch-pound system first and the metric system, known as the International System of Units (SI), following in parenthesis. Beginning with the 2002 edition of the *NEC®*, the metric system SI units will be used with the inch-pound system in parenthesis following. The *NEC®* may soon be renamed the *International Electrical Code* because of its widespread use throughout the world. The majority of the world uses the metric system of measure, and it will be necessary for the United States to follow their lead if we wish to compete in world markets.

Question 2: I heard the term arc-fault interrupter used, but what does it mean?

Answer: An arc-fault circuit-interrupter is a device intended to provide protection from the effects of arc-faults by recognizing the characteristics unique to arcing and by functioning to de-energize the circuit when an arc-fault is detected. All dwelling unit bedroom branch-circuits supplying 125-volt, single-phase, 15- or 20-ampere receptacle outlets will be required to have arc-fault circuit-interrupter protection effective January 1, 2002. See *210.12*.

Question 3: What exactly is an arc-fault?

Answer: An arc-fault is an arc or flash-over from an energized conductor, either to ground or to another conductor of opposite polarity. A flash-over is a short to ground or to another conductor of opposite polarity, and the short circuit may not have enough current flow to open the overcurrent device but will have the capability of igniting any combustibles it may come in contact with.

Question 4: Are the receptacles used on a small-appliance branch-circuit required to be rated 20-amperes?

Answer: No. Receptacles rated 15-amperes can be used on 15- or 20-ampere rated circuits. The maximum load permitted on a 15-ampere receptacle used on a 20-ampere circuit is 12 amperes. See *Table 210.21(B)(2)*. A 120-volt circuit with a 12-ampere load is rated at 1440 VA. There are many kitchen appliances on the market that are rated in excess of 1440 watts. The *NEC®* is a minimum safety requirement and is not intended as a design specification. The design criteria should always consider the adequacy of the electrical wiring installed.

Question 5: Can 20-ampere rated receptacles be installed on 15-ampere circuits?

Answer: No. The rating of the receptacle must clearly indicate to the user the maximum permissible load allowed on the circuit. Using a 20-ampere receptacle on a 15-ampere circuit would incorrectly indicate to the user that a 16-ampere load could be used with that receptacle.

Question 6: Is a wall switch-controlled lighting outlet required in an attic space?

Answer: A lighting outlet is required in attic spaces only if the attic space is used for storage or contains equipment that requires servicing. The lighting outlet may contain a switch or a wall switch may be used. The lighting outlet must be at, or near, the equipment requiring servicing. One point of control for the lighting outlet must be at the usual point of entry to the space [see *210.70(A)(3)*]. This means that if the equipment is not near the usual point of entry, the switch must be remote from the lighting outlet. An adequate wiring design would dictate the number of lighting outlets and switch locations required.

Question 7: What is a general use receptacle?

Answer: All of the 15- or 20-ampere rated receptacles installed in a dwelling unit, except those required by *210.52(B), (C),* and *(F)*. These receptacles are considered general use receptacles and are included in the general lighting calculations of *220.3(A)*.

Question 8: What is meant by demand factor?

Answer: The ratio of the maximum demand of a system to the total connected load of a system is the demand factor. For example, a 1500 sq ft dwelling unit requires 3 VA per sq ft for general lighting according to *Table 220.03(A)*. This is a load of approximately 4500 watts of lighting. It is difficult to assume that the equivalent of 45-100-watt lights would be on at the same time in that dwelling unit. For this reason, a demand factor allowances is permitted by *220.11*.

Question 9: If I have a dryer rated at 4500 watts, why do I have to compute the load at 5000 watts?

Answer: The dwelling unit and the dryer outlet will still be there when you are gone. The load is computed for the largest dryer that may be installed at that location.

Question 10: Where can I find the requirements for the installation of Romex?

Answer: Romex is a manufacturer's trade or brand name for nonmetallic sheathed cable. Type NM Cable can be found in *Article 334*. Wiring methods can be found in *Article 300*.

CHAPTER 4

Grounding

OBJECTIVES

After completing this chapter, the student should understand:
- the definition of grounding.
- the purpose of grounding.
- grounding electrodes.
- grounding-electrode conductors.
- bonding jumpers.
- equipment grounding conductors.
- GFCIs.

KEY TERMS

Bonding: The permanent joining of metallic parts to form an electrically conductive path that will ensure electrical continuity and the capacity to conduct safely any current likely to be imposed

Bonding Jumper: A reliable conductor to ensure the required electrical conductivity between metal parts required to be electrically conductive

Bonding Jumper, Main: The connection between the grounded circuit-conductor and the equipment grounding conductor at the service

Ground: A conducting connection, whether intentional or accidental, between an electrical circuit or equipment and the earth or to some conducting body that serves in place of earth

Grounded: Connected to earth or to some conducting body that serves in place of earth

Grounded Conductor: A system or circuit conductor that is intentionally grounded

Grounding Conductor, Equipment: The conductor used to connect the noncurrent-carrying metal parts of equipment, raceways, and other enclosures to the system grounded conductor, the grounding-electrode conductor (or both) at the service equipment or at the source of a separately derived system

Grounding-Electrode Conductor: The conductor used to connect the grounding electrode to the equipment-grounding conductor, to the grounded conductor (or both) of the circuit at the service equipment, or at the source of a separately derived system

Ground-Fault Circuit-Interrupter (GFCI): A device intended for the protection of personnel that functions to de-energize a circuit or portion thereof within an established period of time when a current to ground exceeds some predetermined value that is less than that required to operate the overcurrent protective device of the supply circuit

CHAPTER 4 Grounding

INTRODUCTION

Grounding and **bonding** are very complex subjects and many books have been written based entirely on this subject. This chapter will explain the basic fundamentals of grounding and clarify the terminology used in grounding applications. Systems, circuits, and equipment required (permitted or not permitted) to be grounded will be explained. The circuit-conductor required to be grounded on grounded systems, location of grounding connections, types and sizes of grounding and bonding conductors and electrodes, methods of grounding and bonding, and conditions under which guards, isolation, or insulation may be substituted for grounding will be explained. Grounding procedures for various types of buildings will be illustrated and described with reference to the *NEC®* requirements and good design characteristics.

FUNDAMENTALS OF GROUNDING AND BONDING

There are two types of **grounds**. One is intentional and the other is accidental or unintentional. Intentional grounding is required for electrical systems and for electrical equipment. Accidental grounds are faults in electrical wiring systems or circuits.

Grounding Electrical Systems

We will look at intentional grounding of electrical systems first. The *NEC®* requires grounding for some electrical systems. Electrical systems that are required to be grounded must be connected to earth in a manner that will limit any current imposed on that system that may be caused by lightning or unintentional contact with higher voltage lines.

NEC® 250.20(B) requires alternating systems of 50 volts to 1000 volts to be **grounded** under any of the following conditions:

1. Where the system can be grounded so that the maximum voltage to ground on the ungrounded conductors does not exceed 150 volts (Figure 4-1). In Figure 4-1, the midpoint (common conductor) of a 120/240 volt, single-phase, 3-wire secondary is grounded, limiting the maximum voltage to ground to 120 volts.

2. Where the system is 3-phase, 4-wire, wye-connected in which the neutral is used as a circuit-conductor (Figure 4-2). In Figure 4-2, the neutral conductor is grounded. In a 120/208 volt system, the voltage to ground would be 120 volts. In a 480/277 volt system, the voltage to ground would be 277 volts. In this configuration, grounding is required even though the voltage to ground exceeds 150 volts.

3. Where the system is 3-phase, 4-wire delta-connected in which the midpoint of one phase winding is used as a circuit-conductor (Figure 4-3). In Figure 4-3, one phase is tapped at midpoint and is grounded at this point. This conductor is 90° out of phase with phases A and C, but is 120° out of phase with B phase. In a 240-volt delta system, the B phase voltage to ground is 120 × 1.73 or 208 volts.

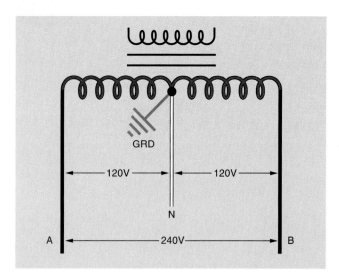

Figure 4-1 Transformer with grounded (neutral) conductor voltage to ground 120 volts

Electrical systems not included in *250.20(B)* are not required to be grounded, however, others may be grounded. An example of a system that is not required to be grounded is a corner-grounded delta system. In this system, the conductor to be grounded is one phase conductor (see Figure 4-4).

In Figure 4-4, B phase is the conductor that is generally grounded. This grounding system results in the voltage to ground being equal to phase voltage.

Systems that cannot be grounded are covered in *250.22* (Figure 4-5).

1. Cranes operating over combustible fibers as shown in *503.13*. A spark caused by a fault to ground could ignite combustible flying fibers.

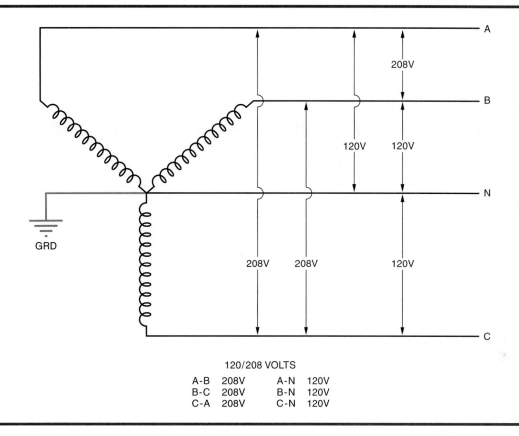

Figure 4-2 Transformer secondary with grounded (neutral) conductor voltage to ground 120 volts

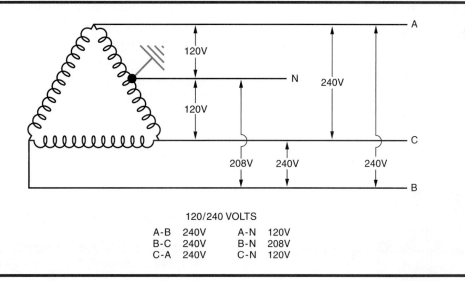

Figure 4-3 Transformer secondary with one phase tapped at midpoint and grounded

2. Circuits in health care facilities as provided in *517.60*. A ground-fault may cause interruption of power to operating rooms and anesthetic induction rooms.

3. Electrolytic cell circuits as provided in *668.3(C)(3)*. A ground-fault could interrupt the dc power resulting in product and equipment damage.

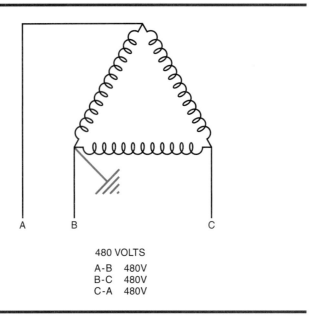

Figure 4-4 Corner grounded 480-volt Delta permitted but not required to be grounded

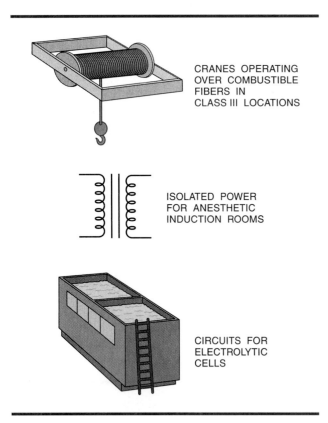

Figure 4-5 Circuits not permitted to be grounded

Intentional grounding of electrical systems is accomplished by connecting one of the circuit or system conductors to earth by means of a conductor called the **grounding-electrode conductor** (Figure 4-6).

As shown in Figure 4-6, the neutral conductor becomes the **grounded conductor** when it is attached to the neutral bar where the grounding-electrode conductor is connected. The neutral conductor, if present, is always the conductor that is intentionally grounded.

The conductor that can be grounded in alternating current systems is covered in *250.26* of the *NEC*® (Figure 4-7).

(A) Single phase, 2-wire
(B) Single phase, 3-wire
(C) Three phase, 4-wire
(D) Three phase, 3-wire

Grounding Electrical Equipment

Now we will look at the intentional grounding of electrical equipment. Electrical equipment is grounded (connected to earth) to limit the voltage to ground on these materials. Electrical equipment grounding is accomplished by means of an **equipment-grounding conductor** and the **main bonding jumper**. The equipment-grounding conductor may be a conductor run with the circuit-conductors, or where metal raceways are used, the raceway may be the equipment-grounding conductor (Figure 4-8).

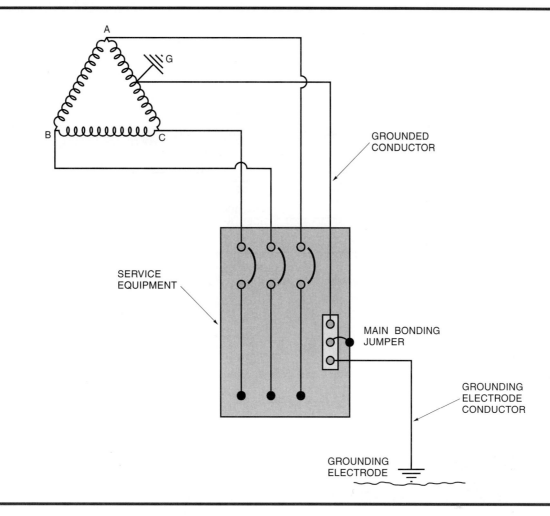

Figure 4-6 Connecting system conductor to earth to establish intentional grounding

The purpose of the equipment-grounding conductor is to establish a path or circuit for the ground-fault current back through the overcurrent device to open the circuit. The complete path for ground-fault current begins at the utility supply transformer and ends through the utility supply transformer (Figure 4-9).

It is important to understand that a low impedance ground-fault path or circuit can never be through the grounding-electrode conductor and back to the utility supply transformer through the earth (Figure 4-9).

The impedance (resistance + reactances) of the grounding-electrode circuit is too great for enough current to flow to open the overcurrent device. The impedance of the circuit consists of the service conductor, the grounding-electrode conductor, the grounding electrode at the service (ground rod), the earth path between the grounding electrodes, the grounding electrode at the utility transformer, that portion of the transformer secondary you use, and whatever inductive reactance is developed by the faulted alternating current circuit.

Because the impedance of the ground path through the grounding-electrode conductor system is too great to allow enough current to flow to open the circuit overcurrent protective device, *250.24(B)* requires that in grounded ac services operating at less than 1000 volts, the grounded conductor shall be run to each service disconnecting means and shall be bonded to each service disconnecting means enclosure (Figure 4-9). This establishes a path of low impedance back to the utility company transformer.

To understand the intentional grounding of electrical equipment, we must look at *250.2(B)* and *(C)* that require: *conductive materials enclosing electrical*

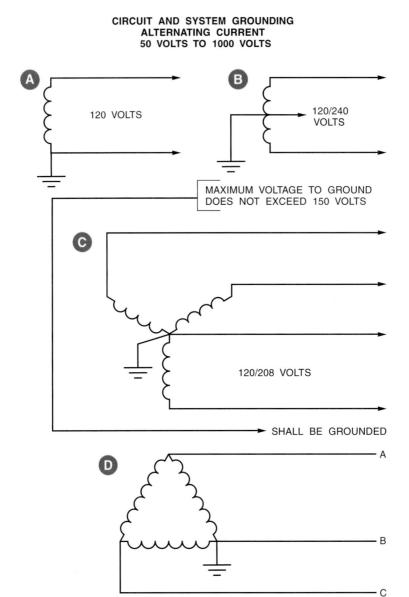

Figure 4-7 Conductor in ac systems required to be grounded (A) single-phase, 2-wire (B) single-phase, 3-wire (C) three-phase, 4-wire (D) three-phase, 3-wire

conductors or equipment shall be connected to earth so as to limit the voltage to ground on these materials and *that electrically conductive materials, such as metal water piping, metal gas piping, and structural steel members that are likely to become energized shall be bonded to the supply system grounded conductor.* "Likely to become energized" is derived from the idea that where wiring is run in close proximity to noncurrent-carrying conductive materials, there is a possibility that a current-carrying conductor is liable to come in contact with these materials. Figure 4-10 shows the grounding of these items.

Why is it so important that we ground this electrical equipment? We have seen where we ground electrical systems or circuits to limit current imposed on the system or circuit by lightning or contact with higher voltage. Now, let us look at the reason for grounding electrical equipment. Generally, you will be working with electrical systems that are supplied

CHAPTER 4 Grounding 75

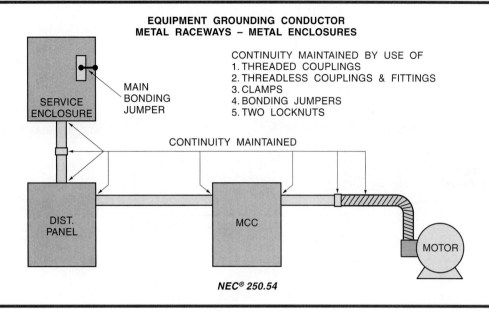

Figure 4-8 Equipment-grounding conductor metal raceway—metal enclosures

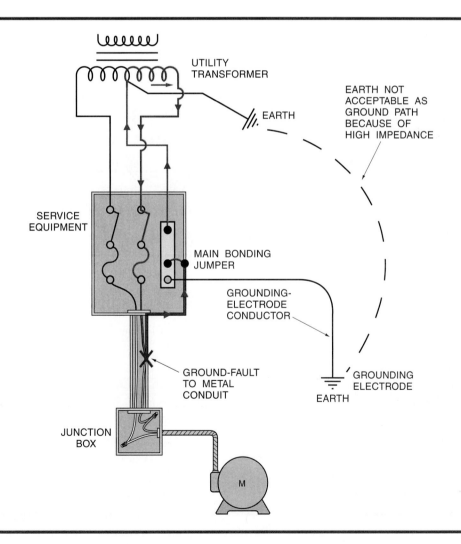

Figure 4-9 Ground-fault current path

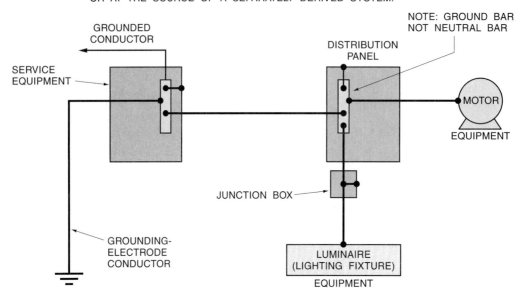

Figure 4-10   Grounding electrical equipment using equipment grounding conductor

by the utility company with the neutral or common conductor grounded at the utility company transformer (Figure 4-9).

This gives us a difference in potential to ground at our wiring system. If an energized conductor should come into contact with a noncurrent-carrying conductive surface, such as a water pipe, a difference in potential would exist between the water pipe and a grounded surface, such as the earth. A person touching the water pipe and earth at the same time would receive an electric shock (Figure 4-11).

By grounding all conductive materials enclosing electrical conductors or equipment by the use of an equipment-grounding conductor and by bonding all electrically conductive materials, such as water piping, an unintentional ground would establish a ground-fault path to *open the overcurrent device* in the faulted circuit and eliminate the danger of electric shock (Figure 4-12). Now that we have covered the fundamentals of grounding, we will look at some practical grounding applications.

PRACTICAL GROUNDING

One-Family Dwelling

We will start our practical grounding with *steps* to ground a one-family dwelling.

Step #1: *Determine the type of grounding electrode that will be used.* In most one-family dwellings, an underground metal water piping system is used as the grounding electrode. *NEC®* 250.50 requires that, if available on the premises, a metal underground water piping system shall serve as the grounding electrode. It was common practice to use the interior metal water piping system as a conductor or bond to the underground metal water piping system, but recent changes in the *NEC®* prohibit interior metal water piping more than 5 ft (1.5 m) from the entrance to the building to be used as a part of the grounding-electrode system or as a conductor to interconnect electrodes that are a part of the grounding-electrode system (Figure 4-13). The reasoning behind this change was that many times nonmetallic fittings or

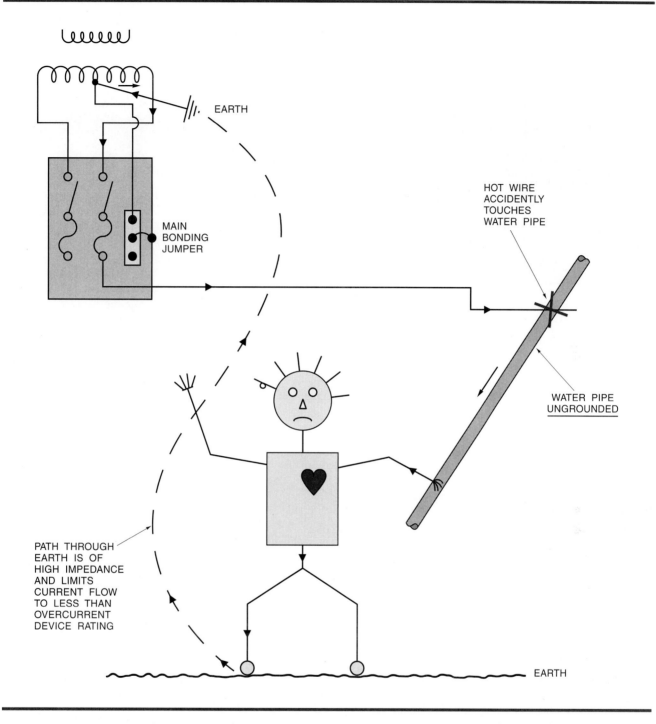

Figure 4-11 Ground-fault path with person completing circuit

piping were introduced into metal piping systems when repairs were made. Much of the interior metal water piping system was not exposed to view and these changes may interrupt the grounding continuity. If available on the premises, each of the following items, a metal underground water pipe, the metal frame of a building, a concrete encased electrode, or a ground ring and any of the following made electrodes, rod, pipe, or a plate shall be bonded together to form the grounding-electrode system.

Because nonmetallic fittings or piping is now being used as the underground water piping system

78 CHAPTER 4 Grounding

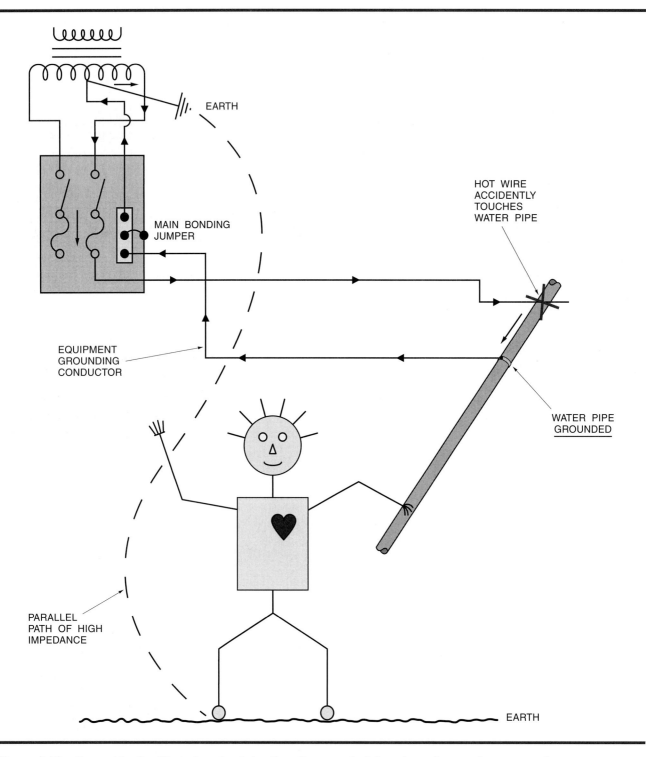

Figure 4-12 Ground-fault with water pipe intentionally grounded. Low impedance allows enough current to open overcurrent device

in many communities, a supplemental electrode must be used with a metal underground water piping system. Generally, a ground rod is used as the supplemental grounding electrode.

A ½ in. (12.7 mm) copper ground rod with not less than 8 ft (2.44 m) in contact with the soil may be used. *NEC® 250.56* requires that a ground rod have a resistance to ground of 25 ohms or less, or it

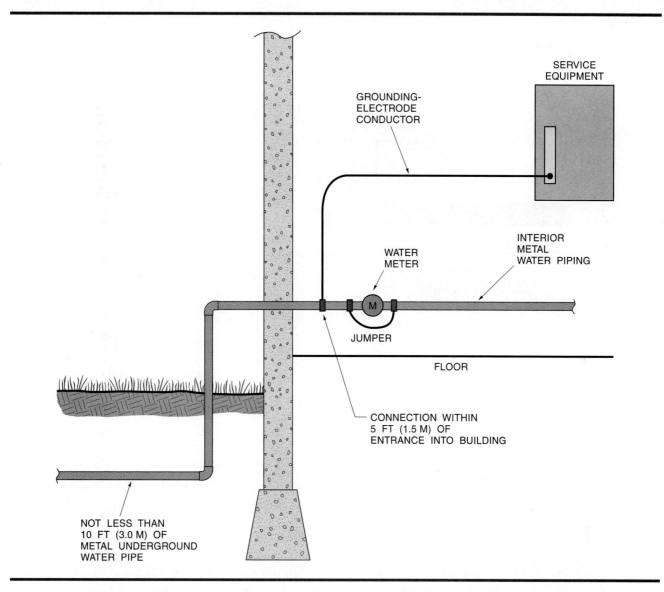

Figure 4-13 Grounding-electrode conductor connection within 5 ft of entrance into building

shall be augmented by one additional ground rod. The supplemental ground rods shall be spaced not less than 6 ft (1.83 m) apart. Paralleling of the ground rods is the same as paralleling resistors. The resistance is reduced by the paralleling of the ground rods. So the result of Step #1 is a metal underground water pipe grounding electrode with two ground rods as supplemental grounding electrodes (Figure 4-14).

Step #2: *Determine the size of the grounding-electrode conductor and bonding jumpers.* The grounding electrode conductor is sized in accordance with the size of the service-entrance conductors as shown in *Table 250.66*. In this example, we will assume a 100-ampere, 120/240-volt, 3-wire, single-phase service. To find what size conductors are required for an ampacity of 100 amperes, we use *Table 310.16*. Using the 75° copper column, we find that a 3 AWG THWN conductor has an ampacity of 100 amperes. We use the 75° column because it fits the conductors generally used today. The 90° column is not used because there are no terminations available for use at 90°. There is another *Table, 310.15(B)(6)* that may be used for dwelling services and feeders of this type. In this *Table*, we find that a 4 AWG THWN copper conductor may be used with a 100-ampere rating for a service or feeder. In either case, *Table 250.66* requires an 8 AWG copper

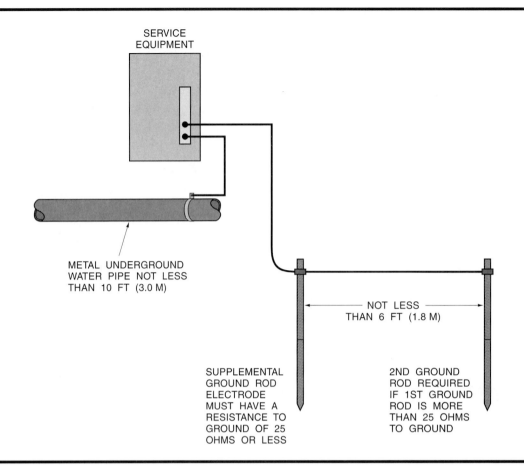

Figure 4-14 Water pipe grounding electrode with ground rod as supplemental electrode and 2nd ground rod required by *250.56*

conductor for the grounding-electrode conductor. *NEC® 250.64* requires that grounding-electrode conductors smaller than 6 AWG shall be in Rigid Metal Conduit, Intermediate Metal Conduit, Rigid Nonmetallic Conduit, Electrical Metallic Tubing, or cable armor (Figure 4-15).

Step #3: *Install the main bonding jumper.* This is the final link for the completion of the ground-fault path back through the grounded circuit-conductor to open the circuit overcurrent protective device. The main bonding jumper is the copper grounding strap that is furnished with the service panel to connect the neutral bar to the metal panel housing, or sometimes it is just a green 8–32 screw that is furnished to install through the neutral bar to the panel housing (Figure 4-16). *NEC® 250.28(D)* requires that the main bonding jumper not be smaller than the grounding-electrode conductor sizes as shown in *Table 250.66*.

In the example (Step #2) we used a one-family dwelling with a 100-ampere, single-phase, 3-wire, 120/240-volt service. A similar service rated at 200-amperes, using *Table 310.16*, would require 3/0 AWG copper service-entrance conductors to be used, or using *Table 310.15(B)(6)*, we are permitted to use 2/0 AWG copper conductors. Referring to *Table 250.66*, both 3/0 and 2/0 AWG copper service-entrance conductors require a 4 AWG copper grounding-electrode conductor. According to *250.64(B)*, a 4 AWG or larger, grounding-electrode conductor shall be protected if exposed to physical damage.

Two-Family Dwelling

A two-family dwelling (commonly called a two-flat) generally has a 2- or 3-gang meter housing, either for overhead or for underground installations (Figure 4-17).

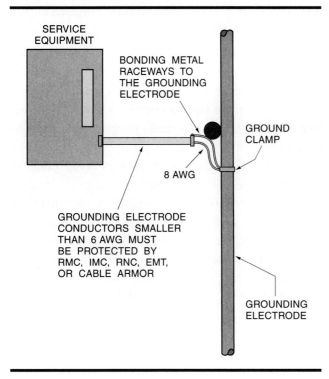

Figure 4-15 Grounding-electrode conductors smaller than 6 AWG

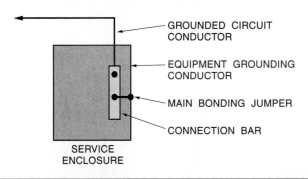

Figure 4-16 Definition of main bonding jumper

The third meter is for the common elements wiring which would include common stair or hall lighting, outside lighting, basement lighting, and in some cases, laundry facility wiring. A commonly used service arrangement for a two-family dwelling is shown in Figure 4-17.

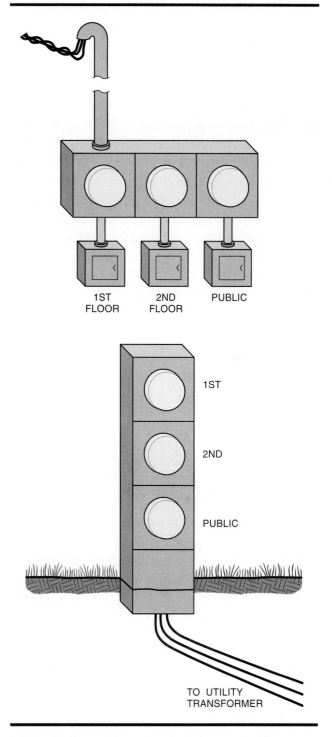

Figure 4-17 Two-family dwelling service with public meter overhead or underground

Step #1: *Determine the grounding electrode that will be used.* This step is generally the same in the installation of a grounding system for any type of building or structure. The grounding electrode used in most buildings will be a metal underground water pipe. The *NEC®* in *250.50* requires that all of the

82 CHAPTER 4 Grounding

electrodes available must be bonded together and used in the grounding-electrode system. In general, only a metal underground water pipe is available and, as required by *250.50(A)(2)*, must be supplemented by an additional electrode, which is generally a ground rod.

Step #2: *Determine the size of the grounding-electrode conductor and bonding jumpers required.* The minimum size of the grounding-electrode conductor is always determined by the size of the service-entrance conductors, as shown in *Table 250.66*. If we assume that a 200-ampere, single-phase, 3-wire, 120/240-volt service is required for this two-flat, we will be required by *Table 310.16* to use 3/0 AWG copper conductors. *Table 250.66* requires a minimum 4 AWG copper grounding-electrode conductor when 3/0 AWG copper service-entrance conductors are used (Figure 4-18).

Step #3: *Install the main bonding jumper.* Again, this is the final link for the completion of the ground-fault path back through the grounded conductor to open the faulted circuit overcurrent protective device. The main bonding jumper would have to be installed in each of the service panels to keep them independent of any alterations being made to the other panels (Figure 4-18).

The procedure for grounding a two-dwelling unit building to a five-dwelling unit building basically follows the same steps. A six-unit building would be the same if there is no common area meter and panel. Each dwelling unit panel must have a main disconnect, and we are limited to six discon-

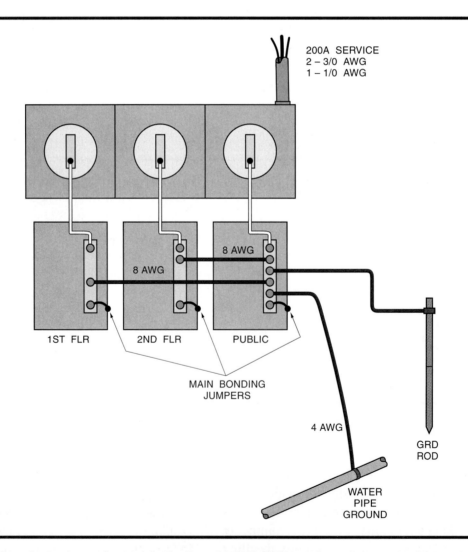

Figure 4-18 Grounding-electrode conductor terminations for 3/0 AWG service-entrance conductors to 3—100 A position meter and service panels

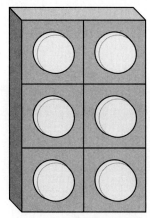

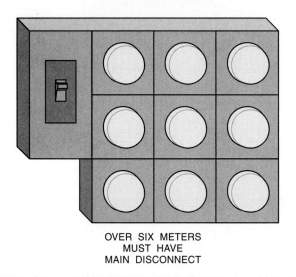

Figure 4-19 Only 6 meters permitted without main disconnecting means

necting means without a main switch (*230.71*) (Figure 4-19).

Multifamily Dwelling

A multifamily dwelling requiring more than six means of disconnect in the main service must have a single main service disconnecting means ahead of the service disconnects. Grounding is then done at this main disconnecting means. We will assume a 12-unit multifamily building requiring a 400-ampere, single-phase, 3-wire 120/240-volt service.

Step #1: *Determine the grounding electrode that will be used.* Generally, a metal underground water pipe will be used as the grounding electrode, and as shown before, must be supplemented by an additional electrode.

Step #2: *Determine the size of the grounding-electrode conductor to be used.* Using *Table 310.16*, we find that for a 400-ampere conductor, we may use a 500-kcmil AWG THW copper conductor. The ampacity of the 500-kcmil AWG THW copper conductor is 380 amperes, but the *NEC®* in *240.3(B)* permits the use of the next higher size overcurrent device, not exceeding 800 amperes, when the ampacity of a conductor does not correspond with the standard ampere rating of a fuse or circuit breaker as shown in *240.6*. *Table 250.66* requires a 1/0 AWG copper conductor for the grounding-electrode conductor where the service-entrance conductors are 500-kcmil AWG copper. This grounding-electrode conductor shall be connected on one end to the neutral bar in the main service disconnect, and the other end shall be connected to a metal underground water pipe within 5 ft (1.50 m) from where it enters the building (Figure 4-20).

For this grounding application, where a grounding-electrode conductor larger than 4 AWG is used, it must be protected if exposed to severe physical damage. A good design would require the 1/0 AWG grounding-electrode conductor to be run in a metric designator 27 (1 in.) nonmetallic conduit if protection is required. The *NEC®* does not define physical damage or severe physical damage, and it would be prudent for the designer to require protection for these conductors.

To determine the size of the enclosure for a grounding-electrode conductor, you must go to Table 1 in Chapter 9 where it requires that a single conductor cannot fill the conduit to more than 53%. Then using Table 5 to obtain the approximate area in sq in. of the conductor and Table 4 to determine what size conduit is required for that area.

Looking at *Table 250.66* we find that there are only 7 copper grounding-electrode conductor sizes. This makes it easy to establish and remember what size conduit enclosures must be used if needed for mechanical protection. An 8 AWG requires a metric designator 16 (½ in.) raceway, 6 and 4 AWG require metric designator 21 (¾ in.) raceways, and 2 AWG, 1/0 AWG, 2/0 AWG and 3/0 AWG require a metric designator 27 (1 in.) raceway (Figure 4-21).

The only raceways permitted for use to protect the grounding-electrode conductor are shown in

84 CHAPTER 4 Grounding

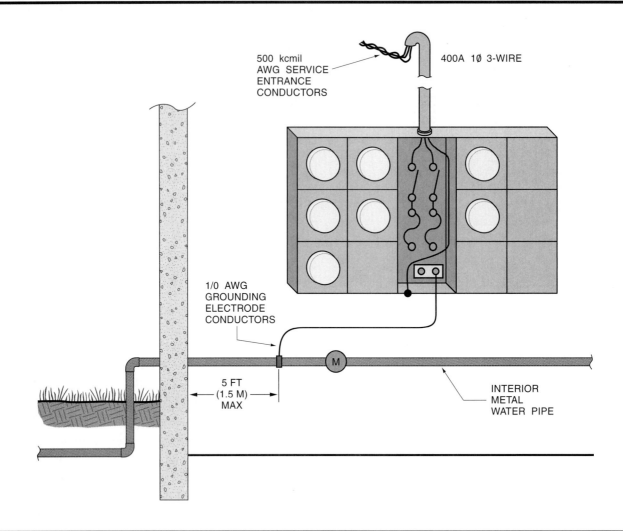

Figure 4-20 Grounding-electrode conductor sized according to service-entrance conductor size

GROUNDING ELECTRODE CONDUCTOR	RACEWAY METRIC	ENGLISH
8 AWG	16	½ in.
6 AWG	21	¾ in.
4 AWG	21	¾ in.
2 AWG	27	1 in.
1/0 AWG	27	1 in.
2/0 AWG	27	1 in.
3/0 AWG	27	1 in.

Figure 4-21 Copper grounding-electrode conductor raceway size where protection is required

250.64(B). They are Rigid Metal Conduit (RMC), Intermediate Metal Conduit (IMC), Electrical Metallic Tubing (EMT), and Rigid Nonmetallic Conduit (RNC). Caution must be used when using metal conduits for physical protection of the grounding-electrode conductor.

A grounding-electrode conductor is a single conductor and if run in a metal enclosure creates a magnetic field. This causes induction in the metal raceway and the metal raceway becomes hot. This induction is due to the rise and fall of the magnetic field of the alternating current flowing in the grounding-electrode conductor. Without a return conductor in the same enclosure, the magnetic fields produced are not cancelled out. The fault current flowing in the grounding-electrode conductor could be very high and the enclosure could become very hot in a short period of time (Figure 4-22).

Because of the problems associated with induction, it is better if the grounding-electrode conductor is not run in a metal enclosure. Only sizes smaller than 6 AWG are required to be run in a raceway.

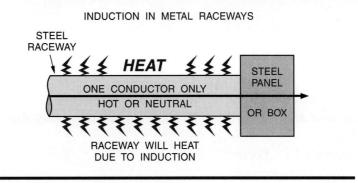

Figure 4-22 Induction heating when one conductor is run in a metal raceway

Grounding-electrode conductors 6 AWG and larger do not need to be run in a raceway or have other protective covering, unless exposed to physical damage according to the requirements of the *NEC,*® but as shown above a nonmetallic conduit will give the protection which may become necessary.

Ground-Fault Circuit-Interrupters

Ground-fault circuit-interrupter (GFCI) protection devices are intended to protect personnel. For many years, the electrical industry held on to the belief that adequate grounding was a sufficient means of protection against electric shock and electric fire hazards. It is now recognized that serious injury and damage may occur by a ground-fault current that is too low to open the overcurrent protective device.

The development of the GFCI enables a circuit or part of a circuit to be opened before there is sufficient ground-fault current to cause injury to personnel or damage to equipment.

The GFCI protective device can be a circuit breaker type or a receptacle type. The GFCI is a device that monitors the current balance between the ungrounded conductor and the grounded (neutral) conductor. These conductors both pass through a current transformer sensor, and as long as there is balance between the current flow in the conductors, the circuit is operable. If a portion of the current returns to the source by means other than the grounded (neutral) conductor, a current is induced in the transformer sensor and the solid-state circuitry opens the protective device, de-energizing the circuit or that portion of the circuit controlled by the GFCI (Figure 4-23).

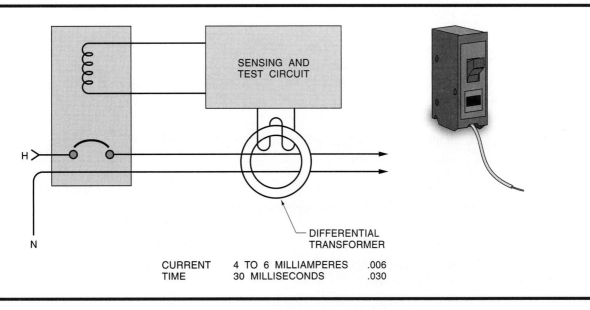

Figure 4-23 Internal wiring of GFCI unit

REVIEW QUESTIONS

MULTIPLE CHOICE

1. A conducting connection, whether intentional or accidental, between an electrical circuit or equipment and the earth, or to some conducting body that serves in place of earth is defined as:
   ____ A. grounded conductor.
   ____ B. grounding electrode.
   ____ C. ground.
   ____ D. bonding conductor.

2. A conductor installed to ensure the required electrical conductivity between metal parts required to be electrically conductive is defined as:
   ____ A. equipment-grounding conductor.
   ____ B. grounded conductor.
   ____ C. bonding jumper.
   ____ D. grounding-electrode conductor.

3. The measurement of total electrical resistance in an ac circuit is called:
   ____ A. impedance.
   ____ B. overcurrent.
   ____ C. amperes.
   ____ D. conductance.

4. Systems or circuits are grounded to:
   ____ A. prevent ground-faults.
   ____ B. measure impedance.
   ____ C. protect against lightning.
   ____ D. prevent overloads.

5. In a corner-grounded delta system, the voltage to ground is:
   ____ A. phase voltage.
   ____ B. 150 volts.
   ____ C. 120 volts.
   ____ D. 240 volts.

6. In a single-phase, 3-wire, 120/240-volt system, the conductor that is intentionally grounded is the:
   ____ A. primary conductor.
   ____ B. neutral conductor.
   ____ C. orange conductor.
   ____ D. phase conductor.

7. The impedance of a circuit is a combination of:
   ____ A. resistance and voltage.
   ____ B. resistance and current.
   ____ C. resistance and reactances.
   ____ D. all of the above.

8. Alternating-current system grounding is required for 50- to 1000-volt premises wiring where the system can be grounded so the maximum voltage to ground on the ungrounded conductors does not exceed:
   ____ A. 208 volts.
   ____ B. 240 volts.
   ____ C. 480 volts.
   ____ D. 150 volts.

9. Where the main bonding jumper is a screw, it is required to have a _____ finish that is visible with the screw installed.
   ____ A. green
   ____ B. orange
   ____ C. white
   ____ D. red

10. Where an underground metal water pipe is used as the grounding electrode, it must be supplemented by another _____ .
    ____ A. main bonding jumper
    ____ B. grounding electrode
    ____ C. equipment-grounding conductor
    ____ D. circuit-conductor

11. Where a single ground rod does not have a resistance to ground of _____ ohms or less, it must be supplemented by one additional electrode.
    ____ A. 50
    ____ B. 75
    ____ C. 30
    ____ D. 25

12. A grounding-electrode conductor is sized based on the size of the:
    ____ A. service-entrance conductors.
    ____ B. grounded conductor.
    ____ C. grounding electrode.
    ____ D. circuit breaker or fuse.

13. Grounding-electrode conductors exposed to severe physical damage that are size _____ AWG or larger require protection.
    ____ A. 4
    ____ B. 6
    ____ C. 8
    ____ D. 1/0

14. The main bonding jumper is always installed in the:
    ____ A. meter housing.
    ____ B. main service switch.
    ____ C. service raceway.
    ____ D. transformer.

15. The minimum size of the equipment-grounding conductor is determined by the size of:
    ____ A. the grounded conductor.
    ____ B. grounding-electrode conductor.
    ____ C. rating of the circuit overcurrent protective device.
    ____ D. service-entrance conductors.

TRUE OR FALSE

1. All electrical systems are required to be grounded.
   ____ A. True
   ____ B. False
   Code reference _____ .

2. The main bonding jumper is installed in the meter housing by the utility company.
   ____ A. True
   ____ B. False
   Code reference _____ .

3. Where a driven ground rod does not have a resistance to ground of 25 ohms or less, it is required to be supplemented by additional rods until this resistance is achieved.
   ____ A. True
   ____ B. False
   Code reference _____ .

4. The neutral conductor, if present, is always the conductor that is intentionally grounded.
   ____ A. True
   ____ B. False
   Code reference _____ .

5. The grounding-electrode conductor must be attached to a metal underground water piping system within 5 ft (1.5 m) of where the water piping enters the premises.
   ____ A. True
   ____ B. False
   Code reference _____ .

6. Multiple ground rods must be spaced not less than 6 ft (1.8 m) apart.
   ____ A. True
   ____ B. False
   Code reference _____ .

7. A metal raceway enclosing an equipment-grounding conductor must be bonded to the conductor at both ends.
   ____ A. True
   ____ B. False
   Code reference _____ .

8. Induction is a result of the rise and fall of the magnetic field of the alternating current flowing in a conductor.
   ____ A. True
   ____ B. False
   Code reference _____ .

9. An alternating system of 50 to 1000 volts that can be grounded so that the maximum voltage to ground on the ungrounded conductors does not exceed 300 volts must be grounded.
   ____ A. True
   ____ B. False
   Code reference _____ .

10. A 3-phase, 4-wire, 480/277-volt system cannot be grounded.
    ___ A. True
    ___ B. False
    Code reference _____ .

FILL IN THE BLANKS

1. The main bonding jumper is the connection between the _____ and the equipment-grounding conductor at the service.

2. A system or circuit-conductor that is _____ grounded is the grounded conductor.

3. The *NEC®* in *250.20(B)* requires alternating systems of 50 volts to 1000 volts to be grounded where the system can be grounded so the _____ to ground does not exceed 150 volts.

4. Electrical systems that are required to be grounded must be _____ earth in a manner that will limit any current imposed on that system which may be caused by _____ or _____ contact with higher voltage lines.

5. The neutral conductor becomes the _____ conductor when it is attached to the neutral bar where the _____ conductor is attached.

6. The conductor that is required to be grounded in alternating systems is shown in _____ of the *NEC®*.

7. The purpose of the equipment-grounding conductor is to establish a path or circuit for the _____ back through the overcurrent protective device to open the circuit.

8. The impedance (resistance + reactances) of the _____ circuit is too great for enough current to flow to open the overcurrent device.

9. *NEC®* 250.56 requires that a ground rod have a resistance to ground of _____ , or it shall be augmented by _____ additional ground rod.

10. *NEC®* 250.64 requires that grounding-electrode conductors smaller than _____ shall be in RMC, IMC, PVC, EMT, or cable armor.

FREQUENTLY ASKED QUESTIONS

Question 1: What is meant when it is said that the Code requires a grounded circuit-conductor to stabilize the voltage to ground?

Answer: An ungrounded conductor can reach any voltage within that system with respect to earth. When one conductor is grounded, a fixed potential to ground is established. For example, if you ground the common conductor of a single-phase, 3-wire, 120/240-volt transformer secondary, the highest voltage to ground that can be attained by any conductor in that system is 50% of the transformer secondary potential or 120 volts because the grounded conductor is tapped at midpoint of the 240-volt secondary winding.

Question 2: What is meant by the *grounding-electrode system*?

Answer: A grounding-electrode system is formed by bonding together all grounding electrodes available on the premises, along with any made electrodes installed on the premises. Electrodes that may be available on the premises include a metal underground water pipe, the metal frame of the building or structure where effectively grounded, a concrete-encased electrode, or a ground ring. Male electrodes are rod and pipe electrodes or plate electrodes. The requirements for a grounding-electrode system may be found in *250.50* of the *NEC*.

Question 3: If the serving utility furnishes a grounded neutral conductor, why is it necessary for us to ground it at the building we are working on?

Answer: Actually, we ground to protect ourselves from the ground that is furnished to us. If we did not ground at our premises, we would not establish a ground-fault path to open the overcurrent protective device in a faulted circuit. If we did not ground and bond all non-current-carrying conductive surfaces and an energized conductor were to accidentally come into contact with them, we would have an energized surface with which we may come into contact with at the same time we are also touching something grounded to earth. Severe shock or electrocution may result. *NEC* 250.2 requires grounding of electrical systems to limit the voltage from lightning, and also requires the grounding of electrical equipment and the bonding of electrically conductive materials in a manner that established an effective path for fault-current.

Question 4: What is a ground-fault circuit or path?

Answer: When an energized conductor touches a grounded surface, the ground-fault circuit starts at the serving transformer, through the energized conductor to the premises main service panel, through the overcurrent device on the faulted circuit, through the circuit conductor to the fault (the point where the energized conductor comes into contact with the grounded surface), through the equipment-grounding conductor back to the service panel enclosure, through the main bonding jumper to the neutral bar and back through the serving transformer. This complete circuit, if of low enough impedance, will cause the overcurrent device to open.

Question 5: Why is the equipment-grounding conductor sized differently than the grounding-electrode conductor?

Answer: The equipment-grounding conductor is run with the circuit-conductors. It may be a bare or insulated conductor run with the circuit-conductors, or it may be the metal raceway enclosing the circuit-conductors. The size of the equipment-grounding conductor is based on the size of the overcurrent protective device for that circuit using *Table 250.122*. If the circuit faults to ground, the maximum amount of current that will flow is equal to the overcurrent protective device (fuse or circuit breaker) rating, and the equipment-grounding conductor will be sized to handle.

The grounding-electrode conductor is sized to carry heavy fault-currents from lightning, line surges, or unintentional contact with higher voltage lines. The requirements for sizing the grounding-electrode conductor are found in *Table 250.66*. The conductor sizes in *Table 250.66* are based on an IR drop not exceeding 40 volts when carrying the short-time current rating for 5 seconds. This short-time rating is based on 100 ft of grounding electrode length. The *NEC* does not limit the length of a grounding-electrode conductor, but proper design would require a calculation to be made for lengths over 100 ft.

Question 6: I do not understand what they mean when they say a water pipe grounding electrode must be supplemented by another type of electrode. Can I use a different type of electrode and only ground to it?

Answer: To be more precise, the water pipe electrode you are referring to is a metal underground water pipe in direct contact with the earth for 10 ft (3.0 m) or more, as described in *250.50*. This section describes the grounding-electrode system and requires that if available on the premises, a metal underground water pipe, the metal frame of the building (where effectively grounded), a concrete-encased electrode, a ground ring, and any made electrodes shall be bonded together to form the grounding-electrode system.

The supplemental grounding-electrode requirement, as shown in *250.50(A)(2)*, prohibits a metal underground water pipe from being the sole grounding electrode for a building or structure.

Question 7: Do I have to go to the street side of the water meter with my grounding electrode conductor, or can I ground to the water pipe near the panel?

Answer: Using the interior metal water piping system as a bonding conductor is permitted by *250.50*, but only within 5 ft (1.5 m) of the point of entrance to the building. Continuity of the interior water piping system would be compromised if someone installed a nonmetallic fitting. This break in the system could be concealed in a wall or ceiling or otherwise not detected.

Question 8: In a 20-unit multifamily building, do I run a grounding-electrode conductor from the service panel in each unit?

Answer: The panel in the unit is not a service panel. It is a sub-panel fed by feeder conductors from the main service. The main service disconnect is where you connect the grounding-electrode conductor. The main service disconnect is fed by the service-entrance conductors and the grounding-electrode conductor is sized according to the size of the service-entrance conductors.

Question 9: Why do we need GFCIs when we have grounded systems?

Answer: Grounded systems provide a path of low impedance for ground-faults. To interrupt the circuit, the ground-fault current must reach the trip level of the overcurrent device. A person who is touching an energized object and is in contact with a grounded surface will be subject to electric shock. The seriousness of the shock is dependent upon the length of time it takes to open the circuit and the voltage level of the circuit. With a 15-ampere overcurrent protective device, the current may never reach that trip level and electrocution may occur.

With GFCI protection, the sensing coil will detect a leakage of current flow not returning to the source through the return conductor, and when it reaches a level of approximately 5 mA, the GFCI will open the circuit. The 5 mA of current flow through the body of the person to ground will still produce a severe shock, but will not cause serious damage.

Question 10: Why don't they just put one big GFCI on the service to a building and protect everything?

Answer: GFCIs can be either a circuit breaker type or a receptacle type. *NEC® 215.9* permits feeders supplying 15- and 20-ampere receptacle branch-circuits to be protected by a GFCI. The more you put on one of these protection devices, the more it is inactivated when it opens. If a complete service were protected by a GFCI, a fault in one thing plugged into a receptacle would shut down the entire house. The less you have connected to a ground-fault device, the easier it is to troubleshoot.

Electrical Inspection—Rough-in

OBJECTIVES

After completing this chapter, the student should understand:
- the meaning of rough-in inspection.
- how to check for proper installation of boxes.
- how to check cable and raceway installation.
- the requirements regarding box fill.
- the requirements for receptacle outlet spacing.
- the requirements for lighting outlets.

KEY TERMS

Authority Having Jurisdiction: Commonly called the AHJ. An employee of a municipality who is responsible for the inspection of electrical installations for conformance with the electrical ordinances of that municipality

Box Fill: The requirements relating to the number of conductors permitted in an outlet or junction box

Raceway Fill: The requirements relating to the number of conductors permitted in a raceway

Reamer: A tool with cutting edges that are inserted into a pipe or raceway and is rotated to remove burrs or sharp edges which may have resulted from cutting the raceway or pipe

Rebars: Steel reinforcing bars placed in concrete to strengthen and reinforce concrete construction

Rough-in: That portion of the work of the electrical installation that is completed before any wall or ceiling covering is applied

Volume Allowance: Free space within a box required for each conductor

INTRODUCTION

The authority having jurisdiction (AHJ) is represented by the electrical inspector. The electrical inspector is responsible for determining and assuring that electrical installations are in conformance with the electrical ordinances of the municipality where the electrical installation is made.

Not all municipalities adopt the entire *NEC*. Some adopt only certain parts of the *NEC* and others adopt the *NEC* with specific revisions. The installing contractor should always verify what the electrical ordinances of a municipality are prior to beginning an installation in that jurisdiction or, in fact, before estimating work in that jurisdiction. Relying on the architect or builder to include on their drawings all work required in conformance with the ordinances of that municipality could turn out to be a costly mistake. Always check with the code enforcement authority of every municipality in which you are about to start work and it will save you from future problems.

INSPECTION PROCEDURES—ROUGH-IN

General

Rough-in inspections are done after all work is completed that must be done before wall or ceiling covering is applied. The items to be inspected may vary from job to job, but residential inspections are generally quite similar because of the many requirements the *NEC®* provides for each room. Inspections of commercial or industrial installations require attention to the specific types of equipment that will be installed (Figure 5-1).

In multistory buildings where concrete slabs are poured, additional rough-in inspections may be required before each concrete pour. It is the responsibility of the installing contractor to request inspections in a timely manner so that building construction schedules will not be adversely affected.

Residential. Rough-in inspections for residential buildings will vary with respect to the wiring method employed. It is difficult for the inspector to determine exactly how the installer has decided to circuit the building when circular raceways such as EMT or Electrical Nonmetallic Tubing (ENT) are being used since the conductors are not pulled in until after the wall covering is applied. For this reason, when inspecting jobs using circular raceways, a check of the number of raceways entering a metal box will help the inspector in determining if the wire fill requirements of *Table 314.16(A)* are being exceeded (Figure 5-2).

Since there will be at least two conductors in every raceway, any metal box with more than three raceways entering it will have to be checked. Generally, boxes with a depth of 1½ in. (38 mm) will be used, and a look at *Table 314.16(A)* in Figure 5-2 shows that when using 14 AWG conductors, only 7 conductors are permitted in a 4 in. (100 mm) octagonal metal box, 10 conductors are permitted in a 4 in. (100 mm) square metal box, and these amounts are reduced when devices or other equipment are installed in the box. Looking at Figure 5-3, illustration (A) we find 3 ½ in. (16 mm) conduits. It appears that 10 conductors are pulled into this box. This conforms to *Table 314.16(A)*. However, looking at illustration (B), we find 5 ½ in. (16 mm) conduits have been installed in this box. In addition, a device ring for the installation of wiring devices is shown. Judging by the conductors already shown in the box, there will be more than the number of conductors permitted by *Table 314.16(A)*.

Figure 5-1 **Commercial electrical equipment**

Table 314.16(A) Metal Boxes

Box Trade Size			Minimum Volume		Maximum Number of Conductors*						
mm	in.		cm³	in.³	18	16	14	12	10	8	6
100 × 32	(4 × 1¼)	round/octagonal	205	12.5	8	7	6	5	5	5	2
100 × 38	(4 × 1½)	round/octagonal	254	15.5	10	8	7	6	6	5	3
100 × 54	(4 × 2⅛)	round/octagonal	353	21.5	14	12	10	9	8	7	4
100 × 32	(4 × 1¼)	square	295	18.0	12	10	9	8	7	6	3
100 × 38	(4 × 1½)	square	344	21.0	14	12	10	9	8	7	4
100 × 54	(4 × 2⅛)	square	497	30.3	20	17	15	13	12	10	6
120 × 32	(4 11/16 × 1¼)	square	418	25.5	17	14	12	11	10	8	5
120 × 38	(4 11/16 × 1½)	square	484	29.5	19	16	14	13	11	9	5
120 × 54	(4 11/16 × 2⅛)	square	689	42.0	28	24	21	18	16	14	8
75 × 50 × 38	(3 × 2 × 1½)	device	123	7.5	5	4	3	3	3	2	1
75 × 50 × 50	(3 × 2 × 2)	device	164	10.0	6	5	5	4	4	3	2
75 × 50 × 57	(3 × 2 × 2¼)	device	172	10.5	7	6	5	4	4	3	2
75 × 50 × 65	(3 × 2 × 2½)	device	205	12.5	8	7	6	5	5	4	2
75 × 50 × 70	(3 × 2 × 2¾)	device	230	14.0	9	8	7	6	5	4	2
75 × 50 × 90	(3 × 2 × 3½)	device	295	18.0	12	10	9	8	7	6	3
100 × 54 × 38	(4 × 2⅛ × 1½)	device	169	10.3	6	5	5	4	4	3	2
100 × 54 × 48	(4 × 2⅛ × 1⅞)	device	213	13.0	8	7	6	5	5	4	2
100 × 54 × 54	(4 × 2⅛ × 2⅛)	device	238	14.5	9	8	7	6	5	4	2
95 × 50 × 65	(3¾ × 2 × 2½)	masonry box/gang	230	14.0	9	8	7	6	5	4	2
95 × 50 × 90	(3¾ × 2 × 3½)	masonry box/gang	344	21.0	14	12	10	9	8	7	2
min. 44.5 depth	FS — single cover/gang (1¾)		221	13.5	9	7	6	6	5	4	2
min. 60.3 depth	FD — single cover/gang (2⅜)		295	18.0	12	10	9	8	7	6	3
min. 44.5 depth	FS — multiple cover/gang (1¾)		295	18.0	12	10	9	8	7	6	3
min. 60.3 depth	FD — multiple cover/gang (2⅜)		395	24.0	16	13	12	10	9	8	4

*Where no volume allowances are required by 314.16(B)(2) through 314.16(B)(5).

Figure 5-2 *NEC® 2002 Table 314.16(A).* (Reprinted with permission from NFPA 70-2002)

Figure 5-3 Examples of box fill requirements

Table 314.16(B) Volume Allowance Required per Conductor

Size of Conductor (AWG)	Free Space Within Box for Each Conductor	
	cm³	in.³
18	24.6	1.50
16	28.7	1.75
14	32.8	2.00
12	36.9	2.25
10	41.0	2.50
8	49.2	3.00
6	81.9	5.00

Figure 5-4 *NEC® 2002 Table 314.16(B).* (Reprinted with permission from NFPA 70-2002)

When using a metal box where the conductors are of different sizes, or when using a nonmetallic box, refer to *Table 314.16(B)* (Figure 5-4). Nonmetallic boxes are marked with their per cubic-in. capacity by the manufacturer.

A one-family dwelling, as we learned in Chapter 3, is a building that consists only of one dwelling unit. In making the rough-in inspection of a one-family dwelling, the inspector has a checklist of items that he must review.

Step #1: *Living rooms, recreation rooms, bedrooms, dens, and similar rooms.* A room-by-room visual inspection is made to determine:

Receptacle Outlets: Proper location of all electrical outlets for receptacles. Receptacle outlets must conform to *210.52(A)* as shown in Chapter 3, "Residential Receptacle Outlets Required" (Figure 5-5).

1. The 6 ft (1.8 m) rule that no point on the wall is more than 6 ft (1.8 m) from a receptacle outlet in that wall space.
2. Each wall space 2 ft (600 mm) or more in width must have a receptacle outlet in that space.

Lighting Outlets: Lighting outlets must be installed in accordance with *210.70* as shown in Chapter 3, "Residential Lighting Outlets Required" (Figure 5-6).

At least one wall switch-controlled lighting outlet must be installed in each habitable room. A switched receptacle may be installed in lieu of a fixed lighting outlet (Figure 5-6).

One method of switching a receptacle outlet for compliance with *210.70(A)(1)* is using a duplex receptacle with breakoff tabs and by removing the tab on the ungrounded terminal. The two receptacles

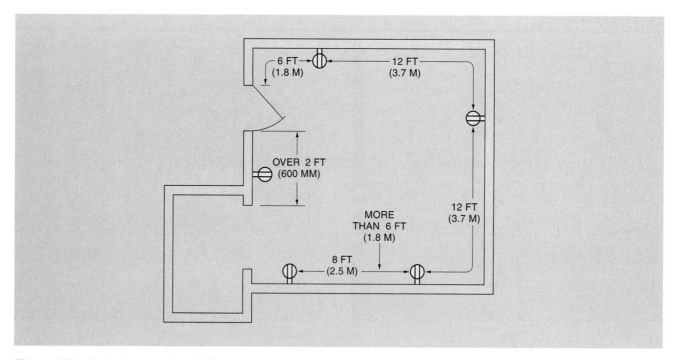

Figure 5-5 Typical wall receptacle layout—no point on wall more than 6 ft (1.8 m) from an outlet in the wall space. Wall space 2 ft (600 mm) or more requires a receptacle outlet

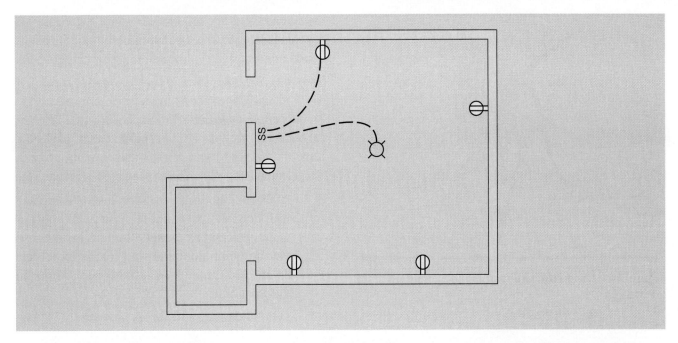

Figure 5-6 Room layout with one wall switch-controlled ceiling outlet and one wall switched receptacle

may be wired with one of the receptacles normally energized (hot), and the other receptacle wired to the switch leg coming from the wall switch (Figure 5-7).

The hot feeding the single-pole switch and continuing to the duplex receptacle is shown as being two conductors under one terminal screw. Generally, not more than one conductor is permitted to be attached to one terminal, and a splice and pigtail to

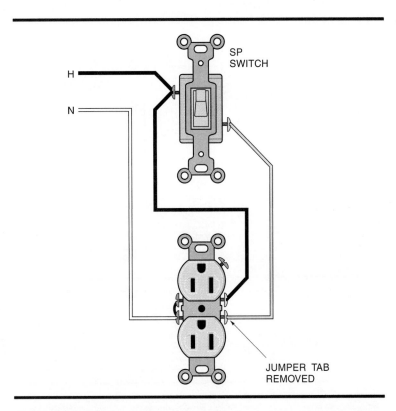

Figure 5-7 Duplex receptacle split-wired with one receptacle controlled by wall switch

98 CHAPTER 5 Electrical Inspection—Rough-in

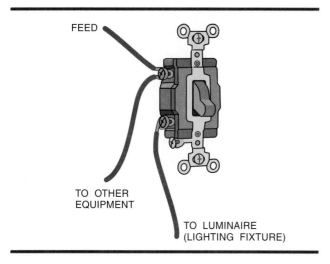

Figure 5-8 **Conductor wrapped around screw terminal without splice**

the switch would be required. In some instances, it is acceptable to skin the insulation in the middle of the conductor and wrap it around the screw (Figure 5-8).

Step #2: *Kitchens and dining rooms.* A room-by-room visual inspection is made to determine:

Receptacle Outlets: Proper location of all electrical outlets for receptacles. Receptacle outlets must conform to *210.52(B)* and *(C)*, as shown in Chapter 3, "Residential Receptacle Outlets Required," (Figure 5-9).

1. A receptacle outlet shall be installed at each wall counter space that is 12 in. (300 mm) or wider. Receptacle outlets shall be installed so that no point along the wall line is more than 24 in. (600 mm) from a receptacle outlet in that space.

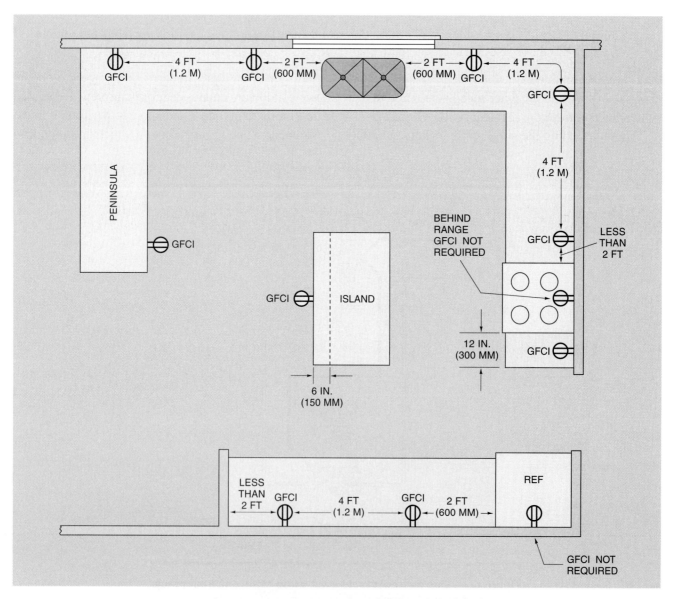

Figure 5-9 **Kitchen room receptacle layout**

2. At least one receptacle outlet must be installed at each island counter space with a long dimension of 24 in. (600 mm), or greater, and a short dimension of 12 in. (300 mm), or greater.
3. At least one receptacle outlet shall be installed at each peninsular counter space with a long dimension of 24 in. (600 mm), or greater, and a short dimension of 12 in. (300 mm), or greater.
4. Countertop spaces separated by range tops, refrigerators, or sinks shall be considered as separate countertop spaces. Receptacle outlets shall be located above, but not more than 18 in. (450 mm) above the countertop. Receptacle outlets installed on walls without countertops and in dining rooms shall comply with the 6 ft (1.8 m) rule. Receptacle outlets, except any used for general lighting and are wall switch-controlled, must be on the small-appliance branch-circuits.

Lighting Outlets: Lighting outlets must be installed in accordance with *210.70* (Figure 5-10).

Either a wall switch-controlled fixed lighting outlet or a receptacle must be installed in the dining room. A wall switch-controlled fixed lighting outlet must be provided for the kitchen. A switched receptacle is not permitted for general lighting in the kitchen. Lighting outlets must be fed from general lighting circuits.

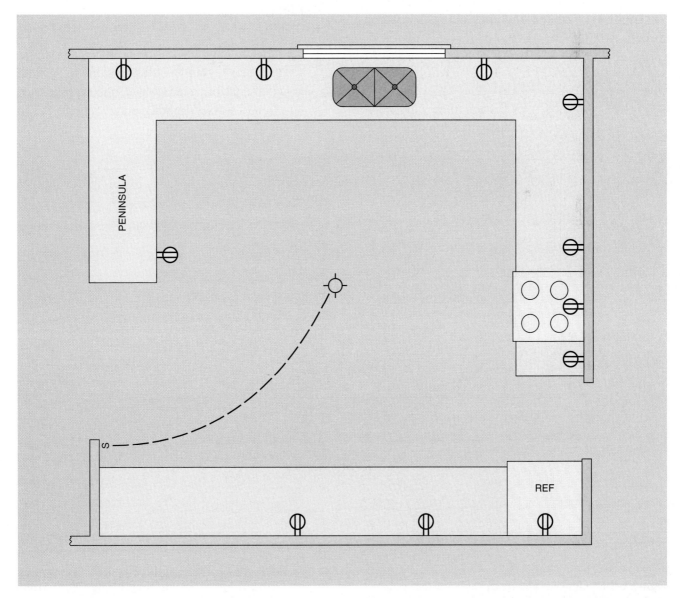

Figure 5-10 Kitchen lighting outlet wall switch-controlled switched receptacle not permitted

Step #3: *Bathrooms and laundry areas.*

Receptacle Outlets: Receptacle outlets must conform to *210.52(D)* and *(F)* (Figure 5-11).

1. At least one receptacle outlet must be installed in each bathroom within 36 in. (900 mm) from the outside edge of each basin. The receptacle outlet must be installed on a wall that is adjacent to the basin.

2. At least one receptacle outlet shall be installed for the laundry. A laundry receptacle outlet need not be installed in a dwelling unit in a multifamily building where laundry facilities are provided as a part of the common elements. A laundry receptacle must be installed within 6 ft (1.8 m) of the intended location of the washing machine.

Lighting Outlets: Lighting outlets must conform to *210.70.*

1. A wall switch-controlled lighting outlet must be installed in the laundry room. A receptacle controlled by a wall switch shall be permitted in lieu of a lighting outlet.

2. A wall switch-controlled lighting outlet must be installed in each bathroom. A wall switch-controlled receptacle may not be used in a bathroom for a lighting outlet.

Step #4: *Hallways, stairways, and closets.*

Receptacle Outlets: Receptacle outlets must conform to *210.52(C)* (Figure 5-12).

1. Receptacle outlets are not required in closets or on stairways.

2. Hallways of 10 ft (3.0 m) or more in length shall have at least one receptacle outlet.

Lighting Outlets: Lighting outlets must be installed in accordance with *210.70*.

1. Hallways must have at least one wall switch-controlled lighting outlet installed. A wall switch-controlled receptacle outlet may be used in lieu of a fixed lighting outlet (Figure 5-12).

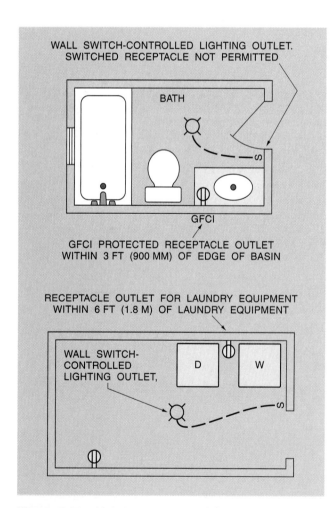

Figure 5-11 Lighting and receptacle outlets required for bathrooms and laundry rooms

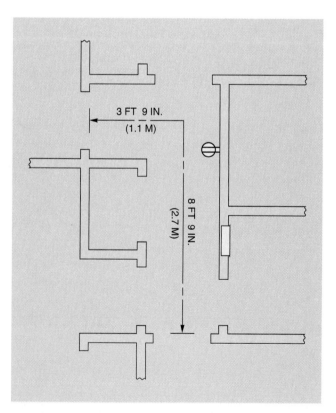

Figure 5-12 Receptacle outlet required in hallways 10 ft (3.0 m) or more in length, lighting outlet controlled by wall switch

Figure 5-13 Stairway lighting outlet with switch at each level for six steps or more

2. At least one wall switch-controlled lighting outlet in stairways. A wall switch-controlled receptacle cannot be used in stairways as the required lighting outlet. Where lighting outlets are installed in interior stairways, there must be a wall switch at each floor level to control the lighting outlet where the difference between floor levels is six steps or more (Figure 5-13). In stairways, remote control, central or automatic control of lighting is permitted.

3. Lighting outlets are not required in closets, but good design would dictate that in many instances fixed lighting outlets should be installed. When installing luminaires (lighting fixtures) in a clothes closet, clearances between the storage spaces and the luminaire (lighting fixture) must be maintained in accordance with *Figure 410.8* of the *NEC.®* Incandescent luminaires (fixtures) with open or partially enclosed lamps and pendant luminaires (fixtures) are not permitted (Figure 5-14). Luminaires (fixtures) permitted in closets are as follows:

   a. Surface-mounted incandescent luminaires (fixtures) installed on the wall above the door or on the ceiling, provided there is clearance of 12 in. (300 mm) between the luminaire (fixture) and the nearest point of a storage space.
   b. Surface-mounted fluorescent luminaires (fixtures) installed on the wall above the door or on the ceiling, provided there is clearance of 6 in. (150 mm) between the luminaire (fixture) and the nearest point of a storage space.
   c. Recessed incandescent luminaires (fixtures) with a completely enclosed lamp installed in the wall or the ceiling, provided there is clearance of 6 in. (150 mm) between the luminaire (fixture) and the nearest point of a storage space.
   d. Recessed fluorescent luminaires (fixtures) installed in the wall or ceiling, provided there is clearance of 6 in. (150 mm) between the luminaire (fixture) and the nearest point of a storage space.

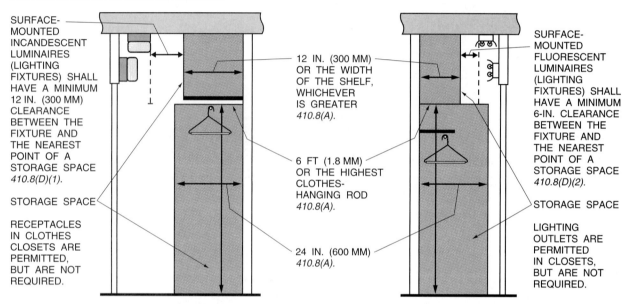
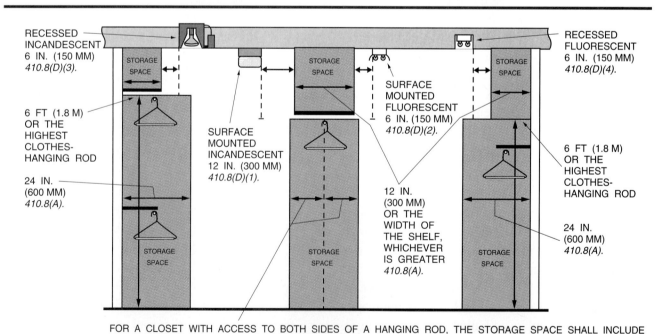

Figure 5-14 Clothes closet

Step #5: *Basements and Garages:*

Receptacle Outlets: Receptacle outlets must conform to *210.52(G)* (Figure 5-15).

1. For a one-family dwelling, at least one receptacle outlet must be installed in each unfinished basement. This receptacle outlet shall be in addition to any provided for the laundry. Finished areas in basements are treated the same as other habitable rooms.

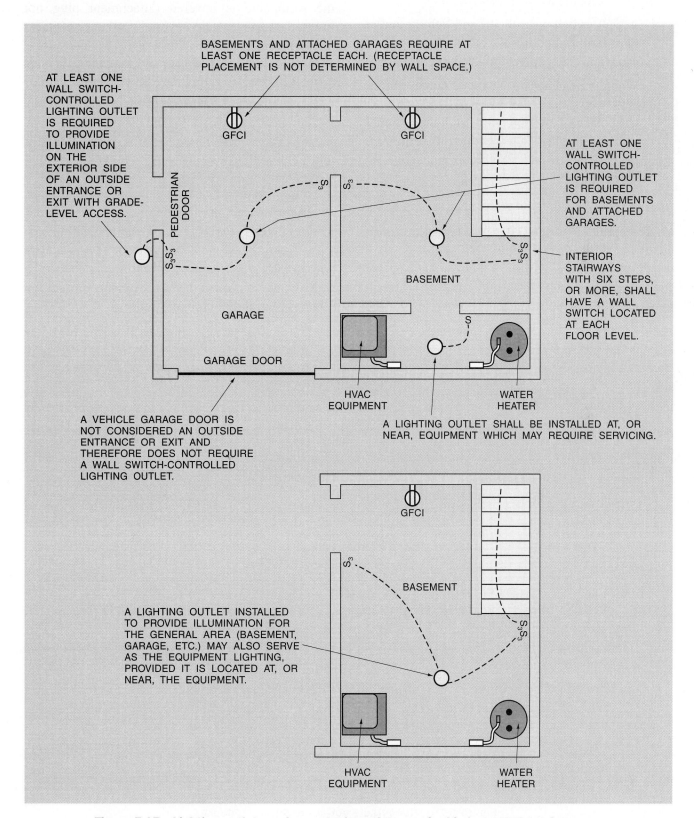

Figure 5-15 Lighting outlets and receptacle outlets required in basements and garages

2. One receptacle outlet must be installed in each attached garage and each detached garage with electric power.

Lighting Outlets: Lighting outlets must be installed in accordance with *210.70(A)(2)*.

1. A wall switch-controlled fixed lighting outlet must be installed in each attached garage and each detached garage with electric power.

2. Where basements are used for storage, or contains equipment that requires servicing, at least one wall switch-controlled lighting outlet must be installed. The lighting outlet must be installed at, or near, the equipment requiring servicing.

Step #6: *Outdoors:*

Receptacle Outlets: Receptacle outlets must conform to *210.52(E)* (Figure 5-16).

Outdoor receptacles must be installed at the front and rear of each one-family dwelling and each two-family dwelling that is at grade level. The receptacle outlets must be accessible at grade level and not more than 6½ ft (2.0 m) above grade. An outdoor receptacle, whether in a location protected from the weather or in other damp locations, requires a weatherproof receptacle enclosure when the receptacle is covered (attachment plug not inserted and receptacle cover closed). A receptacle location protected from the weather includes areas such as under-roofed, open porches, canopies, marquees, etc. that are not subjected to beating rain or water runoff. A wet location receptacle where the product plugs into it and is not attended while in use (e.g., sprinkler system controllers, landscape lighting, holiday lights, etc.), requires an enclosure that is weatherproof, whether the attachment plug is inserted or removed.

Lighting Outlets: Lighting outlets must conform to *210.70* (Figure 5-17). Exterior lighting outlets are required for all outdoor entrances or exits with grade level access. The lighting outlets can be wall switch-controlled or they may be controlled by remote, central, or automatic control.

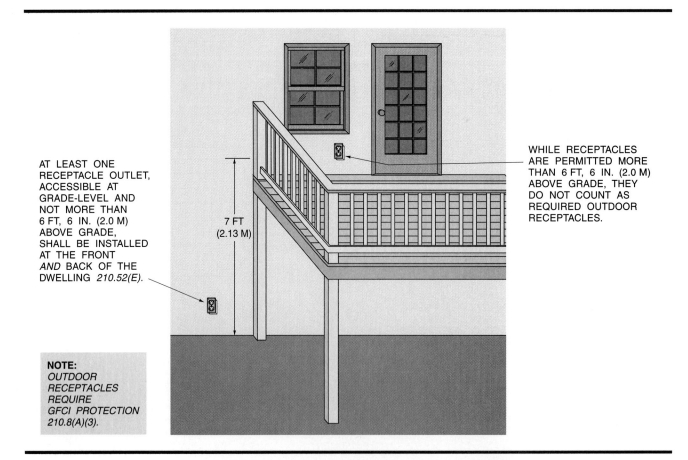

Figure 5-16 Receptacle outlets required outdoors

Figure 5-17 Exterior lighting outlets

Nonmetallic Sheathed Cable. When inspecting a building where Nonmetallic-Sheathed Cable is the wiring method, the Nonmetallic-Sheathed Cable installation is checked during the rough inspection, Figure 5-18.

The installation of the Nonmetallic-Sheathed Cable is done according to the requirements of *Article 334, Nonmetallic Sheathed Cable*. The requirements of *300.4* and *314.17* must also be met. Type NM Cable must be installed as follows:

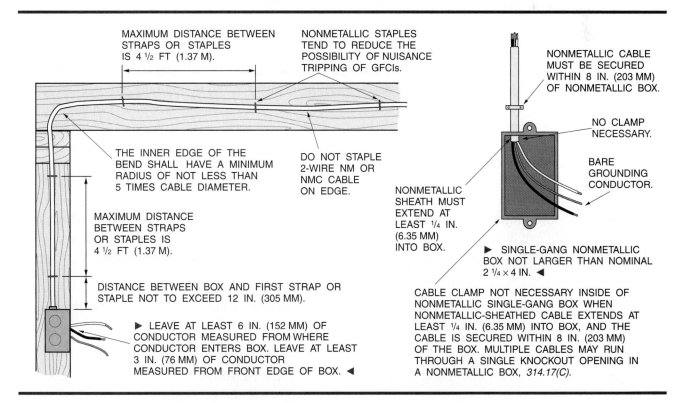

Figure 5-18 Installation of nonmetallic—sheathed cable

1. Secured in place at intervals not exceeding 4½ ft (1.4 m) and within 12 in. (300 mm) from every cabinet, box, or fitting.

2. Bends in cables shall be made so that the radius of the curve shall not be less than five times the diameter of the cable. Bending the cable sharply will put a strain on the insulation on the conductors in the cable.

3. Nonmetallic Sheathed Cable, including the sheath, shall extend into the box not less than ¼ in. (6 mm) through a nonmetallic cable knockout opening. This is to be certain that conductors, not protected by the sheath, are not exposed in the wall where they would be subject to physical damage.

4. In all instances, Nonmetallic-Sheathed Cable shall be secured to the boxes except that where a box not larger than 2¼ in. × 4 in. (57 mm × 100 mm) is used and where the cable is fastened within 8 in. (200 mm) of the box, securing the cable to the box is not required. Multiple cable entries shall be permitted in a single cable knockout opening. This permits cables entering a single gang box to be installed without a cable connector if they are secured closer to the box. Cable connectors can be separate connectors or they may be connector devices installed as a part of a nonmetallic box.

5. Where Nonmetallic-Sheathed Cable passes through either factory or field punched or cut holes in metal members, the cable shall be protected by bushings or grommets covering all metal edges and securely fastened in the opening prior to the installation of the cable. Metal studs or members are rarely installed in a one-family dwelling but are commonly used in multifamily buildings.

6. Where cable is installed through bored holes in wood members, the holes shall be bored so that the edge of the hole is not less than 1¼ in. (32 mm) from the nearest edge of the wood member. Where this distance cannot be maintained, the cable shall be protected by a steel plate at least 1/16 in. (1.6 mm) thick. This protection is necessary to prevent damage to the cable from persons driving nails or screws into the studs after the wall covering is on while unaware that the cable is there.

7. Nonmetallic-Sheathed Cables run parallel to wood framing members must be installed so that the cable is not less than 1¼ in. (32 mm) from the nearest edge of the wood-framing member where nails or screws are likely to penetrate. This means that when running a cable up alongside of a wood stud, you must keep the cable secured away from the edge where it may become damaged.

8. Nonmetallic-Sheathed Cables run across the top of floor joists in accessible attics shall be protected by guard strips. Where this space is not accessible by a permanent ladder or stairs, the protection is only required within 6 ft (1.8 mm) of the nearest edge of the scuttle hole.

9. Inspection of splices and grounding connections. Since the next inspection will be made after the devices have been installed, the inspector must check that proper splicing has been done, and that the equipment grounding conductors have been satisfactorily spliced together and bonded to metal outlet boxes.

When determining box fill, a single volume allowance is counted for each conductor that originates outside of the box and terminates in the box. Conductors that run through the box without splice require a single volume allowance. Conductors that do not leave the box, such as jumper wires between devices, need not be counted. Equipment grounding conductors smaller than 14 AWG and not more than four luminaire (fixture) wires that enter a box from a domed luminaire (fixture) canopy and terminate in that box need not be counted. Additional requirements for determining box fill are shown in *314.16*. These are itemized in the following text.

Clamps: Add a single **volume allowance** (based on the largest conductor in the box) if the box contains one or more internal cable clamps

Support Fittings: Add a single volume allowance if the box contains one or more luminaire (fixture) studs or hickeys

Device or Equipment: Add a double volume allowance for each yoke (based on the largest conductor connected to a device on that yoke) if the box contains one or more wiring devices on a yoke

Equipment-Grounding Conductors: Add a single volume allowance (based on the largest equipment-grounding conductor in the box) if a box contains one or more equipment-grounding conductors

Isolated Equipment-Grounding Conductor: Add a single volume allowance if one or more additional *isolated* equipment-grounding conductors is permitted in the box

Circular Raceways. When inspecting buildings where circular raceways such as RMC, IMC, EMT, ENT, or RNC are used as the wiring method, other inspection procedures must also be done.

Metal Raceways: RMC, IMC, and EMT installations must be checked for proper installation according to *Article 342* (IMC), *Article 344* (RMC), and *Article 358* (EMT)

1. Reaming and threading. Conduits must be carefully threaded so that threads may be made up tight. Conduit ends shall be reamed, using a **reamer**, to remove rough edges so that the conductors will not be damaged when pulled in.

2. Bends. Bends shall be made so that the conduit is not damaged and the internal diameter is not effectively reduced. There cannot be more than the equivalent of four quarter bends (360° total) between pull points (conduit bodies and boxes) (Figure 5-19).

3. Securing and supporting. Each conduit shall be securely fastened within 3 ft (900 mm) of each outlet box. Conduit shall be supported at intervals not exceeding 10 ft (3.0 m). *Table 344.30(B)(2)* may be used for straight runs made up with threaded couplings (Figure 5-20).

4. Bushings. Where a conduit enters a box, fitting, or other enclosure, a bushing shall be provided to protect the wire from abrasion.

Nonmetallic Raceways. RNC and ENT must be checked for proper installation according to *Article 362* (ENT) and *Article 352* (RNC).

1. Bends. Field bends in RNC shall only be done with bending equipment identified for the purpose, and the radius shall not be less than shown in *Table 344.24* (Figure 5-21). Bends in ENT can be made manually and the radius

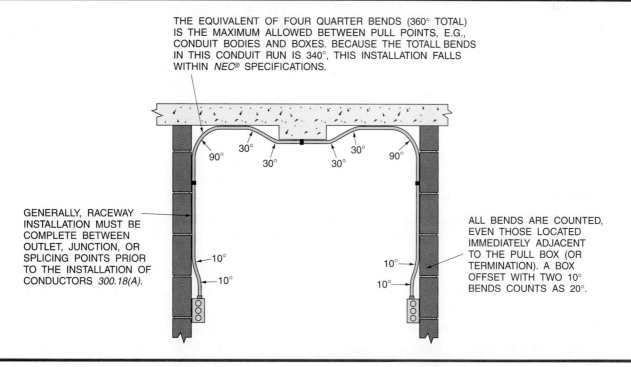

Figure 5-19 Maximum bends in one run

CHAPTER 5 Electrical Inspection—Rough-in

Table 344.30(B)(2) Supports for Rigid Metal Conduit

Conduit Size		Maximum Distance Between Rigid Metal Conduit Supports	
Metric Designator	Trade Size	m	ft
16 – 21	½ – ¾	3.0	10
27	1	3.7	12
35 – 41	1¼ – 1½	4.3	14
53 – 63	2 – 2½	4.9	16
78 and larger	3 and larger	6.1	20

Figure 5-20 NEC® 2002 Table 344.30(B)(2). (Reprinted with permission from NFPA 70-2002)

Table 352.30(B) Support of Rigid Nonmetallic Conduit (RNC)

Conduit Size		Maximum Spacing Between Supports	
Metric Designator	Trade Size	mm or m	ft
16–27	½–1	900 mm	3
35–53	1¼–2	1.5 m	5
63–78	2½–3	1.8 m	6
91–129	3½–5	2.1 m	7
155	6	2.5 m	8

Figure 5-22 NEC® 2002 Table 352.30(B). (Reprinted with permission from NFPA 70-2002)

Table 344.24 Radius of Conduit Bends

Conduit Size		One Shot and Full Shoe Benders		Other Bends	
Metric Designator	Trade Size	mm	in.	mm	in.
16	½	101.6	4	101.6	4
21	¾	114.3	4½	127	5
27	1	146.05	5¾	152.4	6
35	1¼	184.15	7¼	203.2	8
41	1½	209.55	8¼	254	10
53	2	241.3	9½	304.8	12
63	2½	266.7	10½	381	15
78	3	330.2	13	457.2	18
91	3½	381	15	533.4	21
103	4	406.4	16	609.6	24
129	5	609.6	24	762	30
155	6	762	30	914.4	36

Figure 5-21 NEC® 2002 Table 344.24, Reprinted with permission from NFPA 70-2002

shall not be less than shown in *Table 344.24*. There shall not be more than four quarter bends between pull points.

2. Supports. RNC shall be securely fastened within 3 ft (900 mm) of each outlet box. RNC shall be supported as required in *Table 352.30* Horizontal runs of RNC can be supported by openings through framing members at intervals not exceeding those in *Table 352.30* (Figure 5-22). ENT shall be securely fastened within 3 ft (900 mm) of each outlet box or enclosure. ENT shall be secured at intervals not exceeding 3 ft (900 mm), but shall be permitted to be supported by openings in framing members not exceeding 3 ft (900 mm).

Joints in nonmetallic raceways shall be made by an approved method. Expansion fittings for RNC shall be provided to compensate for thermal expansion in accordance with *Table 352.44(A) or (B)*.

Raceways shall be installed complete between outlet boxes or enclosures prior to the installation of conductors.

Concrete Slabs. It is not the responsibility of the **Authority Having Jurisdiction** (AHJ) to determine whether or not the work installed prior to the pouring of the concrete has been installed where it will be in the walls. It is extremely important that care is exercised by the electrical contractor's employees when laying out a slab for the installation of electrical raceways.

The AHJ's responsibility is to check to see that all raceways and fittings used in the concrete pour are listed for use in concrete. That means that a Nationally Recognized Testing Laboratory (NRTL) such as Underwriters Laboratories (UL) tested this equipment for its suitability for use in concrete. He must check the workmanship and the correct usage of all electrical equipment. The AHJ must see that all threaded fittings are made up tight and that all stub-ups are properly capped to prevent entry of concrete into the raceway. He must see that all raceways are properly supported so that they don't give way under the weight of the concrete pour, and that proper spacing is maintained around raceways to permit sufficient concrete to be poured to maintain structural strength. The AHJ must see that proper fittings, such as expansion fittings, are used at expansion joints.

Generally, raceways are installed after the first layer of iron **rebars** are set in place but some installers now place their raceways on the plywood forms and place short pieces of ¼-in. (8 mm) plastic tubing under the raceway to raise it up enough to allow concrete to flow under it. This practice must be approved by the project engineer.

Concrete Slabs on Grade. A slab on grade has all the requirements of slabs on forms, and in addition, care must be exercised to be sure that the raceways are in the concrete and not in the earth below the slab unless the material is judged acceptable for use in that environment. Generally, a vapor barrier sheeting is installed before the electrical installation. Raceways installed in slabs on grade should be installed above the iron mesh or rebars to ensure that concrete completely encloses the raceway. Care must be exercised by the installer not to penetrate the vapor barrier.

It is good practice for the installer to have a representative present when concrete is poured to be sure that raceway stub-ups are not moved or buried by the concrete workers (Figure 5-23).

After slabs are poured, the rough-in inspections follow the same procedure as is required for every type of building or structure. Now all the required applications for **raceway fill**, **box fill**, outlets required, equipment supports, and other wiring methods must be checked and inspected.

INSPECTION PROCEDURES—SERVICE AND GROUNDING

General

Depending upon the job, the service and grounding inspection may be made prior to the rough-in inspection, or it may be done at the same time as the rough-in inspection. On large projects, the service inspection may involve several on-site inspections. For this chapter, we will concentrate on a service and grounding inspection of a residential dwelling.

The drawing shown in Figure 5-24 will be used as a guide, and we will discuss each part of the service drawing starting from the connection point of the service-drop conductors.

Point of Attachment to the Building. The point of attachment shall provide a minimum clearance for the service-drop conductors as required in *230.26*. In no case shall the point of attachment be less than 10 ft (3.0 m) above grade.

Figure 5-23 Concrete slab

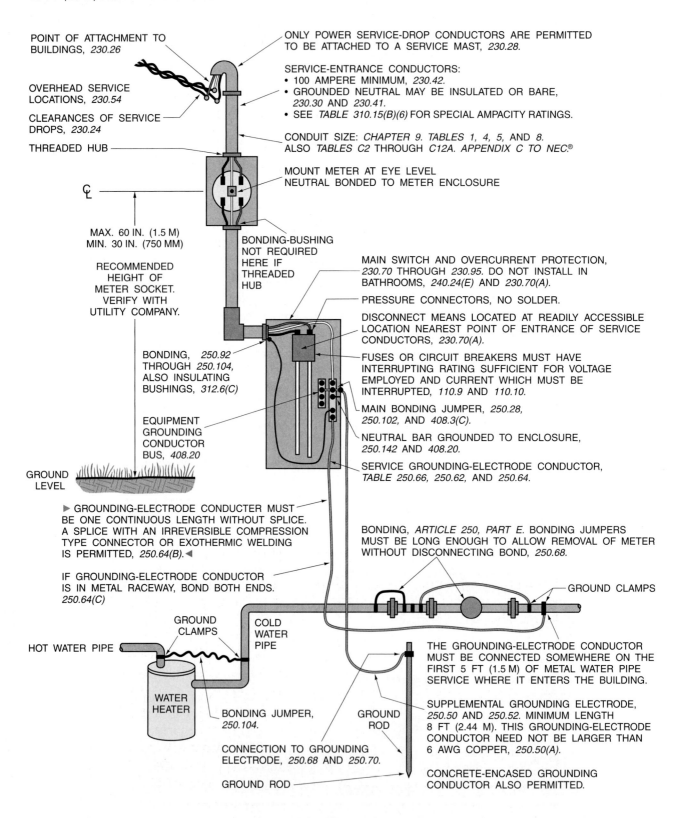

Figure 5-24 Service and grounding

Clearances of Service-Drops. Service-drop conductors shall not be readily accessible (Figure 5-25).

Service-Entrance Conductors. Verify that the proper size service-entrance conductors are used based on the load calculations. For a one-family dwelling, a 100-ampere minimum rating of the service disconnecting means as shown in *230.79*.

Utility Meter Location. Verify height and location of utility meter with utility company.

Disconnecting Means Location. Disconnecting means must be located at a readily accessible location nearest the point of entrance of service-entrance conductors into the building. Service-entrance conductors do not have overcurrent protection on the load side of the utility transformer, and their length in the building must be kept at a minimum.

Main Bonding Jumper. Verify that the main bonding jumper has been properly installed. The main bonding jumper is a part of the ground-fault return path to the grounded (neutral) service conductor that is the path of low impedance back to the utility transformer. This connection is called by many persons *the most important connection in electrical work.*

Grounding-Electrode Conductor. If the grounding-electrode conductor is in a metal raceway, verify that it is bonded at both ends. Failure to bond the conduit at both ends will result in a highly inductive magnetic field that will be in opposition to any ground-fault current, and may result in failure to open the faulted circuit overcurrent protective device.

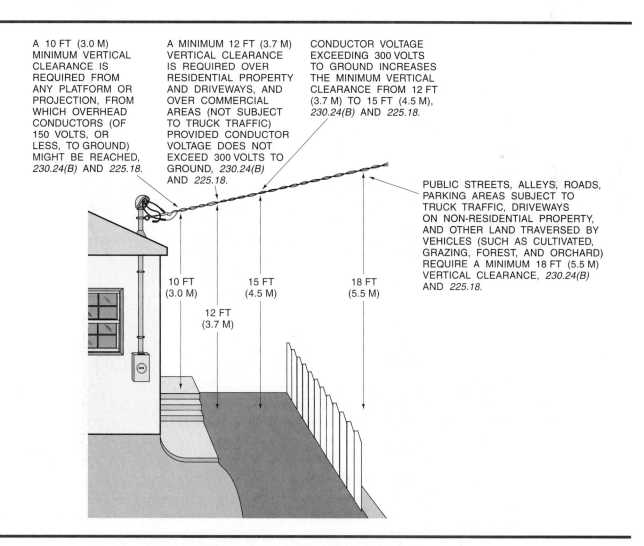

Figure 5-25   Service drop clearances

Verify that the grounding-electrode conductor connection is made within 5 ft (1.5 m) of the point of entry of the water service.

Verify that a supplemental grounding electrode has been installed if a metal underground water pipe has been used as the grounding electrode.

REVIEW QUESTIONS

MULTIPLE CHOICE:

1. No point on a wall space measured horizontally along the floor shall be more than _____ from a receptacle outlet in that space.
   ___ A. 2 ft (600 mm)
   ___ B. 1 ft (300 mm)
   ___ C. 18 in. (450 mm)
   ___ D. 6 ft (1.8 m)

2. Receptacle outlets must be installed at each wall counter space in kitchens that is _____ or wider.
   ___ A. 12 in. (300 mm)
   ___ B. 24 in. (600 mm)
   ___ C. 18 in. (450 mm)
   ___ D. 3 ft (900 mm)

3. A laundry receptacle must be installed within _____ of the intended location of the washing machine.
   ___ A. 6 ft (1.8 m)
   ___ B. 12 ft (3.7 m)
   ___ C. 3 ft (900 mm)
   ___ D. 2 ft (600 mm)

4. Surface-mounted fluorescent fixtures installed on the wall above the door in a closet must have _____ clearance between the luminaire (fixture) and the nearest storage point.
   ___ A. 12 in. (300 mm)
   ___ B. 6 in. (150 mm)
   ___ C. 18 in. (450 mm)
   ___ D. 3 in. (75 mm)

5. The sheath of nonmetallic cable must extend into a box not less than _____ through a nonmetallic cable knockout opening.
   ___ A. ⅛ in. (3.18 mm)
   ___ B. ¼ in. (6 mm)
   ___ C. ½ in. (12.7 mm)
   ___ D. ¹⁄₁₆ in. (1.6 mm)

TRUE OR FALSE

1. All municipalities are required to follow the *NEC*.®
   ____ A. True
   ____ B. False
   Code Reference _____ .

2. A switched receptacle may be installed in lieu of a fixed lighting outlet in a kitchen.
   ____ A. True
   ____ B. False
   Code Reference _____ .

3. A laundry receptacle outlet must be installed adjacent to the washing machine.
   ____ A. True
   ____ B. False
   Code Reference _____ .

4. Each stairway having more than six steps must have a receptacle outlet installed at the bottom landing.
   ____ A. True
   ____ B. False
   Code Reference _____ .

5. Closets over 6 ft (1.8 m) wide must have a fixed lighting outlet installed.
   ____ A. True
   ____ B. False
   Code Reference _____ .

6. Pendant luminaires (fixtures) with enclosed lamps may be installed in closets.
   ____ A. True
   ____ B. False
   Code Reference _____ .

7. All hallways must have a receptacle outlet installed in the center.
   ____ A. True
   ____ B. False
   Code Reference _____ .

8. Lighting outlets are not required in detached garages.
   ____ A. True
   ____ B. False
   Code Reference _____ .

9. Outdoor receptacles must be installed for all dwelling units.
   ____ A. True
   ____ B. False
   Code Reference _____ .

10. Each Nonmetallic Sheathed Cable must be secured to an outlet box.
    ____ A. True
    ____ B. False
    Code Reference _____ .

11. Multiple cable entries shall be permitted in a single cable opening.
    ____ A. True
    ____ B. False
    Code Reference _____ .

12. RMC must be secured within 6 ft (1.8 m) of each box.
    ____ A. True
    ____ B. False
    Code Reference _____ .

13. ENT is permitted to be supported by framing members not exceeding 6 ft (1.8 m).
    ____ A. True
    ____ B. False
    Code Reference _____ .

14. Raceways shall be installed complete between outlet boxes prior to the installation of the conductors.
    ____ A. True
    ____ B. False
    Code Reference _____ .

15. All stairways must have a lighting outlet controlled by 3-way switches.
    ____ A. True
    ____ B. False
    Code Reference _____ .

FILL IN THE BLANKS

1. The requirements relating to the number of conductors _____ in an outlet or junction box is called box fill.

2. The rough-in is that portion of the work of the electrical installation that is _____ before any wall or ceiling covering is installed.

3. An employee of a municipality who is responsible for _____ for conformance with the electrical _____ of that municipality.

4. A receptacle outlet shall be installed _____ that is 12 in. (300 mm) or wider.

5. At least one receptacle outlet shall be installed at _____ with a long dimension of 24 in. (600 mm) or greater, and a short dimension of 12 in. (300 mm) or greater.

6. At least one receptacle outlet must be installed in _____ within 36 in. (900 mm) from the _____ of the basin.

7. A wall switch-controlled lighting outlet must be installed in each bathroom. A wall switch-controlled receptacle _____ in a bathroom for a lighting outlet.

8. Incandescent luminaires (fixtures) with open or partially enclosed _____ luminaires (fixtures) are not permitted in closets.

9. A wall switch-controlled fixed lighting outlet must be installed in each _____ and each detached garage with _____ .

10. Nonmetallic-Sheathed Cables run across the top of floor joists _____ shall be protected by guard strips.

FREQUENTLY ASKED QUESTIONS

Question 1: Why don't all municipalities follow the *NEC*®? Aren't they required to follow it?

Answer: The *NEC*® is sponsored by the National Fire Protection Association. It is made available for adoption by reference to public authorities. The NFPA has no power to police or enforce compliance with the *NEC*.®

Question 2: In the event of a dispute, who has the final say on interpreting the *NEC*®?

Answer: The NFPA has both informal and formal interpretation procedures, see *90.6, Formal Interpretations*. These procedures may be followed, but the NFPA has no authority to enforce their interpretations. The final say may end up being in a court of law.

Question 3: Can an inspector approve materials or equipment that does not have a listing by a Nationally Recognized Testing Laboratory (NRTL)?

Answer: Actually, all approvals are made by the AHJ, see *90.4, Enforcement*. Inspectors generally require that materials and equipment be listed and labeled by a NRTL. Municipalities do not have the resources to test materials and equipment and, therefore, rely on NRTLs.

Question 4: If an inspector finds something wrong, what is his procedure to have it taken care of?

Answer: An inspection defect notice is generally posted on the jobsite and notice is sent to the installing contractor. A time limit is imposed for correction of the defect and if compliance is not obtained the job is shut down. The general contractor is notified that no wall or ceiling covering can be installed until the defect is remedied.

Question 5: Do inspectors make surprise visits to job sites, or do they just come when called?

Answer: Inspection authorities generally have inspection procedures in place for timely inspections that will permit them to inspect all phases of the job.

CHAPTER 6

Circular Raceway Systems

OBJECTIVES

After completing this chapter, the student should understand:
- requirements for rigid metal conduit.
- requirements for intermediate metal conduit.
- requirements for electrical metallic tubing.
- requirements for flexible metallic tubing.
- requirements for flexible metal conduit.
- requirements for liquidtight flexible metal conduit.
- requirements for rigid nonmetallic conduit.
- requirements for electrical nonmetallic tubing.

KEY TERMS

Electrical Metallic Tubing: A metallic tubing of circular cross-section with integral or associated couplings, approved for the installation of electrical conductors, and used with listed fittings to provide electrical continuity

Electrical Nonmetallic Tubing: A pliable, corrugated raceway of circular cross-section with integral or associated fittings. It is composed of a material that is resistant to moisture and chemical atmospheres and is flame retardant

Ferrous: Containing iron

Flexible Metal Conduit: A raceway of circular cross-section made of helically wound, formed, interlocked metal strip

Flexible Metal Tubing: Tubing that is circular in cross-section, flexible, metallic, and liquidtight without a nonmetallic jacket

Intermediate Metal Conduit: A threadable steel raceway of circular cross-section with integral or associated couplings, approved for the installation of electrical conductors, and used with listed fittings to provide electrical continuity

Liquidtight Flexible Metal Conduit: A raceway of circular cross-section having an outer liquidtight, nonmetallic, sunlight-resistant jacket over a flexible metal core with associated couplings, connectors, and fittings, and approved for the installation of electrical conductors

Metallic: Consisting of metal

Nonferrous: Containing little or no iron

Nonmetallic: Containing no metal

Poly-Vinyl Chloride: A type of rigid nonmetallic circular raceway

Rigid Metal Conduit: A threadable steel raceway of circular cross-section with integral or associated couplings, approved for the installation of electrical conductors, and used with listed fittings to provide electrical continuity

Rigid Nonmetallic Conduit: A nonmetallic raceway of circular cross-section with integral or associated couplings, approved for the installation of electrical conductors. It is composed of a material that is resistant to moisture and chemical atmospheres and is flame retardant

INTRODUCTION

In this chapter, the student will learn the *National Electrical Code®* (*NEC®*) requirements for all of the circular raceways used in making electrical installations. Each raceway system has its own advantages. The designer and installer should become familiar with the requirements relating to the use of each type of raceway in order to be able to properly use the raceways to their best advantage. Raceways are the conduits whose function is to facilitate the installation or removal of electrical conductors and to provide protection from mechanical damage. This chapter contains requirements for raceway systems that are frequently used. In addition to defining different raceways, other important information is included, such as minimum and maximum size, installation provisions, and support requirements.

METAL RACEWAYS

Metallic raceway systems include **Rigid Metal Conduit** (RMC), **Intermediate Metal Conduit** (IMC), **Electrical Metallic Tubing** (EMT), **Flexible Metal Tubing** (FMT), **Flexible Metal Conduit** (FMC) and **Liquidtight Flexible Metal Conduit** (LFMC). Metallic systems have the advantage of good physical protection for the conductors, the use of the raceway as an equipment grounding conductor, and the ability to repull conductors in the raceway at a later date.

Rigid Metal Conduit

The *NEC®* requirements for construction, installation, and use of RMC may be found in *Article 344*.

RMC is a listed metal raceway of circular cross-section with integral or associated couplings. It is a **ferrous** metal conduit containing refined iron. It is approved for the installation of electrical conductors and is used with listed fittings to provide electrical continuity (Figure 6-1). The importance of the electrical continuity of a metal raceway intended for use as a grounding means cannot be overemphasized, as we learned in Chapter 4. RMC is generally used as the equipment grounding conductor where grounding is required.

RMC smaller than trade size ½ (16) or larger than trade size 6 (150) shall not be used. The thickness of the wall of RMC is greater than that of IMC resulting in identical outside diameters with a lesser inside diameter for RMC. RMC may be threaded in the field using a standard cutting die with a 1 in 16 taper (1 in. taper per ft) (Figure 6-2). All cut ends must be reamed to remove rough edges.

Figure 6-1 Rigid metal conduit color-coding

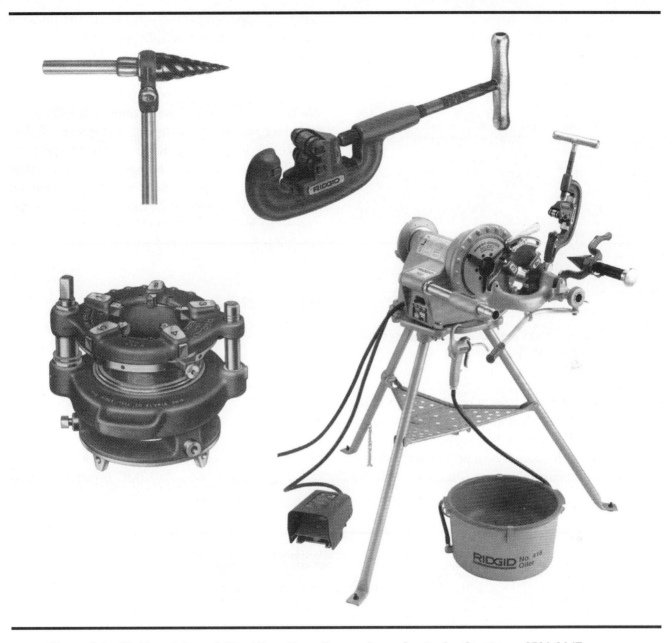

Figure 6-2 Rigid metal conduit cutting, threading, and reaming tools. *Courtesy of* **Ridgid/Emerson**

RMC can be used under all atmospheric conditions and occupancies. When protected solely by enamel, RMC may only be used indoors where occupancy is not subject to severe corrosive conditions. RMC may be used in concrete or in direct contact with the earth when protected against corrosion. RMC may be installed in or under cinder fill (where subject to permanent moisture), if it is protected on all sides by a layer of noncinder concrete not less than 2 in. (50 mm) thick (Figure 6-3).

Bends in RMC must be made so that the conduit will not be damaged and its internal diameter will not be reduced by kinking. Field bends shall comply

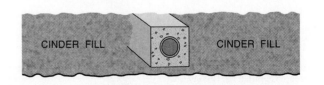

Figure 6-3 RMC in cinder fill protected by 2 in. encasement in non-cinder concrete

with *Table 344.24* (Figure 6-4). There cannot be more than the equivalent of four quarter-bends (360° sum total of bends) between pull points; for example, outlet boxes, junction boxes, and conduit bodies

Table 344.24 Radius of Conduit Bends

Conduit Size		One Shot and Full Shoe Benders		Other Bends	
Metric Designator	Trade Size	mm	in.	mm	in.
16	½	101.6	4	101.6	4
21	¾	114.3	4½	127	5
27	1	146.05	5¾	152.4	6
35	1¼	184.15	7¼	203.2	8
41	1½	209.55	8¼	254	10
53	2	241.3	9½	304.8	12
63	2½	266.7	10½	381	15
78	3	330.2	13	457.2	18
91	3½	381	15	533.4	21
103	4	406.4	16	609.6	24
129	5	609.6	24	762	30
155	6	762	30	914.4	36

Figure 6-4 NEC® 2002 Table 344.24 (Reprinted with permission from NFPA 70-2002)

(Figure 6-5). Offset bends, where made to enter boxes or enclosures, shall be counted in the total.

RMC shall be securely fastened within 3 ft (900 mm) of each outlet box, junction box, or other enclosure or termination point (Figure 6-6). RMC must be supported every 10 ft (3 m). The distance between supports for straight runs of conduit shall be permitted in accordance with *Table 344.30(B)(2)*, provided the conduit is made up with threaded couplings (Figure 6-7). Where structural members do not readily permit fastening within 3 ft (900 mm), fastening can be increased to a distance of 5 ft (1.5 m). When running RMC across bar joists in a building, the distance between bar joists many times exceed 3 ft (900 mm) (Figure 6-6). Horizontal runs of RMC supported by framing members not exceeding intervals of 10 ft (3 m) is permitted if the conduit is securely fastened within 3 ft (900 mm) of terminations.

Intermediate Metal Conduit

The *NEC®* requirements for the construction, installation, and use of intermediate metal conduit may be found in *Article 342*.

IMC is a listed metal raceway of circular cross-section with integral or associated couplings. It is a ferrous metal conduit containing refined iron (Figure 6-8). It is approved for the installation of electrical conductors and is used with fittings to provide electrical continuity. IMC is generally used

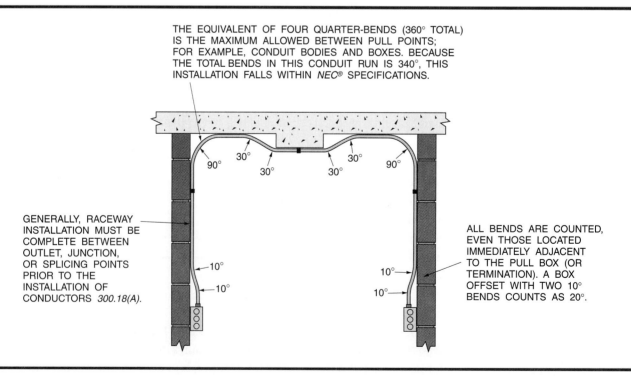

Figure 6-5 Maximum bends in one run

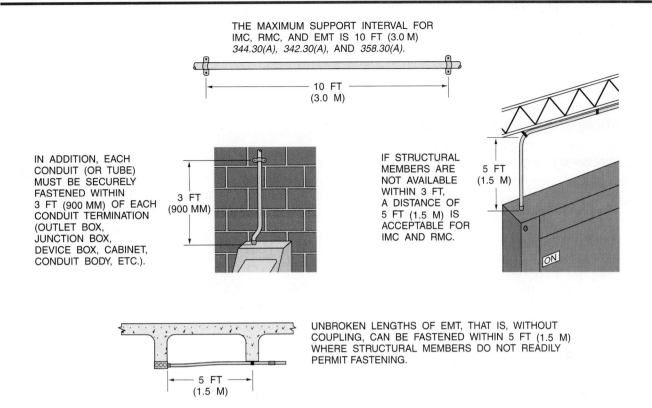

Figure 6-6 RMC, IMC, and EMT support requirements

Figure 6-7 Supports for RMC, *NEC® Table 344.30(B)(2)* (Reprinted with permission from NFPA 70-2002)

Figure 6-8 Intermediate grade conduit color-coding

as the equipment grounding conductor where required.

IMC is accepted wherever RMC is permitted. IMC has the same outside diameter as RMC but has a thinner wall thickness. The reduced weight of IMC as compared to RMC is an important factor when designing electrical systems. The requirements for IMC are generally the same as is required for RMC, except that IMC larger than trade size 4 (103) shall not be used. The requirements for threading, bending, and supporting are the same as those for RMC.

Electrical Metallic Tubing

The *NEC®* requirements for the installation of EMT are found in *Article 358*.

EMT is an unthreaded thinwall raceway of circular cross-section designed for the physical protection and routing of conductors and cables and for use as an equipment grounding conductor. EMT is generally made of steel (ferrous) with protective coatings, but it is also made of aluminum (**nonferrous**) (Figure 6-9).

EMT smaller than trade size ½ (16) or larger than trade size 4 (103) shall not be used. EMT shall not be threaded. All cut ends must be reamed to remove rough edges.

Ferrous (steel) or nonferrous (aluminum) EMT is permitted to be installed in concrete, in direct contact with the earth, or in areas subject to severe corrosive influences where protected by corrosion protection and judged suitable for the condition. Enamel cannot be the sole corrosion protection. In cinder concrete or cinder fill where subject to permanent moisture, EMT must be protected on all sides by a layer of noncinder concrete at least 2 in. (50 mm) thick, unless the tubing is at least 18 in. (450 mm) under the fill.

Bends in EMT must be made so that the conduit will not be damaged and its internal diameter effectively reduced. For trade sizes ½, ¾, and 1 (16, 21, and 27), a hand bender is used (Figure 6-10). Generally, all bends in sizes over trade size 1 (27) are made with hydraulic benders (Figure 6-11). There are also some *power-jack* and ratchet type benders in the trade size 1¼ (35). There must not be more than a total of 360° in bends between pull points, such as outlet boxes and junction boxes.

EMT must be securely fastened in place at least every 10 ft (3.0 m) and shall be securely fastened within 3 ft (900 mm) of each outlet box, junction box, or other tubing termination.

As can be done with RMC and IMC, fastening unbroken lengths of EMT can be increased to a distance of 5 ft (1.5 m) where structural members do not readily permit fastening within 3 ft (900 mm).

Flexible Metallic Tubing

The *NEC®* requirements for the construction, installation, and use of flexible metallic tubing may be found in *Article 360*.

FMT is a raceway that is circular in cross-

Figure 6-9 Electrical metallic tubing

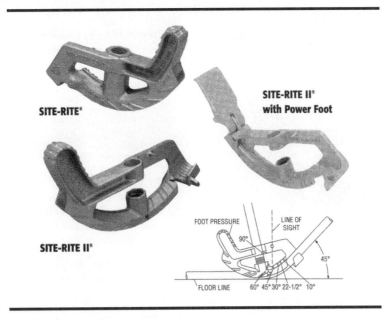

Figure 6-10 RMC and EMT hand benders and hydraulic benders. *Courtesy of* Greenlee® Textron

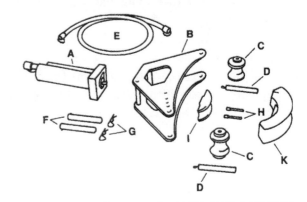

- Strong, portable pin-assembled aluminum components.
- Easy-to-use ram travel scale and bending charts.
- Conduit supports index to suit all sizes.
- Model available to bend PVC coated rigid conduit.

Figure 6-11 Hydraulic bender, power jack. *Courtesy of* Greenlee® Textron

section, flexible, metallic, and liquidtight without a **nonmetallic** jacket (Figure 6-12). It is approved for the installation of electrical conductors and may be used as an equipment grounding conductor when used with fittings listed for grounding and when it does not exceed 6 ft (1.8 m) in length. FMT was developed for use in plenums or ducts to prevent the propagation of smoke from the raceway.

FMT smaller than trade size ½ (16) or larger than trade size ¾ (21) shall not be used, except FMT of trade size ⅜ (12) may be used in ducts or plenums used for environmental air, or it may be used to run from a luminaire (fixture) to an outlet box placed at least 1 ft (300 mm) from the luminaire (fixture), provided this tap is at least 18 in. (450 mm), but not longer than 6 ft (1.8 m) in length.

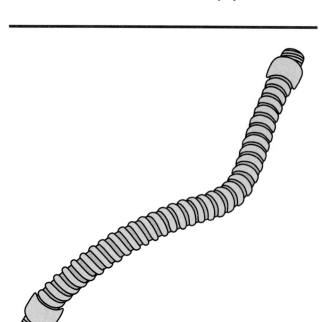

Figure 6-12 Flexible metallic tubing

Table 360.24(A) Minimum Radii for Flexing Use

Metric Designator	Trade Size	Minimum Radii for Flexing Use	
		mm	in.
12	3/8	25.4	10
16	1/2	317.5	12 1/2
21	3/4	444.5	17 1/2

(B) Fixed Bends. Where FMT is bent for installation purposes and is not flexed or bent as required by use after installation, the radii of bends measured to the inside of the bend shall not be less than specified in Table 360.24(B).

Table 360.24(B) Minimum Radii for Fixed Bends

Metric Designator	Trade Size	Minimum Radii for Fixed Bends	
		mm	in.
12	3/8	88.9	3 1/2
16	1/2	101.6	4
21	3/4	127.0	5

Figure 6-13 Minimum radius for FMT bends, *NEC* 2002 Tables 360.24(A) and (B) (Reprinted with permission from NFPA 70-2002)

When bending FMT, care must be exercised not to make the radius of the bends less than is shown in the *Tables* in *360.24(A)* and *(B)*. Bending FMT too sharply will crack the tubing and render its purpose useless (Figure 6-13).

FMT less than trade size 1/2 (16) shall not be used, except trade size 3/8 (12) may be used in lengths not in excess of 6 ft (1.8 m) for utilization equipment, or part of a listed assembly, or for tap connections to luminaires (lighting fixtures) as shown in *410.67(C)*. FMT larger than trade size 4 (103) shall not be used. The maximum number of conductors in trade size 3/8 (12) shall be as shown in *Table 348.22* (Figure 6-14).

Flexible Metal Conduit

The *NEC* requirements for construction, installation, and use of FMC may be found in *Article 348*.

FMC is a raceway of circular cross-section made of helically wound, formed, interlocked metal strip (Figure 6-15).

NEC *250.118(5)* and *(6)* permits FMC to be used as an equipment grounding conductor where both the conduit and the fittings are listed for grounding, or if the conduit is not listed for grounding, then the conduit must be terminated in fittings listed for grounding; the circuit conductors in the conduit must be protected by overcurrent devices rated at not more than 20 amperes; the combined length of FMC, FMT, and LFMC in the same ground path does not exceed 6 ft (1.8 m). When used to connect equipment where flexibility is required, *350.60* requires an equipment grounding conductor to be installed (Figure 6-16).

FMC less than trade size 1/2 (16) shall not be used, except that trade size 3/8 (12) may be used in lengths not in excess of 6 ft (1.8 m) for utilization equipment, as part of a listed assembly, or for tap connections to luminaires (lighting fixtures), as permitted in *410.67(C)*. FMC larger than trade size 4 (103) shall not be used. The maximum number of conductors permitted shall be as shown in *Table 348.22* (Figure 6-14).

FMC cannot be used in wet locations unless the conductors enclosed are approved for the specific conditions: (1) in hoistways; (2) in storage battery

Table 348.22 Maximum Number of Insulated Conductors in Metric Designator 12 (Trade Size ⅜) Flexible Metal Conduit*

Size (AWG)	Types RFH-2, SF-2		Types TF, XHHW, TW		Types TFN, THHN, THWN		Types FEP, FEBP, PF, PGF	
	Fittings Inside Conduit	Fittings Outside Conduit	Fittings Inside Conduit	Fittings Outside Conduit	Fittings Inside Conduit	Fittings Outside Conduit	Fittings Inside Conduit	Fittings Outside Conduit
18	2	3	3	5	5	8	5	8
16	1	2	3	4	4	6	4	6
14	1	2	2	3	3	4	3	4
12	—	—	1	2	2	3	2	3
10	—	—	1	1	1	1	1	2

*In addition, one covered or bare equipment grounding conductor of the same size shall be permitted.

Figure 6-14 Maximum number of conductors in ⅜ in. (metric designator 12) FMT, NEC® Table 348.22 (Reprinted with permission from NFPA 70-2002)

Figure 6-15 Flexible metal conduit

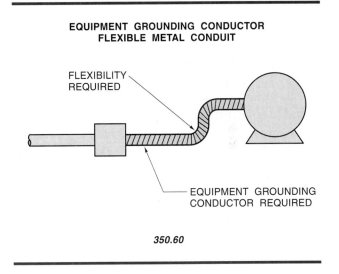

Figure 6-16 Equipment grounding conductor, FMC

rooms; (3) in hazardous classified areas; (4) underground or imbedded in concrete; or (5) where subject to physical damage. FMC can be used in exposed and concealed locations.

There shall not be more than the equivalent of four quarter-bends (360°) between pull points. The radius of any field bend shall not be less than is shown in *Table 344.24* using the column *Other Bends* (see Figure 6-4).

Except where FMC is fished, or for lengths not exceeding 3 ft (900 mm) at terminals where flexibility is required, FMC shall be securely fastened in place within 3 ft (900 mm) of each box or other termination point and shall be supported at intervals not to exceed 4½ ft (1.4 m). Horizontal runs of FMC may be supported by openings through framing members at intervals not greater than 4½ ft (1.4 m).

Liquidtight Flexible Metal Conduit

The *NEC®* requirements for the construction, installation, and use of LFMC are found in *Article 350*.

LFMC is a raceway of circular cross-section having an outer liquidtight, nonmetallic, sunlight-resistant jacket over a flexible metal core with associated couplings, connectors, and fittings for the installation of electrical conductors.

LFMC is permitted to be used as an equipment grounding conductor where the conduit is terminated in fittings listed for grounding; where in trade size 3/8 (12) and trade size 1/2 (16), the circuit conductors are protected by overcurrent devices rated at 20 amperes or less; and where in trade sizes 3/4 through 1 1/4 (21 through 35), the circuit conductors are protected by overcurrent devices rated not more than 60 amperes, and there is no FMC, FMT, or LFMC in trade sizes 3/8 or 1/2 (12 or 16) in the grounding path.

The combined length of FMC, FMT, and LFMC in the same ground path must not exceed 6 ft (1.8 m).

NONMETALLIC RACEWAYS

Nonmetallic raceway systems include **Rigid nonmetallic conduit**, **electrical nonmetallic tubing**, and liquidtight flexible nonmetallic conduit. Nonmetallic raceway systems have the advantage of good physical protection, lighter weight, the ability to repull conductors, and corrosion resistant qualities. Nonmetallic raceways are nonmagnetic, and being nonmetallic, have fewer reactance losses than occur with metallic conduits with the same conductors on ac circuits. Consideration must be given in nonmetallic raceway systems to expansion and contraction.

Rigid Nonmetallic Conduit

The *NEC®* requirements for the construction, installation, and use of RNC will be found in *Article 352*.

RNC (Figure 6-17) is a nonmetallic raceway of circular cross-section with integral or associated couplings, connectors, and fittings for the installation of electrical conductors.

RNC smaller than trade size 1/2 (16) or larger than trade size 6 (155) shall not be used.

RNC may be used in locations subject to severe corrosive influences, in wet locations, in cinder fill, for underground installations, and exposed (where not subject to physical damage).

RNC may *not* be used in hazardous locations, except where specifically permitted or where subject to physical damage, unless identified for such use.

RNC is manufactured in many types, but Rigid Nonmetallic (**Poly-Vinyl Chloride** [PVC]) Conduit, Type Schedule 40 and Schedule 80 are the most commonly used. Schedule 40 Conduit is suitable for underground use by direct burial or encasement in concrete. Unless marked underground use only, Schedule 40 Conduit is also suitable for above-ground use indoors or outdoors exposed to sunlight and weather (where not subject to physical damage). Schedule 80 Conduit is suitable for use wherever Schedule 40 may be used and is suitable for use where exposed to physical damage.

PVC can be cut with a fine-toothed hacksaw (Figure 6-18). After cutting, the PVC should be reamed or deburred. The conduit must be clean and dry with all dirt and plastic shavings removed. To join or assemble PVC, a coat of cement is applied to the end of the conduit and into the fitting, the conduit is then pushed firmly into the fitting. Rotate the conduit in the fitting to evenly distribute the cement.

Figure 6-17 Rigid nonmetallic conduit. *Courtesy of* Carlon, Lamson, and Sessions

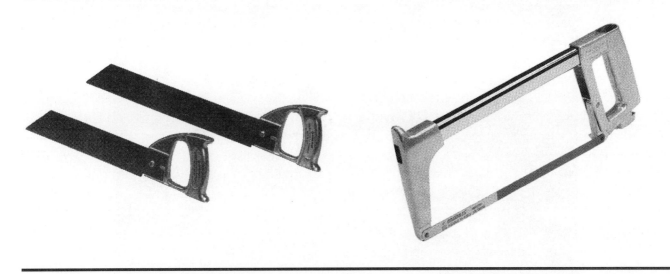

Figure 6-18　PVC hacksaws and PVC saws. *Courtesy of* Greenlee®

Generally, brush-top cans of cement are furnished by the manufacturer for ease in applying the cement.

Bends in PVC Conduit are usually done with factory made elbows (30, 45, or 90°) but using a heat-box for bending the conduit in the field is easily accomplished for trade sizes 1 to 2 (16 to 53). The heat or *hot box*, as it is sometimes called, is a rectangular box approximately 2 ft (600 mm) long with a slot where the conduit to be bent is placed (Figure 6-19). The heat box has a 120-volt, 15-ampere heating element which heats the conduit in a matter of minutes. It is then formed by hand to the exact bend required.

RNC must be provided with expansion fittings to compensate for thermal expansion and contraction. Expansion characteristics of RNC are found in *Tables 352.44(A)* or *(B)*. Expansion fittings are required where the length change is expected to be ¼ in. (6 mm) or greater in a straight run between securely mounted items, such as boxes or other conduit terminations.

RNC must be securely fastened within 3 ft (900 mm) of each outlet box or other conduit termination. RNC shall be supported as required in *Table 352.30*. Horizontal runs of RNC supported by openings through framing members at intervals not exceeding those in *Table 352.30* and securely fastened within 3 ft (900 mm) shall be permitted.

Electrical Nonmetallic Tubing

The *NEC®* requirements for the construction, installation, and use of electrical nonmetallic tubing can be found in *Article 362*.

ENT is a nonmetallic, pliable, corrugated raceway of circular cross-section with integral or associated couplings, connectors, and fittings for the installation of electrical conductors. ENT is composed of a material that is resistant to moisture and chemical atmospheres and is flame retardant (Figure 6-20).

Pliable means that the raceway can be bent by hand without further assistance. As with all circular raceways, not more than the equivalent of four

Figure 6-19　Electric PVC heater/bender. *Courtesy of* Greenlee®

Figure 6-20 Electrical nonmetallic tubing. *Courtesy of* Carlon, Lamson, and Sessions

quarter-bends (360°) between pull points, such as boxes and cabinets, are permitted. Bends must be made so that the tubing is not damaged, and the radius of bends shall not be less than is shown in *Table 344.24* using the column *Other Bends* (see Figure 6-4).

ENT may be used in any building not exceeding three floors above grade for exposed work, and concealed within walls, floors, and ceilings. In any building exceeding three floors above grade, ENT must be concealed within floors, walls, and ceilings that provide a thermal barrier of material that has a 15-minute finish rating. ENT may be used in locations where subject to severe corrosive influences and in concealed, dry, and damp locations. It may be used above suspended ceilings with a 15-minute finish rating, and it may be used encased in concrete, or embedded in a concrete slab on grade, or in wet locations indoors.

ENT may not be used where subject to physical damage; where the voltage is over 600 volts; where exposed to the direct rays of the sun; and except where specifically permitted, in hazardous areas, in theaters and similar areas, and for direct burial.

ENT smaller than trade size ½ (16) or larger than trade size 2 (53) shall not be used.

ENT shall be securely fastened at intervals not exceeding 3 ft (900 mm) and must be securely fastened within 3 ft (900 mm) of each outlet box or other termination. Horizontal runs of ENT supported by openings in framing members at intervals not exceeding 3 ft (900 mm) and securely fastened within 3 ft (900 mm) of termination points shall be permitted.

ENT is permitted as a prewired, manufactured assembly and provided in continuous lengths. Where equipment grounding is required, a separate equipment grounding conductor shall be installed in the raceway.

Liquidtight Flexible Nonmetallic Conduit

The *NEC*® requirements for the construction, installation, and use of liquidtight flexible nonmetallic conduit can be found in *Article 356*.

Liquidtight flexible nonmetallic conduit (LFNC) is a raceway of circular cross-section of various types. There are three types as listed below:

1. Type LFNC-A with a smooth seamless inner core.

2. Type LFNC-B with a smooth inner finish with integral reinforcement within the conduit wall.

3. Type LFNC-C with a corrugated internal and external surface without integral reinforcement within the conduit wall.

LFNC smaller than trade size ½ (16) shall *not* be

used, except trade size ⅜ (12) may be used in lengths not exceeding 6 ft (1.8 m) as part of a listed assembly for tap connections to luminaires (lighting fixtures), for utilization equipment, or for electric sign conductors. LFNC larger than trade size 4 (103) shall not be used.

LFNC may be used in exposed or concealed locations where flexibility is required for installation or maintenance; for direct burial; where protection of the contained conductors is required from vapors, liquids, or solids; for outdoor locations where marked suitable for the purpose. LFNC-B can be installed in lengths longer than 6 ft (1.80 m) where secured within 3 ft (900 mm) of each termination point and supported every 4½ ft (1.4 m). LFNC-B may be installed as a listed prewired assembly, trade size ½ (16) through trade size 1 (27).

Type LFNC-B must be securely fastened at intervals not exceeding 3 ft (900 mm) and within 12 in. (300 mm) on each side of every outlet box or other termination point

Where equipment grounding is required, an equipment grounding conductor shall be run within the conduit.

LFNC cannot be used where subject to physical damage, in lengths longer than 6 ft (1.80 m), or where the voltage of the contained conductors is in excess of 600 volts.

REVIEW QUESTIONS

MULTIPLE CHOICE

1. Bends in circular raceways between pull points shall not exceed the equivalent of _____ quarter-bends.
   - ____ A. three
   - ____ B. four
   - ____ C. six
   - ____ D. none of the above

2. EMT conduit must be securely fastened within _____ of all outlet boxes and other termination points.
   - ____ A. 3 ft (900 mm)
   - ____ B. 6 ft (1.8 m)
   - ____ C. 4 ft (1.2 m)
   - ____ D. 10 ft (3.0 m)

3. The minimum approved size for RMC is:
   - ____ A. trade size ⅜ (12).
   - ____ B. trade size ½ (18).
   - ____ C. trade size ¾ (21).
   - ____ D. none of the above.

4. RMC is generally shipped in standard lengths of _____ .
   - ____ A. 20 ft (6.0 m)
   - ____ B. 4 ft (1.2 m)
   - ____ C. 8 ft (2.5 m)
   - ____ D. 10 ft (3.0 m)

5. The maximum approved size for intermediate grade conduit is:
   - ____ A. trade size 4 (103).
   - ____ B. trade size 6 (155).
   - ____ C. trade size 3½ (91).
   - ____ D. none of the above.

6. The maximum approved size for RMC is:
   ____ A. trade size 4 (103).
   ____ B. trade size 6 (155).
   ____ C. trade size 3½ (91).
   ____ D. none of the above.

7. Unbroken lengths of EMT can be fastened within _____ of an outlet box where structural members do not permit fastening within 3 ft (900 mm).
   ____ A. 8 ft (2.5 m)
   ____ B. 6 ft (1.8 m)
   ____ C. 4 ft (1.2 m)
   ____ D. 5 ft (1.5 m)

8. Which of the following circular metal raceways can be used as the equipment grounding conductor?
   ____ A. RMC
   ____ B. IMC
   ____ C. EMT
   ____ D. All of the above

9. Straight runs of trade size 1¼ RMC and IMC made up with threaded couplings can be supported every _____.
   ____ A. 10 ft (3.0 m)
   ____ B. 12 ft (3.7 m)
   ____ C. 14 ft (4.2 m)
   ____ D. 8 ft (2.5 m)

10. Horizontal runs of FMC can be supported by openings through framing members, at intervals not greater than _____, if securely fastened within 12 in. (300 mm) of each termination point.
    ____ A. 3 ft (900 mm)
    ____ B. 4½ ft (1.4 m)
    ____ C. 6 ft (1.8 m)
    ____ D. 2 ft (600 mm)

11. Liquidtight flexible nonmetallic conduit must be secured within _____ of each termination point.
    ____ A. 6 in. (150 mm)
    ____ B. 12 in. (300 mm)
    ____ C. 18 in. (450 mm)
    ____ D. 24 in. (600 mm)

TRUE OR FALSE

1. Electrical metallic tubing (EMT) can never be used as an equipment-grounding conductor.
   ____ A. True
   ____ B. False
   Code reference _____.

2. Electrical metallic tubing (EMT) can only be threaded in sizes larger than trade size 2 (53).
   ____ A. True
   ____ B. False
   Code reference _____ .

3. Rigid metal conduit (RMC) buried in the earth must have a concrete envelope not less than 4 in. (100 mm) thick.
   ____ A. True
   ____ B. False
   Code reference _____ .

4. Rigid metal conduit (RMC) larger than trade size 6 (155) shall not be used.
   ____ A. True
   ____ B. False
   Code reference _____ .

5. Rigid metal conduit (RMC) shall be securely fastened within 24 in. (600 mm) of each outlet box or termination point.
   ____ A. True
   ____ B. False
   Code reference _____ .

6. Flexible metal tubing (FMT) is not permitted to be used as an equipment grounding conductor.
   ____ A. True
   ____ B. False
   Code reference _____ .

7. When using flexible metal conduit (FMC) where flexibility is required, an equipment grounding conductor must be installed in the raceway.
   ____ A. True
   ____ B. False
   Code reference _____ .

8. Flexible metal conduit (FMC) cannot be used in lengths greater than 6 ft (1.8 m) where grounding is required.
   ____ A. True
   ____ B. False
   Code reference _____ .

9. Flexible metal conduit (FMC) smaller than trade size ½ (16) cannot be fished in walls.
   ____ A. True
   ____ B. False
   Code reference _____ .

10. Electrical nonmetallic tubing (ENT) cannot be run exposed in buildings exceeding three floors above grade.
    ____ A. True
    ____ B. False
    Code reference _____ .

11. Electrical nonmetallic tubing (ENT) may not be used in installations where the voltage exceeds 600 volts.
    ____ A. True
    ____ B. False
    Code reference _____ .

12. Electrical nonmetallic tubing (ENT) is permitted as a prewired, manufactured assembly and provided in continuous lengths.
    ____ A. True
    ____ B. False
    Code reference _____ .

13. Liquidtight flexible nonmetallic conduit (LFNC) larger than trade size 4 (103) shall not be used.
    ____ A. True
    ____ B. False
    Code reference _____ .

14. Special threading equipment must be used for intermediate metal conduit (IMC) because the outside diameter is different than that of rigid metal conduit (RMC).
    ____ A. True
    ____ B. False
    Code reference _____ .

15. When installing flexible metal conduit (FMC), grounding conductors must be installed for circuits over 20 amperes.
    ____ A. True
    ____ B. False
    Code reference _____ .

FILL IN THE BLANKS

1. The thickness of the wall of RMC is _____ of IMC, resulting in identical outside diameters, with a _____ diameter for RMC.

2. Where protected solely by _____ , RMC may only be used indoors in occupancies not subject to _____ corrosive conditions.

3. IMC is a listed metal raceway of _____ cross-section with integral or associated couplings.

4. There must not be more than _____ (360°) between pull points, such as outlet boxes and junction boxes.

5. When bending FMT, care must be exercised not to make the radius of the _____ is shown in the *Tables* in *360.24(A) and (B)*.

6. Where FMC is used to connect equipment where _____ required, an equipment grounding conductor must be installed.

7. FMC cannot _____ , unless the conductors enclosed are approved for the specific conditions.

8. LFMC is permitted to be used as an equipment grounding conductor where the _____ in fittings listed for grounding.

9. RNC _____ in locations subject to severe corrosive influences, in wet locations, _____, for underground installations, and exposed, where not subject to physical damage.
10. RNC must be _____ for thermal expansion and contraction.

FREQUENTLY ASKED QUESTIONS

Question 1: What are the advantages in using RMC, IMC, and EMT in a wiring design?
Answer: The wall thickness and the strength of steel make these wiring methods desirable for the mechanical protection provided to the enclosed conductors. In addition, a properly installed metal conduit system provides a reliable equipment grounding path.

Question 2: What is the purpose of the colored coverings on the threads of RMC?
Answer: Thread protectors are furnished by the manufacturer to keep the threads clean and to prevent damage to the threads. Thread protectors for trade sizes 1, 2, 3, 4, 5, and 6 (27, 53, 78, 103, 129, and 155) are color-coded blue; trade sizes ½, 1½, 2½, and 3½ (16, 41, 63, and 91) are black; and trade sizes ¾ and 1¼ (21 and 35) are red. Different colors are used for the color-coding of intermediate grade conduit.

Question 3: What is the difference between RMC and IMC?
Answer: IMC has a reduced wall thickness and weighs about one-third less than RMC. RMC and IMC are interchangeable and both have the same thread taper. They both use the same fittings and have the same support requirements.

Question 4: What is the reason for the requirement that raceways be installed completely, prior to the installation of conductors?
Answer: The requirement shown in *300.18(A)* states in part *installed complete between outlet, junction, or splicing points, prior to the installation of conductors.* This will protect the conductors from damage if someone tries to slide the conduit over the conductors. It will verify that the raceway system is installed in a manner that permits the conductors to be easily installed in the raceway.

Question 5: What is the color-coding used on thread protectors for IMC?
Answer: Thread protectors for IMC trade sizes 1, 2, 3, and 4 (27, 53, 78, and 103) are color-coded orange; trade sizes ½, 1½, 2½, and 3½ (16, 41, 63, and 91) are yellow; and trade sizes ¾ and 1¼ (21 and 35) are green.

Question 6: Why do they have PVC coated rigid steel conduit? Why not just use PVC Conduit?
Answer: PVC coated rigid metal conduit has the protection of the strength of steel conduit, and the ability to serve as an equipment grounding conductor. In addition, it has all the corrosion resistant qualities of PVC.

Question 7: Can ENT be used as a whip to run tap conductors to recessed fluorescent luminaires (lighting fixtures)?
Answer: Yes. *NEC® 362.30(A) Exception* permits ENT to be run a distance of not more than 6 ft (1.8 m) from a luminaire (fixture) terminal connection, for tap connections, without being secured.

Question 8: Where ENT is run through holes in metal members, are bushings or grommets required in these openings?

Answer: No. Bushings or grommets are not required, but where screws or nails are likely to penetrate the ENT, a steel plate not less than 1/16 in (1.6 mm) shall be used to protect the ENT.

Question 9: Can RMC and IMC be used in the same run of conduit between termination points?

Answer: Yes. Both RMC an IMC can be used in the same locations (they use the same fittings), and they are both suitable for use as equipment grounding conductors. The installer must note, however, that they have different fill requirements due to IMC having a larger internal diameter, and the fill must be based on the lesser of the two—RMC.

Question 10: What is meant by the term thin-wall?

Answer: EMT is commonly called *thin-wall* because of its thinner wall, as compared to RMC or IMC. EMT, because of its thin wall, cannot be threaded and is installed by use of set-screw or compression couplings and connectors. EMT is also permitted to have an integral coupling that is accomplished by having a belled end of the tube with set-screws attached.

CHAPTER 7

Appliances

OBJECTIVES

After completing this chapter, the student should understand:
- the definition of an appliance.
- branch-circuits required for appliances.
- overcurrent protection required for appliances.
- requirements for appliance disconnecting means.
- construction requirements for an appliance.
- requirements for marking an appliance.

KEY TERMS

Authority Having Jurisdiction (AHJ): The entity having responsibility for approving the installation, such as the building department or an inspector

Branch-Circuit Overcurrent Protection, Appliance: Protective device rating marked on an appliance

Branch-Circuit Rating: The rating of the branch-circuit overcurrent device

Central Vacuum Outlet Assemblies: A listed assembly for the connection of a vacuum hose with provision for automatically energizing the power vacuum unit

Hard-Wired: Not cord- and plug-connected

Individual Branch-Circuit, Appliance: Marked rating of an appliance

Infrared Lamp Heating Appliances: Commercial or industrial heating appliances with a lamp that comprises heat through the use of electromagnet radiation of wavelengths

Kitchen Waste Disposer: An electric appliance installed in the drain line of a kitchen sink for grinding up garbage

Live Parts: Energized, electrically connected parts having a source of potential difference, or a potential difference from earth

Resistance Electric Heating Elements: Open- or sheathed-wound resistance coil heating elements

Trash Compactor: An electric appliance that crushes and compresses trash into small bundles

Unit Disconnecting Means: A disconnect switch integral with an appliance used to deenergize an appliance

Water Heaters: Heaters employing resistance type immersion electric heating elements

INTRODUCTION

After completing this chapter, the student will understand and recognize the requirements for the installation and inspection of appliances. Appliances are utilization equipment that use electrical power for electronic, electromechanical, chemical, heating, lighting, and similar purposes. The internal wiring of appliances is based on standards formulated by a Nationally Recognized Testing Laboratory (NRTL). The information in this chapter has been written to be used in conjunction with the requirements of *Article 422* of the *NEC*® The *NEC*® has minimum requirements for safeguarding persons and property related to the installation and use of appliances.

GENERAL

Appliances must not have any **live parts** exposed that can come into contact with the user of the appliance. The only exception to this is open-resistance electric heating elements, such as the heating elements of a toaster, which are necessarily exposed. Generally, cord- and plug-connected appliances, other than **trash compactors**, and **kitchen waste disposers**, are not available on the jobsite for inspection by the **Authority Having Jurisdiction (AHJ)**. The AHJ is responsible for confirming that the **branch-circuit rating** is compatible with the **branch-circuit overcurrent protection** required for each appliance. For appliances that are in place or are **hard-wired**, the inspector must verify that the proper disconnecting means has been installed. In some cases, a **unit disconnecting means** is acceptable as the required disconnecting means.

A disconnecting means must be provided to disconnect each appliance from all ungrounded conductors. Unit disconnect switches having a marked *off* position and that disconnect all ungrounded conductors are permitted as the required disconnecting means where other disconnecting means are also provided as required. See *Article 422, Part III*.

Each electric appliance must be provided with a nameplate that is placed in a position that is visible or easily accessible after installation. The nameplate must give the identifying name and the rating in volts and amperes, or in volts and watts. All **resistance electric heating elements** that are replaceable in the field must have their ratings in volts and amperes, or in volts and watts, or with the manufacturers part number marked on the nameplate. See *Article 422.61*.

INDIVIDUAL APPLIANCES

Toasters

Toasters have an open-resistance heating element that is exposed to the user of the appliance. The user may, and frequently does, use a fork or some other metal device to reach down into the

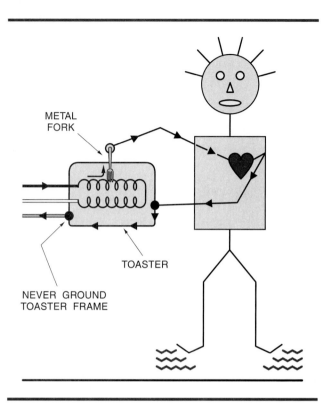

Figure 7-1 Circuit established through grounded toaster

toaster to dislodge a piece of toast. For this reason, the metal frames of toasters are not grounded. If a user were to hold the grounded metal frame of a toaster with one hand and using the other hand, stick a fork or other metal object down into the toaster and make contact with the live parts, a circuit would be established through the users body causing severe electric shock or possibly electrocution (Figure 7-1). Toasters may be plugged into a 15- or 20-ampere receptacle outlet and circuit without individual overcurrent protection, and the plug and receptacle may be used as the disconnecting means for the appliance. See *422.33(A)*.

Kitchen Waste Disposers

Electrically operated kitchen waste disposers are permitted to be cord- and plug-connected with a flexible cord that terminates in a grounding-type attachment plug. A listed kitchen waste disposer that is marked to identify the disposer as being protected by a system of double insulation does not require being terminated with a grounding-type attachment plug. This type of protected unit may be equipped with a two-prong attachment plug (Figure 7-2). See

CHAPTER 7 Appliances 137

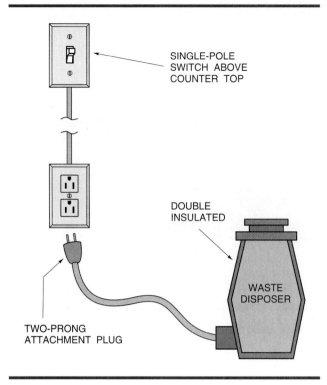

Figure 7-2 Kitchen waste disposer with double insulation, *422.16(B)(1)*

422.16(B)(1). The length of the flexible cord shall not be less than 18 in. (450 mm) or longer than 36 in. (900 mm). The receptacles used for this purpose must be accessible and located to avoid physical damage to the flexible cord (Figure 7-3). The cord and plug connection may also serve as the required disconnecting means.

Kitchen waste disposers may also be hard-wired

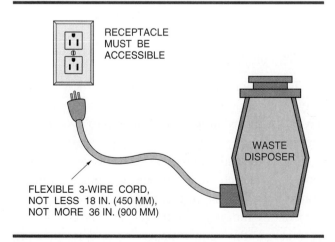

Figure 7-3 Cord- and plug-connected kitchen waste disposer

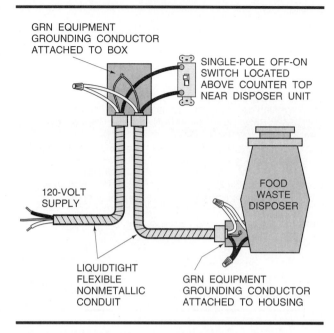

Figure 7-4 Liquidtight flexible nonmetallic raceway with equipment grounding conductor

and in many areas of the country local ordinances *require* that they be hard-wired. A disconnecting means is required for a kitchen waste disposer and generally a toggle switch is located on the wall above the counter top to serve this purpose. FMC is generally the wiring method used to hard-wire a disposer, and the required grounding is accomplished through the metal conduit. If LFNC is used for this purpose, an equipment-grounding conductor must be run with the circuit conductors (Figure 7-4). See *422.16(B)(1)*.

Built-In Dishwashers and Trash Compactors

These appliances may be cord- and plug-connected with a flexible cord that terminates in a grounding-type attachment plug. A listed dishwasher or trash compactor that is marked to identify it as being protected by a system of double insulation does not require being terminated with a grounding-type attachment plug. This type of protected unit may be equipped with a two-prong attachment plug (Figure 7-5). The length of the flexible cord may be from 3 ft to 4 ft (0.9 m to 1.2 m). The receptacles used for this purpose must be accessible and located in the space occupied by the appliance or in a space adjacent to the unit location. See *422.16(B)(2)*.

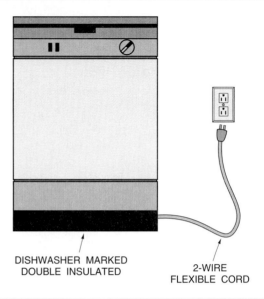

- ▶ THE FLEXIBLE CORD SHALL BE TERMINATED WITH A GROUNDING-TYPE ATTACHMENT PLUG. A LISTED DISHWASHER OR TRASH COMPACTOR, DISTINCTLY MARKED AS PROTECTED BY A SYSTEM OF DOUBLE INSULATION OR ITS EQUIVALENT, DOES NOT REQUIRE TERMINATION WITH A GROUNDING-TYPE ATTACHMENT PLUG.
- ▶ THE CORD MUST BE BETWEEN 3 FT (900 MM) AND 4 FT (1.2 M) IN LENGTH WHEN MEASURED FROM THE FACE OF THE ATTACHMENT PLUG TO THE PLANE OF THE REAR OF THE APPLIANCE.
- ▶ THE RECEPTACLE MUST BE LOCATED SO THE FLEXIBLE CORD WILL NOT BECOME DAMAGED.
- ▶ THE RECEPTACLE SHALL BE ACCESSIBLE.

Figure 7-5   Dishwasher with system of double insulation, 2-wire flexible cord, and attachment cap

Wall-Mounted Ovens and Counter-Mounted Cooking Units

These appliances may be permanently connected or, only for ease in servicing or for installation, they may be cord- and plug-connected. It is important that when installing these appliances with an attachment plug care is taken as to where the cord is located to protect it from physical damage or from interfering with other items that may be occupying that same space (Figure 7-6). See *422.16(B)(3)*.

Household-Type Appliances with Surface Heating Elements

Household-type appliances with surface heating elements having a maximum demand of more than 60 amperes shall have its power supply divided into two or more circuits, each of which must be provided with overcurrent protection not in excess of 50 amperes. When computing the maximum demand for a household electric range, you are permitted to use *Table 220.19* of the *NEC.* Column C is used for ranges not exceeding 12.0 kW, and *Note 1* is used for ranges over 12.0 kW but not more than 27.0 kW. *Note 1* requires that the maximum demand in *Column C* be increased 5% for each additional kilowatt of rating, by which the rating of the individual range exceeds 12.0 kW.

For example, if you have a household electric range with a nameplate rating of 27.0 kW, you may use *Table 220.19* as follows:

- *Column C* is for ranges 12.0 kW and under
- *Note 1* applies to ranges over 12.0 kW to 27.0 kW
- Subtract 12.0 kW from 27.0 kW. 27.0 kW − 12.0 kW = 15 kW
- Multiply 15 by 5% = 75%
- Multiply 8.0 kW (*Column C*) by 75%. 8.0 kW × 75% = 6.0 kW
- Add 6.0 kW to 8.0 kW = 14.0 kW
- 14.0 kW = 14,000 watts divided by 240 volts = 58.3 amperes

This means that ranges rated 27.0 kW or less may be cord- and plug-connected and do not have to have their power supply subdivided into two or more sections.

For cord- and plug-connected household electric ranges, an attachment plug and receptacle connection at the rear base of the range, if accessible from the front by removal of a drawer, shall be considered as meeting the requirements for a disconnecting means for the range. If the attachment plug is not accessible from the front by removal of a drawer,

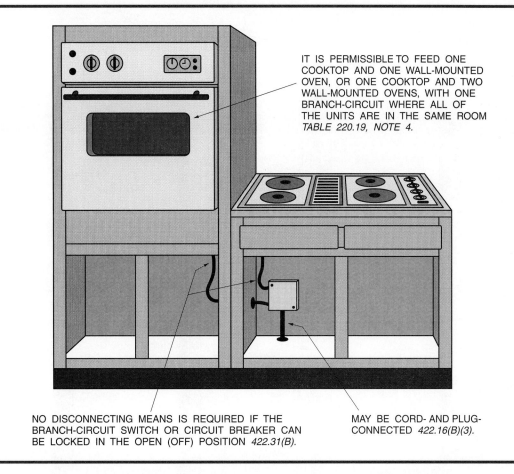

Figure 7-6 Wall-mounted oven and counter-mounted cooktop

then the disconnecting means may be the branch-circuit switch or circuit breaker if it is in sight from the range or is capable of being locked in the open position. See *422.32(B)*.

Central Vacuum Outlets

Central vacuum outlets are the outlets that are located throughout the premises for the attachment of the vacuum hose in that area. These vacuum outlets are equipped with a low-voltage switching mechanism that acts to energize, through a relay, the power outlet at the location of the power vacuum unit. The low-voltage wiring to these outlets is run exposed, generally following the PVC vacuum piping from the power unit to each outlet (Figure 7-7).

Storage-Type Water Heaters

Storage-type **water heaters** that have a capacity of 120 gal (450 L) or less must have a branch-circuit with a rating of not less than 125% of the nameplate reading of the water heater. See *422.13*. The disconnecting means for a household water heater may be the branch-circuit breaker, or in a one-family dwelling, it may also be the main service disconnect, as shown in *422.34(C)*. Generally, the branch-circuit breaker is used as the disconnecting means for a household electric water heater.

The disconnecting means for a water heater, other than in a residential occupancy, may be the branch-circuit switch or circuit breaker if they are readily accessible, within sight from the water heater, or are capable of being locked in the open position. See *422.34(D)*.

The rating of the branch-circuit overcurrent device can be determined by dividing the maximum wattage by the rated voltage shown on the nameplate. For example, a water heater with a nameplate rating of 4500 watts maximum and a voltage of 240 volts would be calculated using the principals

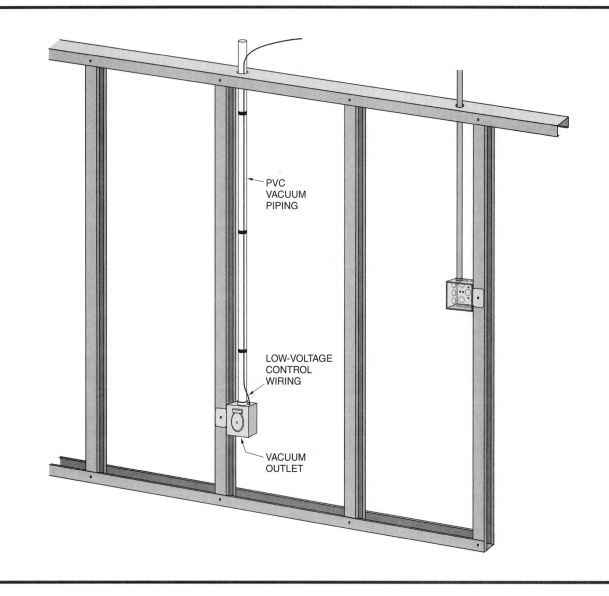

Figure 7-7 PVC vacuum outlet piping with low-voltage wiring

related to Ohms Law, as I = W/E = 4500 divided by 240 = 18.75 amperes. The requirement that the branch-circuit rating must not be less than 125% of the water heaters nameplate rating is: 18.75 × 1.25 = 23.44. This would require a minimum of a 25-ampere fuse or circuit breaker, which is the next highest standard size above the 23.44 calculated amperes.

Water heaters generally have an upper and a lower element (Figure 7-8). The upper element is used for fast recovery, and the lower element maintains water temperature. Each element is thermostatically controlled, but both elements cannot be energized at the same time. The nameplate on the water heater indicates upper wattage, lower wattage, and maximum wattage (Figure 7-9).

Infrared Heating Appliances

Infrared lamp commercial and industrial heating appliances must have overcurrent protection not exceeding 50 amperes as required in *422.11(C)*. In industrial occupancies, infrared lamps shall be permitted to be operated in series on a circuit of over 150 volts to ground provided the voltage rating of the lampholders is not less than circuit voltage. This

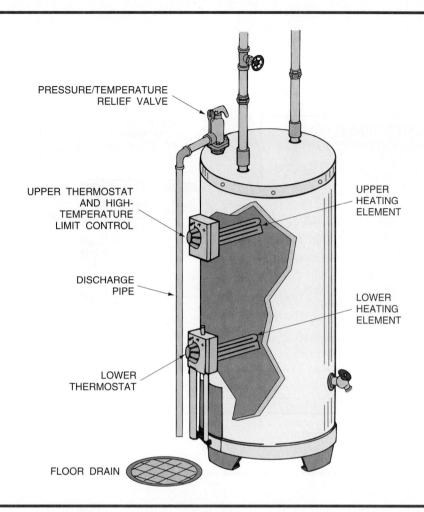

Figure 7-8   Water heater with upper and lower elements

means that in industrial occupancies, four 120-volt infrared lamps may be installed in series on a 480-volt circuit provided the lampholders are rated 600 volts (Figure 7-10). See *422.14*.

Screw-shell lampholders cannot be used with infrared lamps rated over 300 watts unless the lampholders are identified as being suitable for use with infrared heating lamps rated over 300 watts. Infrared lamps generate a great amount of heat at the base where it is mounted in the lampholder. The lampholder construction must be of such design that it will be able to withstand the heat it is subjected to.

Flatirons

Electrically heated smoothing irons must be equipped with a temperature limiting means. Smoothing irons and other cord- and plug- connected electrically heated appliances, intended to be used or applied to combustible material, must be equipped with a stand, as required by *422.45*, as a separate piece of equipment or a part of the appliance. This stand will hold the flatiron in a position where it will not be in contact with combustible material when not in use.

Central Heating Equipment

Central heating equipment includes central gas heating boilers or warm air furnace systems with auxiliary equipment, such as humidifier, electrostatic air cleaner, pump, or valves directly associated with the heating system. Central-heating systems, including their associated equipment, must be supplied by an **individual branch-circuit** (Figure 7-11). See *422.12*.

BASIC WATER HEATER COMPANY
PLAINVILLE, USA
220—240-VOLT A/C 50/60-HERTZ
ELEMENT WATTS—
UPPER 4500 LOWER 4500 MAXIMUM WATTS 4500

SCALD HAZARD: WATER TEMPERATURE OVER 125°F CAN CAUSE SEVERE BURNS INSTANTLY OR DEATH FROM SCALDS. SEE INSTRUCTION MANUAL BEFORE CHANGING TEMPERATURE SETTING.

TEMPERATURE	TIME TO PRODUCE SERIOUS BURN
120°F	Approx. 9½ minutes
125°F	Approx. 2 minutes
130°F	Approx. 30 seconds
135°F	Approx. 15 seconds
140°F	Approx. 5 seconds
145°F	Approx. 2½ seconds
150°F	Approx. 1⁸⁄₁₀ seconds
155°F	Approx. 1 second
160°F	Approx. ½ second

Figure 7-9 Information on water heater nameplates

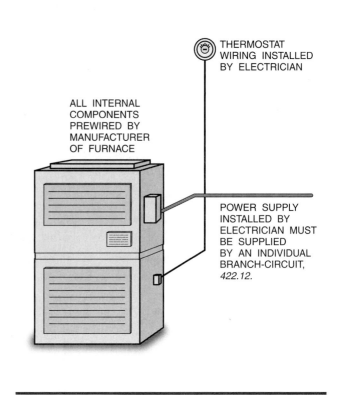

Figure 7-11 Central gas furnace circuit

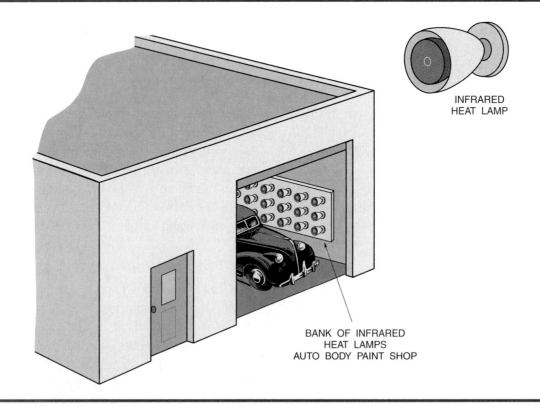

Figure 7-10 Auto repair shop with infrared heat lamps in paint drying booth

REVIEW QUESTIONS

MULTIPLE CHOICE

1. When an appliance is hard-wired it is:
   - ____ A. not cord- and plug-connected.
   - ____ B. connected by means of a rigid metal raceway.
   - ____ C. connected by means of a flexible metal raceway.
   - ____ D. all of the above.

2. A unit disconnect switch is required to:
   - ____ A. open all ungrounded conductors.
   - ____ B. have a marked *off* position.
   - ____ C. be an integral part of the appliance.
   - ____ D. all of the above.

3. Kitchen waste disposers are required to be:
   - ____ A. double insulated.
   - ____ B. connected by means of a flexible metal raceway.
   - ____ C. connected by means of a flexible cord.
   - ____ D. none of the above.

4. Receptacles used for connection dishwashers by an attachment plug must be:
   - ____ A. rated at 20 amperes.
   - ____ B. GFCI protected.
   - ____ C. connected to the kitchen small-appliance circuit.
   - ____ D. none of the above.

5. Flexible cords used to connect trash compactors are limited in length to:
   - ____ A. 24 in. (600 mm).
   - ____ B. 18 in. (450 mm).
   - ____ C. 3 ft to 4 ft (0.9 mm to 1.2 mm).
   - ____ D. none of the above.

6. Wall-mounted ovens may be cord- and plug-connected:
   - ____ A. only in residential occupancies.
   - ____ B. for ease in servicing or for installation.
   - ____ C. only if rated 12.0 kW or less.
   - ____ D. only on circuits rated 20 amperes or less.

7. Electric ranges over 12.0 kW must add _____% for each additional kW by which the rating of the individual range exceeds 12.0 kW.
   - ____ A. 10
   - ____ B. 5
   - ____ C. 3
   - ____ D. none of the above

144 CHAPTER 7 Appliances

8. Storage-type water heaters that have a capacity of 120 gal (450 L) or less, must have a branch-circuit with a rating of not less than _____ % of the nameplate rating of the water heater.
   _____ A. 125
   _____ B. 80
   _____ C. 150
   _____ D. none of the above

9. Screw-shell lampholders cannot be used with infrared lamps rated over _____ watts, unless the lampholder is identified for use with infrared lamps rated over _____ watts.
   _____ A. 150
   _____ B. 250
   _____ C. 300
   _____ D. 500

10. Information regarding the installation of appliances can be found in *Article* _____ of the *NEC.*
    _____ A. 695
    _____ B. 422
    _____ C. 411
    _____ D. none of the above

TRUE OR FALSE

1. Appliances are always cord- and plug-connected.
   _____ A. True
   _____ B. False
   Code reference _____ .

2. Appliances rated over 750 watts must be hard-wired.
   _____ A. True
   _____ B. False
   Code reference _____ .

3. Motor operated appliances cannot be cord- and plug-connected.
   _____ A. True
   _____ B. False
   Code reference _____ .

4. Appliances with heating elements must be on circuits rated at 125% of their nameplate rating.
   _____ A. True
   _____ B. False
   Code reference _____ .

5. The metal frames of toasters are not grounded.
   _____ A. True
   _____ B. False
   Code reference _____ .

6. Kitchen waste disposers may be connected by means of a flexible cord with an attachment plug.
   ____ A. True
   ____ B. False
   Code reference _____.

7. Flexible cords used with trash compactors shall not exceed 4 ft (1.2 m) in length.
   ____ A. True
   ____ B. False
   Code reference _____.

8. Both upper and lower heating elements of a water heater are used for quick recovery of hot water.
   ____ A. True
   ____ B. False
   Code reference _____.

9. Electric ranges rated over 27.0 kW may not be cord- and plug-connected.
   ____ A. True
   ____ B. False
   Code reference _____.

10. Electrically heated smoothing irons must be equipped with a stand to hold the flatiron when not in use.
    ____ A. True
    ____ B. False
    Code reference _____.

FILL IN THE BLANKS

1. Infrared lamp commercial and industrial heating appliances must have overcurrent protection not _____ amperes.

2. The nameplate on a water heater indicates _____ wattage, _____ wattage, and _____ wattage.

3. The rating of the branch-circuit _____ can be determined by dividing the maximum wattage by the rated voltage as shown on the nameplate.

4. *Note 1* to *Table 220.19* requires that the maximum demand in *Column C* be increased _____ for each additional kilowatt of rating, by which the rating of the individual range exceeds _____ kW.

5. Household-type appliances with surface heating elements having a _____ of more than _____ amperes, shall have its power supply divided into two or more circuits.

6. These appliances may be permanently connected or, only for _____ or for installation, they may be cord- and plug-connected.

7. Electrically operated kitchen waste disposers are permitted to be cord- and plug-connected with a flexible cord that terminates in a _____ attachment plug.

8. A _____ is required for a kitchen waste disposer, and generally, a _____ on the wall above the countertop serves this purpose.

9. When computing the _____ demand for a household electric range, you are permitted to use Table _____ of the *NEC*.

10. For cord- and plug-connected household electric ranges, _____ connection at the rear base of the range, if accessible from the front by _____, shall be considered as meeting the requirements for a disconnecting means for the range.

FREQUENTLY ASKED QUESTIONS

Question 1: Does the *NEC* cover the construction of appliances?
Answer: Yes. The *NEC* contains specific requirements relating to appliance internal wiring, unit disconnecting means, and overcurrent protection. These requirements are usually a part of the standard by which a NRTL will evaluate each appliance. The *NEC* in 90.7 makes it clear that it is not the intent of the Code that factory-installed internal wiring, or the construction of equipment be inspected at the time of installation of the appliance, except to detect alterations or damage to the equipment. See *Article 422, Part IV* for information on appliance construction.

Question 2: How can an inspector inspect the appliances in a residence when they usually are not there when he inspects?
Answer: Cord- and plug-connected portable appliances are generally only inspected at the time of their manufacture, and that is when the listing label is applied. The education of the consumer to only purchase listed appliances is the best way to increase safety in a residence. Requirements for the installation of permanently installed appliances can be found in *Article 422, Appliances*.

Question 3: How is the cord to a floor lamp protected from overheating and burning up?
Answer: There are two types of problems that can occur with a flexible cord supplying a floor lamp. *First*, the conductors may be overloaded, but this is a remote possibility because the lamps that may be used will not be of sufficient wattage to exceed the ampacity of the lamp cord. *Second*, a short between conductors or a short to ground could occur, but this condition will generally draw a high magnitude of current in a short period of time, and will open the branch-circuit overcurrent device quickly before any damage is done to the equipment.

Question 4: What is the rule regarding disconnecting means for permanently connected appliances?
Answer: Each permanently connected appliance must be provided with a means for disconnection from all ungrounded conductors.
1. For appliances rated at not more than 300 volt-amperes or ⅛ horsepower, the branch-circuit switch or circuit breaker may serve as the disconnecting means.

2. For appliances rated more than 300 volt-amperes or ⅛ horsepower, the branch-circuit switch or circuit breaker may serve as the disconnecting means if the switch or circuit breaker is within sight of the appliance or is capable of being locked in the open position. Further information regarding the disconnection of permanently connected appliances may be found in *422.31* of the *NEC.*®

Question 5: Are cord- and plug-connected appliances required to have disconnecting means?
Answer: Yes, for cord- and plug-connected appliances, the separable plug and receptacle, if accessible, is permitted as the disconnecting means. For cord- and plug-connected household electric ranges, if the attachment plug and receptacle are accessible from the front by removal of a drawer, it can be considered as being accessible. The placement of the receptacle behind an electric range is critical as being accessible through the drawer. See *422.33(B)*.

Question 6: Can a unit switch on an appliance be considered as the disconnecting means?
Answer: Yes. A unit switch that has a marked *off* position, and disconnects all of the ungrounded conductors, is permitted as the required disconnecting means where other disconnecting means are provided as well. The other disconnecting means required may be found in *422.34*.

Question 7: What is the maximum rating of a single cord- and plug-connected appliance that may be connected to a multioutlet branch-circuit?
Answer: The maximum rating of any one cord- and plug-connected appliance may not exceed 80% of the multioutlet branch-circuit rating. See *210.23(A)(1)*.

Question 8: What is the maximum rating of any one appliance fastened in place, where lighting units and cord- and plug-connected equipment are also supplied on a multioutlet branch-circuit?
Answer: The rating of any one appliance fastened in place cannot exceed 50% of the branch-circuit rating, where lighting units and/or other utilization equipment is supplied. See *210.23(A)(2)*.

Question 9: I have an electric furnace for a condominium unit. The furnace is located in a utility closet on the balcony of the unit. Do I need a disconnecting means at the furnace, or can the branch-circuit circuit breaker be used as the disconnecting means?
Answer: The branch-circuit switch or circuit breaker may be used as the required disconnecting means if it is within sight of the furnace or is capable of being locked in the open position. This information may be found in *422.31(B)*.

Question 10: I have an electric fry pan that is rated 1500 watts. Can I plug it into an outlet on the kitchen small-appliance circuit?
Answer: If the receptacle outlet is rated 20 amperes, you may use this appliance on the 20-ampere small-appliance circuit. If the receptacle is rated 15 amperes, as permitted by *210.21(B)(2)*, you may *not* use this appliance on the 20-ampere small-appliance circuit. *Table 210.21(B)(2)* allows a maximum load of 12 amperes on a 15-ampere receptacle. A 1500-watt appliance on a 120-volt circuit would have an amperage draw of approximately 13 amperes. There is one opinion that if the appliance is rated 1500 watts at 125 volts, then the amperage is 12 amperes and if the voltage is decreased to 120 volts, the wattage is also decreased, and there is no problem. Another thought is that if the appliance does not constitute a continuous load, that is, 3 hours or more of continuous operation, then the 80% rule is not a requirement, but this is contrary to the present requirements of *210.21(B)(2)*.

CHAPTER 8

Optional Load Calculations

OBJECTIVES

After completing this chapter, the student should understand how to make optional load calculations for:
- a dwelling unit.
- an existing dwelling unit.
- a multifamily dwelling.
- two dwelling units.
- schools.
- new restaurants.

KEY TERMS

Central Electric Space-Heating: Space-heating unit other than electric space-heating equipment, such as a gas-fired furnace or a gas-fired boiler

Connected Load: The total load in kVA connected to the power supply after demand factors are applied

Dwelling Unit: One or more rooms for the use of one or more persons as a housekeeping unit, with space for eating, living, and sleeping, and permanent provisions for cooking and sanitation

Electric Space-Heating Equipment: A central furnace with electric resistance-type elements or electric resistance-type baseboard heating units

Feeder: All circuit-conductors between the service equipment or other power supply source and the final branch-circuit overcurrent device

Service: The conductors and equipment for delivering electric energy from the serving utility to the premise wiring system

INTRODUCTION

This chapter will familiarize the student with the available optional methods for calculating loads consistent with the requirements of good design procedures, while maintaining the guidelines for a hazard free installation established by the *NEC*.® The information in this chapter has been written for use in junction with the requirements of *Article 220* of the *NEC*.®

GENERAL

Optional calculations are a result of tests that have been made to determine the actual demand in various types of occupancies. Care must be exercised by anyone using these calculations to be sure that there are not extenuating circumstances that render their installation not suitable for these optional calculations. The *NEC*® is not intended as a design manual, although many of its requirements are of a design nature in its effort to establish guidelines for an installation free from the hazards resulting from the use of electricity. A designer of electrical installations should always be cognizant of the adequacy of his design in providing an efficient and convenient wiring system with the ability for future expansion of electrical use.

DWELLING UNIT

The optional calculation for the **service** or **feeder** conductors for **dwelling units** can be used for those units that have a total **connected load** of 100 amperes or more, and are served by a single 3-wire, 120/240-volt or 208Y/120-volt set of service, or feeder conductors with an ampacity of 100 amperes or greater, as shown in *220.30* (Figure 8-1).

To use this method of calculation, it is required that it be determined that using the standard calculation, service or feeder conductors with an ampacity of 100 amperes or more will be required. The optional calculated load shall include 100% of the first 10 kVA plus 40% of the remainder for the following loads:

- 1500 VA for each 2-wire, 20-ampere appliance circuit
- 1500 VA for each laundry branch-circuit

Figure 8-1 Service conductors on a typical one-family dwelling including telephone, TV, and AC disconnect

- 3 VA per sq ft of floor area
- Nameplate rating of all appliances fastened in place
- Nameplate rating of all motor loads

To this, we must add the largest of the following loads:

- 100% of the nameplate ratings of the air conditioner
- 100% of the nameplate ratings of heat pump compressors
- 65% of the nameplate ratings of **central electric space-heating** equipment
- 65% of the nameplate ratings of less than four, separately controlled, **electric space-heating** units (Figure 8-2)
- 40% of the nameplate ratings of four or more, separately controlled, electric space-heating units

The reason for the qualification of electric space-heating units being separately controlled is because separately controlled units will add to the diversity factor of units cycling on and off separately. Four or more space-heating units are given a greater demand reduction due to the probability of greater diversity in on and off cycling.

To demonstrate the use of the optional calculation procedure in comparison with the standard calculation, we will go through the required steps as follows:

Criteria

Dwelling unit with 1500 sq ft floor area, 12,000 VA electric range, 4500 VA water heater, 1200 VA dishwasher, 4500 VA clothes dryer, ⅓ horsepower kitchen waste disposer, and 10,000 VA of separately controlled electric baseboard in six rooms. The air-conditioning load is smaller than the heat load.

Standard Calculation
EXAMPLE

Floor Area—1500 sq ft @ 3 VA/sq ft   4500 VA
Appliance Circuits—2 @ 1500 VA   3000 VA
Laundry Circuit—1 @ 1500 VA   1500 VA
                              Total: 9000 VA

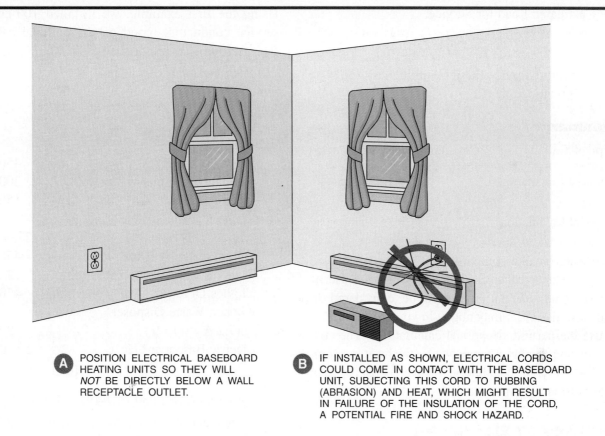

Figure 8-2 Electric baseboard heaters

Using Demand Factors—
3000 VA @ 100% =	3000 VA
6000 VA @ 35% =	2100 VA
Net General Lighting and Small Appliance	5100 VA
12,000 VA Electric Range	8000 VA
Water Heater	4500 VA
Dishwasher	1200 VA
Clothes Dryer (See *220.18*)	5000 VA
Kitchen Waste Disposer	1035*
Baseboard Electric Heat	10,000 VA
Total:	34,835 VA

Calculated Load for Service:

$$\frac{34{,}835}{240\ V} = 145\ \text{amperes}$$

---

*Table *430.148* FLA 7.2 × 115V = 825 VA
825 VA × 1.25 (*430.24*) = 1035 VA

Optional Calculation *(220.30)*

EXAMPLE

Floor Area—1500 sq ft @ 3 VA	4500 VA
Appliance Circuits—2 @ 1500 VA	3000 VA
Laundry Circuit—1 @ 1500 VA	1500 VA
Range @ Nameplate Rating	12,000 VA
Water Heater @ Nameplate Rating	4500 VA
Dishwasher @ Nameplate Rating	1200 VA
Clothes Dryer @ Nameplate	4500 VA
⅓ Horsepower Kitchen Waste Disposer (7.2 A × 115V × 1.25)	1035 VA
Total:	32,235 VA
Application of Demand Factor—	
10 kVA @ 100%	10,000 VA
Remainder of General Load @ 40% (22,235 × .40)	8894 VA
Total of General Load:	18,894 VA
10 kVA of Baseboard Heat @ 40% (10,000 × .40)	4000 VA
Total:	22,894 VA

Calculated Load for Service:

$$\frac{22{,}894}{240 \text{ V}} = 95.4 \text{ amperes}$$

Minimum Service Rating Permitted by *230.79* = 100 amperes

Summary

Type of Calculation	Calculated Load for Service
Standard Calculation	$\frac{34{,}835 \text{ VA}}{240 \text{ V}} = 145$ amperes
Optional Calculation	$\frac{22{,}894 \text{ VA}}{240 \text{ V}} = 95.4$ amperes

Based on this comparison, it is easy to realize the importance of investigating whether the use of the optional calculation should be considered in your installation. It may not be in your best interests to use the permitted optional calculations. The cost savings may be far out weighed by the limitations on the use of your installation's lower capacity as a result of the optional calculation.

EXISTING DWELLING UNIT

This calculation may be used, according to *220.31*, to determine if the existing service or feeder is of sufficient capacity to serve the additional loads. If the dwelling unit is served by a 3-wire, 120/240-volt or 208Y/120-volt service, it shall be permissible to compute the total load as follows, based on additional loads installed.

No Additional Air-Conditioning or Electric Space-Heating Equipment to be Installed

Formula to be used

Load (kVA)	Percent of Load
First 8 kVA of Load	100
Remainder of Load	40

Load calculations shall include the following:

- Floor area @ 3 VA per sq ft
- Appliance circuits @ 1500 VA each
- Laundry circuits @ 1500 VA each
- Electric range, ovens, cooking tops @ nameplate rating
- All appliances fastened in place @ nameplate rating

Using the first example, we installed 100-ampere service conductors based on the optional calculation. Assuming that at a later date, we have decided to install a 27 kW range in place of the 12.0 kW range originally installed, the following calculation may be used:

EXAMPLE

Floor Area—1500 sq ft @ 3 VA/sq ft	4500 VA
Appliance Circuits—2 @ 1500 VA	3000 VA
Laundry Circuit—1 @ 1500 VA	1500 VA
Range @ Nameplate Rating (new range)	27,000 VA
Water Heater @ Nameplate Rating	4500 VA
Dishwasher @ Nameplate Rating	1200 VA
Clothes Dryer @ Nameplate Rating	4500 VA
Kitchen Waste Disposer (7.2 A × 115 V)	828 VA
Electric Baseboard Heat	10,000 VA
Total:	57,028 VA

First 8 kVA of Load @ 100% = 8000 VA

Remainder of Load @ 40% = 57,028 − 8000 = 49,028 × .40 = 19,611 VA

19,611 + 8000 = 27,611 VA

Calculated Load for Service:

$$\frac{27{,}611 \text{ VA}}{240 \text{ V}} = 115 \text{ amperes}$$

By using this optional calculation method, we find that it is necessary to increase the service size from 100 amperes to 125 amperes to accommodate the new 27,000-VA range.

Where Additional Air-Conditioning or Electric Space-Heating Equipment Is Installed

If additional air-conditioning or electric space-heating is to be installed the following formula shall be used.

Air-Conditioning Equipment	100%
Central Electric Space-Heating	100%
Less Than Four Separately Controlled Space-Heating Units	100%
First 8 kVA of All Other Loads	100%
Remainder of All Other Loads	100%

Other Loads Include the Following:
1. General Lighting @ 3 VA sq ft
2. Appliance and Laundry Circuits @ 1500 VA ea.
3. Household Cooking Units @ Nameplate Rating
4. All Appliances Fastened in Place @ Nameplate Rating

Assume now that the increase in load was in electric space-heating equipment. The 10,000 VA of baseboard space-heating equipment is changed to 20,000 VA of central space-heating equipment.

Using the previous formula:

General Lighting @ 3 VA sq ft	4500 VA
2—Appliance Circuits @ 1500 VA	3000 VA
1—Laundry Circuit @ 1500 VA	1500 VA
Water Heater	4500 VA
Dishwasher	1200 VA
Disposal	828 VA
Electric Dryer	4500 VA
Electric Range	12,000 VA
Total:	23,028 VA

First 8000 @ 100% = 8000 VA
Remainder at 40% = 15,028 × .40 = 6011
Central Electric Space Heating   20,000 VA
 (increased from 10,000)
Calculated Load for Service:
 26,011 VA @ 240 V = 108 amperes

Summary

Type of Calculation	Calculated Load for Service
Optional Calculation Existing Dwelling Unit With 15 kVA Additional Range Load Added:	27,694 VA @ 240 V = 115 amperes
Optional Calculation Existing Dwelling Unit with 10 kVA Electric Space-Heating Load Added:	26,011 VA @ 240 V = 108 amperes

MULTIFAMILY DWELLING

The optional calculation for a multifamily dwelling (see Figure 8-3) may be used if each dwelling unit is equipped with electric cooking equipment, and each dwelling unit is also equipped with electric

Figure 8-3   Typical multifamily building

space-heating, or air-conditioning, or both. The *NEC®* provides a *Table (220.32)* that shows the demand factor percentage that may be applied based on the number of dwelling units in the multifamily building. The designer must note that house loads (corridor and stairwell lighting, outside lighting, etc.) must be added to the dwelling unit loads computed in accordance with *Table 220.32* (Figure 8-4).

Optional Calculation

EXAMPLE: 12, 1200 sq ft Dwelling Unit Multi-family Building

Floor Area—12 × 1200 × 3 VA =	43,200 VA
Appliance Circuits	
12 × 2 × 1500 VA =	36,000 VA
Laundry Circuits	
12 × 1 × 1500 VA =	18,000 VA
Electric Ranges 12 × 12,000 VA =	144,000 VA
Water Heaters 12 × 4500 VA =	54,000 VA
Dishwashers 12 × 1200 VA =	14,400 VA
Kitchen Waste Disposers	
12 × 828 (7.2 A × 115 V) =	9936 VA
Clothes Dryers	
(In Common Elements)	
Central Space-Heating	
12 × 10,000 VA =	120,000 VA
Total Connected Load:	439,536 VA
(*Table 220.32*)	× .41
	180,210 VA
(Estimated House Loads)	+21,000 VA
	201,210 VA
@ 240 V = 838 amperes	

Now we will do a standard calculation for the same building so that a comparison may be made.

Standard Calculation

Floor Area—	
12 × 1200 sq ft × 3 VA =	43,200 VA
Appliance Circuits	
12 × 2 × 1500 VA =	36,000 VA
Laundry Circuits	
12 × 1 × 1500 VA =	18,000 VA
Total General Lighting and	
Small Appliance Load:	97,200 VA
First 3000 VA @ 100% 3000 VA	
Remainder @ 35%	
(94,200 × .35)	32,970 VA

Table 220.32 Optional Calculations — Demand Factors for Three or More Multifamily Dwelling Units

Number of Dwelling Units	Demand Factor (Percent)
3–5	45
6–7	44
8–10	43
11	42
12–13	41
14–15	40
16–17	39
18–20	38
21	37
22–23	36
24–25	35
26–27	34
28–30	33
31	32
32–33	31
34–36	30
37–38	29
39–42	28
43–45	27
46–50	26
51–55	25
56–61	24
62 and over	23

Figure 8-4 *NEC® Table 220.32* (Reprinted with permission from NFPA 70-2002)

Lighting and Small Appliance	35,970 VA
Electric Ranges	
(*Table 220.19, Column C*)	27,000 VA
Water Heaters	
12 × 4500 VA = 54,000 VA	
Dishwashers	
12 × 1200 VA = 14,400 VA	
Kitchen Disposals	
12 × 828 =	9936 VA
(220.17)	78,336 VA × .75 = 58,752 VA
Central Space-Heating	
12 × 10,000 VA =	120,000 VA
Common Elements	21,000 VA
Total **Connected Load**:	262,722 VA
@ 240 V = 1095 amperes	

Summary

Type of Calculation	Calculated Load for Service
Standard Calculation	262,722 VA @ 240 V = 1095 amperes
Optional Calculation	201,210 VA @ 240 V = 838 amperes

This summary shows the wide variance of total connected load amperes that can be found when using optional calculations. The designer and installer must understand that the requirements in the *NEC®* are the *minimum* requirements for safety and that following these requirements may not provide an installation that is adequate for good service or will provide for future expansion.

TWO DWELLING UNITS

Where two dwelling units are supplied by a single feeder and the standard computed load using *Article 220, Part II* exceeds that for three identical units computed under *220.32*, the lesser of the two loads may be used (see Figure 8-4).

EXAMPLE

Floor Area—2 × 1200 sq ft @ 3 VA =		7200 VA
Appliance Circuits 2 × 2 × 1500 VA =		6000 VA
Laundry Circuits 2 × 1 × 1500 VA =		3000 VA
Total General Lighting and Small Appliance Load:		16,200 VA
First 3000 VA @ 100% = 3000 VA		
Remainder @ 35% (13,200 × .35) =		4620 VA
Net General Lighting and Small Appliance Load:		7620 VA
Electric Ranges (Table 220.19, Column C)		11,000 VA
Water Heaters 2 × 4500 =		9000 VA
Dishwashers 2 × 1200 =		2400 VA
Kitchen Disposals 2 × 828 =		1656 VA
(220.17)	13,056 VA × .75 =	9792 VA
Central Space-Heating 2 × 10,000 VA =		20,000 VA
House Loads		8500 VA
	Total:	56,912 VA

Calculated Load for Service:

$$\frac{56{,}912}{240 \text{ V}} = 237 \text{ amperes}$$

Three Identical Units Computed Under *220.32*

Floor Area—3 × 1200 × 3 VA =	10,800 VA
Appliance Circuits 3 × 2 × 1500 VA =	9000 VA
Laundry Circuits 3 × 1, 1500 VA =	4500 VA
Electric Ranges 3 × 12,000 VA =	36,000 VA
Water Heaters 3 × 4500 VA =	13,500 VA
Dishwashers 3 × 1200 VA =	3600 VA
Kitchen Waste Disposers 3 × 828 (7.2 A × 115 V) =	2484 VA
Clothes Dryers (In Common Elements)	
Central Space-Heating 3 × 10,000 VA =	30,000 VA
Total Connected Load:	109,884 VA
×	.45
	49,448 VA
House Loads	8500 VA
Total:	57,948 VA

Calculated Load for Service:

57,948 VA @ 240V = 241 amperes

Summary

Number of Units	Calculated Load for Service
Two Dwelling Units	$\frac{56{,}912 \text{ VA}}{240 \text{ V}} = 237$ amperes
Three Dwelling Units	$\frac{57{,}948 \text{ VA}}{240 \text{ V}} = 241$ amperes

The calculation for three units based on *220.32* resulted in 241 amperes compared with the two dwelling unit computation of 237 amperes based on the standard calculation. In accordance with *220.33*, the service may be based on 237 amperes. By using the optional calculation permitted by *220.32*, the three unit service is basically the same rating as the two unit standard calculation service.

SCHOOLS

An optional method for calculating the feeder or service load for a school (see Figure 8-5) can be done using *Table 230.34* when the school is equipped with electric space-heating or air-conditioning or both. The connected load to which the demand factors may be applied include all lighting, power, water heating, cooking, all other loads, and the larger of the air-conditioning load or electric space-heating load (Figure 8-6).

EXAMPLE

Calculations using the standard method shown in *Article 220* of the *NEC®* permits demand factors to be used for kitchen equipment as shown in

156 CHAPTER 8 Optional Load Calculations

Figure 8-5 Typical service for a school

Table 220.34 Optional Method — Demand Factors for Feeders and Service-Entrance Conductors for Schools

Connected Load	Demand Factor (Percent)
First 33 VA/m² (3 VA/ft²) at	100
Plus	
Over 33 to 220 VA/m² (3 to 20 VA/ft²) at	75
Plus	
Remainder over 220 VA/m² (20 VA/ft²) at	25

Figure 8-6 *NEC® Table 220.34* (Reprinted with permission from NFPA 70-2002)

Table 220.20 Demand Factors for Kitchen Equipment — Other Than Dwelling Unit(s)

Number of Units of Equipment	Demand Factor (Percent)
1	100
2	100
3	90
4	80
5	70
6 and over	65

Figure 8-7 *NEC® Table 220.20* (Reprinted with permission from NFPA 70-2002)

Table 220.20 (Figure 8-7), and general lighting loads in accordance with *Table 220.3(A)*. Using the following criteria, we will compare the methods of calculating the service or feeder loads.

Criteria
Service: 3-Phase, 4-Wire, 208/120
   Wye-Connected System
60,000 sq ft Floor Area
5 Units of Kitchen Equipment Totaling
   50,000 VA

100,000 VA Power Load
25,000 VA Water Heating Load
30,000 VA Miscellaneous Loads
125,000 VA Electric Heating Load

Standard
Lighting: 60,000 sq ft @ 3 VA
   [*220.3(A)*]   180,000 VA
Kitchen: 50,000 VA × .70 (*220.20*)   35,000 VA
Power   100,000 VA

Water Heating Load	25,000 VA
Miscellaneous Loads	30,000 VA
Electric Heating Load	125,000 VA
Total:	495,000 VA

Standard Calculation:

$$\frac{495,000 \text{ VA}}{208 \times 1.72} = \frac{495,000 \text{ VA}}{360} = 1375 \text{ amperes}$$

Optional

Lighting: 60,000 sq ft @ 3 VA	
[220.3(A)]	180,000 VA
Kitchen	50,000 VA
Power	100,000 VA
Water Heating Load	25,000 VA
Miscellaneous Loads	30,000 VA
Electric Heating Load	125,000 VA
Total:	510,000 VA

Optional Calculation

$$\frac{510,000 \text{ VA}}{60,000 \text{ sq ft}} = 8.5 \text{ VA per sq ft}$$

60,000 sq ft @ 3 VA sq ft (220.34)	180,000 VA
60,000 sq ft @ 5.5 VA sq ft =	
330,000 VA × .75	247,500 VA
	427,500 VA

$$\frac{427,500 \text{ VA}}{360} = 1187 \text{ amperes}$$

Summary

Type of Calculation	Calculated Load for Service
Standard Calculation	$\frac{495,000 \text{ VA}}{360} = 1375$ amperes
Optional Calculation	$\frac{427,500 \text{ VA}}{360} = 1187$ amperes

NEW RESTAURANTS

The optional calculation for new restaurants (Figure 8-8) may be used to calculate the service and feeder conductors. To use this optional method, you add all the electrical loads, including both heating and cooling loads, and multiply the total connected load by the single demand as shown in *Table 220.36* (Figure 8-9). This calculation is permitted in lieu of the requirements in *Part II* of *Article 220*.

The optional method applies only to feeder and service conductors for the entire building load and not to feeders for individual loads within the building. Feeder conductors are not required to be larger than the service conductors. The grounded (neutral) conductor load is determined by *220.22*.

EXAMPLE

Criteria

Service Characteristics: 3-Phase, 4-Wire, 120/208-Volt, Wye-Connected System.	
Lighting Load [Continuous Loads, see 210.20(A)]	80 kVA
Miscellaneous Power	50 kVA
Refrigeration	20 kVA
Air-Conditioning	300 kVA
Electric Space-Heating	65 kVA
Food Prep and Cleaning (4 units) (*Table 220.20*)	25 kVA

	Standard	Optional
Lighting Loads:	100 kVA	80 kVA
	(80A × 1.25)	
Power	50 kVA	50 kVA
Air-Conditioning	300 kVA	300 kVA
Refrigeration	20 kVA	20 kVA
Food Prep and Cleaning	20 kVA	25 kVA
	(*Table 220.20—80%*)	
Electric Space-Heating		65 kVA
Total:	490 kVA	540 kVA

Summary

Standard Calculation:

$$\frac{490 \text{ kVA}}{(208 \times 1.73)} = \frac{490 \text{ kVA}}{360} = 1361 \text{ amperes}$$

Optional Calculation:
$$540 \text{ kVA} - 325 = 215 \times .50 =$$
$$107.5 + 172.5 = 280 \text{ kVA}$$

$$\frac{280 \text{ kVA}}{360} = 777 \text{ amperes (Figure 8-9)}$$

NEC® Table 220.36 states the permitted load calculations for service and feeder conductors for new restaurants. The *Table* is divided into two categories, one represents calculations for an *all-electric restaurant*, which is a restaurant that uses electricity for cooking, water heating, heating, and

158 CHAPTER 8 Optional Load Calculations

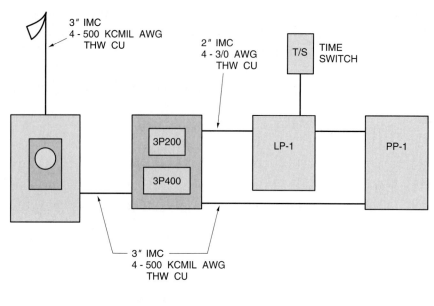

1. 400A, 3 PH, 4W, 208Y/120V C.T. FITTING W/ INSTRUMENT METER HOUSING
2. 400A, 3 PH, 4W, 208Y/120V MAIN DISTRIBUTION
   1-3P200A LP-1
   1-3P400A PP-1
3. 200A, 3 PH, 4W, 208Y/120V LIGHTING PANEL (LP-1)
   INTERIOR LIGHTING AND RECEPTACLES
   TIME SWITCH EXTERIOR LIGHTS
4. 400A, 3 PH, 4W, 208Y/120V POWER PANEL (PP-1)
   ELECTRIC COOKING
   ELECTRIC BOOSTER HTR FOR DISHWASHER
   KITCHEN APPLIANCES
   HVAC

Figure 8-8 Typical restaurant service

Table 220.36 Optional Method — Permitted Load Calculations for Service and Feeder Conductors for New Restaurants

Total Connected Load (kVA)	All Electric Restaurant Calculated Loads (kVA)	Not All Electric Restaurant Calculated Loads (kVA)
0–200	80%	100%
201–325	10% (amount over 200) + 160.0	50% (amount over 200) + 200.0
326–800	50% (amount over 325) + 172.5	45% (amount over 325) + 262.5
Over 800	50% (amount over 800) + 410.0	20% (amount over 800) + 476.3

Note: Add all electrical loads, including both heating and cooling loads, to compute the total connected load. Select the one demand factor that applies from the table, and multiply the total connected load by this single demand factor.

Figure 8-9 *NEC® Table 220.36* (Reprinted with permission from NFPA 70-2002)

air-conditioning. The second category is for a new restaurant that is a *not all-electric restaurant*, that is, one that may use gas-operated equipment.

The foregoing text emphasizes the variance that may be present relating to actual demand and the connected load. The designer and installer must study carefully all aspects of the intended installation before applying the optional calculations. Remember, the *NEC®* is the minimum requirements for safety and does not assure electrical installations that are necessarily convenient to the user or that contain provisions for future expansion.

REVIEW QUESTIONS

MULTIPLE CHOICE

1. If the optional calculation method for a one-family dwelling is used, the other loads above 10 kilowatts are subject to a demand factor of:
   ____ A. 30%.
   ____ B. 40%.
   ____ C. 50%.
   ____ D. 60%.

2. If the optional calculation method for a dwelling unit is used, the load of less than four separately controlled electric space-heating units is subject to a demand factor of:
   ____ A. 45%.
   ____ B. 50%.
   ____ C. 65%.
   ____ D. 100%.

3. The demand load for a 12.0 kW electric range in a dwelling unit is:
   ____ A. 12.0 kW.
   ____ B. 8.0 kW.
   ____ C. 6.0 kW.
   ____ D. 10.0 kW.

4. The optional method of calculation for a multifamily dwelling may be used when _____ or more dwelling units are served by the same service.
   ____ A. four
   ____ B. three
   ____ C. two
   ____ D. ten

5. If the optional calculation is used, the load for a 4.5 kilowatt clothes dryer is:
   ____ A. 4.5 kilowatts.
   ____ B. 5.0 kilowatts.
   ____ C. 6.0 kilowatts.
   ____ D. none of the above.

TRUE OR FALSE

1. When the optional calculation method is used, the minimum rating for a household electric dryer is 5000 watts.
   ____ True
   ____ False
   Code reference _____.

2. If the optional calculation method for a dwelling unit is used, the load of less than four separately controlled electric space-heating units is subject to a demand load of 65%.
   ____ True
   ____ False
   Code reference _____.

3. The optional calculation method may not be used for a dwelling unit served by a 208Y/120-volt service.
   ____ True
   ____ False
   Code reference _____ .

4. When using the optional calculation method for a new restaurant, both the air-conditioning and electric space-heating loads must be added.
   ____ True
   ____ False
   Code reference _____ .

5. The optional calculation method for a one-family dwelling permits all general loads above the initial 10 kilowatts to be subject to a demand factor of 50%.
   ____ True
   ____ False
   Code reference _____ .

FILL IN THE BLANKS

1. Optional calculations are a result of tests that have been made to determine the _____ in various types of occupancies.

2. A designer of electrical systems should always be cognizant of the _____ of his design in providing an efficient and convenient wiring system, with the ability for future expansion of electrical use.

3. The optional calculation for the service or feeder conductors for dwelling units can be used for those units that have a total connected load of _____ or more.

4. Four or more space-heating units are given a greater demand reduction due to the probability of _____ in on and off cycling.

5. The optional calculation for a multifamily dwelling may be used if each dwelling unit is equipped with _____ , and each dwelling unit is also equipped with _____ , or air-conditioning, or both.

FREQUENTLY ASKED QUESTIONS

Question 1: Why do they have these optional methods of calculation in the *NEC*®?

Answer: There are those who feel that the standard calculation methods, as shown in *Article 220*, are too restrictive. Enough documentation was presented to show that the actual demands on many buildings were considerably less than load calculations required in *Article 220*. These calculations do not give proper attention to adequacy for the user's convenience or consideration for future expansion, but they do meet the minimum requirements for safety.

Question 2: When making the optional method calculation, is it correct to use demand factor diversity as shown in *Article 220*?

Answer: No. Actual connected loads and percent of loads as permitted in the optional calculation may only be used. For instance, actual lighting loads must be used without applying demand factors (as shown in *Table 220.11*) or without increasing the rating due to continuous loads, as shown in *210.20(A)*.

Question 3: Can you explain what is meant in *Table 220.36* for the optional method of calculation for new restaurants, when it says 50% (amount over 325) + 172.5 for *All Electric Restaurant Calculated Loads (kVA)*?

Answer: After you find the total service or feeder load in kVA, which is done by adding all of the electrical loads, including both heating and cooling loads as indicated in the Note to *Table 220.36*, you go to the right-hand column of *Table 220.36* that is titled *Total Connected Load (kVA)*. Select the category in which your total load fits and move right to the column under *All Electric Restaurant Calculated Loads (kVA)*. Use the percentage shown on the amount your total kVA exceeds the maximum amount of the previous category in the left-hand column plus the amount shown in the second column. For example, if your total load in kVA is 400 kVA, you would choose the category in the left-hand column of *326—800*. Your total load of 400 kVA exceeds the maximum amount shown in the previous category (325) by 75. You use 50% of this number, plus the amount shown in the center column, $75 \times .50 = 37.5 + 172.5 = 210$ kVA.

Question 4: If I use the optional calculation method, can I also use *Table 310.15(B)(6)* to determine the conductor size required?

Answer: Yes. If the occupancy qualifies under *310.15(B)(6)*, you are permitted to use this table to calculate the conductor size required. There is considerable debate about the use of either of these *optional* calculations, but the use of both of these calculations, individually or together, is not prohibited.

Question 5: If I have a multifamily dwelling where only one-half of the units are equipped with electric cooking equipment, can I use the optional calculation method in some way?

Answer: No. *NEC® 220.32(A)(3)* requires that *each dwelling unit is equipped with electric cooking equipment*.

CHAPTER 9

Transformers

OBJECTIVES

After completing this chapter, the student should understand:
- dry-type transformers.
- voltage transformers.
- current transformers.
- oil-insulated transformers.
- autotransformers 600 volts or less.
- grounding-type autotransformers.
- isolation transformers.

KEY TERMS

Autotransformer: A transformer that has only one winding and is used for both the primary and secondary circuits simultaneously

Constant-Potential Transformer: A transformer that consists of three parts: a primary winding which carries the supply current, the magnetic core where the alternating magnetic flux is produced, and the secondary winding where an electromotive force is generated by the change in magnetism in the iron core

Current Transformer: A doughnut or bar type transformer (toroid) used to measure large currents in conductors that pass through the center opening, reproducing the effect of the current to a scale suited to the measuring instrument used

Dry-Type Transformer: Dry-type transformers are those in which their windings are air-cooled and are not immersed in any liquid for cooling

Isolation Transformer: A transformer of the multiple winding type with the primary and secondary windings physically separated, which inductively couples its secondary winding to the grounded feeder systems that energize its primary winding

Oil-Insulated Transformers: Oil-insulated transformers are those in which their windings are immersed in oil for cooling

Primary Winding: The winding that is connected to the source of power. It can be either the high-voltage winding or the low-voltage winding, depending on the transformer application

Secondary Winding: The winding from which the load is served. It can be either the high-voltage winding or the low-voltage winding, depending upon the transformer application

Voltage Transformer: A transformer with a high turns ration used to measure high voltage's primaries and having a low-voltage secondary connected to a low-scale voltmeter

163

INTRODUCTION

One of the greatest advantages of alternating current over direct current is that ac current can be transformed and dc current cannot be transformed.

This chapter will familiarize the student with the installation, protection, and maintenance of transformers. The serving utility transformer installation, protection, and maintenance is governed by the rules and regulations of the utility company and is not covered by this chapter. The information in this chapter has been written for use in conjunction with the requirements of *Article 450* of the *NEC*.®

GENERAL

Transformers are used in ac systems to raise or lower the voltage level in distribution systems. Figure 9-1 shows a utility sub-station using a large transformer to feed transmission lines to the area where power is needed. Figure 9-2 shows a pole-mounted utility transformer used to feed traffic-control equipment. The utility company uses pole-mounted transformers to feed metering for various equipment at roadside sites (Figure 9-3). Voltages required at distribution locations are derived from transformers, which have the ability to change the voltage and the current by varying the turns ratio of the transformer windings. The turns ratio between the windings determines the amount of increase or decrease in voltage. The volt-ampere rating of the primary of a transformer is equal to the volt-ampere rating of the secondary. This indicates that the ratio of current is inversely proportional to the voltage ratio of primary to secondary. For example, a 50 kVA, 480V/240V transformer has a full-load current of 50,000 divided by 480 = 104 amperes on the primary. The secondary has a full-load current of 50,000 divided by 240 = 208 amperes. The current is increased in direct proportion to the decrease in voltage. One goes up and the other goes down in same amount. A simple way to look at this is to take the numbers 2 and 4 equals 6. If you raise the 2 to 4 and lower the 4 to 2, you still have 4 and 2 equals 6.

Transformer windings are generally marked as high- and low-voltage rather than primary and secondary. The letter X is typically used to mark the low-voltage windings. X1 and X2 are used to identify the opposite terminals of one winding. Additional windings would be marked X3 and X4,

Figure 9-1 Utility sub-station transformer

Figure 9-2 Pole-mounted transformer

Figure 9-3 Metering and equipment fed from a pole-mounted transformer

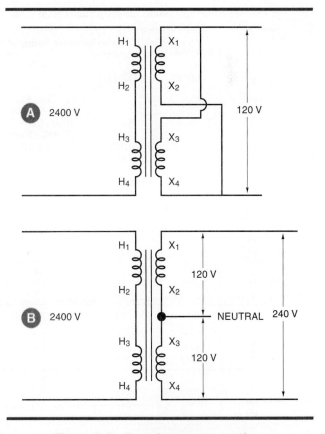

Figure 9-4 Transformer connections

X5 and X6, and so forth. High-voltage windings are marked H1 and H2, H3 and H4, and so on.

All even-numbered terminals will have the same polarity and will be negative or positive at the same time. The even-numbered terminals will be out of phase with the odd-numbered terminals. For example, assume a transformer has two equal **secondary windings**, each producing 120 volts, and are marked X1 and X2, X3 and X4, and each winding is rated 120 V. If the terminals X1 and X3, and X2 and X4 were connected, we would have an output of 120 volts (Figure 9-4[A]). However, if X2 is connected

166 CHAPTER 9   Transformers

to X3, and the output is taken from X1 and X4, the output would be 240 volts (Figure 9-4[B]).

The most important application of a transformer is in the transmission of energy over long distances. Figure 9-5 shows a typical transmission line system.

The voltage is raised and the current is proportionately lowered, allowing the use of smaller conductors to transmit the energy, then when the destination is reached, the voltage is lowered for use at safe levels, and the current is proportionately increased to carry large loads.

For example, to send a current of 400 amperes a distance of 1500 ft (450 m) over a conductor on a 480-volt system, you would be required to use at least a 500-kcmil copper conductor (not including any voltage drop considerations). This information is found in *Table 310.17, Allowable Ampacities of Single Insulated Conductors Rated 0 through 2000 Volts in Free Air* (Figure 9-6).

Using the 75° copper column, we find a 500-kcmil conductor can carry 620 amperes (depending

Figure 9-5   Typical transmission line system

Table 310.17 Allowable Ampacities of Single-Insulated Conductors Rated 0 Through 2000 Volts in Free Air, Based on Ambient Air Temperature of 30°C (86°F)

	Temperature Rating of Conductor (See Table 310.13.)						
	60°C (140°F)	75°C (167°F)	90°C (194°F)	60°C (140°F)	75°C (167°F)	90°C (194°F)	
	Types TW, UF	Types RHW, THHW, THW, THWN, XHHW, ZW	Types TBS, SA, SIS, FEP, FEPB, MI, RHH, RHW-2, THHN, THHW, THW-2, THWN-2, USE-2, XHH, XHHW, XHHW-2, ZW-2	Types TW, UF	Types RHW, THHW, THW, THWN, XHHW	Types TBS, SA, SIS, THHN, THHW, THW-2, THWN-2, RHH, RHW-2, USE-2, XHH, XHHW, XHHW-2, ZW-2	
Size AWG or kcmil	COPPER			ALUMINUM OR COPPER-CLAD ALUMINUM			Size AWG or kcmil
18	—	—	18	—	—	—	—
16	—	—	24	—	—	—	—
14*	25	30	35	—	—	—	—
12*	30	35	40	25	30	35	12*
10*	40	50	55	35	40	40	10*
8	60	70	80	45	55	60	8
6	80	95	105	60	75	80	6
4	105	125	140	80	100	110	4
3	120	145	165	95	115	130	3
2	140	170	190	110	135	150	2
1	165	195	220	130	155	175	1
1/0	195	230	260	150	180	205	1/0
2/0	225	265	300	175	210	235	2/0
3/0	260	310	350	200	240	275	3/0
4/0	300	360	405	235	280	315	4/0

Figure 9-6   *NEC® Table 310.17* (Reprinted with permission from NFPA 70-2002) *(continues)*

Table 310.17 Continued

Size AWG or kcmil	Temperature Rating of Conductor (See Table 310.13.)						Size AWG or kcmil
	60°C (140°F)	75°C (167°F)	90°C (194°F)	60°C (140°F)	75°C (167°F)	90°C (194°F)	
	Types TW, UF	Types RHW, THHW, THW, THWN, XHHW, ZW	Types TBS, SA, SIS, FEP, FEPB, MI, RHH, RHW-2, THHN, THHW, THW-2, THWN-2, USE-2, XHH, XHHW, XHHW-2, ZW-2	Types TW, UF	Types RHW, THHW, THW, THWN, XHHW	Types TBS, SA, SIS, THHN, THHW, THW-2, THWN-2, RHH, RHW-2, USE-2, XHH, XHHW, XHHW-2, ZW-2	
	COPPER			ALUMINUM OR COPPER-CLAD ALUMINUM			
250	340	405	455	265	315	355	250
300	375	445	505	290	350	395	300
350	420	505	570	330	395	445	350
400	455	545	615	355	425	480	400
500	515	620	700	405	485	545	500
600	575	690	780	455	540	615	600
700	630	755	855	500	595	675	700
750	655	785	885	515	620	700	750
800	680	815	920	535	645	725	800
900	730	870	985	580	700	785	900
1000	780	935	1055	625	750	845	1000
1250	890	1065	1200	710	855	960	1250
1500	980	1175	1325	795	950	1075	1500
1750	1070	1280	1445	875	1050	1185	1750
2000	1155	1385	1560	960	1150	1335	2000

CORRECTION FACTORS

Ambient Temp. (°C)	For ambient temperatures other than 30°C (86°F), multiply the allowable ampacities shown above by the appropriate factor shown below.						Ambient Temp. (°F)
21–25	1.08	1.05	1.04	1.08	1.05	1.04	70–77
26–30	1.00	1.00	1.00	1.00	1.00	1.00	78–86
31–35	0.91	0.94	0.96	0.91	0.94	0.96	87–95
36–40	0.82	0.88	0.91	0.82	0.88	0.91	96–104
41–45	0.71	0.82	0.87	0.71	0.82	0.87	105–113
46–50	0.58	0.75	0.82	0.58	0.75	0.82	114–122
51–55	0.41	0.67	0.76	0.41	0.67	0.76	123–131
56–60	—	0.58	0.71	—	0.58	0.71	132–140
61–70	—	0.33	0.58	—	0.33	0.58	141–158
71–80	—	—	0.41	—	—	0.41	159–176

* See 240.4(D).

Figure 9-6 *NEC® Table 310.17 (continued)*

upon ambient temperature considerations). A 500-kcmil conductor is about as thick as your thumb.

If you raised (transformed) the voltage from 480 volts to 2400 volts, an increase of 5 times or 500%, the current would be decreased or lowered proportionately to ⅕ or 124 amperes, and you would then be able to use a 1 AWG copper conductor (see *Table 310.16*), which is about as thick as a straw

Table 310.16 Allowable Ampacities of Insulated Conductors Rated 0 Through 2000 Volts, 60°C Through 90°C (140°F Through 194°F), Not More Than Three Current-Carrying Conductors in Raceway, Cable, or Earth (Directly Buried), Based on Ambient Temperature of 30°C (86°F)

Size AWG or kcmil	Temperature Rating of Conductor (See Table 310.13.)						Size AWG or kcmil
	60°C (140°F)	75°C (167°F)	90°C (194°F)	60°C (140°F)	75°C (167°F)	90°C (194°F)	
	Types TW, UF	Types RHW, THHW, THW, THWN, XHHW, USE, ZW	Types TBS, SA, SIS, FEP, FEPB, MI, RHH, RHW-2, THHN, THHW, THW-2, THWN-2, USE-2, XHH, XHHW, XHHW-2, ZW-2	Types TW, UF	Types RHW, THHW, THW, THWN, XHHW, USE	Types TBS, SA, SIS, THHN, THHW, THW-2, THWN-2, RHH, RHW-2, USE-2, XHH, XHHW, XHHW-2, ZW-2	
	COPPER			ALUMINUM OR COPPER-CLAD ALUMINUM			
18	—	—	14	—	—	—	—
16	—	—	18	—	—	—	—
14*	20	20	25	—	—	—	—
12*	25	25	30	20	20	25	12*
10*	30	35	40	25	30	35	10*
8	40	50	55	30	40	45	8
6	55	65	75	40	50	60	6
4	70	85	95	55	65	75	4
3	85	100	110	65	75	85	3
2	95	115	130	75	90	100	2
1	110	130	150	85	100	115	1
1/0	125	150	170	100	120	135	1/0
2/0	145	175	195	115	135	150	2/0
3/0	165	200	225	130	155	175	3/0
4/0	195	230	260	150	180	205	4/0
250	215	255	290	170	205	230	250
300	240	285	320	190	230	255	300
350	260	310	350	210	250	280	350
400	280	335	380	225	270	305	400
500	320	380	430	260	310	350	500
600	355	420	475	285	340	385	600
700	385	460	520	310	375	420	700
750	400	475	535	320	385	435	750
800	410	490	555	330	395	450	800
900	435	520	585	355	425	480	900
1000	455	545	615	375	445	500	1000
1250	495	590	665	405	485	545	1250
1500	520	625	705	435	520	585	1500
1750	545	650	735	455	545	615	1750
2000	560	665	750	470	560	630	2000
CORRECTION FACTORS							
Ambient Temp. (°C)	For ambient temperatures other than 30°C (86°F), multiply the allowable ampacities shown above by the appropriate factor shown below.						Ambient Temp. (°F)
21–25	1.08	1.05	1.04	1.08	1.05	1.04	70–77
26–30	1.00	1.00	1.00	1.00	1.00	1.00	78–86
31–35	0.91	0.94	0.96	0.91	0.94	0.96	87–95
36–40	0.82	0.88	0.91	0.82	0.88	0.91	96–104
41–45	0.71	0.82	0.87	0.71	0.82	0.87	105–113
46–50	0.58	0.75	0.82	0.58	0.75	0.82	114–122
51–55	0.41	0.67	0.76	0.41	0.67	0.76	123–131
56–60	—	0.58	0.71	—	0.58	0.71	132–140
61–70	—	0.33	0.58	—	0.33	0.58	141–158
71–80	—	—	0.41	—	—	0.41	159–176

* See 240.4(D).

Figure 9-7 *NEC® Table 310.16* (Reprinted with permission from NFPA 70-2002)

(Figure 9-7.) The economy that results from the use of smaller copper wire, together with the lesser cost of supporting this wire over that distance, is a great advantage.

Transformers Over 600 Volts

These transformers are air-cooled and are generally of the constant-potential type (Figure 9-8).

A **constant-potential transformer** is designed to change the voltage of a system. It operates with its primary connected to a constant-potential source and provides a secondary voltage that is substantially constant from no load to full-load. Potential is another way of saying voltage, as we learned in Chapter 1.

Overcurrent protection for transformers over 600 volts varies from that required for transformers 600 volts or under. When we speak of transformers 600 volts and under, or over 600 volts, we are referring to the higher voltage potential, either primary or secondary.

If a transformer primary is rated more than 600 volts, it must be protected by an individual overcurrent device on the primary side, in accordance with *Table 450.3(A)* (Figure 9-9).

If a fuse is used for this protection, it may be sized at a maximum of 300% of the transformer full-load amperes (Figure 9-10).

If a circuit breaker is used, it may be sized at a maximum of 600% of the transformer full-load amperes for transformers with a rated impedance not more than 6%, or a maximum of 400% for transformers with a rated impedance of more than 6% but not more than 10% (Figure 9-11).

If the secondary is also over 600 volts, the secondary fuse protection may be sized at a maximum of 250% for transformers with an impedance not greater than 6%, or 225% for transformers with an impedance greater than 6% but not greater than 10% (Figure 9-10). If circuit breakers are used for the secondary protection, they may be sized at 300% for transformers with a rated impedance of not more than 6%, or they may be sized at 250% for transformers with an impedance more than 6% but not greater than 10% (Figure 9-11).

For transformers over 600 volts with secondary voltage 600 or below, the rating of a fuse or circuit breaker must *not* be more than 125% of the transformer secondary rated current (Figure 9-10 and Figure 9-11).

In supervised locations, where maintenance of the transformer is under engineering supervision, transformers over 600 volts with secondary voltage 600 or below, the rating of a fuse or circuit breaker is permitted to be 250% of the transformer secondary rated current. This information may be found in *Table 450.3(A)* of the *NEC®* (refer to Figure 9-9).

It must be understood that the overcurrent protection, required by *Table 450.3(A)*, is for the protection of the transformer, and will not necessarily protect the primary or secondary conductors, or equipment connected to the secondary. Using overcurrent protection to the full values permitted will require much larger conductors than the full-load current rating of the transformer.

Transformers 600 Volts or Less

The difference in installation procedures between a **dry-type transformer** rated over 600 volts, and a dry-type transformer rated 600 volts or less, is for the most part, in the overcurrent protection requirements.

Figure 9-8 Pole-mounted constant-potential transformer. *Courtesy of* Niagara Mohawk

PROTECTION BY PRIMARY OVERCURRENT DEVICE

Conductors are permitted by *240.21(C)* to be connected to a transformer secondary, without overcurrent protection at the secondary as follows:

For transformer primary rated current of 9 amperes or more, conductors supplied by a single-phase transformer with a 2-wire secondary (single voltage), or a 3-phase, delta-delta connected transformer having a 3-wire (single voltage) secondary, provided this primary protection is in accordance with *450.3*, and does not exceed the value determined by multiplying the secondary conductor ampacity by the secondary to primary transformer voltage ratio.

EXAMPLE

Assume a 50 kVA, 600/240 V, 3-phase, 3-wire transformer.

50 kVA, 3-phase, 3-wire, 240-volt delta secondary

$50{,}000/(240 \times 1.732) = 120$ amperes (secondary conductor ampacity)

600 volts ÷ 240 = 2.5 (secondary to primary ratio)

2.5×120 amperes = 300 amperes

In this example, *Table 450.3(B)* permits primary only protection when this protection does not exceed 125% of the transformer primary rated current. The transformer primary rated current is 50,000 divided

---

Table 450.3(A) Maximum Rating or Setting of Overcurrent Protection for Transformers Over 600 Volts (as a Percentage of Transformer-Rated Current)

Location Limitations	Transformer Rated Impedance	Primary Protection Over 600 Volts		Secondary Protection (See Note 2.)		
				Over 600 Volts		600 Volts or Below
		Circuit Breaker (See Note 4.)	Fuse Rating	Circuit Breaker (See Note 4.)	Fuse Rating	Circuit Breaker or Fuse Rating
Any location	Not more than 6%	600% (See Note 1.)	300% (See Note 1.)	300% (See Note 1.)	250% (See Note 1.)	125% (See Note 1.)
	More than 6% and not more than 10%	400% (See Note 1.)	300% (See Note 1.)	250% (See Note 1.)	225% (See Note 1.)	125% (See Note 1.)
Supervised locations only (See Note 3.)	Any	300% (See Note 1.)	250% (See Note 1.)	Not required	Not required	Not required
	Not more than 6%	600%	300%	300% (See Note 5.)	250% (See Note 5.)	250% (See Note 5.)
	More than 6% and not more than 10%	400%	300%	250% (See Note 5.)	225% (See Note 5.)	250% (See Note 5.)

Notes:
1. Where the required fuse rating or circuit breaker setting does not correspond to a standard rating or setting, a higher rating or setting that does not exceed the next higher standard rating or setting shall be permitted.
2. Where secondary overcurrent protection is required, the secondary overcurrent device shall be permitted to consist of not more than six circuit breakers or six sets of fuses grouped in one location. Where multiple overcurrent devices are utilized, the total of all the device ratings shall not exceed the allowed value of a single overcurrent device. If both circuit breakers and fuses are used as the overcurrent device, the total of the device ratings shall not exceed that allowed for fuses.
3. A supervised location is a location where conditions of maintenance and supervision ensure that only qualified persons monitor and service the transformer installation.
4. Electronically actuated fuses that may be set to open at a specific current shall be set in accordance with settings for circuit breakers.
5. A transformer equipped with a coordinated thermal overload protection by the manufacturer shall be permitted to have separate secondary protection omitted.

Figure 9-9 *NEC® Table 450.3(A) 2002* (Reprinted with permission from NFPA 70-2002)

by 600 V × 1.732 = 48 amperes times 125% = 60 amperes. Using a primary protective device rated at 60 amperes does not exceed the value 300 amperes derived in the example.

Single-phase (other than 2-wire) and multi-phase (other than delta-delta, 3-wire) transformer secondary conductors are not considered protected by the primary overcurrent protective device. Figure 9-12 shows an example of a single-phase, 2-wire transformer and a single-phase, 3-wire transformer. Figure 9-13 shows an example of a delta-delta, 3-wire transformer, and a multi-phase transformer other than delta-delta, 3-wire.

The installer must recognize that although *Table 450.3(B)* does not require secondary overcurrent

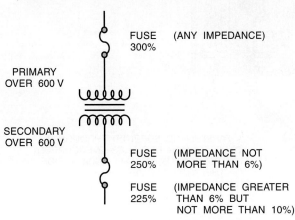

Figure 9-10 Maximum rating or setting of fuse overcurrent protection for transformers over 600 volts (as a percentage of transformer-rated current)

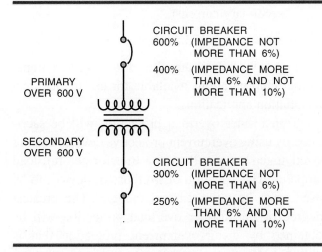

Figure 9-11 Maximum rating or setting of circuit-breaker overcurrent protection for transformers over 600 volts (as a percentage of transformer-rated current)

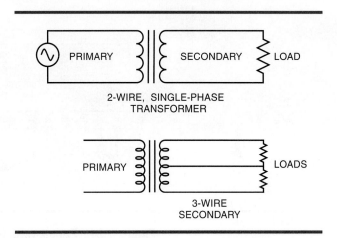

Figure 9-12 2-wire, single-phase transformer; 3-wire, single-phase transformer

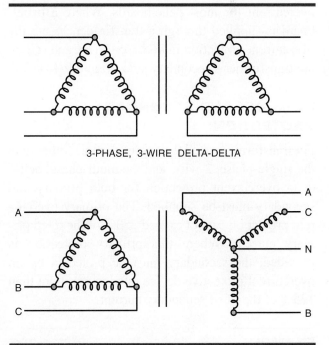

Figure 9-13 3-phase, 3-wire delta-delta; 3-phase, 4-wire delta-wye

172 CHAPTER 9 Transformers

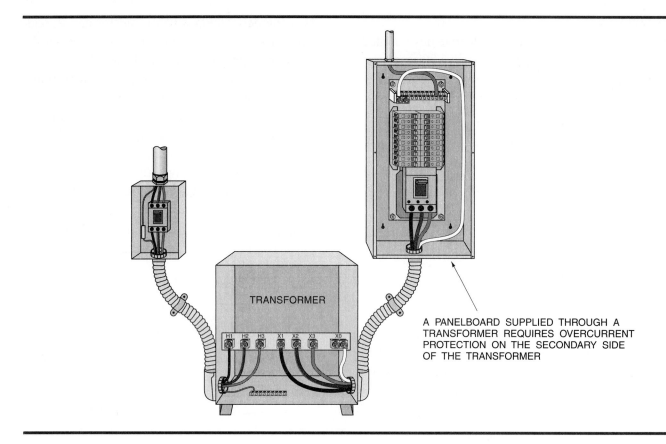

Figure 9-14 Panelboard supplied through transformer

protection if the primary overcurrent protection is limited to 125%, *408.16(D)* requires secondary overcurrent for most panelboards. Where a panelboard is supplied through a transformer, locate the overcurrent protection in *408.16(A)*, *(B)*, and *(C)* on the transformer's secondary side (Figure 9-14).

PRIMARY AND SECONDARY PROTECTION

For transformers rated 9 amperes or more, other than the single-phase, 2-wire, and the multi-phase, delta-delta overcurrent protection for both primary and secondary must be supplied. The primary overcurrent protection cannot exceed 250% of the rated primary current. Where this primary protection is provided, the secondary must be protected by an overcurrent protective device rated at not more than 125% of the rated secondary current.

EXAMPLE

Assume a 50 kVA, 600 V, 208 Y/120-volt, 3-phase, 4-wire transformer.

$50,000/(600 \times 1.732) = 48$ amperes (rated primary current)

48×2.5 (250%) = 120 amperes (maximum primary protection)

$50,000/(208 \times 1.732) = 138.8$ amperes (rated secondary current)

$138.8 \times 1.25 = 173.5$ amperes

Figure 9-15 illustrates the various 3-phase transformer configurations available for use in electrical installation applications.

Transformer overload protection will be sacrificed by using overcurrent protective devices that are sized much greater than the transformer full-load amperes. The selection of overcurrent devices to be used must be given careful thought. The greatest degree of transformer overload protection will be obtained by using overcurrent protection that is within 125% of the transformer full-load amperes.

Where secondary protection is required, the secondary overcurrent device shall be permitted to consist of not more than six circuit breakers or six sets

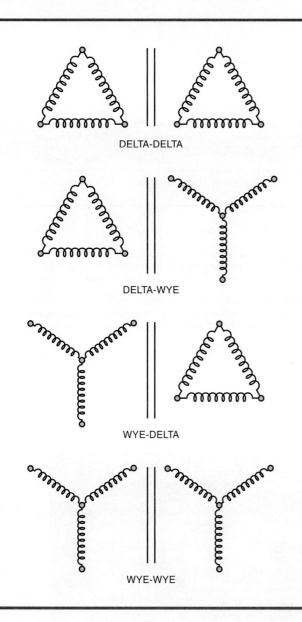

Figure 9-15 Three-phase transformer configuration

of fuses grouped in one location. Where multiple overcurrent devices are used, the total of all the ratings shall not exceed the allowed value of a single device, as shown in *Table 450.3(B), Note 2* (Figure 9-16).

Although *Table 450.3(B)* does not require secondary overcurrent protection, in some applications, where the primary overcurrent protection is limited to 125% of the primary rated current, *408.16(D)* requires secondary overcurrent protection for some panelboards. Where a panelboard is supplied through a transformer, you must locate the overcurrent protection required in *408.16(D)* on the secondary side of the transformer

Oil-Insulated Transformers

Oil-insulated transformers installed indoors must be installed in a vault, unless they fall under one of the *Exceptions 450.26*. The *Exceptions* generally cover installations where suitable protection is taken to ensure that a transformer oil fire will not spread to other parts of the building. Oil-insulated transformers installed outdoors require that precautions be taken to prevent combustible parts of buildings from being ignited by a transformer installed on a roof or attached to a building. Safeguards such as space separations fire resistant barriers or enclosures that confine the oil of a ruptured transformer tank might be applied (Figure 9-17).

Autotransformers

Almost any transformer can be connected as an **autotransformer**. The primary and secondary windings can be connected in series to obtain the desired input and output voltages. When connected as an autotransformer, some of the power is conducted directly from the source to the load, and the autotransformer only transforms part of the load. The voltages available from an autotransformer are equal to the sum of the winding voltages (Figure 9-18).

Assume a transformer with a 480-volt, single-phase, 2-wire primary and a 120/240-volt, 3-wire secondary. The top winding marked A and B represents a 480-volt primary marked H1 and H2. The next two windings represent a 3-wire, 120/240-volt secondary marked X1, X2, and X3. By connecting H2 to X1 you would have the configuration shown in Figure 9-18.

A very common use years ago for an autotransformer was to obtain 120-volt, single-phase power for lighting and small power circuits from a 240-volt, 3-phase system used for motor loads in an industrial plant (Figure 9-19).

In Figure 9-19, the 240-volt supply is obtained to one phase of the 3-phase delta system. The secondary of the transformer used as an autotransformer winding is a center tapped 240-volt winding, typical of those used in 120/240-volt, 3-wire systems. The primary is unused. The system must be grounded as shown.

The *NEC®* allows the use of autotransformers under certain conditions. Branch-circuits cannot be derived from autotransformers, unless the circuit

Table 450.3(B) Maximum Rating or Setting of Overcurrent Protection for Transformers 600 Volts and Less (as a Percentage of Transformer-Rated Current)

	Primary Protection			Secondary Protection (See Note 2.)	
Protection Method	Currents of 9 Amperes or More	Currents Less Than 9 Amperes	Currents Less Than 2 Amperes	Currents of 9 Amperes or More	Currents Less Than 9 Amperes
Primary only protection	125% (See Note 1.)	167%	300%	Not required	Not required
Primary and secondary protection	250% (See Note 3.)	250% (See Note 3.)	250% (See Note 3.)	125% (See Note 1.)	167%

Notes:
1. Where 125 percent of this current does not correspond to a standard rating of a fuse or nonadjustable circuit breaker, a higher rating that does not exceed the next higher standard rating shall be permitted.
2. Where secondary overcurrent protection is required, the secondary overcurrent device shall be permitted to consist of not more than six circuit breakers or six sets of fuses grouped in one location. Where multiple overcurrent devices are utilized, the total of all the device ratings shall not exceed the allowed value of a single overcurrent device. If both breakers and fuses are utilized as the overcurrent device, the total of the device ratings shall not exceed that allowed for fuses.
3. A transformer equipped with coordinated thermal overload protection by the manufacturer and arranged to interrupt the primary current shall be permitted to have primary overcurrent protection rated or set at a current value that is not more than six times the rated current of the transformer for transformers having not more than 6 percent impedance and not more than four times the rated current of the transformer for transformers having more than 6 percent but not more than 10 percent impedance.

Figure 9-16 *NEC® Table 450.3(B)* (Reprinted with permission from NFPA 70-2002)

Figure 9-17 Oil-insulated transformer at utility sub-station. *Courtesy of* Magnatek

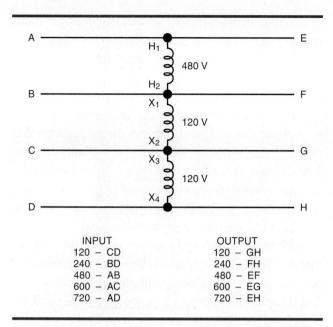

Figure 9-18 Autotransformer voltage connections

Voltage Transformers

Voltage transformers are transformers with a high turns ratio. These transformers are also known as potential transformers. They are used in voltage measuring meters and watthour-meters. The turns ratio is dependent upon the line voltage to be measured and the limits of the meter. For example, a voltage meter with a capability of reading 0-150 volts connected to a 2400-volt primary would need a turns ratio of 16 to 1. The meter scale will be altered to show actual line voltage (Figure 9-20).

Current Transformers

Current transformers are bar or doughnut type transformers. They are used to measure high levels of current with a calibrated meter or to measure the amount of current flow on a conductor to operate some other calibrated device, such as a GFCI. The conductor to be measured is passed through the *doughnut* or ring, and the ammeter or device is connected to windings around the ring. The current

derived has a grounded conductor that is electrically connected to a grounded conductor of the system supplying the autotransformer, as shown in Figure 9-19.

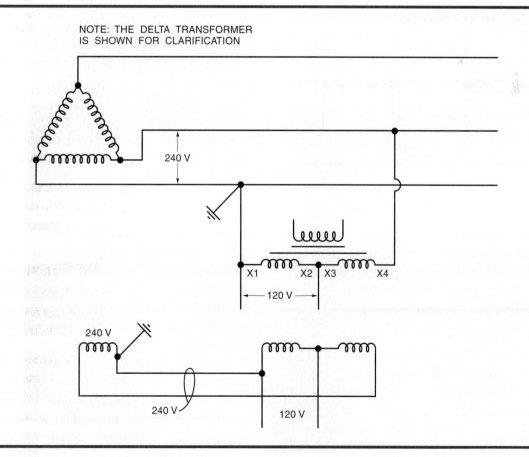

Figure 9-19 Using a 3-phase, 3-wire, 240-volt delta source to obtain 120-volt power

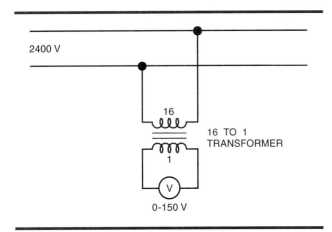

Figure 9-20　Voltage transformer connection

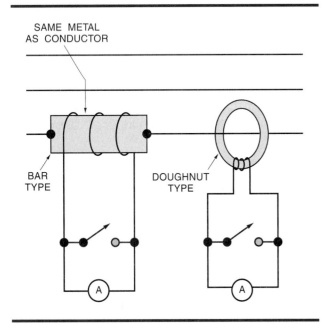

Figure 9-21　Current transformer connection

induced in the doughnut winding is directly proportional to turns ration between the line conductor and the doughnut windings (Figure 9-21).

Grounding-Type Transformers

Grounding-type transformers, or zigzag as they are commonly called, are used to convert a 3-phase, 3-wire, delta ungrounded secondary to a 3-phase, 4-wire, wye-connected system and to derive an artificial neutral which can be grounded. Generally, a zigzag transformer is used to gain the advantages of a grounded system and need only be sized to carry ground-fault current for a short duration.

A zigzag transformer is classified as an autotransformer and has no secondary winding. Two windings for each phase are wound on each core leg, but each pair of windings on the same core leg is wound in the opposite direction (Figure 9-22).

This winding technique creates a high resistance to 3-phase currents, and reduces the impedance that can interfere with the flow of ground-fault current. The derived grounded conductor can be used as the system grounded (neutral) conductor, but in this application, the grounding transformer must be sized for the maximum unbalanced load.

Where a transformer is wound in this zigzag manner, it does two things. It limits the 3-phase current to a level where it only has a magnetizing effect, and it limits the impedance in the windings so that a high level of fault current will flow. This high level of ground-fault current is desirable in most applications where it will serve to open the overcurrent protective device rapidly.

Isolation Transformers

Isolation transformers have the secondary winding physically and electrically isolated from the **primary winding**. There is no electrical connection between the primary and secondary. The basic construction of an isolation transformer is shown in Figure 9-23. Isolation transformers are used extensively in health care facilities as shown in *517.160*. This transformer inductively couples its secondary winding to the grounded feeder systems that energize its primary winding. It is sometimes necessary to use ungrounded circuits and feeders to equipment. The isolation transformer is used for this purpose. The grounded characteristics of the primary are not inductively transferred to the secondary winding of the transformer. The magnetic field produced by ac current induces energy in the secondary winding without the grounding properties (Figure 9-24). If grounding is required in the secondary then grounding procedures must be established.

Isolation transformers are also used in accordance with *680.23(A)(2)*. The transformer is used for underwater luminaires (lighting fixtures). The transformer must be an isolated winding type with an ungrounded secondary that has a grounded metal barrier between the primary and secondary windings.

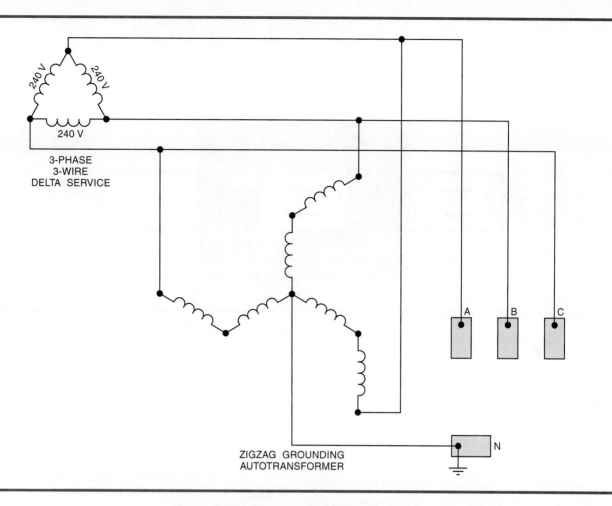

Figure 9-22 Zigzag transformer winding

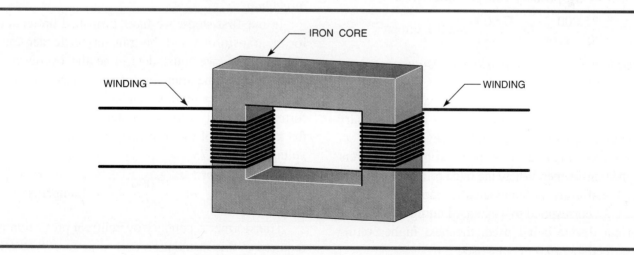

Figure 9-23 Basic construction of an isolation transformer

Five Steps to a Transformer Installation

Assume a 75 kVA, 3-phase, 4-wire, 480-208 Y/120-volt transformer is to be installed in an industrial plant, to obtain 3-phase, 4-wire, 120/208-volt circuits for lighting and small power loads. The transformer is fed from a 480-volt distribution panel located in an electric closet. In order to gain the

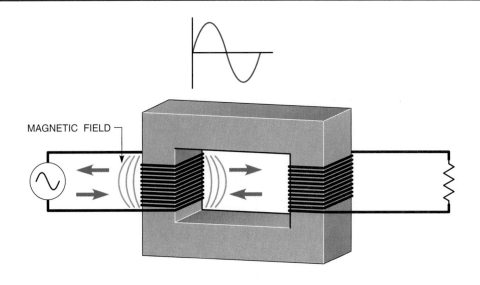

Figure 9-24 Magnetic field produced by ac current

highest use of the transformer, it will be fed with the maximum safe load it will carry. The transformer secondary will feed a 200-ampere main breaker lighting and appliance panelboard, located adjacent to the transformer.

Step #1: *Determine the transformer primary current.*

The safe current the transformer will carry is determined by dividing the kVA rating of the transformer by the primary voltage, times 1.73.

$$\frac{75,000}{480 \times 1.73} \text{ or } \frac{75,000}{830} = 90.3 \text{ amperes}$$

Step #2: *Determine the primary overcurrent protection.*

NEC® 450.3(B)(1) requires that each transformer rated 600 volts or less be protected on the primary side by an overcurrent protective device rated or set at not more than 125% of the rated primary current of the transformer. Where the rated primary current of the transformer is 9 amperes or more, and 125% does not correspond to a standard rating of the overcurrent device being used, the next higher rating described in *240.6* may be used.

90.3 × 1.25 = 112.8 amperes or next higher 125 amperes

Step #3: *Determine transformer feeder conductor size.*

A conductor is required that will safely carry the primary current of the transformer and will be protected by a 125-ampere overcurrent device. *Table 310.16* shows a 2 AWG THWN copper conductor at 115 amperes. *NEC® 240.4(B)* permits this conductor to be protected by the next highest overcurrent device size.

A 2 AWG THWN copper conductor will be used.

Step #4: *Determine the secondary overcurrent protection.*

In our first steps, we have furnished protection for the transformer and the transformer feeder conductors. Now we must determine the overcurrent protection for the transformer secondary tap conductors. The maximum secondary current the transformer will safely supply is determined by dividing the kVA rating of the transformer by the secondary voltage, times 1.73.

$$\frac{75,000}{208 \times 1.73} \text{ or } \frac{75,000}{360} = 208.3 \text{ amperes}$$

Transformer secondary overcurrent protection is required in this installation according to *408.16(D)*, which tells us if a panelboard is supplied through a transformer, the overcurrent protection shall be located on the secondary side of the transformer. In this installation, the 200 ampere main breaker in the lighting and appliance panelboard will serve as the secondary overcurrent protection. This protection conforms to the requirements of *240.21(C)(2)* relating to the location in the circuit of the overcurrent protection.

Step #5: *Transformer grounding.*

This transformer installation is a separately derived system, as defined in *Article 100* of the *NEC*, that says in part: *a premises wiring system whose power is derived from a transformer that has no electrical connection to supply conductors originating in another system.* In effect, this means it is not a part of the service equipment connected to a serving utility.

NEC 250.30 requires a bonding jumper to connect the equipment grounding conductors of the derived system to the grounded conductor. The grounded conductor is the one that is intentionally grounded by connecting it to the neutral bar in the panel. The equipment grounding conductors are the green or bare conductors used in nonmetallic systems and the metal raceways used in metallic systems. These are effectively bonded together by connection to the metal panel enclosure, either through the ground bar or by connection to the enclosure. The main bonding jumper is connected to the neutral bar and to the metal enclosure. There are a variety of other ways to do this, but generally it is done in the panelboard. See *250.30(A)(1)* for alternate methods.

A grounding electrode conductor must be installed to connect the grounded conductor to a grounding electrode as specified in *250.30(A)(3)*. The grounding electrode can be an effectively grounded structural steel member of the structure, an effectively grounded metal water pipe within 5 ft (1.5 m) from the point where it enters the building, or other electrodes as specified in *250.50* and *52* where the electrodes specified in *(1)* or *(2)* are not available.

This means that for separately derived systems, the grounding electrode can be building steel or a metal water pipe, and if they are not available, you can use a ground rod, pipe, ring, plate, or concrete encased electrode.

Miscellaneous Requirements

When installing transformers, attention must be paid to the following:

- Provisions must be made to minimize the possibility of physical damage to the transformer, *450.8(A)*
- Dry-type transformers must be provided with an enclosure that will provide protection against accidental insertion of foreign objects, *450.8(B)*
- All energized parts must be guarded in accordance with *110.27* and *110.34* and accessible to qualified persons only, *450.8(C)*
- The operating voltage of exposed live parts of transformer installations must be indicated by signs or other markings, *450.8(D)*
- Transformers with ventilating openings must be installed so the ventilating openings are not blocked by walls or other obstructions
- Dry-type transformers, not over 112½ kVA, installed outdoors must have a separation of at least 12 in. (305 mm) from combustible material, *450.21(A)*

Transformers are long lasting pieces of equipment that require very little attention other than periodic inspection and maintenance. Transformers should be checked periodically to be sure clearances are maintained and that there is no excessive heating.

REVIEW QUESTIONS

MULTIPLE CHOICE

1. The rated primary current of a 37½-kVA, 480-208/120-volt transformer is:
   - ____ A. 45 amperes.
   - ____ B. 104 amperes.
   - ____ C. 37 amperes.
   - ____ D. 75 amperes.

2. A voltage meter with a voltage transformer having an 16 to 1 ratio, connected to a 2400-volt line would read:
   ____ A. 120 volts.
   ____ B. 240 volts.
   ____ C. 150 volts.
   ____ D. 480 volts.

3. An autotransformer has how many windings?
   ____ A. Two
   ____ B. One
   ____ C. Four
   ____ D. Six

4. A single-phase, 480-volt-240/120-volt transformer, connected as an autotransformer, would have _____ output voltages available.
   ____ A. four
   ____ B. three
   ____ C. five
   ____ D. two

5. The rated secondary current of a 50 kVA, 480-208 Y/120-volt, 3-phase transformer has an approximate primary to secondary ratio of:
   ____ A. 2 to 4.
   ____ B. 1 to 4.
   ____ C. 1 to 2.3.
   ____ D. 2 to 3.

TRUE OR FALSE

1. A zigzag transformer is a constant potential transformer.
   ____ A. True
   ____ B. False

2. Dry-type transformers have their windings immersed in oil.
   ____ A. True
   ____ B. False

3. The primary winding is always the high-voltage winding.
   ____ A. True
   ____ B. False

4. The turns ratio between the windings determines the amount of increase or decrease in voltage.
   ____ A. True
   ____ B. False

5. The letter X is usually used to mark the high-voltage terminal of a transformer.
   ____ A. True
   ____ B. False

6. In a two winding secondary, the windings are connected in series to gain a higher voltage.
   ____ A. True
   ____ B. False

7. A constant-potential transformer operates with its primary connected to a variable voltage source.
   ____ A. True
   ____ B. False

8. The secondary of transformers 600 volts or less does *not* ever need overcurrent protection if the primary overcurrent protection is limited to 125% of the primary rated current.
   ____ A. True
   ____ B. False

9. Where secondary overcurrent protection is required, the secondary overcurrent device is limited to a single device.
   ____ A. True
   ____ B. False

10. Primary overcurrent protection protects only the transformer feeder and the transformer.
    ____ A. True
    ____ B. False

FILL IN THE BLANKS

1. One of the greatest advantages of _____ over direct current is that ac current can be _____, and dc current cannot be transformed.

2. The winding from which the load is served. It can be either the _____ winding or the _____ winding, depending upon the transformer application.

3. The _____ rating of the primary of a transformer is equal to the _____ rating of the secondary.

4. The most important application of a transformer is in the _____ over long distances.

5. It operates with its primary connected to a _____ source and provides a secondary _____ that is substantially constant from no load to full-load.

6. For transformers over 600 volts with secondary voltage _____, the rating of a fuse or circuit breaker must *not* be more than _____ of the transformer secondary rated current.

7. Single-phase (other than 2-wire) and multi-phase (other than delta-delta, 3-wire) transformer secondary conductors are _____ protected by the primary overcurrent protective device.

8. Where this primary protection is provided, the _____ must be protected by an overcurrent protective device rated at not more than _____ of the rated secondary current.

9. The primary and secondary windings can be _____ to obtain the desired _____ and _____ voltages.

10. Branch-circuits cannot be derived from autotransformers unless the _____ derived has a grounded conductor that is _____ to a grounded conductor of the system supplying the autotransformer.

FREQUENTLY ASKED QUESTIONS

Question 1: When a transformer is said to be rated over 600 volts, does that mean the primary or secondary?

Answer: Either one. Transformers can be stepup or stepdown. The primary is where the source voltage is applied, and the secondary is where the load is supplied. For example, we may have a transformer with a 7200-volt primary and a 480-volt secondary or we may have a transformer with a 480-volt primary and a 7200-volt secondary for some electronic application. In either case, the transformer is classified as over 600 volts.

Question 2: If I have primary overcurrent protection not greater than 125% of the rated primary current of the transformer rated at 600 volts or less, what is the maximum rating or setting of overcurrent protection I can use for the secondary?

Answer: According to *Table 450.3(B)*, secondary overcurrent protection is not required if the primary overcurrent protection is not greater than 125% of the rated primary current of the transformer in certain applications. If, however, you are working with a transformer that is not a single-phase, 2-wire, or a 3-phase, 3-wire, delta-delta, then you need a primary overcurrent protection 250% or less than the rated transformer primary current and a secondary overcurrent protection not exceeding 125% of the transformer rated secondary current.

Question 3: Can I install a dry-type transformer rated 75 kVA above a lay-in type ceiling?

Answer: *NEC® 450.13(B)* permits transformers 600 volts or less, and not exceeding 50 kVA, to be installed in hollow spaces of buildings not permanently closed in by building structure provided that the ventilation is adequate to dispose of the heat generated by the transformer without creating a temperature rise that is in excess of the transformer rating. Transformers so installed in hollow spaces of buildings shall not be required to be readily accessible.

Question 3: What are the clearance requirements for dry-type transformers 600 volts or less when installed indoors?

Answer: According to *450.9*, transformers with ventilating openings must be installed so that the ventilating openings are not blocked by walls or other obstructions. The required clearances must be clearly marked on the transformer.

Question 4: When are transformer vaults required?

Answer: Dry-type transformers rated over 112½ kVA must be installed in a fire resistant transformer room having a minimum fire rating of 1 hour. Dry-type transformers over 35,000 volts must be installed in a vault. Oil-insulated transformers installed indoors must be installed in a vault if they do not fall under one of the *Exceptions* to *450.26*.

Question 5: What are the requirements for the construction of a transformer vault?

Answer: The requirements for transformer vaults can be found in *Article 450, Part III*. The walls and roofs of vaults must be made of materials that have a minimum of 3 hours fire resistance. The floor of vaults in contact with the earth must be at least 4 in. (100 mm) thick.

Each doorway must have a minimum of 3 hours fire resistance, door sills must be of sufficient height to contain the oil from the largest transformer within the vault, and in no case shall the height be less than 4 in. (100 mm). Doors shall be kept locked, but must swing out and be equipped with panic hardware.

According to *450.45*, adequate ventilation must be provided. Any piping foreign to the electrical installation must not enter or pass through a transformer vault.

CHAPTER 10

Motors and Controllers

OBJECTIVES

After completing this chapter, the student should understand:
- motor code letter.
- service factor.
- motor full-load current.
- motor controller.
- motor branch-circuit conductors.
- motor branch-circuit short-circuit and ground-fault protective devices.
- motor disconnect switch.
- motor overload protective devices.
- motor overload relay.
- motor-control circuit protective device.
- locked-rotor current.

KEY TERMS

Code Letter: A letter marked on the motor nameplate to show motor input in kilovolt-amperes per horsepower with the rotor locked

Full-Load Current: The running current of a motor under full-load

Locked-Rotor Current: The current taken by the motor with the rotor held stationary

Motor Controller: A device that serves to govern, in some predetermined manner, the electric power to the motor to which it is connected

Motor Disconnect Switch: A disconnecting means installed within sight from a motor to be used for safety precautions during maintenance

Overload Protective Relay: A protective relay that responds to temperature caused by excessive current flow beyond its predetermined setting

Service Factor: The amount a motor may be overloaded by multiplying the service factor times the horsepower of the motor

INTRODUCTION

This chapter will cover basic motor and controller installations. A step-by-step installation procedure will illustrate to the student installation methods and the requirements of the *NEC.* The *NEC* requirements for motors, motor circuits, and controllers can be found in *Article 430*. Motors used in electrical installations are generally furnished by the entity that furnishes the equipment that the motor powers.

SIX STEPS TO A MOTOR INSTALLATION

In this lesson, we will assume a typical motor installation, and following Figure 10-8, we will determine the requirements of the *NEC®* for this installation.

A 10-horsepower, 3-phase, 208-volt motor with a **code letter** F and a **service factor** of 1.15 will be fed from a *main distribution panel* a distance of 60 ft (18 m) from the proposed motor installation. The motor will be manually started by means of a *start-stop* button in the cover of the **motor controller**, and it will have a remote *stop* button a distance of 50 ft (15 m) from the motor controller. The motor controller will be located next to the motor and will contain the motor overload devices.

Step #1: *Determine the motor full-load current (FLC) of the motor to be used in our calculations.*

NEC® 430.6 requires that *Tables 430.147* through *Table 430.150* be used to determine the FLC of motors and *not* the nameplate. *Table 430.150* covers 3-phase, alternating current motors, and using this Table, we find that a 10-horsepower, 208-volt motor has a FLC of 30.8 amperes (Figure 10-1).

Table 430.150 Full-Load Current, Three-Phase Alternating-Current Motors
The following values of full-load currents are typical for motors running at speeds usual for belted motors and motors with normal torque characteristics.
  Motors built for low speeds (1200 rpm or less) or high torques may require more running current, and multispeed motors will have full-load current varying with speed. In these cases, the nameplate current rating shall be used.
  The voltages listed are rated motor voltages. The currents listed shall be permitted for system voltage ranges of 110 to 120, 220 to 240, 440 to 480, and 550 to 600 volts.

Horsepower	Induction-Type Squirrel Cage and Wound Rotor (Amperes)							Synchronous-Type Unity Power Factor* (Amperes)			
	115 Volts	200 Volts	208 Volts	230 Volts	460 Volts	575 Volts	2300 Volts	230 Volts	460 Volts	575 Volts	2300 Volts
½	4.4	2.5	2.4	2.2	1.1	0.9	—	—	—	—	—
¾	6.4	3.7	3.5	3.2	1.6	1.3	—	—	—	—	—
1	8.4	4.8	4.6	4.2	2.1	1.7	—	—	—	—	—
1½	12.0	6.9	6.6	6.0	3.0	2.4	—	—	—	—	—
2	13.6	7.8	7.5	6.8	3.4	2.7	—	—	—	—	—
3	—	11.0	10.6	9.6	4.8	3.9	—	—	—	—	—
5	—	17.5	16.7	15.2	7.6	6.1	—	—	—	—	—
7½	—	25.3	24.2	22	11	9	—	—	—	—	—
10	—	32.2	30.8	28	14	11	—	—	—	—	—
15	—	48.3	46.2	42	21	17	—	—	—	—	—
20	—	62.1	59.4	54	27	22	—	—	—	—	—
25	—	78.2	74.8	68	34	27	—	53	26	21	—
30	—	92	88	80	40	32	—	63	32	26	—
40	—	120	114	104	52	41	—	83	41	33	—
50	—	150	143	130	65	52	—	104	52	42	—
60	—	177	169	154	77	62	16	123	61	49	12
75	—	221	211	192	96	77	20	155	78	62	15
100	—	285	273	248	124	99	26	202	101	81	20
125	—	359	343	312	156	125	31	253	126	101	25
150	—	414	396	360	180	144	37	302	151	121	30
200	—	552	528	480	240	192	49	400	201	161	40
250	—	—	—	—	302	242	60	—	—	—	—
300	—	—	—	—	361	289	72	—	—	—	—
350	—	—	—	—	414	336	83	—	—	—	—
400	—	—	—	—	477	382	95	—	—	—	—
450	—	—	—	—	515	412	103	—	—	—	—
500	—	—	—	—	590	472	118	—	—	—	—

*For 90 and 80 percent power factor, the figures shall be multiplied by 1.1 and 1.25, respectively.

Figure 10-1 *NEC® Table 430.150* showing full-load current for 3-phase alternating current motor (Reprinted with permission from NFPA 70-2002)

Step #2: *Determine the size of the motor branch-circuit conductors.*

NEC® 430.22 requires branch-circuit conductors supplying a single motor to have an ampacity not less than 125% of the motor FLC (Figure 10-2).

$$30.8 \times 1.25 = 38.5 \; (430.22)$$

8 AWG THWN CU - 50 A (*Table 310.16*)

Step #3: *Determine the fuse size (dual element) to be used as motor branch-circuit short-circuit and ground-fault protection.*

NEC® 430.52 refers to *Table 430.52* for the maximum rating or setting of motor branch-circuit short circuit and ground-fault protection devices (Figure 10-3).

$$30.8 \times 1.75 = 53.9 \; (430.52)$$

60 A next higher size (*430.52*) (*240.6*)

Step #4: *Determine the rating required for the motor disconnect switch.*

NEC® 430.110 requires the motor disconnecting means to have an ampere rating of at least 115% of the FLC rating of the motor.

$$30.8 \times 1.15 = 35.42 \; (430.110)$$

60-ampere disconnect switch is required.

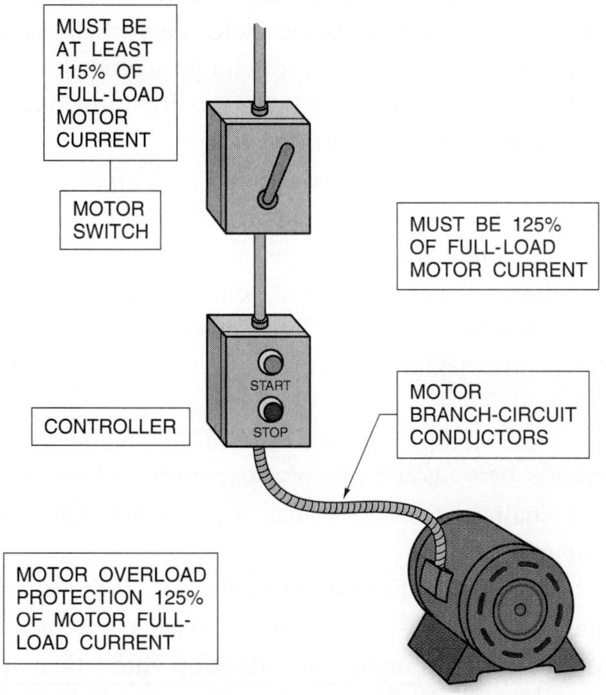

Figure 10-2 Motor branch-circuit conductors. Motor disconnect switch and controller

NEC® 430.102(B) requires a motor disconnecting means shall be located in sight from the motor and the driven machinery location. This disconnecting means shall not be required if the controller disconnecting means is individually capable of being locked in the open position.

NEC® 430.102(A) requires that the controller disconnecting means be located in sight from the controller location. *In sight from* is defined as being visible and not more than 50 ft (15 m) distant from the other (Figure 10-4).

Step #5: *Determine the motor and branch-circuit overload protection required.*

Table 430.52 Maximum Rating or Setting of Motor Branch-Circuit Short-Circuit and Ground-Fault Protective Devices

	Percentage of Full-Load Current			
Type of Motor	Nontime Delay Fuse[1]	Dual Element (Time-Delay) Fuse[1]	Instantaneous Trip Breaker	Inverse Time Breaker[2]
Single-phase motors	300	175	800	250
AC polyphase motors other than wound-rotor				
Squirrel cage — other than Design E or Design B energy efficient	300	175	800	250
Design E or Design B energy efficient	300	175	1100	250
Synchronous[3]	300	175	800	250
Wound rotor	150	150	800	150
Direct current (constant voltage)	150	150	250	150

Note: For certain exceptions to the values specified, see 430.54.
[1]The values in the Nontime Delay Fuse column apply to Time-Delay Class CC fuses.
[2]The values given in the last column also cover the ratings of nonadjustable inverse time types of circuit breakers that may be modified as in 430.52(C), Exception No. 1 and No. 2.
[3]Synchronous motors of the low-torque, low-speed type (usually 450 rpm or lower), such as are used to drive reciprocating compressors, pumps, and so forth, that start unloaded, do not require a fuse rating or circuit-breaker setting in excess of 200 percent of full-load current.

Figure 10-3 Maximum rating or setting of motor branch-circuit protective device, *NEC® Table 430.52* (Reprinted with permission from NFPA 70-2002)

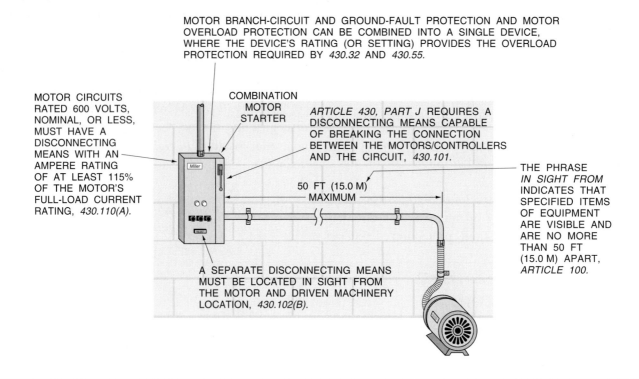

Figure 10-4 Motor disconnecting means within sight from a motor and its driven machinery

NEC® 430.32(A)(1) requires a separate overload device that is responsive to motor current, or a thermal protector integral with the motor that will prevent overheating of the motor due to overload and failure to start. In this installation, the motor overload protective devices will be in the motor controller. For motors with a service factor of not less than 1.15, *430.32(A)(1)* allows 125% of the FLC of the motor. $30.8 \times 1.25 = 38.5$ (*430.32*) (Figure 10-2).

Where the overload relay is not sufficient to start the motor or carry the load, the next higher size overload relay shall be permitted to be used, provided that the trip current of the overload relay does not exceed the percentages of motor FLC shown in *430.34*. For a motor with a service factor of not less than 1.15, a percentage of 140 may be used.

$$30.8 \times 1.40 = 43.12$$

Step #6: *Determine the requirements for the motor-control circuit overcurrent protection.*

NEC® 430.72 outlines the requirements for motor-control circuit overcurrent protection. The motor-control circuit extends beyond the motor controller to a remote stop button. *NEC® 430.72(B) Exception No. 2* permits the motor-control circuit to be protected by the branch-circuit protective devices if they do not exceed the value specified in *Column C* of *Table 430.72(B)* (Figure 10-5).

The branch-circuit overload protective device used in this installation is 60-ampere, and if 12 AWG copper conductors are used for the motor-control circuit conductors, they will be considered as being protected by the branch-circuit protective device and no supplemental overcurrent protection will be required.

Code letters marked on motor nameplates, Figure 10-6 shows motor input with locked rotor and shall be in accordance with *Table 430.7(B)* (Figure 10-7).

Service factor is a margin of safety. If the motor manufacturer gives a motor a "service factor," it means that the motor will develop more than its rated horsepower without damage to the motor. For example, the 10 horsepower motor we have been working with has a service factor of 1.15, allowing

Table 430.72(B) Maximum Rating of Overcurrent Protective Device in Amperes

	Column A Separate Protection Provided		Protection Provided by Motor Branch-Circuit Protective Device(s)			
			Column B Conductors Within Enclosure		Column C Conductors Extend Beyond Enclosure	
Control Circuit Conductor Size (AWG)	Copper	Aluminum or Copper-Clad Aluminum	Copper	Aluminum or Copper-Clad Aluminum	Copper	Aluminum or Copper-Clad Aluminum
18	7	—	25	—	7	—
16	10	—	40	—	10	—
14	(Note 1)	—	100	—	45	—
12	(Note 1)	(Note 1)	120	100	60	45
10	(Note 1)	(Note 1)	160	140	90	75
Larger than 10	(Note 1)	(Note 1)	(Note 2)	(Note 2)	(Note3)	(Note 3)

Notes:
1. Value specified in 310.15 as applicable.
2. 400 percent of value specified in Table 310.17 for 60°C conductors.
3. 300 percent of value specified in Table 310.16 for 60°C conductors.

Figure 10-5 Maximum rating of overcurrent protection device, *NEC*® Table 470.72(B) (Reprinted with permission from NFPA 70-2002)

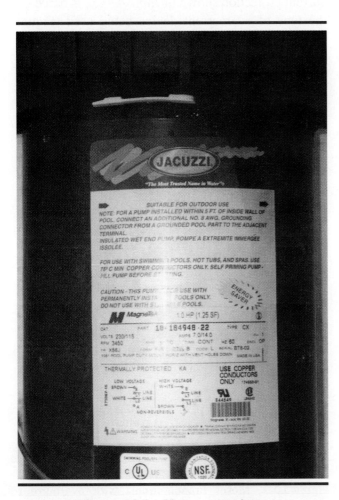

Figure 10-6 Motor nameplate

it to develop 11.5 horsepower current without deterioration of the motor insulation. Most motors have a continuous-duty rating and can operate indefinitely at their rated load.

During the period a motor is starting, it draws a high current, called *inrush current*. This inrush current is called the **locked-rotor current** (or LRC) and could be as high as 4 to 10 times the FLC of the motor. *Table 430.52* allows a percentage increase of FLC so that motors can be successfully started while maintaining full overcurrent protection.

1. Determine FLC
   NEC® 430.6(A)(1)—Table 430.150   30.8 A
2. Motor Branch-Circuit Conductors
   NEC® 430.22(A)   8 AWG CU THWN
3. Main Distribution Fuse
   Table 430.52   60 A
4. Motor Switch Rating
   NEC® 430.110(A)   60 A
5. Motor Overload Protection
   NEC® 430.32(A)(1)   38.5
6. Motor-Control Circuit Protection
   NEC® 430.72(B)   None

Figure 10-8 shows a one-line diagram of the motor installation we have just completed.

Table 430.7(B) Locked-Rotor Indicating Code Letters

Code Letter	Kilovolt-Amperes per Horsepower with Locked Rotor
A	0 – 3.14
B	3.15 – 3.54
C	3.55 – 3.99
D	4.0 – 4.49
E	4.5 – 4.99
F	5.0 – 5.59
G	5.6 – 6.29
H	6.3 – 7.09
J	7.1 – 7.99
K	8.0 – 8.99
L	9.0 – 9.99
M	10.0 – 11.19
N	11.2 – 12.49
P	12.5 – 13.99
R	14.0 – 15.99
S	16.0 – 17.99
T	18.0 – 19.99
U	20.0 – 22.39
V	22.4 and up

Figure 10-7 Locked rotor Code letters, NEC® Table 430.7(B) (Reprinted with permission from NFPA 70-2002)

It is important to note that the fuse protecting the 8 AWG THWN CU motor branch-circuit conductor is 60 amperes, and the ampacity of the 8 AWG conductor is 50 amperes. A ground-fault or short-circuit condition would open the 60-ampere overcurrent device before any damage would occur. An overload condition, which would damage the motor or conductors, is taken care of by the motor **overload protective relay**, which is set at 125% of the motor FLC. The conductors were sized at 125% of the FLC, and the service factor indicates that the motor would withstand a short overload condition.

Now let us go through this again. The conductors from the main distribution panel to the motor, through the motor disconnect switch and the controller, are called the motor branch-circuit conductors. These conductors are sized according to *430.22* that requires that these conductors be 125% of the full-load motor current. The full-load motor current is not taken from the motor nameplate, but in compliance with *430.6*, the FLC is shown in *Table 430.147* through *Table 430.150*.

The overcurrent protection for these motor branch-circuit conductors is figured according to *430.52*, which refers you to *Table 430.52*. The overcurrent protection shown in *Table 430.52* is based on the type of overcurrent device being used (Figure 10-9). Our installation is using a dual element (time delay) fuse and 175% of FLC is permitted.

Every motor installation requires a controller

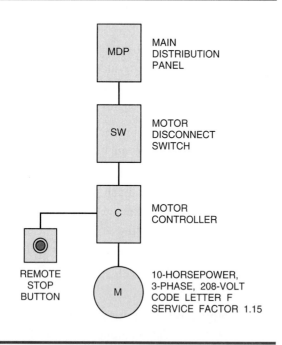

Figure 10-8 Basic one-line diagram of motor installation

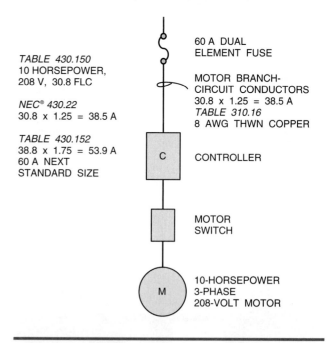

Figure 10-9 Motor branch-circuit with overcurrent protection, sized in accordance with Table 430.52

disconnecting means and a motor disconnecting means. They may be one and the same. There must be a controller disconnecting means in sight from the controller (*See Article 100, Definitions*). If the controller disconnecting means is capable of being locked in the open position, then the motor disconnecting means is not required. The disconnecting means for the motor, whether it be the controller disconnecting means capable of being locked in the open position or a separate motor disconnecting means, must have an ampere rating not less than 115% of the FLC rating of the motor.

The motor overload protection requirements are shown in *430.32*. The percent of motor FLC is determined by the service factor. The motor in this installation has a marked service factor of 1.15, and 125% of the motor FLC is permitted.

This motor has a motor-control circuit, including a remote stop button. The conductors for this control circuit may be tapped from the line terminal of the motor controller, with no supplemental overcurrent protection, if they conform to the requirements of *430.72*. Looking at *Table 430.72(B)*, we find that if we use 12 AWG control circuit conductors extending beyond the controller enclosure, they are considered protected by the motor branch-circuit protective device if this protection does not exceed 60 amperes. This installation conforms to these requirements, and no supplemental protection is required for the motor-control circuit conductors (Figure 10-5).

While this installation has only one motor, there are times when more than one motor will be installed on a single feeder, which in turn is tapped by the individual motor branch-circuit conductors. Information on supplying several motors on a single feeder can be found in *430.24*.

MOTOR CONTROLLERS

We have talked a lot about the controller in our motor installation. *Article 100, Definitions*, defines a

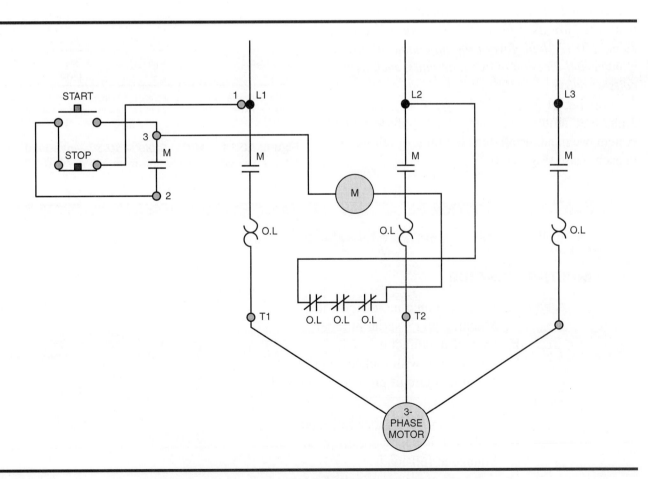

Figure 10-10 Diagram showing overload contacts in series with operating coil. One trip unit in each phase

controller as follows: *A device or group of devices that serve to govern, in some predetermined manner, the electric power delivered to the apparatus to which it is connected.* Simply put, our controller is a magnetic, across the line, motor starter. It has three sets of contacts that are closed by means of a coil. When the coil is energized, it pulls the contacts closed and connects the line to the load. The overload relay contacts are connected in series with the operating coil of the motor controller, and if sufficient current flows through the overload relay, it will open the motor-control circuit, shutting down the motor (Figure 10-10).

Motor Protection

Some tips for motor protection. In accordance with *430.36, Fuses—In Which Conductor,* where fuses are used for motor overload protection, a fuse shall be inserted in each ungrounded conductor and also in the grounded conductor, if the supply system is 3-wire, 3-phase ac with one conductor grounded.

NEC® 430.37, Devices Other than Fuses—In Which Conductor, states that *where devices other than fuses are used for motor overload protection, Table 430.37 shall govern the minimum allowable number and location of overload units, such as trip coils or relays.*

Looking at *Table 430.37*, we find that in any 3-phase ac motor connected to any 3-phase supply system, we must install three trip units or relays, one in each phase (Figure 10-11).

Table 430.37 Overload Units

Kind of Motor	Supply System	Number and Location of Overload Units, Such as Trip Coils or Relays
1-phase ac or dc	2-wire, 1-phase ac or dc ungrounded	1 in either conductor
1-phase ac or dc	2-wire, 1-phase ac or dc, one conductor grounded	1 in ungrounded conductor
1-phase ac or dc	3-wire, 1-phase ac or dc, grounded neutral	1 in either ungrounded conductor
1-phase ac	Any 3-phase	1 in ungrounded conductor
2-phase ac	3-wire, 2-phase ac, ungrounded	2, one in each phase
2-phase ac	3-wire, 2-phase ac, one conductor grounded	2 in ungrounded conductors
2-phase ac	4-wire, 2-phase ac, grounded or ungrounded	2, one per phase in ungrounded conductors
2-phase ac	Grounded neutral or 5-wire, 2-phase ac, ungrounded	2, one per phase in any ungrounded phase wire
3-phase ac	Any 3-phase	3, one in each phase*

*Exception: An overload unit in each phase shall not be required where overload protection is provided by other approved means.

Figure 10-11 *NEC® Table 430.37*—overload units (Reprinted with permission from NFPA 70-2002)

REVIEW QUESTIONS

MULTIPLE CHOICE

1. A code letter marked on a motor indicates:
   ___ A. Manufacturer's serial number
   ___ B. Date of manufacture
   ___ C. Motor input with locked rotor
   ___ D. Maximum circuit protection allowed

2. The FLC of a motor is:
   ___ A. Running current under full-load
   ___ B. Full-load horsepower permitted
   ___ C. Voltage at full-load
   ___ D. Full-load amperes at locked rotor

3. Locked-rotor current represents:
   ____ A. Motor starting current
   ____ B. Motor current with rotor held stationary
   ____ C. Motor current at full-load
   ____ D. Motor current at full horsepower

4. Service factor is:
   ____ A. Lubricant schedule
   ____ B. Motor lead conductor size
   ____ C. Amount motor may be overloaded
   ____ D. Bearing replacement schedule

5. A 10-horsepower, 208-volt, 3-phase motor has a FLC of :
   ____ A. 28.0 A
   ____ B. 30.8 A
   ____ C. 32.2 A
   ____ D. 42.0 A

6. A 3-horsepower, 115-volt, single-phase motor has a FLC of:
   ____ A. 34.0 A
   ____ B. 13.6 A
   ____ C. 11.8 A
   ____ D. 25.0 A

7. A motor overload condition is caused by:
   ____ A. Motor short-circuit
   ____ B. Motor ground-fault
   ____ C. Connected load
   ____ D. None of the above

8. For motors with a service factor of not less than 1.15, the overload relays may sized at _____% of motor FLC.
   ____ A. 125
   ____ B. 140
   ____ C. 175
   ____ D. 125

9. Locked rotor motor input can be found in:
   ____ A. *Table 430.150*
   ____ B. *Table 430.7(B)*
   ____ C. Motor nameplate
   ____ D. None of the above

10. The number of overload trip units required for a 3-phase motor is:
    ____ A. Four
    ____ B. Three
    ____ C. Two
    ____ D. One

CHAPTER 10 Motors and Controllers

TRUE OR FALSE

1. Service factor is the amount a motor may be overloaded by multiplying the service factor by the horsepower of the motor.
   ____ A. True
   ____ B. False
   Code reference: _____ .

2. FLC is taken from the motor nameplate.
   ____ A. True
   ____ B. False
   Code reference: _____ .

3. The motor disconnecting means must have an ampere rating of at least 125% of the motor FLC.
   ____ A. True
   ____ B. False
   Code reference: _____ .

4. *In sight from* is defined as being visible and not more than 50 ft (15 m) distant from each other.
   ____ A. True
   ____ B. False
   Code reference: _____ .

5. Code letters marked on motor nameplates show motor input with locked rotor.
   ____ A. True
   ____ B. False
   Code reference: _____ .

6. There must be a controller disconnecting means in sight from the controller.
   ____ A. True
   ____ B. False
   Code reference: _____ .

7. Motor branch-circuit conductors must be sized at 140% of motor full current.
   ____ A. True
   ____ B. False
   Code reference: _____ .

8. Using *Table 430.52*, a squirrel-cage design E motor can use dual element fuses rated at 300% of the motor FLC for the motor branch-circuit short-circuit ground-fault protective device.
   ____ A. True
   ____ B. False
   Code reference: _____ .

9. A motor starting current (or inrush current) is also called locked-rotor current.
   ____ A. True
   ____ B. False
   Code reference: _____ .

10. A motor disconnecting means shall be located in sight from the motor and the driven machinery location.
    ____ A. True
    ____ B. False
    Code reference: _____ .

FILL IN THE BLANKS

1. *NEC® 430.22* requires the branch-circuit conductors, supplying a _____, to have an ampacity not less than _____ of the motor FLC.

2. *NEC® 430.102(B)* requires a motor disconnecting means shall be located _____ the motor and the driven machinery location.

3. If the motor manufacturer gives a motor a _____, it means that the motor will develop more than its rated horsepower without damage to the motor.

4. During the period a motor is starting, it draws a high current, _____ current.

5. An _____ condition which would damage the motor or conductors, is taken care of by the motor _____, which is set at 125% of the motor FLC.

6. A _____ is a device or group of devices that _____, in some predetermined manner, the electric power delivered to the apparatus to which it is connected.

7. The overload contacts are connected in series with the _____ of the motor controller, and if sufficient current flows through the overload relay, it will open the motor-control circuit.

8. Where fuses are used for _____, a fuse shall be inserted in each ungrounded conductor and also in the _____ if the supply system is 3-wire, 3-phase ac with one conductor grounded.

9. A motor disconnect switch is a disconnecting means installed _____ to be used for safety precautions during maintenance.

10. *NEC® 430.6* requires that the _____ through _____ be used to determine the FLC of motors and *not* the nameplate.

FREQUENTLY ASKED QUESTIONS

Question 1: I'm confused about sizing motor circuit wiring and fuses. How come we can over fuse the motor with *Table 430.52*?

Answer: *Table 430.52* does not provide for the overcurrent protection of a motor. This table, as the title states, provides a percentage of FLC of motors to establish the *Maximum Rating or Setting of Motor Branch-Circuit Short-Circuit and Ground-Fault Protective Devices*. These overcurrent protective devices (fuses or circuit breakers) protect the conductors feeding the motor from overcurrents caused by short-circuits or ground-faults. These conductors are also protected from overload by the motor overload protection specified in *430.32*. The key to understanding conductor and motor protection is to know the meaning of ground-fault, short-circuit, and overload.

Question 2: How are motor branch-circuit conductors protected?

Answer: Motor branch-circuit conductors are electrically protected from two possible problems—ground-fault and short-circuit protection, and overload. First let's get a few definitions in order:

Short-Circuit: Two or more conductors of opposite polarity coming into contact with each other with relatively low resistance between them or contact between them made *short of the load*.

Ground-fault: One or more ungrounded conductors coming into contact with a grounded conductor or a grounded surface.

Overload: Operation of equipment or a conductor with a current value in excess of its rated ampacity that would cause damage or dangerous overheating. A fault such as a short-circuit or ground-fault is not an overload.

These are the things against which we must protect our motor branch-circuit conductors. Short-circuit and ground-fault protection of motor branch-circuit conductors is covered in *430.52*. The first rule is *The motor branch-circuit and ground-fault protective device shall be capable of carrying the starting current of the motor*. NEC® *430.52* refers us to *Table 430.52*, which, depending upon the type of motor and type of overcurrent device used, permits a maximum percentage of full-load current that may be used to size the overcurrent device. This overcurrent device is either a fused motor disconnect switch or a circuit breaker. Short-circuits and ground-faults develop currents of a high magnitude and will open the overcurrent protective devices, which are sized according to *Table 430.52*.

Overload protection for motor branch-circuit conductors are provided by the motor overload devices. This also protects the motor from an overload condition. These overload devices are located in the motor controller or are an integral part of the motor. They are sized according to *430.32*. Because these overload devices can be sized as high as 125% of the motor FLC, the motor branch-circuit conductors are required by *430.22* to be sized at 125% of the motor FLC.

Question 3: Why must motor branch-circuit conductors be sized at 125% of the motor FLC?

Answer: Motor overload devices are permitted by *430.32(A)(1)* to be sized according to the marked service factor of the motor. These values are generally 115% or 125% of the motor FLC. There are some limited exceptions to this, as shown in *430.32(A)(2)*. If the motor is allowed to operate with an overload of 125% of motor FLC, then we must protect the motor branch-circuit conductors by sizing them at 125% of motor FLC also.

There is one school of thought that believes if the service factor is 1.15, then the motor can run continuously at 115% of motor FLC, and the sizing of the motor branch-circuit conductors to 125%, according to *430.22*, should actually be calculated based on the 115% of motor FLC. This is incorrect because the overload protective device is based on 125% of motor FLC and not on 125% of the manufacturer's *service factor*.

Question 4: What is the service factor of a motor?

Answer: Service factor is a margin of safety. If the motor manufacturer designs a service factor into his motor, it means that the motor will develop more than its rated horsepower without damage to the motor. For example, a 10-horsepower motor with a service factor of 1.15 can be allowed to develop 11.5-horsepower current without deterioration of the motor winding insulation. The *NEC®* permits a motor with a service factor of 1.15 to use overload protection of 1.25. This is found in *430.32(A)(1)*. If the motor were to develop an overload of 115%, the motor branch-circuit conductors are not affected because *430.22* has required these conductors to be sized at 125% of motor FLC.

Question 5: Are motor branch-circuit conductors sized at 125% of motor FLC to handle the motor starting current?

Answer: No. The motor starting or *inrush* currents, which are also called *locked-rotor current*, are only present during the acceleration period at the moment a motor is started. The inrush current decreases rapidly as the motor begins to rotate. The motor branch-circuit overcurrent protection, as calculated from *Table 430.52*, easily handles these currents within the limitations of the motor branch-circuit conductors. These motor branch-circuit overcurrent protective devices, which are permitted to be sized much higher than the rated ampacity of the motor branch-circuit conductors, are also able to protect the motor branch-circuit conductors from short-circuit or ground-fault currents because of the magnitude of the currents produced so rapidly by these types of faults.

Question 6: Is the controller disconnecting means required to be in sight from the controller?

Answer: In installations of motor circuits under 600 volts, each controller must have an individual disconnecting means, and it shall be in sight from the controller location [see *430.102(A)*]. A new *Exception No. 2* permits a single disconnecting means for a group of coordinated controllers that drive several parts of a machine or apparatus. The disconnecting means and controllers must be located in sight from the machine or apparatus.

Question 7: Is a motor disconnecting means required within sight of the motor?

Answer: Yes, with one exception. *NEC® 430.102(B)* requires a separate disconnecting means to be located in sight from the motor and the driven machinery. This disconnecting means shall not be required if the controller disconnecting means is individually capable of being locked in the open position.

Question 8: How is the overcurrent protective device sized in the motor disconnecting means?

Answer: No overcurrent protective devices are required in the motor disconnecting means. This disconnecting means is solely for the protection of anyone maintaining the motor or its driven machinery and is not required if the controller disconnecting means is capable of being locked in the open position [See *430.102(B)*].

Question 9: How is the motor disconnecting means sized?

Answer: *NEC® 430.110(A)* requires the disconnecting means for motor circuits to have an ampere rating of at least 115% of the motor FLC rating.

Question 10: Are motor-control circuits, feeding remote control devices, required to have overload protection?

Answer: Overcurrent protection for motor-control circuit wiring is covered in *430.72(B)*. The requirements for conductors that extend beyond the enclosure (remote) can be found in *Table 430.72(B)* in column C. For example, if your motor branch-circuit protective device is rated at 60 amperes and you are using copper control circuit conductors, then you find 60 in the copper column and move to the left to control circuit conductor size where you find 12. This means that you need to install control circuit conductors not smaller than 12 AWG copper. A smaller conductor would require supplemental overcurrent protection to protect that conductor.

Question 11: Explain how after using Column C in *Table 430.72(B)* the electrician can say a 60-ampere overcurrent device will protect a 12 AWG copper control circuit conductor?

Answer: The motor-control circuit conductors are protected by the raceway in which they are installed. The danger of physical damage is minimal. The coil circuit current is small and overload is not a problem. Short-circuit or ground-fault currents rise rapidly and would open the motor branch-circuit protective devices before any damage could occur.

Question 12: If I tap a motor feeder supplying a group of motors, how far can I run that tap without overcurrent protection for the tap conductors?

Answer: If you are tapping to a single motor and are not reducing the conductor size, overcurrent is not a problem. If you are reducing the conductor size, *430.52(D)(2)* requires that the tap conductors have physical protection, the conductors to the overcurrent device be not more than 25 ft in length, and no conductor shall have an ampacity less than ⅓ that of the feeder with a minimum in accordance with *430.22*. This means that the tap conductor must be sized at 125% of the motor FLC and be not less than ⅓ the ampacity of the feeder conductor.

CHAPTER 11

Commercial Load Calculations

OBJECTIVES

After completing this chapter, the student should understand load calculations for:

- **store buildings.**
- **office buildings.**
- **farms.**
- **marinas and boatyards.**
- **recreational vehicle parks.**
- **motels.**

KEY TERMS

Continuous Loads: Electrical loads that are expected to remain in operation or be energized for a period of three hours or more

Demand Factor: The ratio of the maximum demand of a system to the total connected load of a system

Overcurrent Device: A device, usually a fuse or a circuit breaker, that is designed to protect equipment or conductors from an excess of the rated current of equipment or an excess of current beyond the rated ampacity of a conductor

Service Feeder Load: The load on the conductors of a feeder between the power source and the final branch-circuit overcurrent device

Service Load: The total calculated load on the service conductors from the utility company serving the premises

Voltage Drop: A loss of voltage at the load caused by the length and size of the conductors supplying the load and influenced by the resistance of the conductors and the amount of the load

INTRODUCTION

This chapter will familiarize the reader with the methods for making proper load calculations for commercial occupancies. Calculations for commercial occupancies are based on the design criteria and specific Code requirements that relate to these occupancies. Lighting and sign loads are generally considered to be **continuous loads**. Continuous loads are loads that are expected to operate continuously for three hours or more. Some loads are continuous by Code application but others may be by design. Limited **demand factors** are permitted by the Code, but the user's needs must be considered by the designer to assure sufficient capacity for an efficient installation with room for future expansion of electrical usage. **Voltage drop** calculations are not a requirement of the *NEC,*® but should always be a part of the designer's calculation.

GENERAL

In general, Code requirements for load calculations are found in *Article 220* with specific Code requirements for various electrical equipment to be found in the article covering that equipment. For example, if transformers are going to be used in the installation, specific requirements for their use will be found in *Article 450, Transformers*.

STORE BUILDING

Criteria

Figure 11-1 shows a typical store building. The outside dimensions of the store building are 100 ft × 300 ft. The service to the store building will be 3-phase, 4-wire, 208Y/120 volts. The actual lighting load is 87 kVA. The requirements of *Table 220.3(A)* for the general lighting load multiplied by 125% for continuous loading exceeds the actual lighting load at 125% and will be used. A load of 200 volt-amperes per linear ft shall be used for the show window computation, as permitted in *220.3(B)(7)*. The receptacles will not be used as continuous loads and will be computed at 180 VA per outlet in accordance with *220.3(B)(9)*. Two circuits will be required for the sign outlets to comply with *600.5(A)*. The air-conditioning load is larger than the heating load and will be computed at 100%, as permitted in *220.21*.

Lighting Load—sq ft Area—30,000 sq ft 30,000 × 3 VA [*Table 220.3(A)*] × 1.25 (*215.3*) =	112.5 kVA
Show Window—60 ft 60 × 200 VA [*220.12(A)* and *220.3(B)(7)*] =	12.0 kVA
General Purpose Receptacles (Non-Continuous)—20 20 × 180 VA (*220.13*) =	3.6 kVA
Sign Lighting—2 Circuits 2 × 1200 VA [*220.3(B)(6)*] =	2.4 kVA
Lighting Track—40 ft 40 ÷ 2 = 20 × 150 VA [*220.12(B)*] =	3.0 kVA
Air-Conditioning—35 kW	30.0 kVA
Total:	163.5 kVA

Service Feeder Load:

$$\frac{163{,}500}{208 \times 1.732} = \frac{163{,}500}{360} = 454 \text{ amperes}$$

Figure 11-1 Commercial store building

Neutral Load:

Lighting	112.5 kVA
Show Window	12.0 kVA
Receptacles	3.6 kVA
Sign Lighting	2.4 kVA
Lighting Track	3.0 kVA
Total:	133.5 kVA

Neutral Feeder Load:

$$\frac{133.5 \text{ kVA}}{208 \times 1.732} = \frac{133.5 \text{ kVA}}{360} = 370.8 \text{ amperes}$$

Using *Table 310.16* (Figure 11-2).

Service Feeder—4/0 AWG THWN CU

Allowable Ampacity—

230 amperes × 2 = 460 amperes

Table 310.16 Allowable Ampacities of Insulated Conductors Rated 0 Through 2000 Volts, 60°C Through 90°C (140°F Through 194°F), Not More Than Three Current-Carrying Conductors in Raceway, Cable, or Earth (Directly Buried), Based on Ambient Temperature of 30°C (86°F)

Size AWG or kcmil	60°C (140°F) Types TW, UF	75°C (167°F) Types RHW, THHW, THW, THWN, XHHW, USE, ZW	90°C (194°F) Types TBS, SA, SIS, FEP, FEPB, MI, RHH, RHW-2, THHN, THHW, THW-2, THWN-2, USE-2, XHH, XHHW, XHHW-2, ZW-2	60°C (140°F) Types TW, UF	75°C (167°F) Types RHW, THHW, THW, THWN, XHHW, USE	90°C (194°F) Types TBS, SA, SIS, THHN, THHW, THW-2, THWN-2, RHH, RHW-2, USE-2, XHH, XHHW, XHHW-2, ZW-2	Size AWG or kcmil
	COPPER			ALUMINUM OR COPPER-CLAD ALUMINUM			
18	—	—	14	—	—	—	—
16	—	—	18	—	—	—	—
14*	20	20	25	—	—	—	—
12*	25	25	30	20	20	25	12*
10*	30	35	40	25	30	35	10*
8	40	50	55	30	40	45	8
6	55	65	75	40	50	60	6
4	70	85	95	55	65	75	4
3	85	100	110	65	75	85	3
2	95	115	130	75	90	100	2
1	110	130	150	85	100	115	1
1/0	125	150	170	100	120	135	1/0
2/0	145	175	195	115	135	150	2/0
3/0	165	200	225	130	155	175	3/0
4/0	195	230	260	150	180	205	4/0
250	215	255	290	170	205	230	250
300	240	285	320	190	230	255	300
350	260	310	350	210	250	280	350
400	280	335	380	225	270	305	400
500	320	380	430	260	310	350	500
600	355	420	475	285	340	385	600
700	385	460	520	310	375	420	700
750	400	475	535	320	385	435	750
800	410	490	555	330	395	450	800
900	435	520	585	355	425	480	900
1000	455	545	615	375	445	500	1000
1250	495	590	665	405	485	545	1250
1500	520	625	705	435	520	585	1500
1750	545	650	735	455	545	615	1750
2000	560	665	750	470	560	630	2000

	CORRECTION FACTORS						
Ambient Temp. (°C)	For ambient temperatures other than 30°C (86°F), multiply the allowable ampacities shown above by the appropriate factor shown below.						Ambient Temp. (°F)
21–25	1.08	1.05	1.04	1.08	1.05	1.04	70–77
26–30	1.00	1.00	1.00	1.00	1.00	1.00	78–86
31–35	0.91	0.94	0.96	0.91	0.94	0.96	87–95
36–40	0.82	0.88	0.91	0.82	0.88	0.91	96–104
41–45	0.71	0.82	0.87	0.71	0.82	0.87	105–113
46–50	0.58	0.75	0.82	0.58	0.75	0.82	114–122
51–55	0.41	0.67	0.76	0.41	0.67	0.76	123–131
56–60	—	0.58	0.71	—	0.58	0.71	132–140
61–70	—	0.33	0.58	—	0.33	0.58	141–158
71–80	—	—	0.41	—	—	0.41	159–176

* See 240.4(D).

Figure 11-2 *NEC® Table 310.16* (Reprinted with permission from NFPA 70-2002)

Neutral Feeder—3/0 AWG THWN CU

Allowable Ampacity—

200 amperes × 2 = 400 amperes

Parallel two sets of three 4/0 AWG THWN CU and one 3/0 AWG THWN CU

To determine the size of the IMC to use, we must go to *Table 5* and find the approximate sq in. area of the conductors we are using. Using the table that includes Type THWN, we find 4/0 AWG @ .3237 and 3/0 AWG @ .2679. Adding .3237 + .3237 + .3237 + .2679 = 1.239 (Figure 11-3)

Using *Table 4* under *Article 342, Intermediate Metal Conduit*, we find that a 2 in. IMC conduit has a usable area of 1.452 for a 40% fill with over two wires (Figure 11-4).

Table 5 Dimensions of Insulated Conductors and Fixture Wires

Type	Size (AWG or kcmil)	Approximate Diameter mm	Approximate Diameter in.	Approximate Area mm²	Approximate Area in.²
Type: RHH*, RHW*, RHW-2*, THHN, THHW, THW, THW-2, TFN, TFFN, THWN, THWN-2, XF, XFF					
THHN, THWN, THWN-2	14	2.819	0.111	6.258	0.0097
	12	3.302	0.130	8.581	0.0133
	10	4.166	0.164	13.61	0.0211
	8	5.486	0.216	23.61	0.0366
	6	6.452	0.254	32.71	0.0507
	4	8.230	0.324	53.16	0.0824
	3	8.941	0.352	62.77	0.0973
	2	9.754	0.384	74.71	0.1158
	1	11.33	0.446	100.8	0.1562
	1/0	12.34	0.486	119.7	0.1855
	2/0	13.51	0.532	143.4	0.2223
	3/0	14.83	0.584	172.8	0.2679
	4/0	16.31	0.642	208.8	0.3237
	250	18.06	0.711	256.1	0.3970
	300	19.46	0.766	297.3	0.4608

Figure 11-3   NEC® *Table 5* (Reprinted with permission from NFPA 70-2002)

Article 342 — Intermediate Metal Conduit (IMC)

Metric Designator	Trade Size	Nominal Internal Diameter mm	Nominal Internal Diameter in.	Total Area 100% mm²	Total Area 100% in.²	2 Wires 31% mm²	2 Wires 31% in.²	Over 2 Wires 40% mm²	Over 2 Wires 40% in.²	1 Wire 53% mm²	1 Wire 53% in.²	60% mm²	60% in.²
12	3/8	—	—	—	—	—	—	—	—	—	—	—	—
16	1/2	16.8	0.660	222	0.342	69	0.106	89	0.137	117	0.181	133	0.205
21	3/4	21.9	0.864	377	0.586	117	0.182	151	0.235	200	0.311	226	0.352
27	1	28.1	1.105	620	0.959	192	0.297	248	0.384	329	0.508	372	0.575
35	1 1/4	36.8	1.448	1064	1.647	330	0.510	425	0.659	564	0.873	638	0.988
41	1 1/2	42.7	1.683	1432	2.225	444	0.690	573	0.890	759	1.179	859	1.335
53	2	54.6	2.150	2341	3.630	726	1.125	937	1.452	1241	1.924	1405	2.178
63	2 1/2	64.9	2.557	3308	5.135	1026	1.592	1323	2.054	1753	2.722	1985	3.081
78	3	80.7	3.176	5115	7.922	1586	2.456	2046	3.169	2711	4.199	3069	4.753
91	3 1/2	93.2	3.671	6822	10.584	2115	3.281	2729	4.234	3616	5.610	4093	6.351
103	4	105.4	4.166	8725	13.631	2705	4.226	3490	5.452	4624	7.224	5235	8.179

Figure 11-4   NEC® *Table 4*, IMC (Reprinted with permission from NFPA 70-2002)

In summary, we will parallel two 1½ in. IMC conduits, each with three 4/0 AWG THWN CU, and one 3/0 AWG THWN CU conductors.

OFFICE BUILDING

Criteria

Figure 11-5 shows a typical office building service. The outside dimensions of the office building are 150 ft × 100 ft (45 m × 30 m). The service to the office building will be 480/277-volt, 3-phase, 4-wire. The actual lighting load is 50 kVA, 277-volt. The actual lighting load is greater than the load requirement of *Table 220.3(A)* for general lighting. All of the lighting is considered to be a continuous load. There are 70 receptacles used for continuous loads and 100 general-purpose receptacles. There is a 20 kVA, 480-volt electric water heater and 20 kW of electric cooking equipment in a kitchen. Outside lighting consists an 8.0-kVA load. There will be a 5.0-kVA sign load. The electric heating load of 150 kVA exceeds the air-conditioning load and will be used in accordance with *220.15*.

Service Feeder Load

Continuous Loads

Lighting Load—sq ft Area—15,000 sq ft
15,000 × 3½ VA [*220.3(A)*] = 52.5 kVA
Actual Lighting Load is 50.0 kVA
Lighting—According to *220.3(A)* use: 52.5 kVA
Receptacles—70 × 180 = 12.6 kVA
Outside Lighting 8.0 kVA
Sign Lighting 5.0 kVA
Total: 78.1 kVA

Total Continuous Loads:
78.1 × 1.25 [*215.2(A)*] = 97.6 kVA

Non-Continuous Loads

Receptacles—100 × 180 VA = 18.0 kVA
First 10.0 kVA @ 100% = 10.0 kVA
Remainder (*Table 220.13*)
8.0 kVA @ 50% = 4.0 kVA
Total: 14.0 kVA
Electric Water Heater 20.0 kVA
Electric Cooking Equipment (*220.20*) 20.0 kVA
Electric Heating Load (*220.15*) 150.0 kVA
Total Non-Continuous Load: 204.0 kVA

Figure 11-5 Typical office building service

Continuous Loads	97.6 kVA
Non-Continuous Loads	204.0 kVA
Total:	301.6 kVA

Total **Service Feeder Load**:

$$\frac{301.6 \text{ kVA}}{480 \times 1.732} = \frac{301.6 \text{ kVA}}{830} = 363.3 \text{ amperes}$$

Neutral Feeder Load

Lighting	52.5 kVA
Outside Lighting	8.0 kVA
Sign Lighting	5.0 kVA
Receptacles	12.6 kVA
	78.1 kVA × 1.25 = 97.6 kVA
Receptacles Non-Continuous	14.0 kVA
Total:	111.6 kVA

Total Neutral Feeder Load:

$$\frac{111.6 \text{ kVA}}{830} = 134.4 \text{ amperes}$$

The neutral feeder load is computed for only those loads that require a neutral conductor and are expected to carry the maximum unbalanced load of the circuits for which they are used.

Summary

Service Feeder Load: 363.3 amperes
Neutral Feeder Load: 134.4 amperes

Using *Table 310.16* (see Figure 11-2).

Service feeder—500 kcmil AWG THWN CU allowable ampacity—380 amperes

Neutral feeder—1/0 AWG THWN CU allowable ampacity—150 amperes

To determine the size of the IMC to use, go to *Table 5* and find the approximate sq in. area of the conductors we are using. Using the part of *Table 5* that includes Type THWN, we find 500 kcmil THWN CU @ 0.7073 and 1/0 AWG THWN CU @ .1855. Adding .7073 + .7073 + .7073 + .1855 = 2.31 (see Figure 11-4).

Using *Table 4* under *Article 342, Intermediate Metal Conduit*, we find that a 3-in. IMC conduit has a usable area of 3.169 for a 40% fill with over two wires (see Figure 11-4).

Summary: One 3-in. IMC conduit with three 500 kcmil THWN CU and one 1/0 AWG THWN CU conductor.

In this office installation, we will use a transformer to obtain the 3-phase, 4-wire, 208Y/120-volt system required for part of the load.

Receptacles	12.6 kVA
Outside Lighting	8.0 kVA
Sign Lighting	5.0 kVA
	25.6 kVA × 1.25 = 32.0 kVA
Receptacles—Non-Continuous	14.0 kVA
Total:	46.0 kVA

The power supply for the 208Y/120-volt portion of this load will be a 50-kVA, 480-208Y/120-volt, 3-phase, 4-wire transformer.

A typical service and distribution would be as shown in Figure 11-6.

Figure 11-6 shows a trade size 3 (78) intermediate grade (IMC) raceway containing three 500 kcmil AWG THWN CU (copper) conductors and one 1/0 AWG THWN CU conductor to provide a 400-ampere service entrance to the building. Actually, as shown in Figure 11-6, *Table 310.16* indicates that 500 kcmil AWG THWN CU conductors have an ampacity of 380 ampere. Since 380 amperes is not shown as a standard **overcurrent device** rating in *240.6, 240.4(B)* permits using the next higher standard overcurrent device rating.

These service-entrance conductors are brought into a 400-ampere, 3-phase, 4-wire utility current transformer meter housing, which in turn feeds a 400-ampere, 3-phase, 4-wire distribution panel that is protected by a 400-ampere rated overcurrent protective device, *240.4(B)*. The main distribution panel contains the overcurrent protective devices for the 480-volt and the 277-volt circuits and equipment. In addition, the main distribution panel has a 3-pole, 70-ampere rated overcurrent protective device to feed the 50-kVA transformer.

The 50-kVA transformer has a primary rated current of 60.2 amperes (50,000 ÷ 480 × 1.732). Although *Table 450.3(B)* permits using a protective device rated at 250% of the transformer primary rated current (60.2 × 2.5 = 150 A—See Chapter 9), it is not necessary to do so for the load being served and would result in requiring larger conductors to feed the transformer. Transformer overload protection will be sacrificed by using overcurrent devices sized much greater than the transformer full-load current rating, and keeping the overcurrent protection within

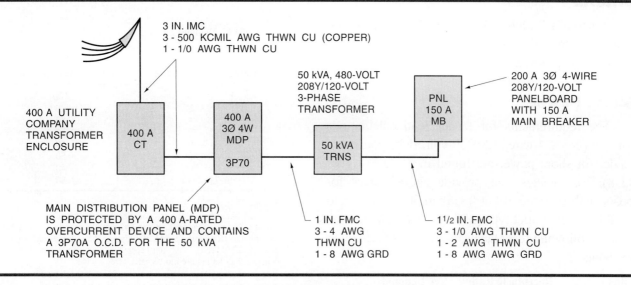

Figure 11-6 Office building service distribution

125% of the transformer full-load amperes is recommended. Based on this recommendation, a trade size 1 (27) flexible metal conduit containing three 4 AWG THWN CU conductors, and one 8 AWG THWN CU grounding conductor will be used to feed the 50-kVA transformer.

The 50-kVA transformer has a secondary rated current of 139 amperes (50,000 ÷ 208 × 1.732). *Table 450.3(B)* permits secondary protection of 125% of the secondary rated current of the transformer. Using this permitted application, 139 A × 1.25 = 178 A, a trade size 1½ (41) FMC containing three 1/0 AWG THWN CU conductors, one 2 AWG THWN CU conductor, and one 6 AWG THWN CU conductor will be used to feed the 200-ampere, 3-phase, 4-wire, 208Y/120-volt panelboard. The panelboard will have a 150-ampere factory-installed overcurrent device in the panel.

MARINAS AND BOATYARDS

Figure 11-7 shows a typical marina and boatyard.

Figure 11-7 Typical marina and boatyard

Criteria

Assume a 3-phase, 4-wire, 120/240-volt delta-connected feeder is used for shore power. Example is shown as follows.

Shore Power

The requirements for feeder load calculations for a 3-phase, 4-wire, 120/240-volt, delta-connected feeder for shore power are found in *555.12* (Figure 11-8). Receptacles that provide shore power for boats shall be rated at not less than 30 amperes and shall be single outlet type, *555.19(A)(4)*. The marina has the following receptacles to provide shore power for boats.

Slips	Receptacle Rating	kW Demand
30 Slips	30-ampere/120-volt	30 × 3600 = 108.0 kW
15 Slips	50-ampere/240-volt	15 × 12,000 = 180.0 kW
10 Slips with 2 receptacles	30-ampere/120-volt 50-ampere/240-volt	(see *555.12 Note 1*) 10 × 12,000 = 120.0 kW
5 Slips with 2 receptacles	30-ampere/120-volt 100-ampere/240 volt	10 × 24,000 = 240.0 kW

Total Connected Load: 648.0 kW

540.0 kW @
   240 volts × 1.732 (415) =   1301 amperes
108.0 kW @ 120/240 volts =   450 amperes
                                                         1751 amperes

Demand Factor (*Table 555.12*)
   65 receptacles: .40
Service Requirements for
   Shore Power: 700 amperes

The calculation in the previous text is for shore power receptacles only. Where two receptacles for shore power are provided for an individual slip and these receptacles have different voltages, only the one with the higher kilowatt demand shall be required to be calculated. A distribution panel to feed shore power receptacles would require a 3-phase, 4-wire, 120/240-volt, delta-connected feeder with an 800-ampere capacity. For other than shore power, see *555.19(B)*.

A 3-phase, 4-wire, 120/240-volt delta-connected system has a grounded (neutral) conductor that has a higher voltage-to-ground on one phase, usually the B phase. One of the phase windings in the transformer is center tapped to obtain the grounded or neutral conductor (see Figure 11-9). For this reason,

Table 555.12 Demand Factors

Number of Receptacles	Sum of the Rating of the Receptacles (percent)
1 – 4	100
5 – 8	90
9 –14	80
15 –30	70
31 –40	60
41 –50	50
51 –70	40
71-plus	30

Notes:
1. Where shore power accommodations provide two receptacles specifically for an individual boat slip and these receptacles have different voltages (for example, one 30 ampere, 125 volt and one 50 ampere, 125/250 volt), only the receptacle with the larger kilowatt demand shall be required to be calculated.
2. If the facility being installed includes individual kilowatt-hour submeters for each slip and is being calculated using the criteria listed in Table 555.12, the total demand amperes may be multiplied by 0.9 to achieve the final demand amperes.

Figure 11-8 *NEC® Table 555.12, Demand Factors* (Reprinted with permission from NFPA 70-2002)

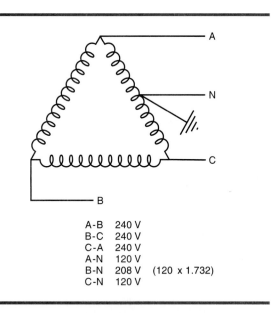

Figure 11-9 Three-phase, 4-wire delta transformer

only the receptacles rated 240 can be balanced over all three phases, while the receptacles rated 120 volts cannot be used with the phase that has the higher voltage-to-ground. This high phase has many names, "red-leg," "wild-leg," or "high phase."

RECREATIONAL VEHICLE PARKS

Figure 11-10 shows a typical recreational vehicle park.

Criteria

A recreational vehicle park is a plot of land on which two or more recreational vehicle sites are located. Recreational vehicles are generally a travel trailer, camping vehicle, truck camper, or motor home. They are primarily designed as temporary living quarters for recreational, camping, or travel use.

The proposed recreational vehicle park will have 60 recreational vehicle sites available for use. Every recreational vehicle site shall be equipped with one 20-ampere, 125-volt receptacle and one 30-ampere, 125-volt receptacle. Thirty sites (50%) shall be equipped with a 50-ampere, 125/250-volt receptacle. This criteria exceeds the minimum requirements of *551.71*.

Basis of Calculations for *551.73(A)*

30 Sites (20-ampere and 30-ampere site facilities) @ 3600 VA =	108 kVA
30 Sites (50-ampere, 120/240-volt site facilities) @ 9600 VA =	288 kVA
Total Connected Load:	396 kVA
Demand Factor (*Table 551.73*):	.41
Total Demand Load:	162 kVA

$$\frac{162{,}000 \text{ VA}}{240 \text{ volts}} = 675 \text{ amperes}$$

Demand factors are shown in Figure 11-11.

Recreational vehicle secondary distribution must be a single-phase, 3-wire, 120/240-system where 50-ampere receptacles are used. The reason for this is that 50-ampere receptacle configurations rated for 208-volt systems are not available at this time. In accordance with *551.72*, the grounded (neutral) conductors cannot be reduced in size below the size of the ungrounded conductor for the site distribution. Each secondary distribution system must be grounded at the transformer.

Recreational vehicle site equipment shall be located as shown in *551.77*. For back-in sites, the supply equipment shall be located on the left (driver's) side on a line that is 5 ft to 7 ft (1.5 m

Figure 11-10 Recreational vehicle park

206 CHAPTER 11 Commercial Load Calculations

Table 551.73 Demand Factors for Site Feeders and Service-Entrance Conductors for Park Sites

Number of Recreational Vehicle Sites	Demand Factor (percent)
1	100
2	90
3	80
4	75
5	65
6	60
7 – 9	55
10 –12	50
13 –15	48
16 –18	47
19 –21	45
22 –24	43
25 –35	42
36 plus	41

Figure 11-11 *NEC® Table 551.73* (Reprinted with permission from NFPA 70-2002)

to 2.1 m) from the edge of the vehicle and shall be located at any point on this line from the rear of the stand to 15 ft (4.5 m) forward of the rear of the stand. For pull-through sites, the supply equipment shall be located at any point along a line that is 5 ft to 7 ft (1.5 m to 2.1 m) from the drivers side and from 16 ft (4.9 m) forward of the rear of the stand to the center point between the two roads that give access to and egress from the pull-through sites.

Receptacles used to supply electric power to a recreational vehicle shall be one of the configurations shown in Figure 11-12.

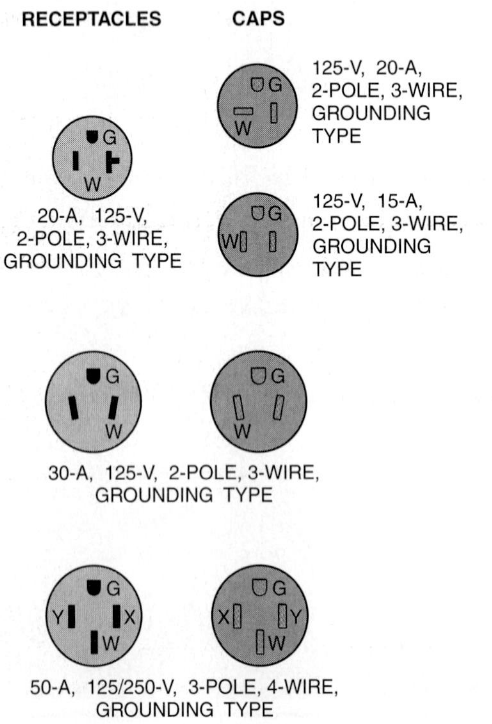

Figure 11-12 Receptacle configurations

FARMS

Figure 11-13 shows a typical farm with outbuilding. This farm consists of two buildings other than the farm dwelling and in accordance with *220.41(B)*.

Criteria Building No. 1

Lighting Load—5.0 kVA
5-horsepower, Single-Phase Motor Load
Other Loads (Power) 18.0 kVA

The method for computing farm loads recognizes a two-part calculation. One part is the calculation of the farm dwelling unit and the other consists of equipment buildings and other stationary electric equipment. The farm dwelling and the farm load in this calculation are supplied by a common service that will be overhead service conductors to a pole

CHAPTER 11   Commercial Load Calculations   207

Figure 11-13   Typical farm with buildings

located on the farm property. The utility company meter and a main breaker distribution panel will be mounted on the pole. Feeders will be run overhead to the dwelling unit and to each of the two equipment buildings from the distribution panel located at the private property pole. The service characteristics will be 400 amperes, single-phase, 3-wire, 120/240 volts (Figure 11-14).

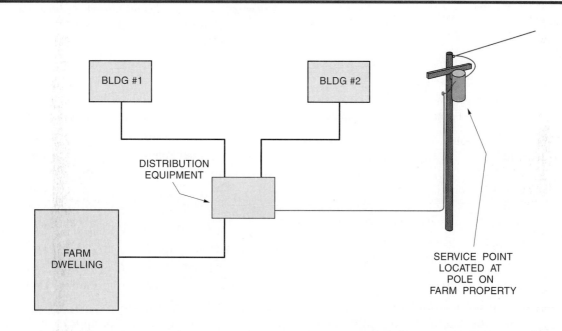

Figure 11-14   Diagram of farm service

Projected Load

Building No. 1

The current shown in the neutral is the maximum unbalance of the load as required in *220.22*. This means that if everything connected to A were turned off, there could be a current flow of 26 amperes in the neutral acting as the return conductor for B. This is the same reaction for A, if B were disconnected.

	A	B	N
Lighting Load:			
5000 VA ÷ 240 volts × 1.25 =	26.0	26.0	26.0
5-horsepower Motor Load @			
240 volts = 28 amperes × 1.25 =	35.0	35.0	0
Total Non-Diverse Loads	61.0	61.0	26.0
Other Loads—(Power)			
18,000 VA ÷ 240 volts =	75.0	75.0	75.0

Demand Factors (see Figure 11-15):
Non-Diverse Loads @ 100%   61.0
Other Loads: First 60 amperes @ 50% = 30.0
Remainder 75 − 60 = 15 @ 25% =   4.0
                    Feeder Load: 95.0 amperes

Criteria Building No. 2

	A	B	N
Lighting Load:			
3000 VA ÷ 240 volts × 1.25 =	15.6	15.6	15.6
Heaters 15.0 kVA @			
240 volts = 62.5 amperes × 1.25 =	78.0	78.0	
Total Non-Diverse Loads	93.7	93.7	15.6
Other Loads—(Power)			
20,000 VA ÷ 240 volts =	83.3	83.3	83.3

Demand Factors:
Non-Diverse Loads @ 100%   93.7
Other Loads: First 60 amperes @ 50% = 30.0
Remainder 83.3 − 60 = 23.3 @ 25% =   5.8
                    Feeder Load: 129.5 amperes

Total Farm Load (see Figure 11-16):
Largest load (Bldg #2)
  129.5 @ 100% =   129.5
Second Largest Load (Bldg #1)
  95.0 @ 75% =   71.0
    Farm Load Less Dwelling: 200.5
  Farm Dwelling Load (Estimate)   100.0
          Total Farm Load: 300.5 amperes

The computation for the farm dwelling unit is the same as a calculation for a dwelling unit and is not repeated here. Where the dwelling and the farm load are supplied by a common service and the dwelling has electric heat and the farm has electric grain-drying systems, the optional calculation shown in *Part III* shall not be used.

The total **service load** of 300.5 amperes for the entire farm load requires that the ungrounded conductors be at least 350 kcmil AWG THWN CU and the grounded (neutral) conductor be rated the same. Farm load calculations are the only service calculations required by the *NEC®* to be done using amperage rather than using volt-amperes. There is no direction given for applying demand factors to the grounded (neutral) conductor, and using a full-size grounded (neutral) conductor appears to be the best method to use.

Table 220.40 Method for Computing Farm Loads for Other Than Dwelling Unit

Ampere Load at 240 Volts Maximum	Demand Factor (Percent)
Loads expected to operate without diversity, but not less than 125 percent full-load current of the largest motor and not less than the first 60 amperes of load	100
Next 60 amperes of all other loads	50
Remainder of other load	25

Figure 11-15 *NEC® Table 220.40, Method for Computing Farm Loads for Other than Dwelling Unit* (Reprinted with permission from NFPA 70-2002)

Table 220.41 Method for Computing Total Farm Load

Individual Loads Computed in Accordance with Table 220.40	Demand Factor (Percent)
Largest load	100
Second largest load	75
Third largest load	65
Remaining loads	50

Note: To this total load, add the load of the farm dwelling unit computed in accordance with Part II or III of this article. Where the dwelling has electric heat and the farm has electric grain-drying systems, Part III of this article shall not be used to compute the dwelling load.

Figure 11-16 *NEC® Table 220.41, Method for Computing Total Farm Load* (Reprinted with permission from NFPA 70-2002)

MOTELS

Figure 11-17 shows a typical motel building.

Criteria

The calculation will be done for a motel with 60 guest rooms, motel office, laundry facilities, exterior lighting, and signs. The service characteristics will be 3-phase, 4-wire, 208Y/120 volts.

Guest Rooms
300 sq ft in area
5.0 kW combination heat pump/air conditioner
1.5 kW auxiliary heat source

Motel Office
400 sq ft in area (office lighting continuous load)
5.0 kW combination heat pump/air conditioner
1.5 kW auxiliary heat source
10 duplex receptacles

Laundry Room
300 sq ft in area
5.0 kW combination heat pump/air conditioner
1.5 kW electric heat source
2, 5.0 kW electric dryers
1, 1.5 kW ironer
10 duplex receptacles

Outside Lighting and Signs
Outside lighting 8.0 kW
Signs 6.0 kW with 5, 20-ampere circuits

Calculations:

Guest Rooms
300 sq ft × 2 VA [220.3(A)] =	
600 VA × 60 Units =	36,000 VA
First 20,000 @ 50% =	10,000
Remainder	
(36,000 − 20,000) @ 40% =	6400
Total (*Table 220.11*):	16,400 VA
Heat Pump and Auxiliary Heat	
(Figure 11-18) 6500 VA × 60 =	
390,000 × .40 =	156,000 VA
Total:	172,400 VA

Motel Office
400 sq ft × 3½ VA [220.3(A)] =	1400 VA
Heat Pump and Auxiliary Heat =	6500 VA
10 Receptacles @ 180 VA	
[220.3(B)(9)] =	1800 VA
Total:	9700 VA

Figure 11-17 Typical motel

Figure 11-18 Motel heat pump unit in a guest room

Figure 11-20 Motel outside sign and lighting

Outside Lighting—8000 VA × 1.25 = 10,000 VA
Signs—6000 VA × 1.25 = 7500 VA
Total: 17,500 VA

Total Service Load 220,000 VA

$$\frac{220{,}000 \text{ VA}}{208 \times 1.732} = \frac{220{,}000 \text{ VA}}{360 \text{ volts}} = 611 \text{ amperes}$$

In calculating the demand load for the heat pump and auxiliary heat units, the only reference to fixed electric space-heating calculations is shown in *220.15* where it requires these loads to be computed at 100% of the total connected load. The exception, however, permits the AHJ to grant permission for feeder and service conductors to have an ampacity less than 100%. This permission is granted for installations where reduced loading of the conductors results from all units not operating at the same time. A demand factor of 40% is used in this calculation, but each calculation must be based on evidence of reduced loading. This evidence may be obtained from the utility company or from other publications dealing with design calculations for motel occupancies where creditable statistics have been compiled relating to this design.

Figure 11-19 Motel laundry room

Laundry Room (See Figure 11-19.)
300 sq ft × 2 VA [*220.3(A)*] = 600 VA
Heat Pump and Auxiliary Heat = 6500 VA
Electric Dryers—2 × 5000 VA = 10,000 VA
Electric Ironer = 1500 VA
10 Receptacles @ 180 VA
  [*220.3(B)(9)*] = 1800 VA
Total: 20,400 VA

Outside Lighting and Signs
(See Figure 11-20.)

REVIEW QUESTIONS

MULTIPLE CHOICE

1. Continuous loads are loads that are expected to remain in operation for a period of _____ hours or more.
   ___ A. two
   ___ B. three
   ___ C. four
   ___ D. five

2. The unit lighting load per sq ft for a store occupancy is:
   ___ A. 5 VA.
   ___ B. 10 VA.
   ___ C. 3 VA.
   ___ D. 2 VA.

3. Show windows may be computed by a unit load per outlet or by applying _____ volt-amperes per linear ft of show window.
   ___ A. 100
   ___ B. 50
   ___ C. 150
   ___ D. 200

4. The feeder load for circuits supplying shore power to receptacles in a marina is based on:
   ___ A. the rating of the circuit.
   ___ B. the rating of the receptacles.
   ___ C. applied voltage.
   ___ D. boat horsepower.

5. For motel occupancies, a demand factor for the first 20,000 volt-amperes of general lighting load is:
   ___ A. 100%.
   ___ B. 50%.
   ___ C. 25%.
   ___ D. 40%.

6. General purpose receptacles in a store occupancy must be calculated at _____ volt-amperes each.
   ___ A. 250
   ___ B. 150
   ___ C. 180
   ___ D. 100

7. The second largest load for a farm service is permitted using a demand factor of _____%.
   ___ A. 50
   ___ B. 75
   ___ C. 60
   ___ D. 80

8. The minimum rating of receptacles that provide shore power for boats shall not be less than:
   ___ A. 20.
   ___ B. 30.
   ___ C. 50.
   ___ D. 60.

9. Demand factors for 30 shore power receptacles shall be based on _____% of the sum of the rating of the receptacles.
   ___ A. 90
   ___ B. 80
   ___ C. 60
   ___ D. 70

10. Lighting that is expected to be in operation for a period of three hours or more is subject to a demand factor of _____%.
    ___ A. 100
    ___ B. 125
    ___ C. 150
    ___ D. 115

TRUE OR FALSE

1. In an office building occupancy, the lighting load is always computed using *Table 220.3(A)* instead of the actual lighting load.
   ___ A. True
   ___ B. False
   Code reference _____.

2. Where the actual number of general purpose receptacles for an office occupancy is unknown, a unit load of 2½ VA per sq ft must be used.
   ___ A. True
   ___ B. False
   Code reference _____.

3. Voltage drop calculations for office building feeders and circuits is not a mandatory requirement of the *NEC*.
   ___ A. True
   ___ B. False
   Code reference _____.

4. Motel lighting calculations are not subject to any demand factors.
   ___ A. True
   ___ B. False
   Code reference _____.

5. Fixed electric space heating shall be computed at 100% of the total connected load.
   ___ A. True
   ___ B. False
   Code reference _____.

6. No separate receptacle load is required for guest rooms in a motel.
   ___ A. True
   ___ B. False
   Code reference _____.

FILL IN THE BLANKS

1. The ratio of the maximum demand of a system to the _____ of a system is the demand factor.

2. Receptacles that provide shore power for boats shall be rated at not less than 30 amperes and shall be of the _____ type.

3. Where two receptacles for shore power are provided for an individual slip and these receptacles have different voltages, only the one with the _____ demand shall be required to be calculated.

4. The high phase has many names, _____, _____, or _____.

5. A recreational vehicle park is a plot of land on which _____ recreational vehicles are located.

FREQUENTLY ASKED QUESTIONS

Question 1: If the actual lighting load is known, should it be used instead of the unit load shown in *Table 220.3(A)*?

Answer: *NEC® 220.3(A)* says a unit load of not less than that shown in *Table 220.3(A)* shall constitute the minimum lighting load. That means that you must use the higher of the two amounts.

Question 2: Why is the lighting load in commercial occupancies always multiplied by 125%?

Answer: Generally, commercial lighting loads are considered as being continuous loads. A continuous load is one that is expected to remain in operation for three hours or more. Store lighting and office lighting is generally on for much longer periods than three hours and therefore must be computed at 125% to compensate for the heat generated during that time.

Question 3: Are voltage drop considerations required when making load calculations?

Answer: The *NEC®* does not generally require voltage drop calculations. However, it does recommend in fine print notes to *210.19(A)* and *215.2(D)* that voltage drop considerations be made. These FPNs are not mandatory and are not enforceable. The *NEC®* has not been furnished with sufficient documentation that a problem within the purpose of the Code exists with branch-circuits or feeder relating to voltage drop that would merit mandatory requirements for voltage drop. Voltage drop is a requirement in *647.4(D), Sensitive Electronic Equipment* and *695.7, Fire Pumps* where sufficient substantiation was provided to establish this requirement. Remember, the fact that the *NEC®* does not require voltage drop to be taken into account does not mean that it is not a necessary part of a proper electrical design.

Question 4: The requirements for show-window load calculations are confusing. What do they mean in *220.3(B)(7)* when they say you can compute using a unit load per outlet or you can compute using 200 volt-amperes per linear ft? Then in *220.12(A)* it says you must use 200 volt-amperes per linear ft of show window.

Answer: The requirements you see in *220.7(B)(7)* refer to making a computation for the show-window branch-circuit. You can compute by unit load per outlet in the show window, or you can use 200 volt-amperes per linear ft of show window. The requirement you see in *220.12(A)* refers to making a computation to be added to the feeder or service load.

Question 5: What is meant by *diverse loads and non-diverse loads* in the methods of making calculations?

Answer: Diverse loads are those that have a demand factor allowance permitted and are allowed to be computed at less than 100% of their rating in volt-amperes. An example of diverse loads that are permitted to be calculated at less than their rated load are shown in *Table 220.13* for non-dwelling receptacle loads and those shown in *Table 220.20* for non-dwelling kitchen equipment. These loads do not include non-diverse loads such as motor loads and other continuous loads that are required to be calculated at 125%. Other non-diverse loads include lighting loads and grain-drying equipment loads that are generally continuous loads, operating continuously for three hours or more.

CHAPTER 12

Overcurrent Protection

OBJECTIVES

After completing this chapter, the student should understand:
- fault-current calculations.
- purpose of overcurrent devices.
- selection of overcurrent devices.
- selective coordination.
- series rated system.
- short-circuit current rating.
- withstand capabilities.

KEY TERMS

Continuous-Current Rating: The total amperes that a device or conductor can carry continuously without interrupting the circuit

Interrupting Rating: The highest current at rated voltage that a device is intended to interrupt under standard test conditions

Short-Circuit Current Rating: Short-circuit current rating is based on the amount of current that must flow for a given period of time to open the protective device

Speed of Response: The time required for the overcurrent device to open in relation to the magnitude of current that flows through the device

Voltage Rating: The maximum voltage that may be applied to a fuse or circuit breaker in accordance with the manufacturer's rating

Withstand Rating: The ability to withstand a certain amount of current for a given time

INTRODUCTION

In previous chapters, we have discussed the definition of overcurrent protective devices and some of the requirements for their use. This chapter will explain the function of overcurrent devices and their role in the protection of electrical circuits and equipment. *Article 240* contains the requirements for overcurrent protection for conductors and equipment. *NEC® 110.9* and *110.10* contain the requirements for interrupting rating and circuit protection against fault-current conditions. This chapter has been written for use in conjunction with those *NEC®* articles. The use of fuses and circuit breakers as overcurrent protective devices will be discussed in this chapter. Motor overload protective devices are covered in Chapter 10, "Motors and Controllers."

GENERAL

Fuses and circuit breakers are both used as the required overcurrent protective devices (Figure 12-1). The designer of the electrical installation has the responsibility of determining which type of

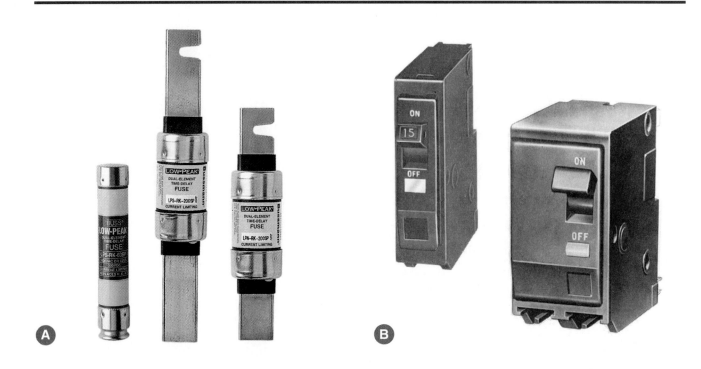

Figure 12-1 (A) Typical fuses. *Courtesy of* Cooper Bussman. (B) Typical circuit breakers

overcurrent device will best handle the requirements for circuit and equipment protection in the system being designed. This chapter will help the user to determine the protection required and will show the methods that may be used to accomplish this protection. Reliable protective devices prevent or minimize damage to electrical equipment and conductors.

PURPOSE OF OVERCURRENT DEVICES

An overcurrent device in an electrical circuit opens the electrical circuit whenever an overload or short-circuit occurs that causes a current to flow which is greater than the rating of the overcurrent device. An overload current does not leave the normal conducting path of the circuit. An overload current is caused by a change in the amount of load that increases the current in relation to the impedance of the load. The amount of current is directly proportional to the magnitude of the load. Most loads are rated directly in watts (power), and using the power formula $P = E \times I$, an illustration can be shown. Assuming a load of 1200 watts and a potential of 120 volts, we can by using the formula I (current) = P (watts) divided by E (volts) establish a current of 10 amperes.

$$I = \frac{P}{E} \text{ or } I = \frac{1200}{120} = 10 \text{ amperes}$$

If we were to increase the load to 3600 watts, our formula would look like this:

$$I = \frac{P}{E} \text{ or } I = \frac{3600}{120} = 30 \text{ amperes}$$

Figure 12-2 shows a circuit properly protected and a circuit not properly protected. If the conductors we had chosen for this circuit were 20-ampere rated, the 1200-watt load would have been properly designed for this installation, and a 20-ampere overcurrent device would be installed to protect the conductors. If the load were increased to 3600 watts, as shown in the second example, the 30-ampere current flow would be too much for the 20-ampere conductors, and they would be damaged if they were not protected by a 20-ampere overcurrent

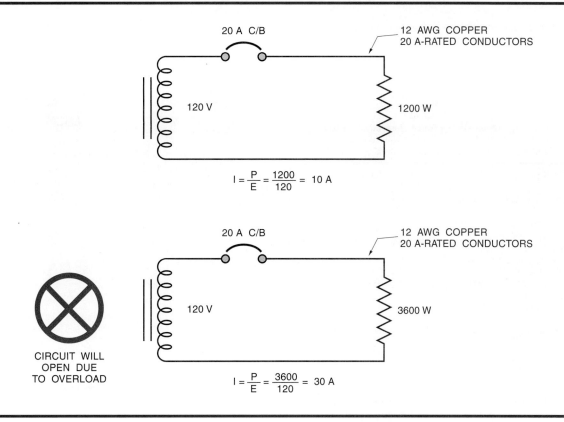

Figure 12-2 One-line diagram showing power calculation

device that would open because of the overload on the conductors.

All overloads are not caused by too much of a designed load connected on the circuit. Other overloads may be caused by defective motors (such as worn bearings) or by overloaded equipment. These are the destructive overloads that can cause severe damage to conductors and equipment, and they must be shut down before the damage can occur (Figure 12-3).

Overloads are sometimes caused by temporary currents that occur when motors are started up or when transformers are energized. These currents are normal and their brief duration is not harmful to conductors or equipment. It is important that the overcurrent devices do not open in these cases. This is why *Table 430.52* permits a higher percentage of overcurrent protection for motor installations (Figure 12-4).

A short-circuit, as we learned in Chapter 1, is unintentional contact between one or more conductors of opposite polarity. In most cases, the contact

Figure 12-3 Damage to electrical equipment not properly protected. *Courtesy of* Cooper Bussman

is between a phase conductor and the metal raceway that is serving as the equipment-grounding conductor. Short-circuit current can be several hundred times larger than the normal current and must be cut

Table 430.52 Maximum Rating or Setting of Motor Branch-Circuit Short-Circuit and Ground-Fault Protective Devices

Type of Motor	Percentage of Full-Load Current			
	Nontime Delay Fuse[1]	Dual Element (Time-Delay) Fuse[1]	Instantaneous Trip Breaker	Inverse Time Breaker[2]
Single-phase motors	300	175	800	250
AC polyphase motors other than wound-rotor				
Squirrel cage — other than Design E or Design B energy efficient	300	175	800	250
Design E or Design B energy efficient	300	175	1100	250
Synchronous[3]	300	175	800	250
Wound rotor	150	150	800	150
Direct current (constant voltage)	150	150	250	150

Note: For certain exceptions to the values specified, see 430.54.
[1]The values in the Nontime Delay Fuse column apply to Time-Delay Class CC fuses.
[2]The values given in the last column also cover the ratings of nonadjustable inverse time types of circuit breakers that may be modified as in 430.52(C), Exception No. 1 and No. 2.
[3]Synchronous motors of the low-torque, low-speed type (usually 450 rpm or lower), such as are used to drive reciprocating compressors, pumps, and so forth, that start unloaded, do not require a fuse rating or circuit-breaker setting in excess of 200 percent of full-load current.

Figure 12-4 *NEC® Table 430.52* (Reprinted with permission from NFPA 70-2002)

off in a few thousands of a second, or severe damage will occur to conductor insulation and electrical equipment. Short-circuit currents can cause severe magnetic field stresses which may be beyond the **withstand rating** capabilities of the bus bars and the equipment bracing.

SELECTION OF OVERCURRENT DEVICES

In general, most applications require consideration of **voltage rating**, the **continuous-current rating**, the **interrupting rating**, and the time required to open the overcurrent device or **speed of response** (Figure 12-5). The voltage rating selection is governed by *110.4*, which requires the voltage rating of

Type SC
(Fast-Acting) 1/2-6A
(Time-Delay) 8-60A
SC 100,000AIR AC
1/2-20A (600VAC)
25-60A (480VAC)

A high performance general-purpose branch-circuit fuse for lighting appliance, and motor branch-circuits of 480 volts (or less).

Figure 12-5 Fast acting fuse—quick rate of response. *Courtesy of* Cooper Bussman

electrical equipment to be not less than the nominal voltage of a circuit to which it is connected.

Voltage Rating—Fuses

Most low-voltage power distribution fuses have 250-volt or 600-volt ratings (other ratings are 125, 300, and 480 volts). The voltage rating of a fuse must be at least equal to or greater than the circuit voltage. It can be higher but never lower. The voltage rating of a fuse is a function of its capability to open a circuit under an overcurrent condition (Figure 12-6).

Voltage Rating—Circuit Breakers

For circuit breakers, the voltage rating is usually available at every voltage. All molded-case circuit breakers are marked with a voltage rating. The ratings for ac are 120, 120/240, 240, 277, 347, 480Y/277, 480, 600Y/347 and 600 volts. Circuit breakers with a single voltage rating are intended for use in circuits where the circuit voltage and the voltage-to-ground do not exceed the voltage rating of the circuit breaker. Dual or slant-rated breakers such as

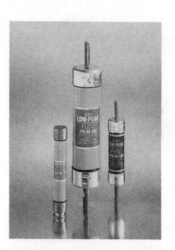

(A) One-time, single-element, noncurrent-limiting fuses. 1/8–60 amperes Class K5 — 50,000 amperes interrupting rating. 250– and 600–volt ratings. 70–600 amperes Class H — 10,000 amperes interrupting rating. 250– and 600–volt ratings.

(B) Single-element, current-limiting, fast-acting fuses. 1–600 amperes Class RK1 — 200,000 amperes interrupting rating. 250– and 600–volt ratings.

(C) Dual-element, time-delay, current-limiting fuses. 1/10–600 amperes Class RK5 — 200,000 amperes interrupting rating. 250– and 600–volt ratings.

(D) Dual-element, time-delay, current-limiting fuses. 1/10–600 amperes Class RK1 — 300,000 amperes interrupting rating. 250– and 600–volt ratings.

Figure 12-6 Voltage rating—fuses. *Courtesy of* Cooper Bussman

480Y/277 are intended for use in circuits where the circuit voltage does not exceed the higher of the two voltages and the voltage-to-ground does not exceed the lower of the two voltages. Based on this, dual-rated circuit breakers (120/240-volt) are not suitable for use on dual-voltage delta systems. This is because the high leg voltage to neutral is 208 volts and should not be higher than the lower of the two voltages, which is 120 volts. A 3-pole, 240-volt breaker should be used for this application (Figure 12-7).

Continuous-Current Rating

The continuous-current rating of an overcurrent protective device is the total amount of current the device can carry continuously without interrupting the circuit. The standard ratings of fuses and circuit

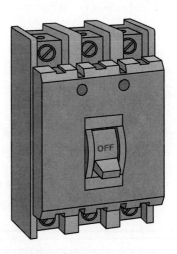

Figure 12-7 Three-pole, 240-volt circuit breaker

breakers are listed in *240.6*. The current rating of the device is selected to match the circuit-conductor ampacity but there are some exceptions to this. For instance, a motor may have the branch-circuit, short-circuit, and ground-fault protection sized higher than the conductor ampacity, as permitted in *Table 430.52*. As we learned in Chapter 10, this is permitted because the conductors are sized at 125% of the motor FLC, and the motor overload protection is generally sized at 125% of the motor FLC. This, together with the fact that the short-circuit current rises very quickly to a high magnitude and will open the motor branch-circuit overcurrent device quickly to protect the motor-circuit conductors, allows the overcurrent device to be sized larger than the rated ampacity of the motor-circuit conductors (Figure 12-8).

Another application where the overcurrent device is permitted to be larger than the conductor current rating is shown in *240.4(B)*. This section permits the next higher standard overcurrent device rating above the rating of the conductors being protected to be used provided (1) that the conductors being protected are not part of a multioutlet branch-circuit supplying receptacles for cord- and plug-connected portable loads; (2) that the ampacity of the conductors does not correspond with the standard ampere rating of a fuse or circuit breaker without overload trip adjustments above its rating, or (3) that the next higher standard rating selected does not exceed 800 amperes.

Interrupting Rating

The interrupting rating is defined in *Article 100* as *the highest current at rated voltage that a device is intended to interrupt under standard test conditions*. This means that the interrupting capabilities are manufactured into the device and tested for confirmation of the interrupting capability. Interrupting ratings for fuses shall be marked where other than 10,000 amperes. Interrupting ratings for circuit breakers other than 5000 amperes shall be shown on the circuit breaker. The interrupting rating of a fuse or circuit breaker is the highest amount of current that the device will interrupt without damage to the device (Figure 12-9).

Speed of Response

The time it takes to open an overcurrent device from when the current value exceeds the current rating of the device is called the speed of response. The speed of response is generally measured in thousandths of a second. The speed of response is the governing factor in the amount of *let-through* current that will pass through the overcurrent device before it opens. The let-through current, as its name implies, is the amount of current that will pass through an overcurrent device before it successfully opens to interrupt current flow. The time required to open a fusible element decreases with the increase of current flow through the fuse. For circuit breakers, this reduction of opening time is also dependent upon the inertia of the moving parts of the circuit breaker. Time current characteristic curves are available from manufacturers of fuses and circuit breakers (Figure 12-10).

For example, if a 30-horsepower, 3-phase, 230-volt motor with a Code letter F is being installed, the full-load amperes are determined from *Table 430.150* as 80 amperes. This amperage is multiplied by 125% (80 × 1.25 = 100) to determine the maximum running current permitted by *430.32*. The locked- rotor amperage can be determined by *Table 430.7(B)* using the Code letter F (5000 VA × 30

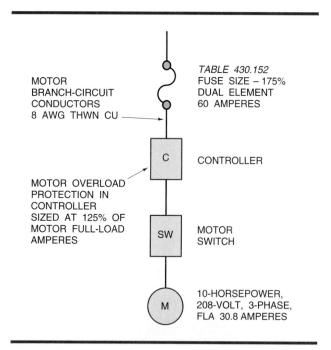

Figure 12-8 One-line diagram showing motor branch-circuit with overcurrent protection larger than ampacity of the motor branch-circuit conductors

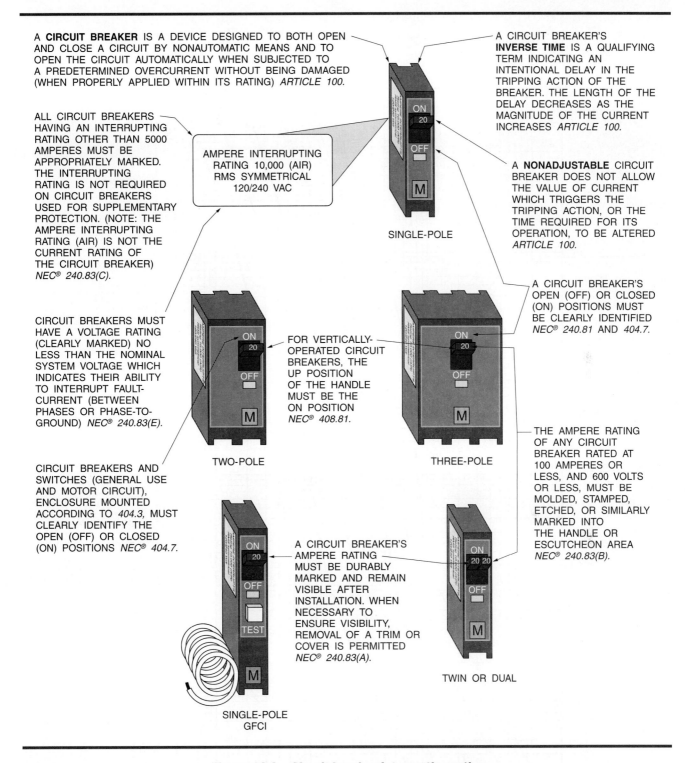

Figure 12-9 Circuit-breaker interrupting rating

horsepower ÷ 360 = 416 LRA). Now look at the chart. Go up the left side to 400, then go horizontal until you intersect the 100-ampere line, now go vertical to the bottom of the chart where it shows 30 seconds. The dual element fuse will hold for 30 seconds to allow the motor to start.

Short-Circuit Current Rating

Electrical equipment is sometimes marked with **short-circuit current ratings**. *NEC® 110.10* requires that electrical components must not be connected to systems capable of delivering a higher fault-current than the equipment's short-circuit current rating.

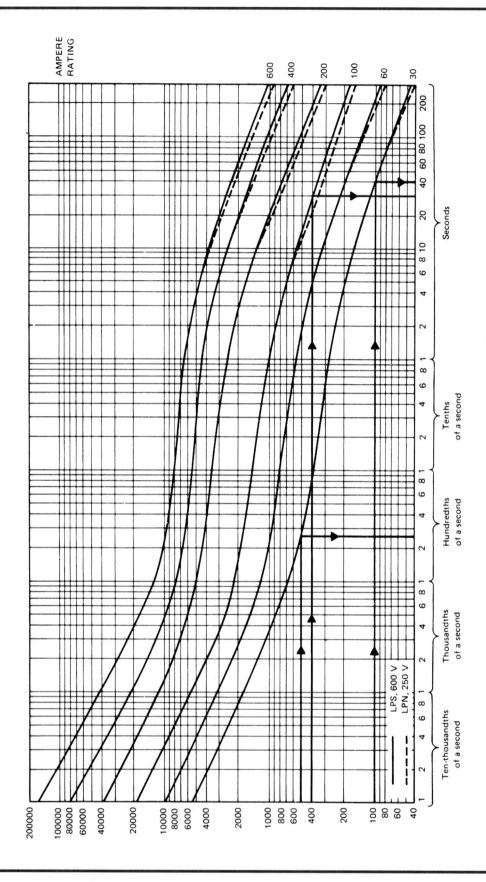

Figure 12-10 Time current characteristic chart for dual element fuse

The short-circuit current rating of electrical equipment is established and marked on the equipment by the manufacturer. This rating is developed in conformance with standards developed by NRTL.

When determining the use of fuses or circuit breakers in a circuit or system, it is necessary to determine the short-circuit current available at the line terminals of these devices so that overcurrent devices with the proper interrupting rating can be selected.

When operating under normal conditions, the current in a circuit is determined by the amount of load and the applied voltage. If a short-circuit occurs, there is an abnormally high amount of current flow until the overcurrent device opens. If the short-circuit current is higher than the interrupting rating of the overcurrent device, the overcurrent device may be damaged or destroyed with the possibility of severe damage to other equipment.

Fault-Current Calculations

Determining the available fault-current at the line terminals of an overcurrent protective device is accomplished by use of formulas. These formulas need several factors to be used in the formula. One of these is the impedance of the transformer furnishing power to the installation. Generally, to determine available fault-current at a service, the serving utility can be contacted, and they can provide you with this information. The information provided by the utility company is generally inclusive of the available short-circuit current at the secondary terminals of the installed transformer and could be determined as the amount available on the largest transformer they may use in that location. Using this information will keep you from being in trouble if the utility should make a transformer change.

Where the power source is from a utility service or a separately derived system, the short-circuit current can be determined as follows:

Assume that a 3-phase transformer is used as a power source and has a secondary rating of 150 kVA at 208Y/120 volts. The nameplate impedance of the transformer shows an impedance of 2%. It is assumed that the primary of the transformer has infinite bus (no limitation on the amount of short-circuit current that can be delivered to the primary of the transformer). Using the formula: Short-circuit current is equal to the FLC times the multiplier M.

Where: FLC = kVA divided by line-to-line voltage (3-phase × 1.73)

M = 100 divided by transformer impedance

Step #1: Determine the *FLC* rating of the transformer

$$FLC = \frac{150{,}000}{E \times 1.73} = \frac{150{,}000}{208 \times 1.73} = 416.6 \text{ FLC}$$

Step #2: Determine the multiplier M.

$$M = \frac{100}{\text{Impedance}} = \frac{100}{2} = 50$$

Step #3: Determine the short-circuit current (Isc)

$$Isc = FLC \times M = 416.6 \times 50 = 20{,}830$$

At this point, we have determined the available short-circuit current at the transformer secondary terminals. To determine the short-circuit current available at the terminals of the main switchgear, distribution panelboard, or other points at various distances from the transformer, we use the point-to-point method shown as follows:

Find the available short-circuit current at the terminals of the main switch that is located 50 ft from the transformer. The main switch is supplied by four 600-kcmil AWG copper conductors in steel conduit. Using the formula:

Isca = short-circuit current at transformer (Isca) × M_2 (Multiplier 2)

Step #1: Determine f

$$f = \frac{1.73 \times L \times I}{C \times E_{L\text{-}L}} = \frac{1.73 \times 50 \times 20{,}830}{22{,}965 \times 208} = .377$$

Step #2: Determine M_2

$$\text{Where: } M_2 = \frac{1}{1 + f} = \frac{1}{1 + .377} = \frac{1}{1.377} = 0.726$$

Where:
L = Length of circuit to fault in ft
I = Isca at transformer
C = Constant table (Figure 12-11)
E = Line-to-Line

Step #3:

Isca = Isca (at transformer) × M_2 (Multiplier 2)

20,830 × .726 = 15,127 (Isca at main switch)

CHAPTER 12 Overcurrent Protection

AWG or kcmil	Copper Conductors Three Single Conductors						Copper Conductors Three Conductor Cable					
	Steel Conduit			Nonmagnetic Conduit			Steel Conduit			Nonmagnetic Conduit		
	600V	5KV	15KV	600V	5KV	15KV	600V	5KV	15KV	600V	5KV	15KV
14	389	—	—	389	—	—	389	—	—	389	—	—
12	617	—	—	617	—	—	617	—	—	617	—	—
10	981	—	—	981	—	—	981	—	—	981	—	—
8	1557	1551	1557	1556	1555	1558	1559	1557	1559	1559	1558	1559
6	2425	2406	2389	2430	2417	2406	2431	2424	2414	2433	2428	2420
4	**4779**	3750	3695	3825	3789	3752	3830	3811	3778	3837	3823	3798
3	4760	4760	4760	4802	4802	4802	4760	4790	4760	4802	4802	4802
2	5906	5736	5574	6044	5926	5809	5989	5929	5827	6087	6022	5957
1	7292	7029	6758	7493	7306	7108	7454	7364	7188	7579	7507	7364
1/0	8924	8543	7973	9317	9033	8590	9209	9086	8707	9472	9372	9052
2/0	10755	10061	9389	11423	10877	10318	11244	11045	10500	11703	11528	11052
3/0	12843	11804	11021	13923	13048	12360	13656	13333	12613	14410	14118	13461
4/0	15082	13605	12542	16673	15351	14347	16391	15890	14813	17482	17019	16012
250	16483	14924	13643	18593	17120	15865	18310	17850	16465	19779	19352	18001
300	18176	16292	14768	20867	18975	17408	20617	20051	18318	22524	21938	20163
350	**19529**	17385	15678	22736	20526	18672	**22646**	21914	19821	**24904**	24126	21982
400	20565	18235	16365	24296	21786	19731	24253	23371	21042	26915	26044	23517
500	22185	19172	17492	26706	23277	21329	26980	25449	23125	30028	28712	25916
600	22965	20567	17962	28033	25203	22097	28752	27974	24896	32236	31258	27766
750	24136	21386	18888	28303	25430	22690	31050	30024	26932	32404	31338	28303
1000	25278	22539	19923	31490	28083	24887	33864	32688	29320	37197	35748	31959

AWG or kcmil	Aluminum Conductors Three Single Conductors						Aluminum Conductors Three Conductor Cable					
	Steel Conduit			Nonmagnetic Conduit			Steel Conduit			Nonmagnetic Conduit		
	600V	5KV	15KV	600V	5KV	15KV	600V	5KV	15KV	600V	5KV	15KV
14	236	—	—	236	—	—	236	—	—	236	—	—
12	375	—	—	375	—	—	375	—	—	375	—	—
10	598	—	—	598	—	—	598	—	—	598	—	—
8	951	950	951	951	950	951	951	951	951	951	951	951
6	1480	1476	1472	1481	1478	1476	1481	1480	1478	1482	1481	1479
4	2345	2332	2319	2350	2341	2333	2351	2347	2339	2353	2349	2344
3	2948	2948	2948	2958	2958	2958	2948	2956	2948	2958	2958	2958
2	3713	3669	3626	3729	3701	3672	3733	3719	3693	3739	3724	3709
1	4645	4574	4497	4678	4631	4580	4686	4663	4617	4699	4681	4646
1/0	5777	5669	5493	5838	5766	5645	5852	5820	5717	5875	5851	5771
2/0	7186	6968	6733	7301	7152	6986	7327	7271	7109	7372	7328	7201
3/0	8826	8466	8163	9110	8851	8627	9077	8980	8750	9242	9164	8977
4/0	10740	10167	9700	11174	10749	10386	11184	11021	10642	11408	11277	10968
250	12122	11460	10848	12862	12343	11847	12796	12636	12115	13236	13105	12661
300	13909	13009	12192	14922	14182	13491	14916	14698	13973	15494	15299	14658
350	15484	14280	13288	16812	15857	14954	15413	15490	15540	16812	17351	16500
400	16670	15355	14188	18505	17321	16233	18461	18063	16921	19587	19243	18154
500	18755	16827	15657	21390	19503	18314	21394	20606	19314	22987	22381	20978
600	20093	18427	16484	23451	21718	19635	23633	23195	21348	25750	25243	23294
750	21766	19685	17686	23491	21769	19976	26431	25789	23750	25682	25141	23491
1000	23477	21235	19005	28778	26109	23482	29864	29049	26608	32938	31919	29135

Figure 12-11 Table of *C* values

The calculation just completed involved 3-phase faults. For single-phase faults, use the following formulas to determine the *f* factor.

For single-phase, line-to-line (L-L) or line-to-neutral (L-N) faults on single-phase, center-tapped transformers:

$$f = \frac{2 \times L \times I}{C \times E_{L-L}} \qquad f = \frac{2 \times L \times I}{C \times E_{L-N}}$$

A chart used to convert *f* values to *M* values is shown in Figure 12-12.

Series-Rated System

Another method of protecting circuit breakers, where the available short-circuit current is higher than the rating of the branch-circuit overcurrent devices, is by using a listed series-rated system.

Chart to Convert "f" Values to "M" Values When Using the Point-to-Point Method				
f	M	f	M
0.01	0.99	1.20	0.45
0.02	0.98	1.50	0.40
0.03	0.97	2.00	0.33
0.04	0.96	3.00	0.25
0.05	0.95	4.00	0.20
0.06	0.94	5.00	0.17
0.07	0.93	6.00	0.14
0.08	0.93	7.00	0.13
0.09	0.92	8.00	0.11
0.10	0.91	9.00	0.10
0.15	0.87	10.00	0.09
0.20	0.83	15.00	0.06
0.30	0.77	20.00	0.05
0.40	0.71	30.00	0.03
0.50	0.67	40.00	0.02
0.60	0.63	50.00	0.02
0.70	0.59	60.00	0.02
0.80	0.55	70.00	0.01
0.90	0.53	80.00	0.01
1.00	0.50	90.00	0.01
		100.00	0.01

$$M = \frac{1}{1+f}$$

Figure 12-12 Conversion chart for *f* values to *M* values

NEC® 240.86 permits a circuit breaker to be used on a circuit having an available fault-current higher than its marked interrupting rating by being connected on the load side of an acceptable overcurrent protective device having the higher interrupting rating. There are conditions attached to this permission concerning testing and motor contribution.

Electric motors that are running when a fault condition occurs contribute additional short-circuit current on an electrical system. Motors should never be connected between the load side of the higher-rated protective device and the line terminals of the lower-rated protective device.

The combination of line-side overcurrent devices and load-side circuit breakers must be tested and marked on the end-use equipment, such as switchboards and panelboards. A common way of using the series rated combination is shown in Figure 12-13.

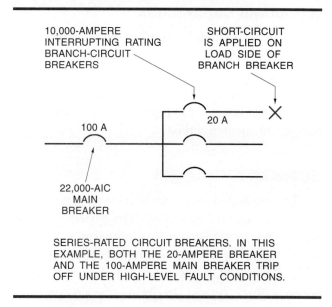

Figure 12-13 Series-rated circuit breakers

Selective Coordination

Selective coordination is a method of selection of overcurrent devices so that in the event of a fault, only the protective device nearest to the fault is opened. Selective coordination will limit the power outage to only those circuits downstream from the overcurrent device protecting those circuits (Figure 12-14). Methods of establishing selective coordination involve the use of time current curves of fuses and circuit breakers. These time current characteristics are obtained from the manufacturers of the overcurrent protective devices. A properly engineered system will allow only the protective device nearest the fault to open.

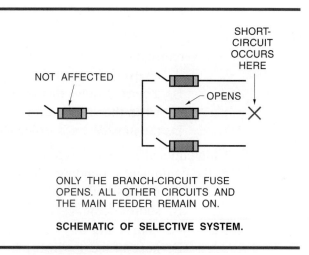

Figure 12-14 Selective coordination with fuses

Withstand Capabilities

The term component short-circuit current rating is used in *110.10*. This term means the same as withstand rating. It is the ability to withstand a certain amount of current for a certain amount of time or what is known as I–squared–T, which stands for current squared x time.

SUMMARY

The protection for an electrical system must provide for continuity of service. The system must be electrically coordinated to ensure that if a problem occurs, only the faulted portion of the system is taken out or isolated without disturbing any other portions of the system. To obtain reliable operation and make sure that the system components are protected from damage, the available fault-current must be calculated at all critical points in the system. After short-circuit current levels are determined, the designer can properly specify interrupting rating, selectively coordinate the system, and provide component protection. Most fuses and some circuit breakers are current limiting. They restrict fault-currents to low levels so that a high degree of protection is given to circuit components against even very high short-circuit currents. Fuses selectively coordinated with circuit breakers can permit circuit breakers with lower interrupting ratings to be used. It is important to remember that just using a fuse in front of a breaker without determining the let-through current does not solve any problem. Coordinating the rating of the device after the fuse is just as important.

REVIEW QUESTIONS

MULTIPLE CHOICE

1. The total amperes that a device or conductor can carry continuously without interrupting the circuit is:
   ____ A. voltage rating.
   ____ B. withstand rating.
   ____ C. interrupting rating.
   ____ D. continuous-current rating.

2. The ability to withstand a certain amount of current for a given amount of time is:
   ____ A. short-circuit rating.
   ____ B. overload rating.
   ____ C. impedance rating.
   ____ D. withstand rating.

3. An overcurrent device in an electrical circuit opens the electrical circuit when:
   ____ A. an overload occurs.
   ____ B. a current flows that is greater than the device rating.
   ____ C. when the voltage drop exceeds 5%.
   ____ D. the impedance is exceeded.

4. Unintentional contact between a phase conductor and the metal raceway that is serving as the equipment-grounding conductor is a:
   ____ A. ground-fault.
   ____ B. short-circuit.
   ____ C. both A and B.
   ____ D. none of the above.

5. Fuse voltage ratings are:
   ____ A. 250 volts.
   ____ B. 600 volts.
   ____ C. both A and B.
   ____ D. none of the above.

6. The next higher standard overcurrent device rating above the rating of the conductors being protected may be used if:
   ____ A. the conductors are part of a multiwire circuit.
   ____ B. the conductors are part of a multioutlet circuit.
   ____ C. the voltage is not greater than 250 volts.
   ____ D. the next higher standard rating selected does not exceed 800 amperes.

7. The time it takes to open an overcurrent device from when the current value exceeds the current rating of the device is called:
   ____ A. interrupting rating.
   ____ B. speed of response.
   ____ C. voltage rating.
   ____ D. fault-current.

8. The total opposition to the flow of current in a transformer is called:
   ____ A. resistance.
   ____ B. conductance.
   ____ C. impedance.
   ____ D. reactance.

9. A 3-phase transformer secondary being used as a power source for an office building is rated 500 kVA, 600V–208Y/120 volts. What is the full-load secondary current?
   ____ A. 208.3 FLC
   ____ B. 416.6 FLC
   ____ C. 1388 FLC
   ____ D. 602 FLC

10. A transformer with an impedance of 1.5% would require a multiplier of _____ to determine short-circuit current.
    ____ A. 66.6
    ____ B. 55
    ____ C. 100
    ____ D. 1.73

TRUE OR FALSE

1. Three-phase, line-to-line voltages must be multiplied by 1.73 for use in short-circuit current formulas.
   ____ A. True
   ____ B. False

2. Selective coordination is a method of selection of overcurrent devices so that in the event of a fault, only the protective device nearest the fault will open.
   ____ A. True
   ____ B. False

3. The total amperes that a device or conductor can carry continuously without interrupting the circuit is called interrupting rating.
   ____ A. True
   ____ B. False

4. The time required for the overcurrent device to open in relation to the magnitude of current is called the speed of response.
   ____ A. True
   ____ B. False

5. An overload current does not leave the normal conducting path of the circuit.
   ____ A. True
   ____ B. False

6. Overloads are sometimes caused by temporary currents that occur when motors are started or when transformers are energized.
   ____ A. True
   ____ B. False

7. Fuses have basically two voltage ratings, 250 volts or 600 volts.
   ____ A. True
   ____ B. False

8. Dual-rated circuit breakers are not suitable for use on dual-voltage delta systems.
   ____ A. True
   ____ B. False

9. On a 3-phase delta system, the high leg voltage-to-ground cannot be used in 120-volt circuits.
   ____ A. True
   ____ B. False

10. Selective coordination is not required when fuses are used as the overcurrent protective device.
    ____ A. True
    ____ B. False

FILL IN THE BLANKS

1. An overcurrent device in an electrical circuit _____ whenever a sustained overload or short-circuit occurs that causes a current to flow that is greater than the rating of the overcurrent device.

2. Overloads are sometimes caused by _____ that occur when motors are started up or when transformers are energized.

3. All molded-case circuit breakers are marked with a _____ rating.

4. Circuit breakers with a single _____ are intended for use in circuits where the circuit voltage and the voltage-to-ground do not exceed the _____ of the circuit breaker.

5. The _____ of an overcurrent device is the total amount of current the device can carry continuously without interrupting the circuit.

6. The time required to open a fusible element _____ with the _____ of current flow through the fuse.

7. If a short-circuit occurs, there is an abnormally _____ of current flow until the overcurrent device opens.

8. Selective coordination is a method of selection of overcurrent devices so that in the event of a fault, only the _____ to the fault is opened.

9. Selective coordination will limit the _____ to only those circuits downstream from the overcurrent device protecting those circuits.

10. Fuses _____ with circuit breakers can permit circuit breakers with lower interrupting ratings to be used.

FREQUENTLY ASKED QUESTIONS

Question 1: Can I just simplify everything and use a dual-element fuse in my main switch and protect everything downstream?

Answer: I'm afraid not. To properly and selectively coordinate, it is necessary that you determine the amount of let-through current that will pass through the dual-element fuse, and how much will reach the next overcurrent device. It may be necessary to employ some other type of current-limiting fuse in your coordination procedure. A fast acting current-limiting (non-time delay) fuse has a very fast response on both the low-overload and short-circuit ranges. This type of fuse has the lowest let-through current value and is frequently used to provide protection to mains, feeders, and sub-feeders that lack an adequate interrupting rating. So check your short-circuit current availability through the entire system, and choose your components carefully.

Question 2: Why don't they have current-limiting circuit breakers?

Answer: Current-limiting circuit breakers are available. The label on this type of circuit breaker will have the words *current-limiting* on it. The let-through data will either be on the label, or it will indicate where this information may be obtained. As it is with fuses, this information is important to properly protect all the downstream devices and components. All downstream components must be capable of withstanding the let-through current of the circuit breaker.

Question 3: There must be an easier way to do all these point-to-point calculations. Where can I get help?

Answer: Representatives from either fuse or circuit-breaker manufacturers will be glad to help you make these calculations. You can be sure that each one will show you the advantages of using the type of device he manufactures. That is why you must understand how and why these calculations are done in order to make your own determination as to how to properly coordinate your installation. Time/current graphs are readily available from these manufacturers, and computer software is also available to assist you in this most important aspect of electrical installations.

Question 4: What do they mean when they say *inverse-time* circuit breaker?

Answer: An inverse-time circuit breaker is a circuit breaker with a time delay feature built into it. Inverse-time circuit breakers are thermal-magnetic type and are very good for use as motor branch-circuit protection. The inverse-time circuit breaker will allow the heavy starting current to pass and still offer good protection against sustained overloads. Overload protective devices with inverse time characteristics are practical for use in circuits since they will prevent interruption of the circuit during short overloads, which would be harmless to the circuit components.

Question 5: Why do some short-circuit calculations use a .9 multiplier for the transformer impedance figure shown on the transformer nameplate?

Answer: UL listed transformers 25 kVA and greater have a plus or minus 10% tolerance on their nameplate. Applying a .9 multiplier to the given impedance rating reduces the impedance to 90%, and thereby increases the multiplier M used in short-circuit calculations and increases the available short-circuit current figure. This is taking a worst-case approach, which is added to the worst-case approach of using infinite bus on the primary of the transformer and is also added to the worst-case approach of assuming a bolted fault; however, with all this beefing up, you probably cannot go wrong. The staggering losses in equipment and downtime from an inadequate design of overcurrent protective devices warrants serious consideration of using the worst-case approach in short-circuit calculations.

Question 6: Why are rejection clips put on fuseholders?

Answer: Rejection clips will reject any attempt to install a fuse that is not a Type R. Type R fuses are current-limiting fuses having a high degree of current limitation and a short-circuit interrupting rating of 300,000 amperes. Class H fuses are not current-limiting and are recognized as having only an interrupting rating of 10,000 amperes. Serious damage could result if a Class R fuse were replaced with a Class H fuse. The use of Class R fuseholders is very important. *Article 240.60(B)* requires that fuseholders for current-limiting fuses shall not permit insertion of fuses that are not current-limiting.

Question 7: What is the difference between a non-time-delay fuse and a time-delay fuse?

Answer: A non-time-delay fuse is a *single element* fuse. The basic component of a fuse is the link. The single element fuse may have more than one link in the element, depending upon the ampere rating of the fuse. Under normal operation, the fuse functions as a conductor. If an overload current occurs, the link reaches a temperature level that will melt a restricted segment of the link. As a result, a gap is formed and an electric arc is formed. After the arc is extinguished by the design of the fuse, the circuit is open. Suppression of the arc is accelerated by filler material. Single element fuses have a very high speed of response to overloads and open with very little delay.

The dual-element time-delay fuse contains two separate types of elements that are connected in series. The fuse links are similar to those in non-time-delay fuses and they provide short-circuit protection, the overload element provides protection against overloads. Further detailed information about these fuses may be obtained from the manufacturers of fuses.

Question 8: Are circuit breakers required to have line and load identification?

Answer: A circuit breaker may or may not be marked line and load. Where a circuit breaker is not marked line and load, it is acceptable to use reverse connections.

Question 9: What does it mean if the nameplate on an air conditioner says *use HACR breakers only*?

Answer: An HACR circuit breaker is a circuit breaker that has been tested and found acceptable for use in heating, air-conditioning, and refrigeration equipment comprising group motor installations. This marking alone, however, does not indicate the acceptability of the circuit breaker in these installations. For an acceptable installation, the end-use equipment must be marked to indicate that *HACR Type* circuit breakers may be used for branch-circuit overcurrent protection.

Question 10: Can circuit breakers marked *SWD* be used for switching *HID* luminaires (fixtures)?

Answer: A circuit breaker rated 50 amperes maximum, 480 volts, or less and are intended to switch high-intensity discharge (HID) luminaires (lighting fixtures) on a regular basis must be marked *HID*. Circuit breakers marked HID may be used to switch fluorescent luminaires (fixtures) but circuit breakers marked SWD can only be used to switch fluorescent luminaires (fixtures).

CHAPTER 13

Hazardous (Classified) Locations

OBJECTIVES

After completing this chapter, the student should understand:
- Class I locations.
- Class II locations.
- Class III locations.
- intrinsically safe systems.
- Zone 0, 1 and 2 locations.

KEY TERMS

Class I Locations: Locations in which flammable gases or vapors are, or may be, present in the air in quantities sufficient to produce explosive or ignitable mixtures

Class II Locations: Locations that are hazardous because of the presence of combustible dust

Class III Locations: Locations that are hazardous because of the presence of easily ignitable fibers or flyings

Conduit Seals: Fittings into which a sealing compound is poured to surround the conductors to minimize the passage of gases or vapors from one portion of the electrical installation to another through the conduit

Dusttight: Enclosures constructed so that dust will not enter

Fire Triangle: Three conditions for fire or explosion, fuel, ignition, and oxygen

Flash Point: Minimum temperature required to initiate self-sustained combustion

Flat Ground Joint: A machined joint between two surfaces that are tightly bolted together

Intrinsically Safe System: Low energy system designed to eliminate the ignition source leg of the fire triangle

Labyrinth Joint: A joint that forces the expanding hot gases to make several right-angle turns before they can exit the enclosure

INTRODUCTION

Hazard (classified) locations are defined as those locations where flammable gases or vapors, flammable liquids, combustible dust, or ignitable flyings or fibers are present in sufficient amounts to create fire or explosion hazards.

Although flammable gases and vapors and combustible dusts are present everywhere, the amounts are so minute that fire and explosion hazards do not exist. To constitute a hazard, these materials must be present in concentrations sufficient to sustain gas or vapor ignition and explosion problems.

This chapter will explain the equipment, materials, and wiring methods required for safe electrical installations in hazardous (classified) locations. This text has been written to be used in conjunction with *Articles 500* through *516*.

GENERAL

In order for a fire or explosion to occur, three conditions must be present.

1. There must be a fuel (flammable gas or vapor, or combustible dust).
2. There must be an ignition source in the form of heat or a spark.
3. There must be oxygen.

These three conditions are referred to as the **fire triangle** (Figure 13-1).

Eliminate one or more of these conditions, and fire or explosion cannot occur. This is the basis for protection in hazardous (classified) locations. This chapter will acquaint the reader with the various types of electrical equipment, and electrical design, used in hazardous locations and *NEC®* classified locations in which a fire or explosion hazard may exist. The requirements in the *NEC®* are intended to confine sparking or arcing and to prevent passage of hazardous vapors, dusts, or fibers to non-hazardous locations.

It is mistakenly assumed that explosionproof equipment is gas-tight; however, there is no way equipment can be made gas-tight. Any time equipment is opened for any reason, gas enters and becomes trapped inside and is ready to explode when the equipment is used again. The requirement is that enclosures be strong enough to contain an explosion and to prevent the escape of hot gases, flame, or heat that could ignite the surrounding atmosphere.

It is assumed that flammable gas or vapors will enter the enclosure and be ignited by a spark within the enclosure. The threaded joints provided by RMC are not gas-tight and gas can enter, but it is through these same threads that the hot gases, caused by an explosion in the enclosure, leave the enclosure and are sufficiently cooled by passage through five full threads before entering the atmosphere in the hazardous area.

Explosionproof equipment uses several methods of permitting the hot gases to escape, but provides by design a method to cool these gases before they enter the surrounding atmosphere. **Flat ground joints** between two tightly bolted together machined surfaces and **labyrinth joints** are methods used other than threaded joints (Figure 13-2).

CLASS I LOCATIONS

Class I locations are areas where flammable gases and vapors are, or may be, present in the air in quantities sufficient to produce explosive or ignitable mixtures. All flammable liquids have what is called a **flash point**. Substances with low flash points are more susceptible to spark ignition sources. The requirements for installations in Class I locations are contained in *Article 501*.

Class I, Division 1

Class I, Division 1 is an area where the hazard exists under normal operating conditions. These situations occur when transferring flammable or combustible liquids from one container to another, such as gasoline dispensing pumps, open vats, paint

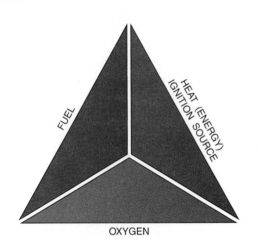

IN ORDER FOR A FIRE OR EXPLOSION TO OCCUR THREE CONDITIONS MUST EXIST.

(1) THERE MUST BE A FUEL (THE FLAMMABLE GAS OR VAPOR, OR COMBUSTIBLE DUST) IN IGNITIBLE QUANTITIES;

(2) THERE MUST BE AN IGNITION SOURCE (ENERGY IN THE FORM OF HEAT OR A SPARK) OF SUFFICIENT ENERGY TO CAUSE IGNITION; AND

(3) THERE MUST BE OXYGEN, USUALLY THE OXYGEN IN THE AIR.

THESE THREE CONDITIONS ARE CALLED THE FIRE TRIANGLE AS SHOWN ABOVE. REMOVE ANY ONE OR MORE OF THESE THREE AND A FIRE OR EXPLOSION CANNOT OCCUR. THIS IS THE BASIS OF VARIOUS PROTECTION SYSTEMS FOR ELECTRICAL EQUIPMENT PERMITTED IN THE *NEC®* FOR USE IN HAZARDOUS LOCATIONS.

Figure 13-1   The fire triangle

CHAPTER 13 Hazardous (Classified) Locations   235

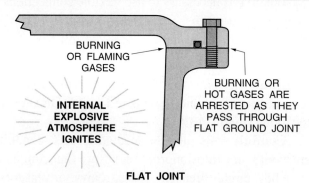

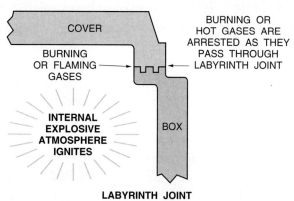

Figure 13-2 Flat joint and labyrinth joint

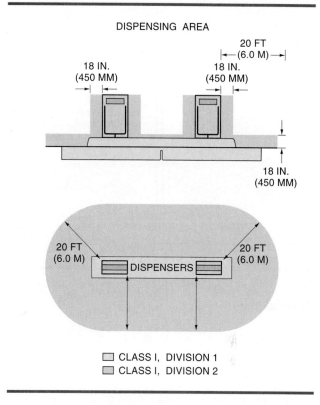

Figure 13-3 Gasoline dispenser classified locations

Figure 13-4 Class I, Division 1 Luminaires (lighting fixtures). *Courtesy of* Killark Manufacturing Company

spray booths, or any other location where ignitable mixtures are used (Figure 13-3).

Wiring Methods. *NEC® 501.4(A)* contains the requirements for wiring methods in Class I, Division 1 locations. Either threaded RMC, threaded steel IMC, or Type MI cable must be the wiring method used. All luminaires (fixtures), boxes, fittings, and joints must be threaded for connection to conduit or cable terminations (Figure 13-4).

RNC may be used when encased in a concrete envelope a minimum of 2 in. (50 mm) thick and buried under at least 2 ft (600 mm) of earth. Where RNC is used, threaded RMC or threaded steel IMC must be used for the last 2 ft (600 mm) of the underground run. An equipment grounding conductor must be included to provide for the electrical continuity of the raceway system and for the grounding of noncurrent-carrying metal parts.

236 CHAPTER 13 Hazardous (Classified) Locations

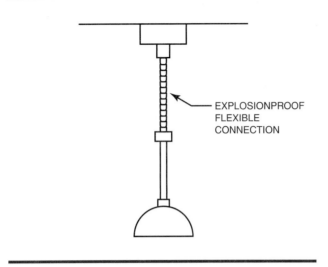

Figure 13-5 Flexible connection at luminaire (lighting fixture)

Where it is necessary to use flexible connections, as at motor terminals, listed flexible fittings or flexible cord of the extra-hard usage-type must be used (Figure 13-5).

Conduit Seals. *NEC® 501.5(A)* contains the requirements for conduit seals in Class I, Division 1 locations (Figure 13-6).

Conduit seals must be located in each conduit entry into an explosionproof enclosure that contains switches, circuit breakers, fuses, relays, or resistors that may produce arcs or sparks or if the entry is 2 in. (50 mm) or larger and the enclosure contains terminals, splices, or taps (Figure 13-7).

In each conduit run leaving a Class I, Division 1 location, a sealing fitting must be located on either side of the boundary of the location within 10 ft (3 m) of the boundary. There must be no union, coupling, box, or fitting between the conduit seal and the point at which the conduit leaves the Division 1 location (Figure 13-8).

Metal conduit containing no unions, couplings, boxes, or fittings that passes completely through a Class I, Division 1 location with no fittings less than 12 in. (300 mm) beyond each boundary shall not require a sealing fitting if the termination points of the unbroken conduit are in unclassified locations (Figure 13-9).

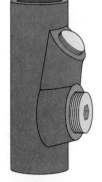

Figure 13-6 Sealing fittings

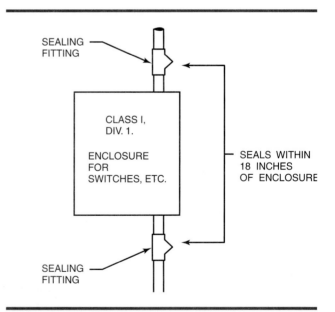

Figure 13-7 Conduit seals for enclosure containing switches

CHAPTER 13 Hazardous (Classified) Locations

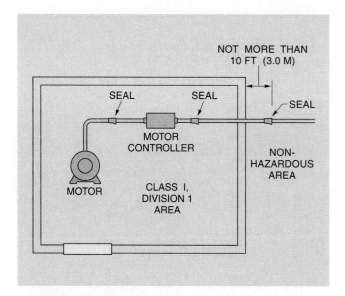

Figure 13-8 Sealing fitting where conduit leaves Class I, Division 1 area

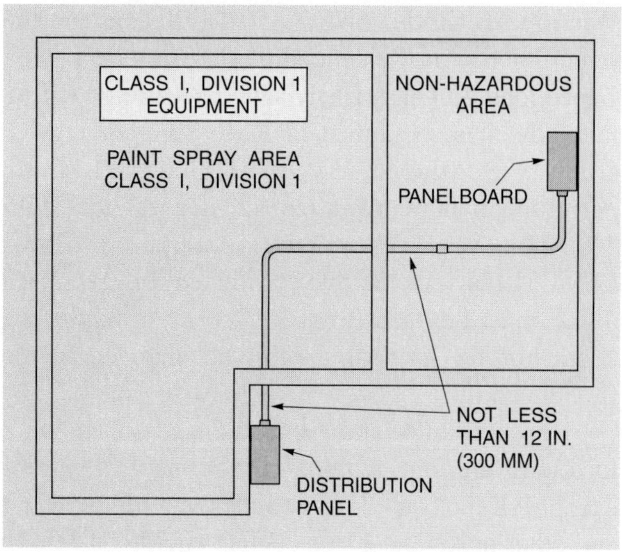

Figure 13-9 Unbroken run of conduit within Class I, Division 1 area and 12 in. (300 mm) beyond

Class I, Division 2

Class I, Division 2 is an area where ignitable gases or vapors are handled, processed, or used but are normally in closed containers from which they can only escape through accidental rupture of the containers. Class I, Division 2 also includes areas that are adjacent to Class I, Division 1 where occasionally gases or vapors might be communicated unless positive ventilation is provided. Class I, Division 2 areas include locations where gasoline or other flammable liquids are contained in storage tanks or other closed containers (Figure 13-10).

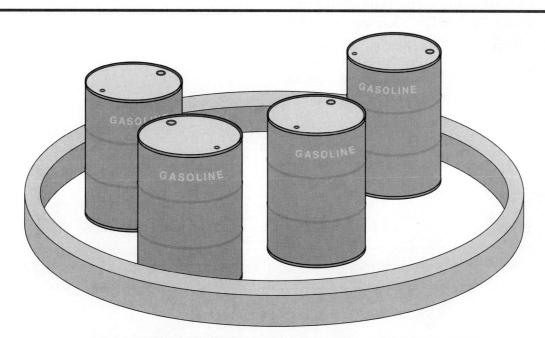

Figure 13-10 Class I, Division 2 gasoline storage with concrete retaining wall

Wiring Methods. *NEC® 501.4(B)* contains the requirements for wiring methods in Class I, Division 2 locations. In Class I, Division 2 locations, wiring methods must be threaded RMC, threaded IMC. Cable types MI, MC MV, TC, ITC, and PLTC are permitted where installed in accordance with *501.4(B)*. Where flexibility is required, FMC, LFMC, LFNC, or flexible cord listed for extra-hard usage must be used (Figure 13-11). An additional conductor for grounding must be included in the flexible cord.

Boxes, fittings, and joints in Class I, Division 2 locations are not required to be explosionproof except for enclosures for switches, circuit breakers, and other make-and-break contact equipment.

Conduit Seals. *NEC® 501.5(B)* shows the requirements for conduit seals in Class 1, Division 2 locations. A conduit seal is required for each enclosure that is required to be explosionproof (Figure 13-12).

A conduit seal is required in each conduit run passing from a Class I, Division 2 location into an unclassified location. The sealing fitting can be located on either side of the boundary within 10 ft (3 m) of the boundary. Threaded rigid metal or threaded steel intermediate metal must be used between the sealing fitting and the point at which the conduit leaves the Division 2 location, and a threaded connection shall be used at the sealing fitting. There must be no unions, couplings, or boxes between the conduit seal and the point at which the conduit leaves the Division 2 location (Figure 13-13).

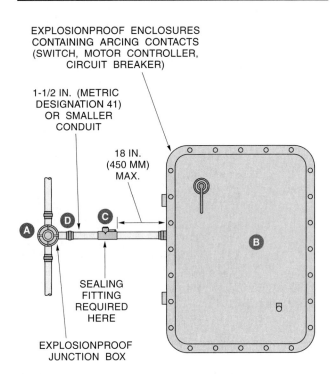

(A) GE SERIES EXPLOSIONPROOF FITTING
(B) B702 SERIES PRISM EXPLOSIONPROOF COMBINATION MAGNETIC MOTOR STARTER
(C) EYS OR ENY SERIES EXPLOSIONPROOF CONDUIT SEALING FITTING
(D) GUF OR GUM SERIES EXPLOSIONPROOF UNIONS

Figure 13-12 Conduit seal for explosionproof enclosure

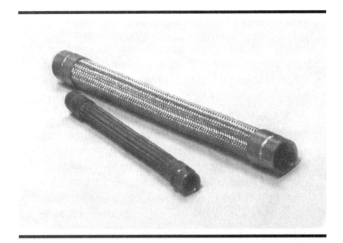

Figure 13-11 Flexible connection. *Courtesy of* Killark Manufacturing Company

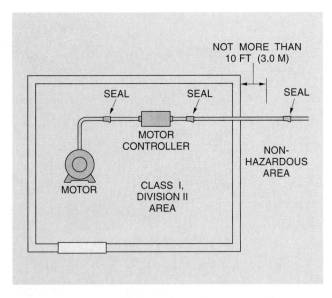

Figure 13-13 Sealing fitting where conduit leaves Class I, Division 2 area

CLASS II LOCATIONS

Class II locations are an area where presence of combustible dust presents a fire or explosion hazard.

Class II, Division 1

Class II, Division 1 is an area where combustible dust is normally in the air in sufficient quantities to produce ignitable mixtures. These areas include grain silos and grain-processing plants (Figure 13-14).

Wiring Methods. *NEC® 502.4(A)* contains the requirements for wiring methods in Class II, Division 1 locations. In Class II, Division 1 locations, threaded RMC, threaded steel IMC or Type MI cable must be the wiring method used.

Fittings and boxes must be provided with threaded bosses for connection to conduit or cable terminations and must be **dusttight** (Figure 13-15).

Where necessary to employ flexible connections, dusttight flexible connectors, LFMC, LFNC, or flexible cord listed for extra-hard usage must be used.

Sealing, Class II, Divisions 1 and 2. Suitable means must be provided to prevent the entrance of dust into a dust ignition-proof enclosure through the raceway. These sealing methods include a permanent and effective seal, a horizontal raceway not less than 10 ft (3 m) long, and a vertical raceway not less than 5 ft (1.5 m) long and extending downward from the dust ignition-proof enclosure. Sealing fittings must be accessible, and seals are not required to be explosionproof. Electrical sealing putty is a method of sealing.

Class II, Division 2

Class II, Division 2 is an area where combustible dust-producing materials are stored or handled only in bags or containers (Figure 13-16).

Wiring Methods. In Class II, Division 2 locations, RMC, IMC, EMT, dusttight wireways, Type MC or MI cable, Type PLTC or ITC in cable trays or Type MC, MI, or TC cable in approved cable trays with spacing as shown in *502.4(B)* may be used. All boxes and fittings must be dusttight (Figure 13-17).

Figure 13-14 Class II, Division 1 location—grain elevator. *Courtesy of* Killark Manufacturing Company

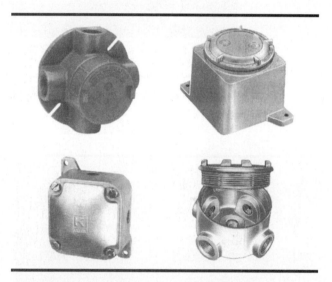

Figure 13-15 Dusttight enclosures Class II, Division I. *Courtesy of* Killark Manufacturing Company

CLASS III LOCATIONS

Class III locations are those that are hazardous because of the presence of easily ignitable fibers or flyings, but in which such materials are not likely to be suspended in air in sufficient quantities to produce ignitable mixtures.

Class III, Division 1

Class III, Division 1 locations are those where easily ignitable fibers or materials, producing combustible flyings, are handled or used. These

CLASS II, DIVISION 2 LOCATION—GRAIN STORAGE COMBUSTIBLE DUST, SEED, OR GRAIN IS NOT NORMALLY IN THE AIR IN QUANTITIES SUFFICIENT TO PRODUCE EXPLOSIVE OR IGNITABLE MIXTURES

Figure 13-16   Class II, Division 2 location

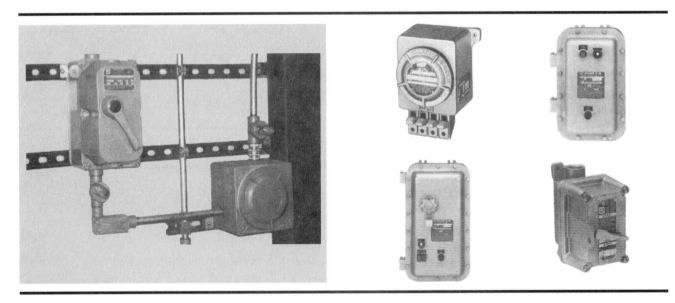

Figure 13-17   Class II, Division 2 dusttight enclosures. *Courtesy of* Killark Manufacturing Company

locations include rayon, cotton, and other textile mills (Figure 13-18).

Wiring Methods. In Class III, Division 1 locations, the wiring method is RMC, IMC, RNC, EMT, dusttight wireways, or Type MC or MI cable. All boxes and fittings shall be dust-tight.

Where necessary to employ flexible connections, LFMC, LFNC or flexible cord of the hard-usage type must be used (Figure 13-19).

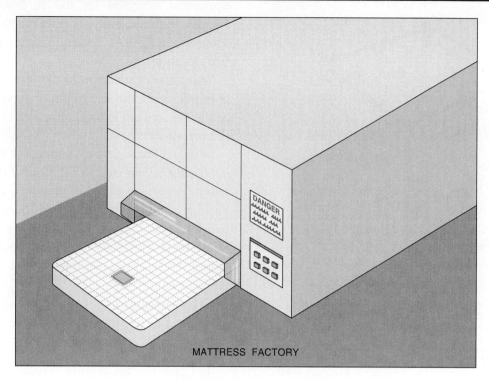

CLASS III, DIVISION 1 LOCATION — WHERE EASILY IGNITABLE FIBERS OR MATERIALS PRODUCING COMBUSTIBLE FLYINGS ARE HANDLED, MANUFACTURED OR USED.

Figure 13-18   Class III, Division 1 location

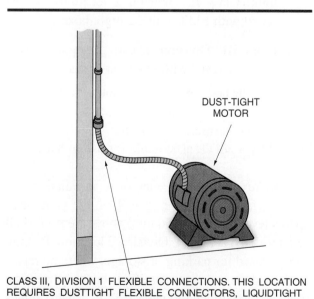

CLASS III, DIVISION 1 FLEXIBLE CONNECTIONS. THIS LOCATION REQUIRES DUSTTIGHT FLEXIBLE CONNECTORS, LIQUIDTIGHT FLEXIBLE METAL CONDUIT WITH LISTED FITTINGS, LIQUID-TIGHT FLEXIBLE NON-METALLIC CONDUIT WITH LISTED FITTINGS, OR FLEXIBLE CORD IN CONFORMANCE WITH *503.10*.

Figure 13-19   Class III, Division 1 flexible connections

Class III, Division 2

Class III, Division 2 locations are those where easily ignitable fibers are stored or handled (Figure 13-20).

Wiring Methods. The wiring methods required for Class III, Division 2 are identical to those required for Division 1 as shown in *503.3(A)*. Boxes and fittings must be dusttight. Flexible connections must be like those shown in *503.3(A)*.

INTRINSICALLY SAFE SYSTEMS

The installation requirements for **intrinsically safe systems** are in *Article 504*. These low-energy systems are designed to assure safety by eliminating the ignition source of the fire triangle. The energy in the system is maintained below that necessary to ignite flammable atmospheres.

The low energy in the system is based on the intrinsically safe power-limited supplies. Intrinsically

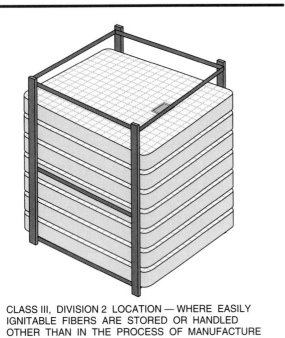

CLASS III, DIVISION 2 LOCATION — WHERE EASILY IGNITABLE FIBERS ARE STORED OR HANDLED OTHER THAN IN THE PROCESS OF MANUFACTURE

Figure 13-20 Class III, Division 2 storage area

safe wiring is already protected through the intrinsically safe system, and ordinary location wiring methods are permitted. Seals must be provided to prevent the transmission of flammable gases, vapors, and combustible dusts through conduits or other raceways from Division 1 to Division 2 locations, or from hazardous locations to unclassified locations.

ZONE 0, 1 AND 2 LOCATIONS

This method of area classification follows the International Electrotechnical Commission (IEC) method of area classification. The requirements are shown in *Article 505*. *Article 505* is applicable only to Class I locations. The divisions are called zones.

SUMMARY

The installation of electrical wiring and equipment in hazardous classified locations is a very complex area of electrical work. This chapter can only acquaint the student with the basic requirements of the *NEC®* on this subject. Further information can be acquired by the student from manufacturers of equipment designed for use in hazardous areas, together with a careful study of all the information contained in the *NEC®* relating to hazardous (classified) locations.

All Class I areas require conduit seals to prevent or minimize the amount of gas or vapors from being communicated to the conduit system beyond the seal (Figure 13-21).

Class II areas require sealing, but the seals need not be explosionproof. *NEC® 502.5, FPN* recognizes sealing putty as a means of sealing in Class II areas. Sealing is not required in Class III areas. All classified locations require that luminaires (lighting fixtures) be designed to minimize the possibility of falling sparks or hot metal from the luminaires (fixtures) or lamps (Figure 13-22).

The wiring methods for each class of hazardous locations can be found in their respective articles, but the general rules are as follows:

1. Class I areas require threaded rigid conduit with explosionproof boxes, with all threaded joints made up with at least five full threads fully engaged. Explosionproof boxes are not generally required in Class I, Division 2 areas (Figure 13-23).

2. Class II, Division 1 locations require threaded RMC with boxes that are approved as dusttight. Class II, Division 2 locations can be wired with EMT with dusttight boxes.

3. Class III, Division 1 and 2 require RMC, RNC, or EMT with dusttight boxes.

While inspecting any hazardous location. it is imperative that all equipment be examined thoroughly to determine that it has been listed and labeled by a NRTL as suitable for use in hazardous locations.

Article 511 covers electrical installations in commercial garages. For each floor, the entire area up to a level of 18 in. (450 mm) above the floor shall be considered a Class I, Division 2 location. Parking garages used for parking or storage, where no repair work is done, are not classified, but they must be adequately ventilated to carry off the exhaust fumes of the engines.

Article 514 covers electrical installations in motor fuel dispensing facilities. *Table 514.3(B)(1)* outlines the extent of the Class I area in these

CHAPTER 13 Hazardous (Classified) Locations   243

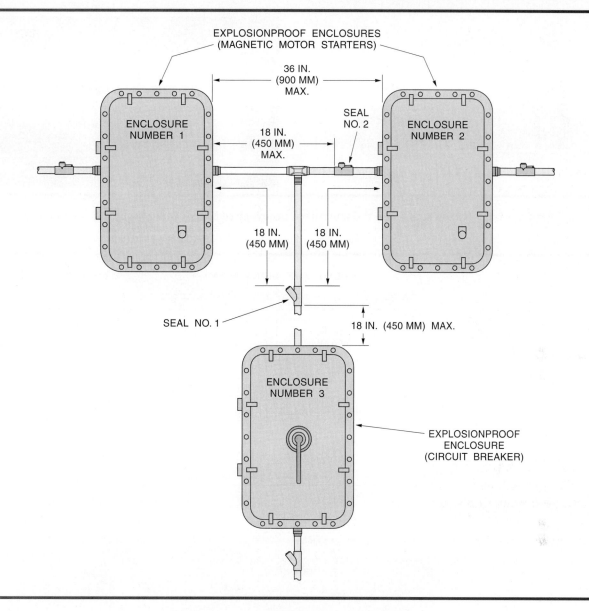

Figure 13-21   Class I, area conduit seals

Figure 13-22   Luminaires (lighting fixtures) for classified locations. *Courtesy of* Killark Manufacturing Company

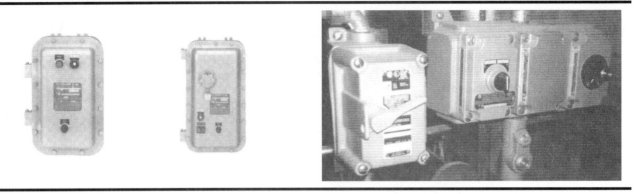

Figure 13-23 Class I, Division 2 enclosures. *Courtesy of* Killark Manufacturing Company

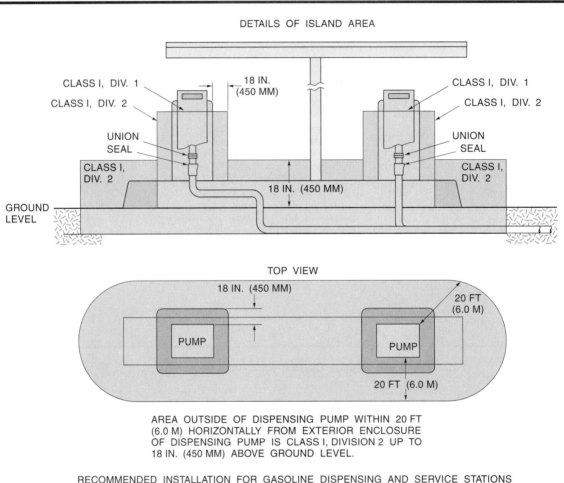

Figure 13-24 Hazardous areas near gasoline dispensing pumps

locations. Generally, it is 18 in. (450 mm) above grade level, extending 20 ft (6.0 mm) horizontally around the dispensing units (Figure 13-24).

Article 516 covers *Spray Application, Dipping, and Coating Processes.* For paint spraying booths with open tops, the area 10 ft (3.0 m) vertically and 20 ft (6.0 m) horizontally adjacent to the booth are considered Class I, Division 2 areas (Figure 13-25). The area within the booth is considered Class I, Division 1. *Article 516* contains diagrams showing all classes according to the nature of the hazardous areas (Figure 13-26).

CHAPTER 13 Hazardous (Classified) Locations 245

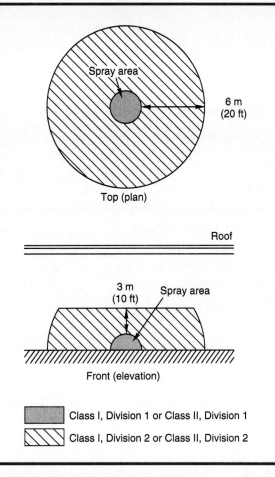

Figure 13-25 *NEC® 516.3(B)(1)*; Electrical area classification for open spray areas (Reprinted with permission from NFPA 70-2002)

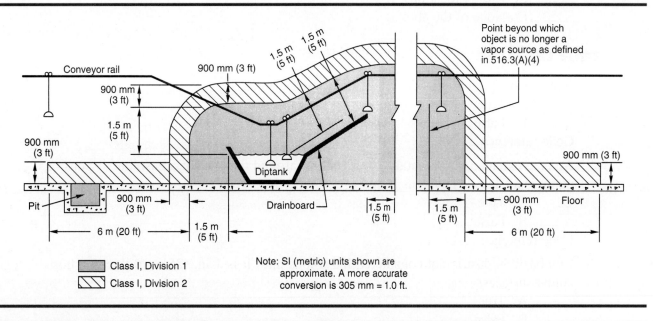

Figure 13-26 *NEC® 516.3(B)(5)*; Electrical area classification for open processes without vapor containment or ventilation (Reprinted with permission from NFPA 70-2002)

REVIEW QUESTIONS

MULTIPLE CHOICE

1. In order for a fire or explosion to occur there must be:
   - ____ A. fuel.
   - ____ B. ignition.
   - ____ C. oxygen.
   - ____ D. all of the above.

2. Threaded conduit in hazardous locations must be made up wrenchtight to:
   - ____ A. prevent gas or vapor leakage.
   - ____ B. prevent arcing or sparking.
   - ____ C. to prevent entry of water.
   - ____ D. to protect the threads.

3. A location where cotton fibers are stored is:
   - ____ A. Class I, Division 1.
   - ____ B. Class III, Division 1.
   - ____ C. Class II, Division 2.
   - ____ D. Class III, Division 2.

4. The Class I, Division 2 area for a gasoline dispenser extends horizontally from the base:
   - ____ A. 20 ft (6.0 m).
   - ____ B. 10 ft (3.0 m).
   - ____ C. 15 ft (4.5 m).
   - ____ D. 18 in. (450 mm).

5. RNC is permitted in:
   - ____ A. Class III locations.
   - ____ B. Class II locations.
   - ____ C. Class I locations.
   - ____ D. None of the above.

TRUE OR FALSE

1. Interiors of spray booths are Class I, Division 2 areas.
   - ____ A. True
   - ____ B. False

   Code reference _____.

2. An area extending horizontally 20 ft (6.0 m) around the base of a gasoline dispensing pump is a Class 1, Division 1 area.
   - ____ A. True
   - ____ B. False

   Code reference _____.

3. Combustible dust is not considered hazardous until it is 1 in. (25 mm) thick on horizontal surfaces.
   - ____ A. True
   - ____ B. False

   Code reference _____.

4. Sealing putty can be used in place of a sealing fitting in Class II locations.
   ___ A. True
   ___ B. False
   Code reference _____ .

5. Explosionproof enclosures must be used in Class I, Division 2 areas.
   ___ A. True
   ___ B. False
   Code reference _____ .

FILL IN THE BLANKS

1. It is assumed that _____ will enter the enclosure and be ignited by a spark within the enclosure.

2. Flat ground joints between two tightly bolted together _____ surfaces and _____ joints are methods used other than threaded joints.

3. Threaded joints in Class I, Division 1 locations must have at least _____ threads fully engaged.

4. The installation requirements for _____ safe systems are found in *Article 504*.

5. All Class I areas require _____ to prevent or minimize the amount of gas or vapors from being communicated to the _____ beyond the seal.

FREQUENTLY ASKED QUESTIONS:

Question 1: Why is it necessary to have five threads fully engaged in all threaded connections?
Answer: In the event of an internal explosion in an enclosure, the hot gases must be contained until they are sufficiently cooled so as not to ignite gases or vapors around the enclosure. By forcing the gases to travel at least five threads before escaping to the surrounding areas allows the gases to cool.

Question 2: What is the purpose of sealing conduit runs in hazardous areas?
Answer: Seals are required in each conduit run entering or leaving an enclosure that contains equipment, such as switches, circuit breakers, or fuses that may produces sparks or arcs. Seals are provided in conduit runs to minimize the passage of gases or vapors, and to prevent passage of flames from one portion of the electrical installation to another.

Question 3: Are sealing fittings required in all hazardous areas?
Answer: Explosionproof sealing is required in Class I areas. Sealing is required in Class II areas, but it need not be explosionproof. Sealing putty may be used for sealing in Class II areas. No sealing is required in Class III areas.

Question 4: Is RMC permitted in hazardous areas?
Answer: RMC is permitted in Class I, Division 1 where encased in 2 in. (50 mm) of concrete with not less than 24 in. (600 mm) of cover. RMC is also permitted in Class III locations.

Question 5: What are *groups* when discussing hazardous locations?

Answer: For the purpose of testing, approval, and area classification, various air mixtures shall be grouped in accordance with *500.6(A)* and *500.6(B)*. Groups A-B-C-D are Class I groups. Groups E-F-G are Class II groups. There are no groups established for Class III. Identification of the materials in each type group may be found in NFPA 497.

Question 6: What would constitute flyings or fibers as shown in Class III locations?

Answer: Easily ignitable particles larger than dust, made up of rayon, cotton or cotton waste, sisal or henequen, jute, hemp, cocoa fiber, oakum, baled waste kapok, Spanish moss, excelsior, and other materials of a similar nature. These materials are likely to be found in textile mills, combustible fiber manufacturing and processing plants, cotton gins, cotton-seed mills, flax-processing plants, clothing manufacturing plants and establishments, and industries involving similar hazardous processes or conditions.

Question 7: Where a conduit leaves a Class I, Division 1 location, should the sealing fitting be installed within the hazardous location or outside of the hazardous location?

Answer: The sealing fitting may installed on either side of the boundary of such location within 10 ft (3 m) of the boundary. *NEC® 501.5(A)(4)* contains these requirements and further states that no union, coupling, box, or fitting can be installed between the conduit seal and the point at which the conduit leaves the Division 1 location.

Question 8: Can I run a RMC through a Class I, Division 1 location without sealing the conduit where it leaves the location?

Answer: Yes. If the conduit does not contain any unions, couplings, boxes, or fittings and terminates in an unclassified location on both ends, that is not less than 12 in. (300 mm) beyond each boundary, a conduit seal is not required.

CHAPTER 14

Swimming Pools

OBJECTIVES

After completing this chapter, the student should understand:
- requirements for swimming pools, spas, hot tubs, and hydromassage tubs.
- the difference between a spa and hot tub.
- wiring methods for swimming pools, spas/hot tubs, and hydromassage tubs.
- types of swimming pool luminaires (lighting fixtures).
- pool and equipment bonding.
- electric shock hazard associated with pools and tubs.

KEY TERMS

Bonding: The permanent joining of metallic parts to form an electrically conductive path that will ensure electrical continuity, and the capacity to conduct safely any current likely to be imposed

Bonding Grid: The structure-reinforcing steel of a concrete pool where the reinforcing rods are bonded together by the usual steel ties

Dry-Niche Luminaire (Fixture): A luminaire (lighting fixture) intended for installation in the wall of a pool or fountain in a niche that is sealed against the entry of pool water

Forming Shell: A structure designed to support a wet-niche luminaire (lighting fixture) assembly, and intended for mounting in a pool or fountain structure

Ground-Fault Circuit-Interrupter (GFCI): A device intended for the protection of personnel that functions to de-energize a circuit or portion within an established period of time when a current to ground exceeds some predetermined value that is less than that required to operate the overcurrent protective device of the supply current

Hydromassage Bathtub: A permanently installed bathtub equipped with a recirculating piping system, pump, and associated equipment. It is designed to accept, circulate, and discharge water after each use

No-Niche Luminaire (Fixture): A luminaire (lighting fixture) intended to be installed below the water without a niche

Spa or Hot Tub: A hydromassage pool or tub for recreational or therapeutic use, not located in health care facilities, designed for the immersion of users and usually having a filter, heater, and motor-driven blower. It can be installed indoors or outdoors, in-ground or aboveground. Generally, a spa or hot tub is not designed or intended to have its contents drained or discharged after each use

Stranded Conductors: A group of wires, usually twisted or braided, forming a conductor

Transformer (Isolation): A transformer with windings that are isolated from each other and that has a grounded metal barrier between the primary and secondary windings

Wet-Niche Luminaire (Fixture): A luminaire (lighting fixture) intended for installation in a forming shell and mounted in a pool or fountain structure where the luminaire (fixture) will be completely surrounded by water

INTRODUCTION

The provisions of *Article 680, Swimming Pools, Fountains, and Similar Installation*, apply to the construction and installation of electric wiring for equipment serving all decorative, swimming, therapeutic, and wading pools, fountains, hot tubs, spas and hydromassage bathtubs, filters, and similar equipment.

Except as modified by *Article 680, Pools and Fountains*, installations must comply with applicable requirements of *NEC*® Chapters 1 through 4.

GENERAL

Swimming pools, fountains, and similar installations must be done with great care. The potential for electric shock is extremely great if the requirements of *Article 680* are not closely followed. The installer and the inspector should be certain that only listed equipment is used in the installation and the grounding procedures are strictly adhered to. The main purpose is to keep energized conductors and equipment that are not properly protected by grounding, bonding, or ground-fault protective methods out of reach from persons using the pool or spa and to ensure that energized conductors or equipment will not contact the water in the pool or spa.

Potential Hazards

The electric shock associated with someone immersed in a pool is quite different from the common *touch* electric shock. The effect of an electric current through the body is determined by how much current is flowing and for how long a time. The magnitude of current is controlled by the body's resistance, which is, among other parameters, dependent on the contact with the electrical potential. The greater the contact area, the lower the resistance, creating the potential for larger current levels. The difference between *touch* shock and *immersion* shock is analogous to the difference between series and parallel circuits. The *touch* shock could be considered a simple path, and the *immersion* shock could be considered multiple paths, resulting in a larger current flow and increased danger from the effects of the electric shock. The effect of electric shock is shown in Figure 14-1.

EFFECT OF ELECTRIC SHOCK		
	Current in milliamperes @ 60 hertz	
	Men	Women
• cannot be felt	0.4	0.3
• a little tingling—mild sensation	1.1	0.7
• shock—not painful—can still let go	1.8	1.2
• shock—painful—can still let go	9.0	6.0
• shock—painful—just about to point where you can't let go—called "threshold"—you may be thrown clear	16.0	10.5
• shock—painful—severe—can't let go—muscles immobilize—breathing stops	23.0	15.0
• ventricular fibrillation (usually fatal) length of time . . . 0.03 sec. length of time . . . 3.0 sec.	1000 100	1000 100
• heart stops for the time current is flowing—heart may start again if time of current flow is short	4 amperes	4 amperes
• burning of skin—generally not fatal unless heart or other vital organs burned	5 or more amperes	5 or more amperes
• cardiac arrest, severe burns, and probable death	10,000	10,000

Figure 14-1 Effect of electrical shock

Grounding

NEC® 680.22 requires **GFCI** protection through the use of ground-fault receptacles or with the use of a ground-fault circuit breaker. In addition, to reduce the hazard of an unprotected electrical source, conductors on the load-side of a GFCI are not to occupy raceways, boxes, or enclosures containing other conductors unless the other conductors are GFCI protected.

Cord and Plug Equipment

Fixed or stationary equipment, other than an underwater luminaire (lighting fixture), for a permanently installed pool is permitted by *680.7* to be connected with a flexible cord. The flexible cord for other than storable pools must not exceed 3 ft (900 mm) in length. The flexible cord must have an equipment-grounding conductor sized in accordance with *Table 250.122*, but not smaller than 12 AWG,

Table 680.8 Overhead Conductor Clearances

	Clearance Parameters	Insulated Cables, 0–750 Volts to Ground, Supported on and Cabled Together with an Effectively Grounded Bare Messenger or Effectively Grounded Neutral Conductor		All Other Conductors Voltage to Ground			
				0 through 15 kV		Over 15 through 50 kV	
		m	ft	m	ft	m	ft
A.	Clearance in any direction to the water level, edge of water surface, base of diving platform, or permanently anchored raft	6.9	22.5	7.5	25	8.0	27
B.	Clearance in any direction to the observation stand, tower, or diving platform	4.4	14.5	5.2	17	5.5	18
C.	Horizontal limit of clearance measured from inside wall of the pool	This limit shall extend to the outer edge of the structures listed in A and B of this table but not less than 3 m (10 ft).					

Figure 14-2 *NEC® Table 680.8* (Reprinted with permission from NFPA 70-2002)

and the cord must terminate in a grounding-type attachment plug.

Overhead Conductor Clearances

The requirements for the clearance of overhead conductors in a pool area are shown in *Table 680.8* (Figure 14-2). In general, no parts of the pool should be placed under existing service-drop conductors or other overhead wiring unless the installation of overhead conductors meet certain clearances above the pool and associated equipment (Figure 14-3). Coaxial cables used for community antenna systems and conforming to the *NEC®* are permitted above swimming pools and associated structures provided a clearance above such equipment is at least 10 ft (3.0 m).

Underground Wiring Location

To reduce electrical hazards, *680.10* requires underground wiring not be installed under the pool or within an area extending 5 ft (1.5 m) horizontally from the inside wall of the pool unless this wiring is

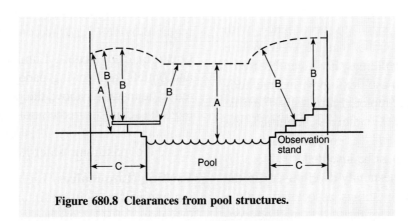

Figure 680.8 Clearances from pool structures.

Figure 14-3 *NEC® Figure 680.8, Clearances from Pool Structures* (Reprinted with permission from NFPA 70-2002)

252 CHAPTER 14 Swimming Pools

Table 680.10 Minimum Burial Depths

Wiring Method	Minimum Burial mm	Minimum Burial in.
Rigid metal conduit	150	6
Intermediate metal conduit	150	6
Nonmetallic raceways listed for direct burial without concrete encasement	450	18
Other approved raceways*	450	18

*Raceways approved for burial only where concrete encased shall require a concrete envelope not less than 50 mm (2 in.) thick.

Figure 14-4 *NEC® Table 680.10, Minimum Burial Depths* (Reprinted with permission from NFPA 70-2002)

necessary to supply pool equipment. Where space limitations prevent wiring from being installed a distance of 5 ft (1.5 m) or more from the pool, wiring will be permitted where installed in RMC, IMC, or a nonmetallic raceway system. The minimum burial depth must be as shown in *Table 680.10* (Figure 14-4).

PERMANENTLY INSTALLED POOLS

Motors

The requirements for motor installations associated with permanently installed pools can be found in *680.21*. The branch-circuits must be installed in RMC, IMC, RNC, or Type MC cable. All wiring methods used must contain a copper equipment-grounding conductor, sized in accordance with *Table 250.122*, but not smaller than 12 AWG. Where installed on or within buildings, EMT can be used.

Where necessary to use flexible wiring methods for connections at the motor, Liquidtight Flexible Metal or Nonmetallic Conduit is permitted. As required in *680.21(A)(1)*, a copper equipment-grounding conductor must be installed.

In the interior of one-family dwellings or in the interior of accessory buildings associated with a one-family dwelling, any of the wiring methods recognized in *NEC® Chapter 3* may be used. As required in *680.21(A)(1)*, a copper equipment-grounding conductor must be run with the circuit-conductors.

Where run in a raceway, the equipment-grounding conductor must be insulated. Where run in a cable assembly, the equipment-grounding conductor may be uninsulated, but must be enclosed within the outer sheath of the cable assembly.

Area Receptacles and Equipment

The placement of receptacles near the pool where a person immersed in the water has access to their use is considered hazardous. *NEC® 680.22* requires that receptacles providing power for water-pump motors or other loads directly associated with the circulation and sanitation system be located at least 10 ft (3.0 m) from the inside walls of the pool or not less than 5 ft (1.5 m) from the inside walls of the pool if they are grounding-type single receptacles with a locking configuration and have GFCI protection.

If a permanently installed pool is installed at a dwelling unit, at least one receptacle must be installed not less than 5 ft (1.5 m) and not more than 20 ft (6 m) from the inside wall of the pool. The receptacle must not be more than 6 ft, 6 in. (2.0 m) above the floor, or grade-level serving the pool. All 125-volt receptacles installed within 20 ft (6.0 m) of the inside walls of a pool must be protected by a GFCI. Receptacles supplying pool pump motors rated 15 or 20 amperes, 120-volt through 240-volt, single-phase must be provided with GFCI protection.

Luminaires (Lighting Fixtures), Lighting Outlets

To reduce the risk of electrical hazards, luminaires (lighting fixtures) and lighting outlets are required by *680.22(B)* to conform to the following:

Outdoor Pool Area. Luminaires (lighting fixtures and outlets, and ceiling paddle fans are not permitted to be installed above the pool or the area extending 5 ft (1.5 m) from the pool's inside wall unless no luminaire (fixture) or fan is less than 12 ft (3.7 m) above the maximum water level (Figure 14-5).

Existing (Luminaires) Lighting Fixtures. Existing luminaires (lighting fixtures) and outlets located less than 5 ft (1.5 m) horizontally from the inside pool wall and at least 5 ft (1.5 m) above the maximum water level, rigidly attached, and GFCI protected, are permitted as shown in Figure 14-5.

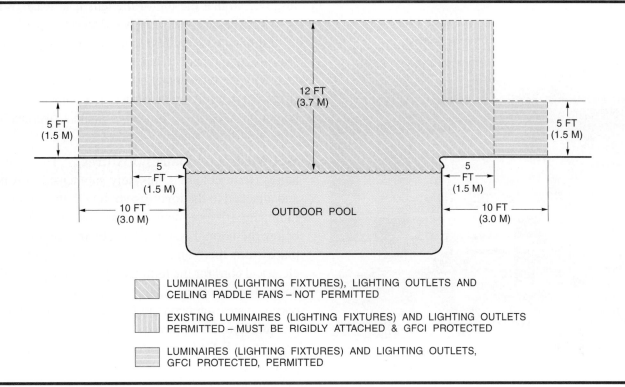

Figure 14-5 Outdoor pool

Indoor Pool Area. The requirements of the indoor pool area are the same as for outdoor pools, except that if the branch-circuit is GFCI protected, totally enclosed luminaires (lighting fixtures) and ceiling suspended paddle fans shall be permitted at a height not less than 7 ft, 6 in. (2.3 m) above the maximum water level.

Luminaires (lighting fixtures) and lighting outlets installed in the pool area, defined as 5 ft (1.5 m) to 10 ft (3.0 m) measured horizontally from the pool's inside wall, shall be protected by a GFCI unless installed not less than 5 ft (1.5 m) above the maximum water level, and rigidly attached to the structure adjacent to or enclosing the pool (Figure 14-6).

Switching Devices. Switching devices within the pool area are required to be located at least 5 ft (1.5 m) from the inside wall of the pool (Figure 14-7).

Transformers. **Transformers** used in electrical wiring installations for swimming pools are required to be isolation transformers with an ungrounded secondary that has a grounded metal barrier between secondary and primary winding. This will reduce the possibility of the higher primary voltage appearing on the secondary windings. An isolation

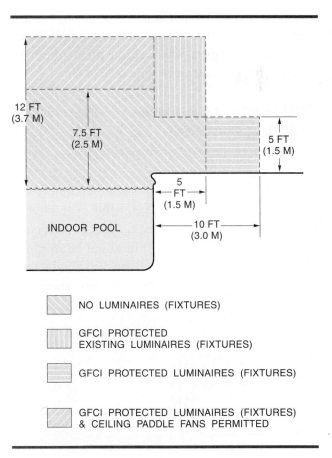

Figure 14-6 Indoor pool

Figure 14-7 Switching devices for pool equipment

transformer is a transformer where the primary and secondary windings are physically and electrically isolated from one another. There is no electrical connection between the primary and secondary winding. The windings are wound on the same core, causing mutual inductance (Figure 14-8).

Underwater Luminaires (Lighting Fixtures)

NEC® 680.23 covers the requirements for underwater luminaires (lighting fixtures). Branch-circuits, which operate at 15 volts or more and supply the luminaires (lighting fixtures) directly or through a transformer, are required to be GFCI protected. Luminaires (lighting fixtures) must not be installed on circuits over 150 volts between conductors. Underwater luminaires (lighting fixtures) should be installed so that the top of the lens is at least 18 in. (450 mm) below the normal water level (Figure 14-9).

Luminaires (fixtures) that are mounted with the lens horizontal need to have the lens adequately protected to prevent contact by any person. Luminaires (fixtures) that rely on the water environment for safe operation need to be protected against overheating when not in the water.

Wet-Niche Luminaires (Fixtures). These luminaires (fixtures) are completely submersed in water and attached to the **forming shell** with the top of the luminaire (fixture) lens at least 18 in. (450 mm) below the normal water level. The re-lamping takes place from the front, so there should be adequate electrical cord for the luminaire (fixture) to reach the deck. The electrical supply conductors are run in corrosion-resistant conduit from the forming shell to the junction box (Figure 14-10). For nonmetallic conduit, an 8 AWG insulated copper must be used to provide continuity between the forming shell and the junction box.

Dry-Niche Luminaires (Fixtures). These luminaires (fixtures) are waterproof, designed to be re-lamped from the rear, and mounted in a dry-niche with the luminaire's (fixture's) lens at least 18 in. (450 mm) from the normal water level. The wiring method is conduit from the luminaire (fixture) to the panelboard. Unlike **wet-niche luminaires (fixtures)**, a junction box for **dry-niche luminaires (fixtures)** is not required. However, if a junction box is used, it is not required to be elevated or located as specified for the wet-niche luminaire (fixture).

No-Niche Luminaires (Fixtures). These luminaires (fixtures) are also waterproof, designed to be re-lamped from the front, and mounted on the inside wall of the pool on a bracket with the top of the luminaire (fixture) lens 18 in. (450 mm) below the normal water level. The wiring method is non-corrosive conduit from the luminaire (fixture) mounting bracket to the above-grade junction box. A junction box that connects conduit from a forming shell or **no-niche luminaire (fixture)** mounting bracket must be labeled and listed for the purpose, made from corrosion-resistant material, have electrical continuity, be located horizontally at least 4 ft (1.2 m) from the inside pool wall, and the vertical dimension of 4 in. (100 mm) above the ground and 8 in. (200 mm) above the maximum pool water level (Figure 14-11).

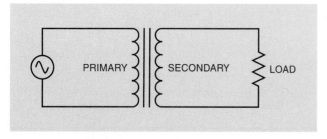

Figure 14-8 Isolation transformer

CHAPTER 14 Swimming Pools 255

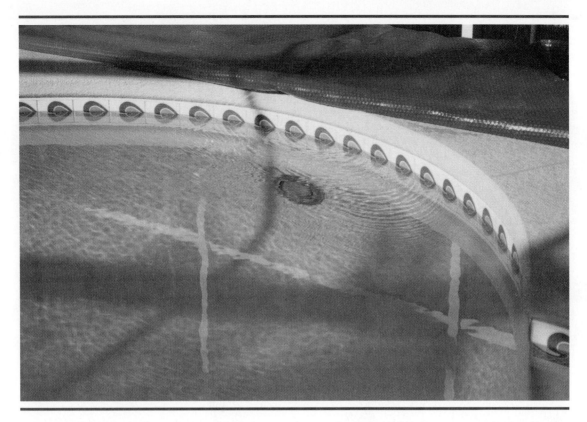

Figure 14-9 Swimming pool underwater luminaire (lighting fixture) on top of lens 18 in. (450 mm) below water level

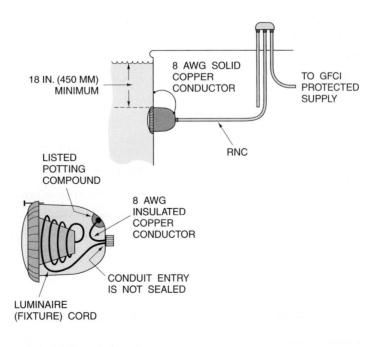

Figure 14-10 Underwater luminaires (lighting fixtures)

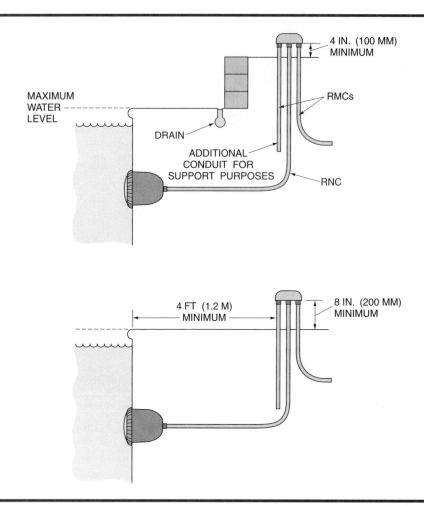

Figure 14-11 Junction boxes for underwater lighting

GROUNDING AND BONDING SWIMMING POOLS

Proper grounding and **bonding** of the metal parts in the pool area reduces the risk of electric shock hazard. This is accomplished by maintaining all the noncurrent metal parts and electrical enclosures at ground or zero potential, and establishes a path to facilitate the operation of the overcurrent protective device.

Grounding

NEC® 680.6 contains the requirements for grounding swimming pools. The following equipment must be grounded:

- Through-wall lighting assemblies and underwater luminaires (lighting fixtures), other than those low-voltage systems listed for application without a grounding conductor
- All electrical equipment located within 5 ft (1.5 m) of the inside wall of the pool or fountain
- All electrical equipment associated with the recirculating system of the pool or fountain
- Junction boxes
- Transformer enclosures
- GFCIs
- Panelboards supplying swimming pool electrical equipment

The grounding of the various metal parts of a permanently installed swimming pool is shown in Figure 14-12.

Bonding

NEC® 680.26 requires that all metal parts of swimming pools be bonded together using 8 AWG or larger solid copper wire insulated, covered, or

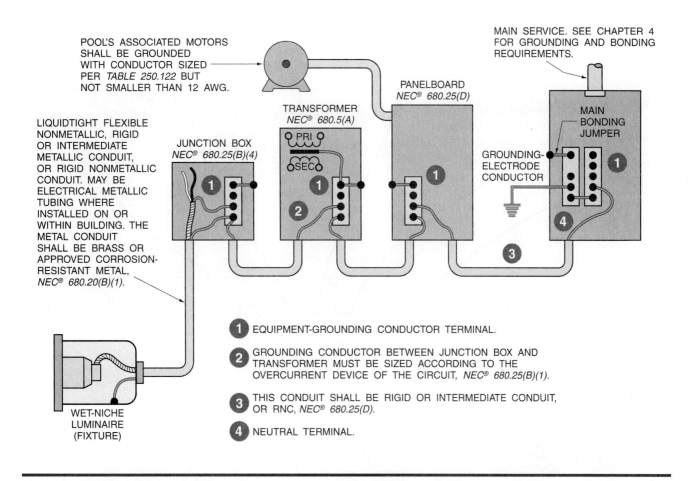

Figure 14-12 Grounding of metal parts of permanently installed pool

bare (Figure 14-13). The use of solid conductors protects against loss of conductor cross-sectional area, whereas **stranded conductors**, which are made by twisting smaller wires together, have the potential for reducing the overall cross-sectional area. Individual strands break because of corrosion or misuse.

The purpose of the swimming pool **bonding grid** is to eliminate voltage gradients in the pool area (Figure 14-14). Voltage gradients or rings range in magnitude from the source voltage (120 volts in Figure 14-14) to zero at the pool wall and floor. Bonding conductors are connected directly to the required metal equipment which needs bonding by means of a non-corrosive clamps.

Electric Deck Area Heating

NEC® 680.27 contains the requirements for unit heaters and permanently connected radiant heaters in the pool area. Unit heaters must not be mounted over the pool or within the area extending 5 ft (1.5 m) horizontally from the inside walls of the pool (Figure 14-15). Radiant heaters must be at least 12 ft (3.7 m) above the deck and must be securely mounted and guarded. Due to the possible deterioration of insulation, radiant heat cables embedded in the deck are not permitted.

Pool Covers

Electrically operated pool covers are a popular means of retaining the pool water heat when the pool is not in use; however, precautions must be exercised to keep electricity at a distance from the pool. The electric motor, controller, and wiring must be located at least 5 ft (1.5 m) from the inside pool wall. The circuit supplying the electric motor and controller must be protected by a GFCI.

258 CHAPTER 14 Swimming Pools

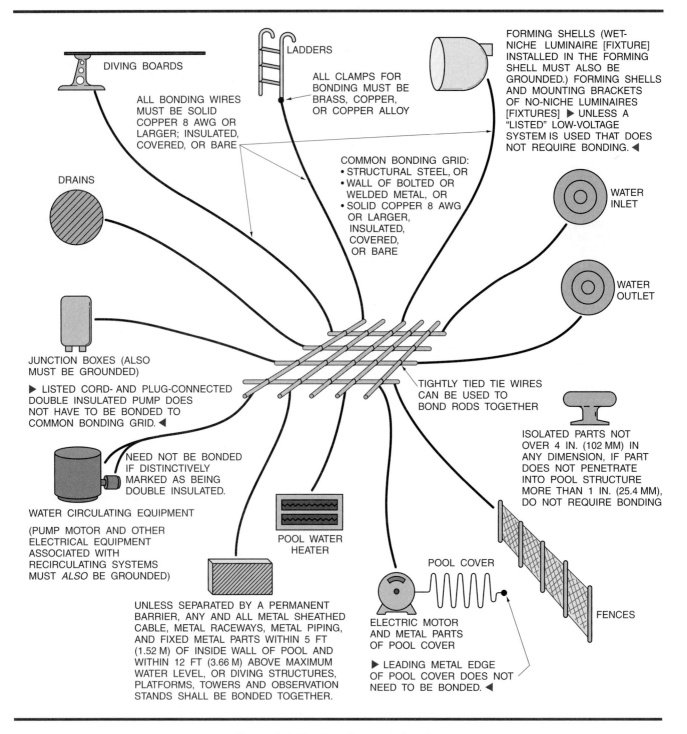

Figure 14-13 Bonding metal parts

Storable Pools

Storable pools that are constructed on or aboveground and are capable of holding water to a maximum depth of 42 in. (1.0 m) or are nonmetallic, molded polymeric walls or inflatable fabric walls, regardless of dimension, are considered to be temporary structures without special electrical wiring and a need for site work. They are likely to be removed and stored for the winter months.

The storable pool does not require bonding; however, the portable filter pump is required to be double insulated and have a grounding conductor

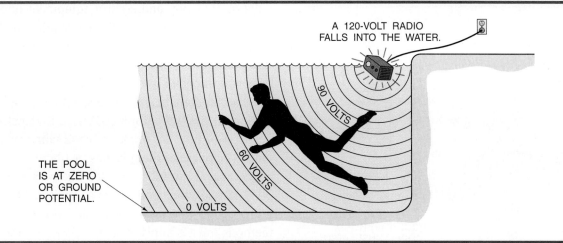

Figure 14-14 Voltage gradients

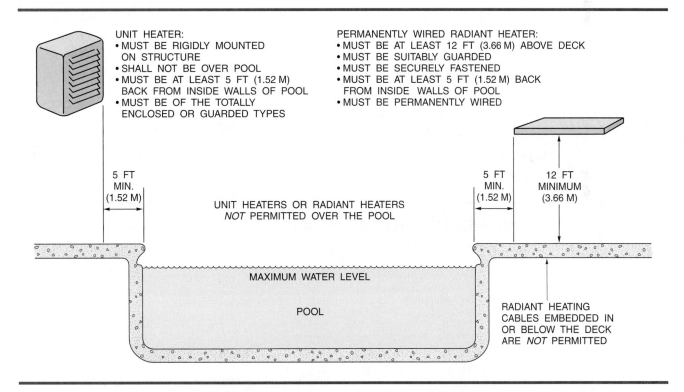

Figure 14-15 Deck area electric heating

integral with the flexible cord. The receptacle that supplies the power is required to be located at least 10 ft (3.0 m) from the pool. All electrical equipment is required to use a GFCI protection (Figure 14-16). Luminaires (lighting fixtures) are required to be cord- and plug-connected lighting assembly with no exposed metal parts, impact resistant, and listed for the purpose.

Spas and Hot Tubs

Spas and hot tubs have an additional health hazard caused by high temperature and duration of

260 CHAPTER 14 Swimming Pools

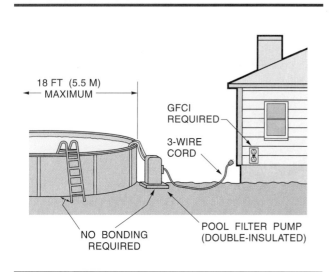

Figure 14-16 Storable pool pump wiring (Reprinted with permission from NFPA 70-2002)

exposure. These conditions, if taken beyond the limit, can cause drowsiness and increase the risk of drowning. Instructions for the safe use of **spas and hot tubs** specify the maximum temperature (about 104°F) and a maximum time (generally 15 minutes).

Hot tubs and spas can be classified as (1) self-contained or (2) packaged assembly. The self-contained spa or hot tub is a factory fabricated unit consisting of a spa or hot tub vessel with water cir-culating, heating, and control equipment integral to the unit. The packaged spa or hot tubs are factory fabricated units consisting of water circulating, heating, and control equipment mounted on a common base and intended to operate the hot tub or spa.

The fundamental difference between the hot tub and spa is the type of material used to construct the vessel. Hot tubs are made of wood whereas a spa is made of concrete, tile, plastics or fiberglass.

Spas and hot tubs installed outdoors must comply with the same Code rules as specified for swimming pools. For indoor spas and hot tubs, wiring methods must comply with *Chapter 3* of the *NEC*.® At least one GFCI receptacle must be installed between 5 ft (1.5 m) and 10 ft (3.0 m) of the inside wall of the spa or hot tub. Any receptacle within 10 ft (3.0 m) of the spa or hot tub that supplies power to a spa or hot tub must be GFCI protected. Figure 14-17 depicts some of the Code requirements. Bonding requirements for spas and hot tubs are similar to those for swimming pools.

It should be noted that spas and hot tubs generally are not drained after each use.

Hydromassage Bathtubs

A permanently installed bathtub equipped with a recirculating system, pump, and other associated

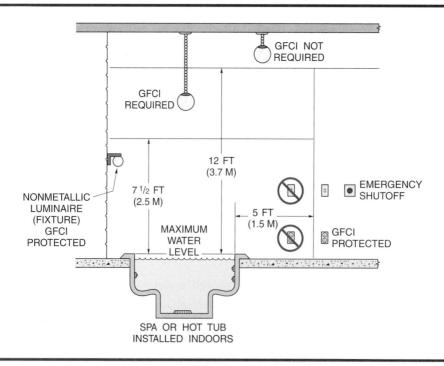

Figure 14-17 Spa or hot tub

equipment is a **hydromassage bathtub**. It is designed to accept water, circulate, and discharge at the end of use.

The hydromassage bathtubs are GFCI protected. GFCI protection is also required for any circuit and any receptacle within 5 ft (1.5 m) from the inside wall. The Code requires accessibility to the electrical equipment associated with the hydromassage tub (Figure 14-18).

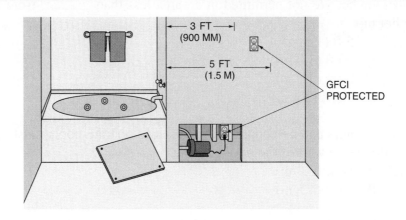

Figure 14-18 Hydromassage tub

REVIEW QUESTIONS

MULTIPLE CHOICE

1. The electric shock hazard to a person swimming in a pool as compared to a normal touch shock is:
   ____ A. similar.
   ____ B. lower.
   ____ C. greater.
   ____ D. less.

2. *Article 680* requires grounding of all equipment within how many feet of the inside wall of the pool?
   ____ A. 10 ft (3.0 m)
   ____ B. 3 ft (900 mm)
   ____ C. 5 ft (1.5 m)
   ____ D. 18 in. (450 mm)

3. The maximum voltage between conductors for underwater luminaires (light fixtures) is:
   ____ A. 120 volts.
   ____ B. 240 volts.
   ____ C. 150 volts.
   ____ D. 15 volts.

4. Deck heaters that are electrically operated are not to be installed over the pool or over the area that is _____ ft from the inside wall of the pool.
   ____ A. 20 ft (6.0 m)
   ____ B. 5 ft (1.5 m)
   ____ C. 10 ft (3.0 m)
   ____ D. 12 ft (3.7 m)

5. Wall switches are not permitted in an area less than _____ from the inside wall of the hot tub.
   ____ A. 5 ft (1.5 m)
   ____ B. 10 ft (3.0 m)
   ____ C. 20 ft (6.0 m)
   ____ D. 7.5 ft (2.4 m)

6. Electric motors and associated wiring used to operate pools must be located _____ from the inside wall of the pool.
   ____ A. 10 ft (3.0 m)
   ____ B. 15 ft (4.5 m)
   ____ C. 5 ft (1.5 m)
   ____ D. 20 ft (6.0 m)

7. The primary purpose for bonding pools is to:
   ____ A. reduce corrosion.
   ____ B. strengthen pool structure.
   ____ C. prevent current leakage.
   ____ D. eliminate voltage gradients.

8. Pools that have nonmetallic inflatable walls, regardless of dimensions, are classified as what kind of pools?
   ____ A. Storable
   ____ B. Permanent
   ____ C. Portable
   ____ D. Temporary

9. Existing luminaires (lighting fixtures) are permitted to be less than 5 ft (1.5 m) measured horizontally from the inside wall of the pool if they are:
   ____ A. GFCI protected.
   ____ B. rigidly fastened in place.
   ____ C. at least 5 ft (1.5 m) above the maximum water level.
   ____ D. all of the above.

10. Luminaires (lighting fixtures) and ceiling paddle fans are permitted over outdoor pools provided they are:
    ____ A. GFCI protected.
    ____ B. at least 7.5 ft (2.4 m) above the maximum water level.
    ____ C. at least 12 ft (3.7 m) above the maximum water level.
    ____ D. totally enclosed.

TRUE OR FALSE

1. The support steel in a pool may be used as a common grid to bond metal parts of a pool and the pool equipment.
   ____ A. True
   ____ B. False
   Code reference _____ .

2. Radiant heat cables are permitted to be embedded in the pool deck.
   ____ A. True
   ____ B. False
   Code reference _____ .

3. All electrical equipment used with storable pools is required to be GFCI protected.
   ____ A. True
   ____ B. False
   Code reference _____ .

4. A wall switch must be located at least 10 ft (3.0 m) from the inside wall of the hot tub.
   ____ A. True
   ____ B. False
   Code reference _____ .

5. Dry-niche luminaires (lighting fixtures) are re-lamped from the front.
   ____ A. True
   ____ B. False
   Code reference _____ .

6. Bonding and grounding of electrical equipment of spas or hot tubs are not required.
   ____ A. True
   ____ B. False
   Code reference _____ .

7. *Article 680* requires that all electrical equipment located within 10 ft (3.0 m) of the inside wall of the pool be grounded.
   ____ A. True
   ____ B. False
   Code reference _____ .

8. EMT is an acceptable wiring method for installing the wet-niche luminaires (fixtures).
   ____ A. True
   ____ B. False
   Code reference _____ .

9. *Article 680* requires an 8 AWG or larger copper conductor, stranded or solid, bare, insulated, or covered as the bonding conductor for bonding the swimming pool and associated equipment.
   ____ A. True
   ____ B. False
   Code reference _____ .

10. Receptacles that provide power to a spa or hot tub are required to be GFCI protected.
    ____ A. True
    ____ B. False
    Code reference _____ .

11. The overhead conductor clearances for pools applies to both the pool itself and the area extending 5 ft (1.5 m) horizontally from the inside wall of the pool.
    ____ A. True
    ____ B. False
    Code reference _____ .

FILL IN THE BLANKS

1. The effect of electric current through the body is determined by _____ current is flowing and for _____ a time.

2. The flexible cord for other than _____ must not exceed 3 ft (900 mm) in length.

3. Where run in a _____ , the equipment-grounding conductor must be insulated.

4. Where necessary to use flexible wiring methods for connections at the motor, liquidtight flexible metal or _____ conduit is permitted.

5. If a permanently installed pool is installed at a dwelling unit, at least one receptacle must be installed _____ and _____ from the inside wall of the pool.

6. Transformers used in electrical wiring installations for swimming pools are required to be _____ transformers with an ungrounded secondary that has a grounded metal barrier between secondary and primary winding.

7. Branch-circuits, which operate at _____ or more and supply the luminaires (lighting fixtures) directly or through a transformer, are required to be GFCI protected.

8. Proper grounding and bonding of the metal parts in the pool area _____ of electric shock hazard.

9. *NEC®* 680.26 requires that all metal parts of swimming pools be bonded together using _____ solid copper wire insulated, covered, or bare.

10. The self-contained spa or hot tub is a _____ unit consisting of a spa or hot tub vessel with water circulating, heating, and control equipment integral to the unit.

FREQUENTLY ASKED QUESTIONS

Question 1: I have an outdoor hot tub installed 2 ft from a single-family dwelling with aluminum siding. *NEC® 680.26(B)(5)* seems to indicate bonding of the aluminum siding is required. Is the aluminum siding required to be bonded? If the answer is yes, what would be an approved method?

Answer: *NEC® 680.26(B)(5)* is in *Part II, Permanently Installed Pools*. So we probably should go to *Part IV, Spas and Hot Tubs*. *NEC® 680.42* refers us back to *Parts I* and *II*. *NEC® 680.26(B)(5)*, as you mentioned, does require all fixed metal parts that are within 5 ft horizontally of the inside walls of the pool, in this case the hot tub, to be bonded. An approved method would be as shown in *680.26(C)*. One method would be to install an 8 AWG solid copper bonding jumper to bond the aluminum siding to any threaded metal piping on the hot tub. Be sure to use a CU-AL termination device at the aluminum siding and provide protection from physical damage for the bonding jumper. You might also check to see if the aluminum siding is already bonded to the electrical system. The local building code may have provisions for adequate bonding or interlocking of the individual sections of siding to each other and for bonding to the electrical system.

Question 2: When sizing conductors for a hot tub rated 50 amperes on the nameplate, do I need to increase the conductors 125% because hot tubs are continuous duty (on for 3 hours or more)?

Answer: If your hot tub is rated at 50 amperes, the electric heater load is included in that rating. A hot tub in itself is not considered a continuous load. However, *680.9* requires that all pool water heaters shall have branch-circuit conductors and overcurrent devices rated at not less than 125% of the total load of the nameplate rating.

If you check *680.2*, you will find three definitions of various hot tubs. There are packaged units, self-contained units, and units which are not packaged or self-contained. I would think that units which are not packaged or self-contained could have the electric water heaters fed separately, and only these conductors would be required to be rated at 125%.

Question 3: Are underwater luminaires (lighting fixtures) in permanently installed swimming pools permitted to be rated at 120 volts?

Answer: Yes, but *680.23(A)(3)* requires that a GFCI be installed in a branch-circuit supplying luminaires (fixtures) operating at more than 15 volts. *NEC® 680.23(A)(4)* requires that no luminaires (lighting fixtures) shall be installed for operation on supply circuits over 150 volts between conductors.

Question 4: For an aboveground pool, I believe a bare 8 AWG solid copper conductor looped around the pool and connected to every other upright, connected to the motor frame, connected to the ladder, and then spliced to an uninsulated 12 AWG equipment-grounding conductor should be sufficient, or does the 8 AWG have to go all the way back to the panel? What if the pool is fed from a sub-panel in the garage?

Answer: The requirements for bonding a pool are contained in *680.26(A)*. This section reads, *the bonding required by this Section shall be installed to eliminate voltage gradients in the pool area as prescribed.* The *FPN* to *680.26(A)* states *this Section does not require that the 8 AWG or larger solid copper bonding conductor be extended or attached to any remote panel board, service equipment, or any electrode.* If the metal uprights are securely fastened together, it is not necessary to bond to them more than once, but it would probably be better to do so. The pool's pump motor and all metal parts over 4 in. in any dimension and penetrating more than 1 in. into the pool must be bonded to the common bonding grid. *NEC® 680.26(C)* identifies the 8 AWG conductor as the common bonding grid. This conductor shall be a solid copper conductor, insulated, covered, or bare, not smaller than 8 AWG.

The requirements are the same if the pool is fed from a sub-panel in the garage. Be sure that all the other pool requirements relating to receptacle locations, ground-fault circuit interrupters, and any lighting you may be using are followed.

Question 5: Does the equipment in a decorative fountain have to be fed from a GFCI protected circuit?

Answer: If the fountain is permanently installed and is larger than 5 ft in any dimension, and the equipment is not listed for operation at 15 volts or less, then *680.51(A)* requires that a GFCI must be installed in the branch-circuit.

Question 6: Is Type UF cable permitted as a wiring method for an outside residential pool motor?

Answer: *NEC® 680.21* requires that wiring to pool-associated motors must be installed in RMC, IMC, RNC, or type MC cable listed for the application. Wiring to pool motors installed in the interior of a one-family dwelling may be done with any of the wiring methods recognized by Chapter 3 that contain an insulated or covered copper equipment-grounding conductor that is not smaller than 12 AWG.

Question 7: *NEC® 680.25(B)* states that when installing a new feeder and panel for a pool/tub/spa, the grounding conductor must be insulated. Do you know the reasoning behind this requirement? Why can't a *covered bare conductor* (as in SER cable) be used?

Answer: Insulated equipment-grounding conductors are required in *sensitive areas*, such as health care facilities and swimming pool areas to lessen the chance of the equipment-grounding conductor coming into accidental contact with the grounded conductor and establishing a parallel path for circuit current. This may cause a difference in potential between grounded surfaces and expose someone to a shock hazard. Bare or covered equipment-grounding conductors are permitted to be used in many applications, but in swimming pool areas, it is considered that the additional protection of insulation on all conductors is most practical. SER cable is not permitted to be used as a feeder to a swimming pool panelboard for the reasons shown previously.

CHAPTER 15

Emergency Systems

OBJECTIVES

After completing this chapter, the student should understand:
- emergency systems.
- legally required standby systems.
- optional standby systems.
- interconnected electric power production sources.

KEY TERMS

Generator Set: A device designed to convert mechanical energy into electrical energy. Usually consisting of a rotating portion (rotor) that is turned by a prime mover, such as a diesel engine or an electric motor, through a magnetic field caused by the windings in the stationary portion (stator) and producing electric power by the induction principle

Integrated Electric System: An integrated electrical system is a unitized segment of an industrial wiring system where an orderly shutdown is required to minimize personnel hazard and equipment damage

Portable Generator Equipment: A generator that is not permanently connected at a specific location and can be easily moved from site to site to furnish power as needed. Portable generators are used on optional standby systems as a temporary source of power

Prime Mover: A mechanical means used to turn the rotating assembly of a generator

Separate Service: A separate service to be used for a source of power for emergency lighting

Storage Battery: A battery comprised of one or more rechargeable cells of the lead-acid, nickel-cadmium, or other rechargeable electrochemical types. The cells may be sealed and have no provision for the addition of water or electrolyte or for the external measurement of electrolytic specific gravity. Sealed cells may be equipped with a pressure-release vent to prevent excessive accumulation of gas pressure. The nominal battery voltage is computed on the basis of 2 volts per cell for the lead-acid type and 1.2 volts per cell for the alkali type

Transfer Equipment: Automatic or nonautomatic devices designed to transfer one or more load conductor connections from one power source to another

Uninterruptible Power Supply (UPS): An alternate source of power, usually a battery and control devices, that is automatically connected parallel to a normal source of power when the normal source of power is cut off. A UPS is a means to keep a power supply uninterrupted for essential equipment, such as data-processing equipment

Unit Equipment: Unit equipment is a source of power for emergency illumination, generally consisting of a rechargeable battery with provision for lamps to be mounted on the equipment and a relaying device to energize the lamps automatically upon failure of the supply to the unit equipment

CHAPTER 15 Emergency Systems

INTRODUCTION

Emergency systems can be classified into two categories, Required Systems and Optional Systems. Required Systems are those mandated by governmental agencies having jurisdiction. There are two types of systems required. There is the **Emergency System** (*Article 700*), which is intended to automatically supply illumination or power (or both) to designated areas and equipment in the event of failure of the normal supply or in the event of failure of a system intended to supply, distribute, and control power and illumination essential to the safety of human life, and the **Legally Required Standby System** (*Article 701*), which is intended to automatically supply power to selected loads (other than those classed as emergency systems) in the event of failure of the normal source of power. Legally Required Standby Systems are typically installed to serve loads, such as heating and refrigeration systems, communication systems, ventilation and smoke removal systems, sewerage and disposal, lighting systems, and industrial processes, that when stopped during any interruption of the normal electrical supply could create hazards or hamper rescue or fire-fighting operations.

The **Optional Standby Systems** (*Article 702*) are intended to protect public or private facilities or property where life safety does not depend on the performance of the system. Optional standby systems are intended to supply on-site generated power to selected loads, either automatically or manually. Optional standby systems are typically installed to provide an alternate source of power to serve loads, such as heating and refrigeration systems, data-processing and communication systems, and **integrated electric systems** serving industrial processes that, when stopped during any power outage, could cause discomfort, serious interruption of the process, damage to the product or process, or the like. This chapter will show the reader the required or intended means of actuating these systems through required or otherwise desired automatic transfer devices or by manual operation.

Additionally, this chapter will acquaint the reader with **Interconnected Electric Power Production Sources** (*Article 705*). These power sources are designed to operate in parallel with and be capable of delivering energy to an electric primary source supply system.

EMERGENCY SYSTEMS

In the event of a failure of the normal source of power in buildings or structures designated in NFPA 101-2000, *Life Safety Code*, emergency systems are required to use automatic transfer switches to transfer the emergency load to the auxiliary source of power to maintain emergency lighting where considered essential to life safety (Figure 15-1).

Transfer equipment must be designed and installed to prevent the inadvertent interconnection of normal and emergency sources of supply in any operation of the transfer equipment. The purpose of this requirement is to protect line workers from backfeed while repairing supply lines and to prevent damage to a standby system if the normal supply is reconnected while simultaneously connected to the auxiliary source of emergency power.

Remember that transformers can work in both directions. If you were to inadvertently connect your source of power to the normal source during an

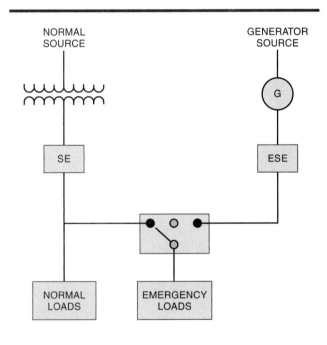

Figure 15-1 Normal power source—Emergency power source using a generator set

Figure 15-2 Automatic transfer control panel

outage, you would feed 240 volts from your generator source back into the service-drop and feed the utility transformer backwards, resulting in a possible 600-volt potential on the supposedly dead lines and putting the utility line repairmen in grave jeopardy (Figure 15-2).

There are several methods or sources of power, other than generators, available to maintain the lighting levels required. In selecting the emergency source of power, consideration must be given to the occupancy and the type of service to be rendered. Emergency lighting may be required for a short duration, such as for evacuation of a theater, or it may be of a longer duration, as for a large building where evacuation will take a long time due to the size of the building. The following are alternate sources of power used for emergency lighting.

Storage Battery

Storage batteries, used as a source of power for emergency systems, must be of suitable rating and capacity to supply and maintain the total load for a minimum of 1½ hours without the voltage applied to the load falling below 87½% of normal. Automotive type batteries cannot be used for this purpose. For a sealed battery, the container shall not be required to be transparent, but for the lead-acid type of battery that requires the addition of water at intervals, the case or jar must be transparent. In all cases, an automatic means of charging shall be furnished.

Generator Sets

Generator sets are the most common alternative source of power for emergency systems that require substantial loads. Generator sets require **prime movers** to drive the generators (Figure 15-3).

Means must be provided to automatically start the prime mover in the event of a failure of the normal source of power and for the automatic transfer of all required electrical circuits. Where internal

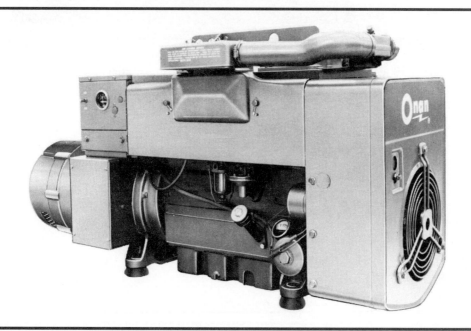

Figure 15-3 Generator for emergency power supply

combustion engines are used as the prime movers, an on-site fuel supply of not less than 2 hours full-demand shall be provided on the premises. Prime movers shall not be solely dependent on a public utility gas system for their fuel supply or municipal water supply for their cooling systems (Figure 15-4).

Generator sets are required to develop power for the system within 10 seconds, unless an auxiliary power supply energizes the emergency system until the generator can pick up the load.

Uninterruptible Power Supplies (UPS)

UPS are packaged battery systems with controls designed to maintain power to critical equipment, such as data-processing equipment or industrial process equipment where an interruption of power could cause catastrophic damage to equipment or personnel. UPS systems are not generally used to handle large loads for long periods of time (Figure 15-5).

Separate Service

A **separate service** is sometimes used as the auxiliary source of emergency power. This type of service is installed in accordance with the requirements of *Article 230*. The separate service must have a separate service-drop or service-lateral and must be widely and physically separated from the normal

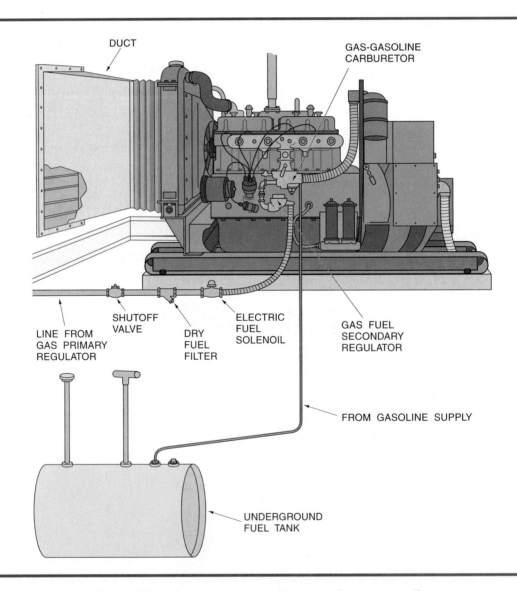

Figure 15-4 Generator powered by natural gas or gasoline

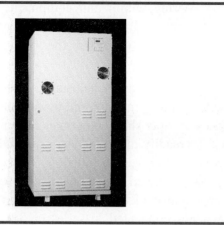

Figure 15-5 Uninterruptible power supply. *Courtesy of* LightGuard

service to minimize the possibility of simultaneous interruption of service. This type of emergency service is not generally acceptable to the AHJ unless fed from a separate utility source (Figure 15-6).

Unit Equipment

Unit equipment consists of a rechargeable battery and a charging means. It generally has one or more lamps mounted on the equipment with a relaying device arranged to energize the lamps automatically upon failure of the supply to the unit equipment. Unit equipment or battery EM lights, as they are commonly called, must be of suitable rating and capacity to supply and maintain not less than 87½% of the nominal battery voltage for the total lamp load for a period of at least 1½ hours. Unit equipment must be permanently fastened in place. The branch-circuit feeding the unit equipment must be the same branch-circuit of the normal supply serving the lighting in that area.

Emergency illumination must include all required means of egress lighting, illuminated exit signs, and all other lights specified as necessary to provide required illumination. Emergency lighting systems shall be installed so that the failure of

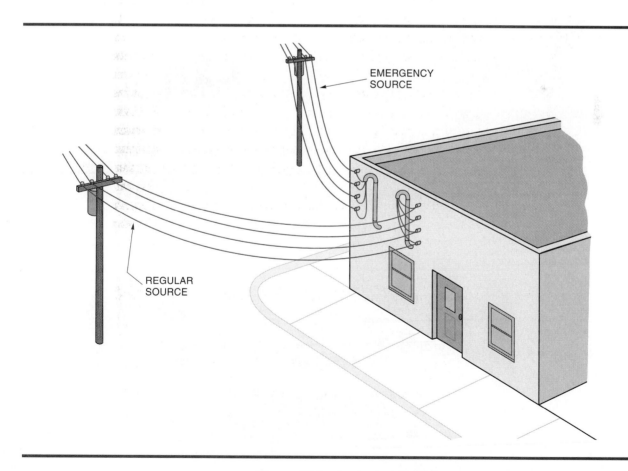

Figure 15-6 Separate services

any individual lighting element, such as the burning out of a lightbulb, cannot leave any space that requires emergency illumination in total darkness. This means that unit equipment with one head (lamp) can only be used to supplement areas where other means of emergency lighting is also present. Unit equipment with two heads cannot be used to provide emergency lighting for more than one area (Figure 15-7).

Emergency Wiring

Wiring from an emergency source shall be kept entirely independent of all other wiring except as follows:

1. Wiring within transfer switches.
2. Wiring within luminaires (lighting fixtures) fed from two sources.
3. Wiring within unit equipment. This requirement means that separate raceways systems must be used for the emergency system (Figure 15-8).

All boxes and enclosures, including transfer switches and generator and power panels, for emergency circuits shall be permanently marked so they will be readily identified as a part of the emergency system. Switches installed in emergency circuits must be arranged so that only authorized persons will have control of emergency lighting (Figure 15-9). Switches connected in series or 3-way and 4-way switches must not be used. In places of assembly, such as theaters, the switch controlling the emergency lighting system shall be located in the lobby or at a place conveniently accessible thereto.

Figure 15-7 Unit battery equipment. *Courtesy of LightGuard*

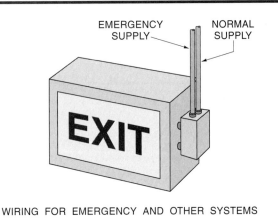

Figure 15-8 Emergency wiring and normal supply in the same exit sign

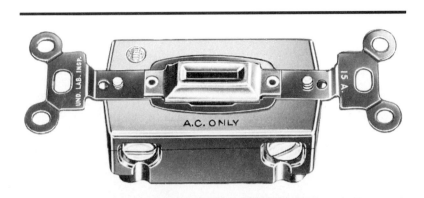

Figure 15-9 Key-operated switch for emergency lighting control

LEGALLY REQUIRED STANDBY SYSTEMS

Legally required standby system requirements parallel those requirements for emergency equipment. These systems are intended to automatically supply power to selected loads (other than those classified as emergency systems) in the event of failure of the normal source. The legally required standby system wiring shall be permitted to occupy the same raceways, cables, boxes, and cabinets as other general wiring.

Where acceptable to the AHJ, connections located ahead of and not within the same cabinet as the service-disconnecting means shall be acceptable. Connections done in this manner are usually made at the service point (Figure 15-10). A separate service raceway is run to the standby disconnecting means. The legally required standby system shall be sufficiently separated from the normal main service disconnecting means to prevent simultaneous interruption of supply through an occurrence within the building. The requirements for transfer equipment for legally required standby systems are the same as for emergency systems.

Optional Standby Systems

The systems covered by *Optional Standby Systems* consist of those that are permanently installed in their entirety, including prime movers and those that are arranged for a connection to a premises wiring system from **portable generator equipment**.

These systems are intended to protect facilities or property where life safety does not depend on the performance of the system. The requirements for transfer equipment are different from the requirements for emergency or legally required standby systems in that the transfer need not be done by automatic means. The other requirements remain the same. Transfer equipment shall be required for all permanently installed standby systems for which an electric utility supply is either the normal or standby source (Figure 15-11).

The optional standby wiring shall be permitted to occupy the same raceways, cables, boxes, and cabinets as other general wiring.

INTERCONNECTED ELECTRIC POWER PRODUCTION SOURCES

Interconnected electric power sources are on-site electric supply sources, such as cogeneration systems, solar photovoltaic systems, and fuel cells. Interconnected electric power production systems must comply with all the applicable requirements of emergency systems, legally required standby systems, and optional standby systems. If the system includes solar photovoltaic systems, the requirements of *Article 690* must also be followed.

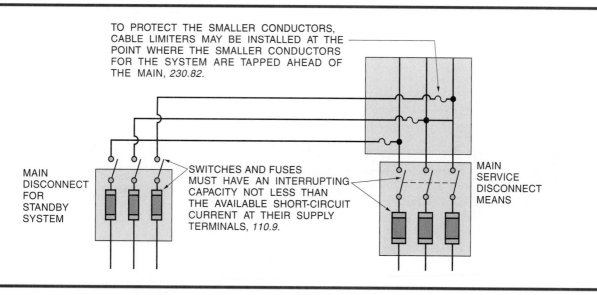

Figure 15-10 Connection ahead of service-disconnecting means

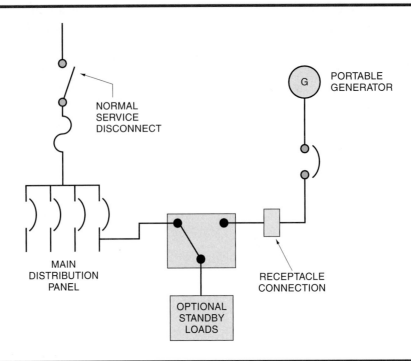

Figure 15-11 Optional standby system with portable generator

An interconnected power production system is one or more systems that are connected in parallel with and capable of delivering energy to an electric primary source supply system.

The point of connection of electric power production systems is at the premise's service-disconnecting means (Figure 15-12).

A permanent plaque or directory must be installed at each service equipment location and at all electric power production sources. A disconnecting means must be provided to disconnect all ungrounded conductors of each electric power production source from all other conductors. It is important that each electric power production source have an individual

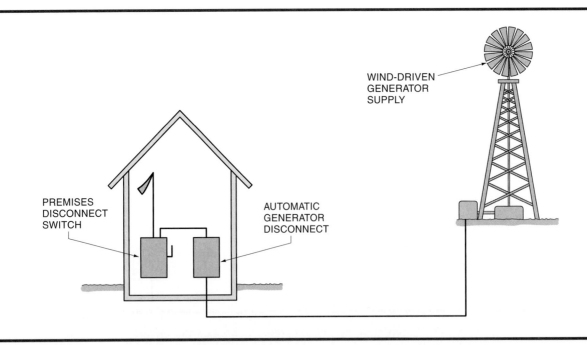

Figure 15-12 Point of connection of electric power production source

disconnecting means to comply with the utility's safe work practices. Upon loss of the primary source of power, an electric power production source shall be automatically disconnected from all ungrounded conductors of the primary source and shall not be reconnected until the primary source is restored (*705.40*).

Solar Photovoltaic Systems

Solar photovoltaic systems may be *interactive* with other power production sources, they may be *hybrid systems* comprised of multiple power sources, such as photovoltaic, wind-driven, water-driven (micro-hydro-electric), and engine-driven generators, but they do not include electric production systems (such as utility systems), or they may *stand alone* (Figure 15-13).

Solar photovoltaic systems consist of a mechanically integrated assembly of modules or panels and other components to form a direct-current power-producing unit or a complete unit of solar cells and other components designed to generate ac power when exposed to sunlight (Figure 15-14).

The ac inverter output from a stand-alone photovoltaic system is permitted to supply ac power to a building's service-disconnecting means at current levels below the rating of the building's service-disconnecting means (Figure 15-15).

If the system were to be reconnected to a utility system at a later date, it would be necessary to replace the service conductors. Additional information may be found in *Article 690*.

Fuel Cells

A fuel cell is an electrochemical system that consumes fuel to produce an electric current. The main chemical reaction that produces electrical power is not combustion. Fuel cells use a chemical process that directly converts a fuel's energy into electrical energy. The fuel cell works similar to a battery, using electrodes in an electrolyte. By supplying hydrogen to the negative electrode and oxygen to the positive electrode, a voltage is generated between the two electrodes. Although the process is similar to the way a battery produces energy, a battery has to be recharged, while a fuel cell keeps running as long as it is supplied with hydrogen and oxygen.

The hydrogen can be extracted from natural gas with a small amount of carbon dioxide as the by-product; however, fuel cells still emit much less

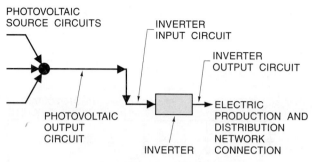

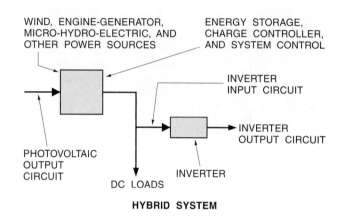

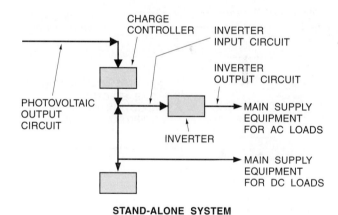

NOTE 1: THESE DIAGRAMS ARE INTENDED TO BE A MEANS OF IDENTIFICATION FOR PHOTOVOLTAIC SYSTEM COMPONENTS, CIRCUITS, AND CONNECTIONS.

NOTE 2: DISCONNECTING MEANS AND OVERCURRENT PROTECTION REQUIRED BY *ARTICLE 690* ARE NOT SHOWN.

NOTE 3: SYSTEM GROUNDING AND EQUIPMENT GROUNDING ARE NOT SHOWN. SEE *ARTICLE 690, PART E*.

NOTE 4: CUSTOM DESIGNS OCCUR IN EACH CONFIGURATION AND SOME COMPONENTS ARE OPTIONAL.

IDENTIFICATION OF SOLAR PHOTOVOLTAIC SYSTEM COMPONENTS IN COMMON SYSTEM CONFIGURATIONS.

Figure 15-13 Solar photovoltaic systems

276 CHAPTER 15 Emergency Systems

ALTERNATING-CURRENT MODULE (ALTERNATING-CURRENT PHOTOVOLTAIC MODULE). A COMPLETE, ENVIRONMENTALLY PROTECTED UNIT CONSISTING OF SOLAR CELLS, OPTICS, INVERTER, AND OTHER COMPONENTS, EXCLUSIVE OF TRACKER, DESIGNED TO GENERATE AC POWER WHEN EXPOSED TO SUNLIGHT.

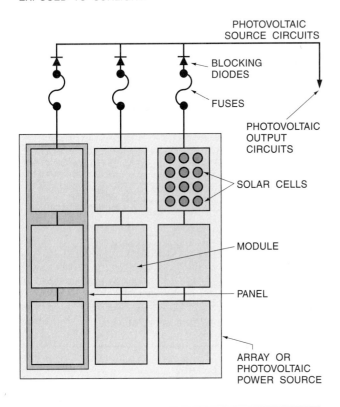

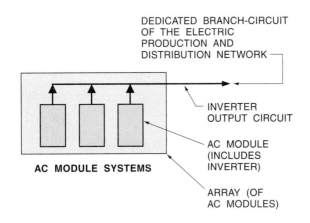

NOTE 1: THESE DIAGRAMS ARE INTENDED TO BE A MEANS OF IDENTIFICATION FOR PHOTOVOLTAIC SYSTEM COMPONENTS, CIRCUITS, AND CONNECTIONS.
NOTE 2: DISCONNECTING MEANS REQUIRED BY ARTICLE 690, PART C ARE NOT SHOWN.
NOTE 3: SYSTEM GROUNDING AND EQUIPMENT GROUNDING ARE NOT SHOWN. SEE ARTICLE 690, PART E.

Figure 15-14 AC photovoltaic module

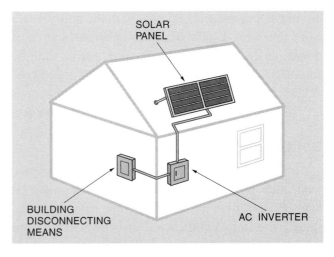

Figure 15-15 AC inverter output permitted to supply power at a current level less than premise's service-disconnecting means

pollution than traditional energy sources. Most people have natural gas lines or propane at their house, so these fuels are most likely to be used for home fuel cells. Because hydrogen is not always readily available, fuel cells use fuels that are more readily available and use a device called a reformer that turns hydrocarbons or alcohol fuels into hydrogen, which is then fed to the fuel cell. Fuel cells are a developing technology, and it is expected that home power generation using fuel cells will be readily available by the year 2002.

Fuel cell power systems may be stand-alone or interactive with other electrical production sources. A fuel cell system typically consists of a reformer, stack, power inverter, and auxiliary equipment. A fuel cell system must not feed the electrical production system when the electrical production system has become de-energized. This means the fuel cell system cannot feed into a de-energized utility power source. Additional information may be found in *Article 692*.

SUMMARY

It would be impossible to discuss every detail related to the requirements for the installation of every type of emergency system and all optional sources of power now available. This brief introduction to these power sources will acquaint the student with the options available and how they must be designed for use. References to the *NEC*® are shown to indicate where additional information may be obtained regarding the installation of these systems.

REVIEW QUESTIONS

MULTIPLE CHOICE

1. Switches for emergency lighting must be:
   - ____ A. 20-ampere rated.
   - ____ B. accessible only to authorized persons.
   - ____ C. located on the first floor.
   - ____ D. key-operated.

2. Emergency systems must be:
   - ____ A. served by two services.
   - ____ B. served through automatic transfer switches.
   - ____ C. served by 120/240-volt systems.
   - ____ D. none of the above.

3. Emergency storage battery systems must provide power for:
   - ____ A. 3 hours.
   - ____ B. 4 hours.
   - ____ C. 1 hour.
   - ____ D. 1½ hours.

4. Optional standby systems are required by:
   - ____ A. municipal governments.
   - ____ B. the *National Electrical Code.*®
   - ____ C. no one.
   - ____ D. fire department.

5. Unit equipment must be:
   - ____ A. fed by an emergency circuit.
   - ____ B. securely fastened in place.
   - ____ C. capable of maintaining power for 4 hours.
   - ____ D. equipped with 2 lamps.

TRUE OR FALSE

1. In the event of a power outage, automatic transfer to emergency power sources must be completed in 20 seconds.
   - ____ A. True
   - ____ B. False

   Code reference: _____ .

2. No appliances and no lamps, other than those specified as required for emergency use, shall be supplied by emergency lighting circuits.
   - ____ A. True
   - ____ B. False

   Code reference: _____ .

3. Optional standby systems are required to be automatically transferred when the normal source of power is disconnected.
   ___ A. True
   ___ B. False
   Code reference: _____.

4. Uninterruptible power supplies are not required to be automatically connected when the normal source of power is cut off.
   ___ A. True
   ___ B. False
   Code reference: _____.

5. Transfer equipment must be designed and installed to prevent the inadvertent interconnection of normal and emergency sources of supply in any operation of the transfer equipment.
   ___ A. True
   ___ B. False
   Code reference: _____.

FILL IN THE BLANKS

1. Where internal combustion engines are used as prime movers, an on-site fuel supply of not less than _____ shall be provided on the premises.

2. Storage batteries that are used as an alternate source of power for emergency systems must be of suitable rating and capacity to supply and maintain the total load for a minimum of _____ without the voltage falling below _____ of normal.

3. Generator sets are required to develop full power for the system within _____, unless an auxiliary power supply energizes the emergency system until the generator can pick up the load.

4. UPS is a packaged _____ system with controls designed to maintain power to critical equipment.

5. The branch-circuit feeding the unit equipment must be the _____-circuit of the normal supply serving the lighting in that area.

FREQUENTLY ASKED QUESTIONS

Question 1: If a main fuse blows, should the emergency system be energized?

Answer: Emergency systems are designed to be energized through the action of the automatic transfer switch whenever the normal source of power, or a part of the normal source of power, is cut off. An automatic transfer switch is a requirement of *700.6(A)*. Automatic transfer switches are electrically operated and mechanically held. This means that the transfer switch is pulled in electrically and is then mechanically latched in so that an electrical failure will not cause the transfer switch to drop out. The mechanical latch can be then electrically unlatched or manually unlatched.

Question 2: When a generator is used as the emergency source of power, how quickly must it come on line when the normal source of power is cut off?

Answer: In the event of failure of the normal supply, emergency power must be available in a time not to exceed 10 seconds. Generators that require more than 10 seconds to develop power shall be permitted if an auxiliary power supply energizes the emergency system until the generator picks up. The auxiliary power supply can be any of the sources of power permitted in *Article 700, Part III, Sources of Power*.

Question 3: If a building is unoccupied at night and the normal power is cut off, what keeps the emergency lights from coming on?

Answer: Buildings that are closed and not occupied for a portion of the day may, in accordance with *700.20*, have the exit and emergency lighting system controlled by a manually operable non-locking switch located in the main lobby or at a place conveniently accessible thereto. These switches shall be arranged so that only authorized persons will have control of emergency lighting. Emergency systems using unit battery equipment do not have this capability. When the supply to this equipment fails, the lamps are energized automatically.

Question 4: Can a portable generator be used as the emergency source of power?

Answer: A portable alternate power supply can be used in an optional standby system. A portable or temporary alternate source of power must be available whenever the emergency generator is out of service for major maintenance or repair. A **portable generator** cannot be used as the permanent alternate source of power for an emergency system.

Question 5: Can solar photovoltaic or fuel cell sources of power be used as the alternate source of power for an emergency system?

Answer: Solar photovoltaic systems and fuel cell systems are designed to work in parallel with normal sources of power, or they may operate as a stand-alone system. They are not acceptable sources for emergency power supply, as shown in *Article 700, Part III, Sources of Power*.

CHAPTER 16

Carnivals, Circuses, Fairs, and Similar Events Temporary Installations

OBJECTIVES

After completing this chapter, the student should understand:
- power sources.
- emergency lighting.
- egress lighting.
- emergency service to selected attractions.
- wiring methods.
- rides, tents, and concessions.
- overhead clearances.
- clearance to rides and attractions.
- GFCI protection required.
- temporary wiring installations.
- grounding conductor continuity assurance.

KEY TERMS

Egress Lighting: Electric lighting intended to provide illumination for safe exit from closed-in areas

Electric Power Equipment Trailers: Equipment trailers used to carry electric power equipment, including generators, from site to site

Portable Distribution or Termination Boxes: Electrical distribution panels or termination boxes intended to be moved from location to location for the installation of temporary wiring

Portable Wiring and Equipment: Electrical equipment and wiring for installations where the installation is intended to be moved or changed frequently or installed on a temporary basis

Rides, Tents, and Concessions: Buildings or structures used in carnivals, circuses, or fairs not intended as permanent structures or buildings for use at a particular location for purposes other than temporary use

Utility Supplied Services: A source of electric power delivered from a local utility to a service point on the premises

INTRODUCTION

The information contained in this chapter applies to the **portable wiring and equipment** for carnivals, circuses, fairs, and similar functions, including wiring in or on all structures. This chapter will show the proper way to install the many cables required to feed the booths, rides, and attractions while maintaining the safety necessary for the attending public. In addition, this chapter will cover temporary wiring installations. This chapter is to be used in conjunction with *Articles 525* and *527*. Information found here is not necessarily a requirement of the *NEC.*®

GENERAL

Before requirements for the installation of electrical work in carnivals, circuses, and fairs was accepted as a new *Article 525* of the 1996 Edition of the *NEC,*® the installation of electrical work was done in a relatively haphazard fashion with little or no inspection and no standards of installation to follow.

Additionally, many areas of the country have not adopted building codes and have no basis for electrical inspection. These installations were done solely at the discretion of the event's promoter and owner. Inspections done in areas where inspection procedures were a part of the ordinances of local municipalities generally followed the requirements for temporary wiring, and they were inadequate for a responsible inspection.

Recognizing the potential liabilities inherent in unsafe electrical equipment and wiring in the area of circuses, carnivals, and fairs, responsible persons involved in these events proposed that a new article be included in the *NEC.*® This article would outline requirements that would ensure safe installations. With the acceptance of new *Article 525*, the hazards associated with these events has been greatly reduced, and inspection authorities have a strong basis for determining the acceptance of wiring installations done in these areas.

This chapter will outline installation methods that should be followed to obtain the degree of safety that the public is entitled to while attending the events covered herein. These methods are not necessarily a part of the *NEC,*® but the reader should

Figure 16-1 Power equipment trailer containing diesel generator

understand that the requirements of the *NEC®* are the minimum requirements for safety and do not ensure an installation that is adequate for the user's needs or convenient for the user.

Power Sources

The power sources for circuses, carnivals, and fairs are generally a separately derived system consisting of a generator with a prime mover. The prime mover is generally a diesel or gasoline driven engine. This equipment is transported from site to site in large **electric power equipment trailers**, and it is the central electrical power source when these events are set up (Figure 16-1). Positioning or locating the power equipment trailers must be done in a way that makes it convenient for storage of fuel and re-supply fuel (Figure 16-2). These fuel storage areas may be classified as Class I, Division 2 areas, and all electric wiring should not be run in these areas without conformance to the appropriate article relating to hazardous classified areas.

The location of the power trailers should be as central as possible to limit the length of the runs of flexible cords and cables to rides, tents, attractions, and to **portable distribution or terminal boxes** (Figure 16-3). The location of portable distribution or terminal boxes should be determined with consideration given to restricting access to them by the general public, while eliminating long runs of flexible cables and cords to minimize voltage drop. *NEC®* 525.22 requires that portable distribution or terminal boxes be designed so that no live parts are exposed to accidental contact. Where installed outdoors, the box must be of weatherproof construction and mounted so that the bottom of the box is not less than 6 in. (150 mm) above the ground. Bus bars must have an ampere rating not less than the overcurrent device supplying the feeder that supplies the box (Figure 16-4). Circuits and feeders supplied by the distribution or terminal box must have overcurrent protection located in the distribution or terminal box.

It was the practice in years gone by for circuses and carnivals to carry wooden termination boxes that would be used as distribution points for the various power demands with little regard for overcurrent protection. With *Article 525* in place, these

Figure 16-2 Truck with diesel generator for power to carnival equipment backed up to fence enclosure for easy refueling away from public areas

Figure 16-3 Portable distribution box

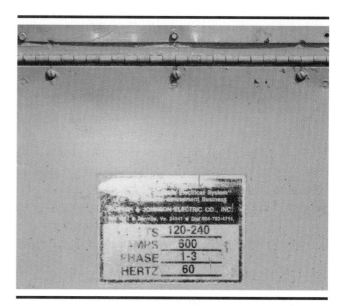

Figure 16-4 Portable distribution box rated at 600 amperes

wooden boxes are no longer acceptable, and they have been replaced by portable distribution centers that comply with *525.22* (Figure 16-5).

In many areas, a **utility supplied service** provided from the local utility company is used as an alternate or a normal power supply for **egress lighting** in case of an interruption of the separately derived power source. This is not a required emergency power source unless it is required by local ordinances.

Requirements for these events vary greatly from municipality to municipality, but generally, an alter-

Figure 16-5 Portable distribution center showing plug-in receptacle ports

nate power source is required to be available with sufficient capacity to power any rides or attractions that may leave a user in a dangerous position due to height or enclosure. *NEC® 525.11* requires that service equipment shall be mounted on a solid backing and be installed in a way that protects it from the weather. Service equipment shall not be installed in a location that is accessible to unqualified persons.

Wiring Methods

When setting up the rides and concession stands, flexible cords are generally run on the ground from the electrical distribution panels in the electrical power trailers to each ride or concession stand. Where used outdoors, flexible cables and cords shall be listed for wet locations and shall be sunlight resistant. Flexible cords and cables shall be listed for extra-hard usage.

Every effort should be made to run these cables in locations where the public has little chance of making contact with them. Where accessible to the public, and where placed on pavement or other hard and smooth surfaces, the cables shall be permitted to be covered with a non-conductive matting. Matting is *not permitted* to be used to cover cables run in soft dirt or grass. Flexible cords shall be run continuously without splice or tap between boxes or fittings. In these areas, it is not required, but it is suggested, that the cables be buried in a shallow trench. The requirements of *305.5* shall not apply (Figure 16-6).

Rides, Tents, and Concessions

NEC® 525.21 requires that all **rides, tents, and concessions** shall be provided with a disconnecting means located within sight and within 6 ft (1.8 m) of the operator's station (Figure 16-7). The disconnecting means shall be readily accessible to the operator, including when the ride is in operation. Where accessible to unqualified persons, the enclosure for the disconnecting means shall be of the lockable type (Figure 16-8). A shunt trip device that opens the disconnecting means when a switch in the ride operator's console is closed, shall be a permitted method of opening the disconnecting means.

NEC® 525.21 also requires that all lamps for illumination inside tents or concessions shall be protected from accidental breakage by a suitable luminaire (fixture) or a lampholder with a guard (Figure 16-9).

Overhead Clearances. Conductors must have a vertical clearance to ground in accordance with *225.18*. These clearances apply only to wiring installed outside of tents and concessions.

Clearance to Rides and Attractions. *NEC® 525.5(B)* requires that amusement rides and attractions shall be maintained not less than 15 ft (4.5 m) in any direction from overhead conductors operating at 600 volts or less, except for the conductors supplying the amusement ride or attraction. Amusement rides or attractions shall not be located under or

Figure 16-6 Cables buried in a shallow trench

286 CHAPTER 16 Carnivals, Circuses, Fairs, and Similar Events Temporary Installations

Figure 16-7 Disconnecting means for ride within sight of operator's station

Figure 16-8 Disconnecting means for ride behind fence to restrict access to qualified persons

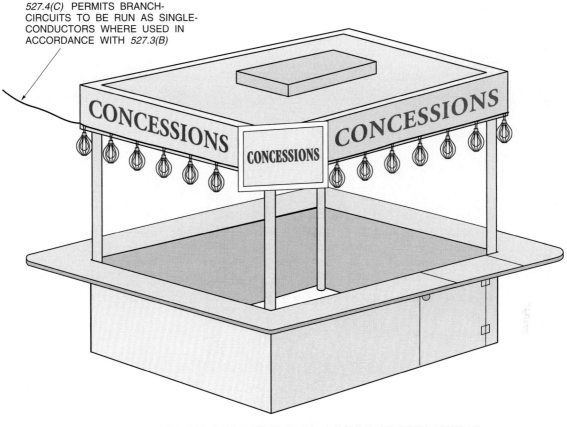
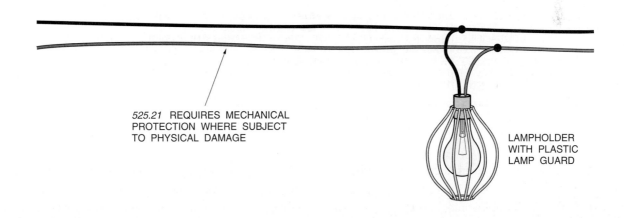

Figure 16-9 Lamp protection at concession stand

within 15 ft (4.5 m) horizontally of conductors operating in excess of 600 volts (Figure 16-10).

GFCI Protection Required. *NEC® 525.23* requires that all 125-volt, 15- or 20-ampere receptacles shall have GFCI protection if they are readily accessible to the general public. Egress lighting shall not be connected to the load-side of a GFCI receptacle.

Grounding Conductor Continuity Assurance. The effectiveness of the grounding conductors to reduce electrical hazards shall be tested each time that portable equipment is reconnected. Equipment

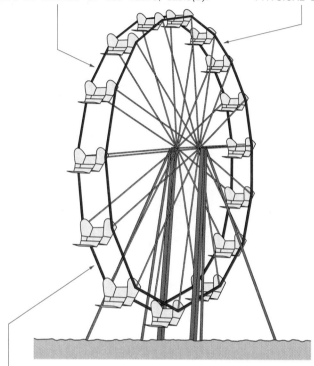

AMUSEMENT RIDES/ATTRACTIONS MUST BE AT LEAST 15 FT (4.5 M) IN ANY DIRECTION AWAY FROM OVERHEAD CONDUCTORS OPERATING AT 600 VOLTS OR LESS, EXCEPT, OF COURSE, FOR THE RIDE/ATTRACTION SUPPLY CONDUCTORS. AMUSEMENT RIDES/ATTRACTIONS MUST NOT BE SITUATED UNDER, NOR WITHIN 15 FT (4.5 M) HORIZONTALLY, OF CONDUCTORS OPERATING IN EXCESS OF 600 VOLTS, *525.5(B)*.

ASSOCIATED ELECTRICAL EQUIPMENT AND WIRING METHODS OF RIDES, CONCESSIONS, OR SIMILAR UNITS MUST BE MECHANICALLY PROTECTED WHERE SUBJECT TO PHYSICAL DAMAGE.

EACH RIDE AND CONCESSION MUST HAVE A VISIBLE FUSED DISCONNECT SWITCH OR CIRCUIT BREAKER WITHIN 6 FT (1.8 M) OF THE OPERATOR'S STATION. THE DISCONNECTING MEANS MUST BE READILY ACCESSIBLE TO THE OPERATOR, EVEN DURING RIDE OPERATION. IF ACCESSIBLE TO UNQUALIFIED PERSONS, THE SWITCH OR CIRCUIT-BREAKER ENCLOSURE MUST BE LOCKABLE. A SHUNT TRIP DEVICE, WHICH OPENS THE FUSED DISCONNECT DEVICE OR CIRCUIT BREAKER WHEN A SWITCH LOCATED IN THE RIDE OPERATOR'S CONSOLE IS CLOSED, IS AN ACCEPTABLE METHOD OF OPENING THE CIRCUIT.

Figure 16-10 Overhead clearance requirements

grounding shall be installed in accordance with the requirements of *Article 250*. Metal raceways, metal-sheathed cable, metal enclosures, and metal parts of rides and concessions shall be bonded.

TEMPORARY INSTALLATIONS

Information relating to the requirements for temporary electrical power and lighting installations can be found in *Article 527*. Temporary wiring methods shall be acceptable only if approved for the conditions of use and any special requirements of the temporary installation. Where temporary power is provided by portable wiring and equipment or vehicle-mounted generator and is connected to the permanent wiring system, a suitable means of transfer or isolation shall be provided to prevent the inadvertent interconnection of normal and temporary power sources.

Temporary electrical power and lighting is permitted by *527.3(A)* during the period of construction, remodeling, maintenance, repair, or demolition of buildings, structures, equipment, or similar activities. Temporary electrical power and lighting installations are permitted by *527.3(B)* for a period of 90 days for holiday lighting, unless provided with AFCI protection.

Branch-Circuits

Branch-circuit conductors shall be within cable assemblies or within multiconductor cables identified for hard or extra-hard usage. Single-insulated conductors are not permitted except for uses shown in *527.3(B)* and in *527.3(C)* for emergencies and tests.

Receptacles

All receptacles must be of the grounding-type. All branch-circuits feeding receptacles on temporary circuits shall contain a separate equipment grounding conductor, and all receptacles shall be electrically connected to the equipment grounding conductor. Receptacles on construction sites shall not be installed on branch-circuits that supply temporary lighting. Receptacles shall not be connected to the same ungrounded conductor of multiwire circuits that supply temporary lighting. This means that receptacles on temporary circuits cannot share a neutral with a lighting branch-circuit.

Splices

On construction sites, a box shall not be required for splices where the circuit-conductors are multiconductor cord or cable assemblies, provided that the equipment grounding continuity is maintained with or without the box. This means splices may be made in the open, but be sure that they are well insulated and are not subject to physical damage.

SUMMARY

Carnivals, circuses, and fairs remain one of the most difficult areas of inspection for the inspection authorities. Because of the temporary nature of these installations and the frequency with which the electrical equipment is moved from one location to another, there is a tendency for the electrical installation to be done in a haphazard manner. Unless there is a utility service required for some special use, such as for emergency egress lighting, the workmen who make the installation leave much to be desired in the way of qualifications.

Generally there is a maintenance foreman who is in charge of the equipment setup when the carnival or circus moves onto a site. The foreman directs the location of the trailers that house the generators that provide power for the concessions and rides. The workmen are usually "roustabouts" and owners of the concessions and rides. There are no electrical qualifications required and everything is cord- and plug-connected.

In the power trailers, distribution panelboards are fed from generators. The distribution panels have integrally mounted receptacles of various ratings from which power cords are run to portable distribution or terminal boxes. From these portable distribution boxes, which contain receptacles of various ratings, power cords are run to concession stands and rides. Many larger rides, which require more available power, have power cords run directly to the power trailer.

Generally a power cord is run to a disconnect switch in each concession stand and lighting is taken from this source. Proper overcurrent protection should be used at this point but all too frequently it is not. All luminaires (lighting fixtures) installed need lamp protection as required by *527.4(F)* or by *525.21* and all branch-circuits shall, for either lighting or receptacles, be run with conductors in a cord or cable identified for hard or extra-hard usage. Receptacles must not be installed on branch-circuits that supply lighting and receptacles must not be connected to the same ungrounded conductor of a multiwire circuit that supplies temporary lighting.

Temporary flexible power cords or cables, where accessible to the public, must be run to afford protection to the public. *NEC® 525.20(G)* contains the requirements for protection of cords or cables. These cables can be placed in a shallow trench and, in accordance with *525.20(G)*, the depth requirements of *300.5* shall not apply. Nonconductive matting may be used to cover cables installed on a hard surface but care must be taken so that a tripping hazard is not created.

All disconnecting means for rides, concessions, and tents must be installed in accordance with *525.21*, which requires that all disconnecting means provided for rides must be readily accessible to the operator and within 6 ft (1.8 m) of the operator's station.

Grounding must be done in accordance with *525.31*. All equipment requiring grounding must be grounded by use of an equipment grounding conductor. The equipment grounding conductor must be sized according to *250.122* and be of a type recognized by *250.118*. The generators providing power are classified as separately derived systems, and the vehicle-mounted generators must be grounded in accordance with *250.34(B)*. Supplemental grounding electrodes in the form of driven ground rods, although not required by the *NEC*,® should be provided at each vehicle-mounted generator.

In order to properly install or inspect carnivals, circuses, fairs, and similar events, both *Article 525* and *Article 527* must be used. The installer must always remember that the *NEC*® contains only the minimum requirements for a safe installation and that many other additional procedures can be used to add to the safety of the public attending carnivals, circuses, fairs, and similar events.

REVIEW QUESTIONS

MULTIPLE CHOICE

1. Overhead conductors shall not pass over an amusement ride or concession stand at a distance less than _____ in any direction.
   - ____ A. 15 ft (4.5 m)
   - ____ B. 6 ft (1.8 m)
   - ____ C. 10 ft (3.0 m)
   - ____ D. 20 ft (6.0 m)

2. Portable terminal boxes must be installed so that the bottom of the box is not less than _____ from the ground.
   - ____ A. 18 in. (450 mm)
   - ____ B. 12 in. (300 mm)
   - ____ C. 6 in. (150 mm)
   - ____ D. none of the above

3. Concession stands may be wired with:
   - ____ A. hard usage cable.
   - ____ B. extra-hard usage cable.
   - ____ C. cable assemblies.
   - ____ D. all of the above.

4. The disconnecting means for a ride must be located within _____ of the ride operator's station.
   - ____ A. 3 ft (900 mm)
   - ____ B. 4 ft (1.2 m)
   - ____ C. 6 ft (1.8 m)
   - ____ D. none of the above

5. Bus bars in a portable distribution or terminal box must have an ampere rating not less than:
   - ____ A. 600 volts.
   - ____ B. 100 amperes.
   - ____ C. the rating of the overcurrent device feeding the unit.
   - ____ D. none of the above.

CHAPTER 16 Carnivals, Circuses, Fairs, and Similar Events Temporary Installations

TRUE OR FALSE

1. Distribution or terminal boxes shall be permitted to be mounted on the earth.
   ____ A. True
   ____ B. False
   Code reference _____ .

2. Bus bars in a portable distribution or terminal box must have an ampere rating not less than the rating of the overcurrent device supplying the distribution or terminal box.
   ____ A. True
   ____ B. False
   Code reference _____ .

3. A shunt trip device may be used as a means to open the disconnecting means for a ride or concession.
   ____ A. True
   ____ B. False
   Code reference _____ .

4. Flexible cords or cables may be spliced with listed splicing kits between boxes or fittings.
   ____ A. True
   ____ B. False
   Code reference _____ .

5. Flexible cords or cables, when buried, must conform to the requirements of *300.5*.
   ____ A. True
   ____ B. False
   Code reference _____ .

FILL IN THE BLANKS

1. Flexible cords or cables shall be continuous _____ or tap between boxes or fittings.

2. Bus bars must have an ampere rating _____ the overcurrent device supplying the feeder in the terminal box.

3. Each ride and concession shall have a disconnecting means with a fused switch or circuit breaker located within sight and within _____ of the operator's station.

4. Service equipment shall not be installed where accessible to _____ unless the equipment is lockable.

5. Egress lighting shall not be connected to the _____ terminals of a GFCI receptacle.

FREQUENTLY ASKED QUESTIONS

Question 1: What do they mean when they talk about portable terminal boxes?

Answer: In the installation of wiring for circuses, carnivals, or fairs, it is common practice to use a *terminal box* as a distribution point for wiring to rides and attractions. This terminal box used to be of wooden construction, and had bus bars mounted in it for connection of the cables running to the rides and attractions. The terminal box was supplied by a feeder that originated at the power source. After the inclusion of *Article 525* in the 1996 *NEC*,® wooden boxes were virtually eliminated, and metal enclosures or distribution panels are now being used. Without regulation or inspection, carnival or fair workmen with little or no electrical knowledge would connect cables to these boxes that had an ampere rating much less then the feeder overcurrent protective device supplying these boxes.

Question 2: What are the grounding requirements for a carnival?

Answer: The grounding requirements for a carnival are not really described in great detail in the *NEC*.® It is important to remember that most of the wiring is done in a temporary manner and that most of the equipment is portable. *NEC*® *525.32* is an extremely good requirement since it shows definite concern for the effectiveness of the grounding conductor. This section requires that portable electrical equipment be tested for effectiveness of the grounding capabilities each time the portable equipment is reconnected. *NEC*® *525.30* also requires bonding of metal raceways and sheathed cable, metal enclosures of electric equipment, metal frames and metal parts of rides, concessions, tents, trailers, and trucks, or other equipment that contains or supports electrical equipment. There are no specific requirements as to how this bonding must be done other than *525.31*, which requires that an equipment grounding conductor of a type and size recognized by *250.118* must be installed in accordance with *Article 250*.

Question 3: Can open single conductors be run as temporary circuit-conductors for lighting circuits in concessions or tents at a carnival or on a construction site?

Answer: According to *527.4(C)*, temporary branch-circuit conductors must be run in cable assemblies or within multiconductor cord or cable of a type identified for hard or extra-hard usage. Single-insulated conductors may be run for temporary electrical power and lighting installations for a period not to exceed 90 days for holiday decorative lighting unless provided with AFCI protection. Single-insulated conductors may also be run for temporary electrical lighting and power during emergencies and for tests, experiments, and development work. In tents, concessions, or on rides at carnivals, circuses, or fairs, open single-insulated conductors cannot be used for temporary lighting or power.

Question 4: Can I run Nonmetallic-Sheathed Cable as temporary branch-circuit wiring on a construction site?

Answer: *NEC*® *527.4(C)* permits the use of cable assemblies as branch-circuit wiring for temporary circuits. Type NM Cable is a listed cable assembly and may be used for this purpose. Type NM Cable is permitted to be used as temporary wiring in any building, dwelling, or structure without any height limitation.

Question 5: Are metal lamp guards on a string of temporary lights required to be grounded?

Answer: Metal lamp guards must be grounded by the means of an equipment grounding conductor run with the circuit-conductors within the power-supply cord.

CHAPTER 17

Health-Care Facilities

OBJECTIVES

After completing this chapter, the student should understand:

- critical branch system.
- essential electrical system.
- life safety branch.
- general care areas.
- critical care areas.
- isolated power systems.
- isolation transformers.
- line isolation monitor.

KEY TERMS

Critical Branch: The critical branch of the emergency system supplies power for task illumination, fixed equipment, selected receptacles, and special power circuits serving the critical care areas

Critical Care Areas: Critical care areas are those special care units, intensive care units, delivery rooms, operating rooms, and similar areas in which patients are intended to be subjected to invasive procedures and connected to line-operated, electro-medical devices

Essential Electrical System: An electrical system capable of supplying a limited amount of lighting and power service, which is considered essential for life safety and orderly continuation of procedures during the time normal electrical service is interrupted for any reason

General Care Areas: General care areas are patient bedrooms, examining rooms, treatment rooms, clinics, and similar areas where it is intended that patients shall come in contact with ordinary appliances, such as a nurse call system, electrical beds, examining lamps, telephones, and entertainment devices

Isolated Power Systems: Systems where the transformers supplying power have no electrical connection between primary and secondary windings or a system supplied by motor generator sets or by means of a suitably isolated battery system

Patient Care Area: A patient care area is any portion of a health-care facility where patients are intended to be examined or treated

X-ray: Electromagnetic radiation capable of penetrating solids and producing a radiograph made of X-rays

CHAPTER 17 Health-Care Facilities

INTRODUCTION

The information in this chapter applies to hospitals, nursing homes, residential custodial care facilities, and other health-care facilities serving patients who are unable to provide for their own care and safety (Figure 17-1).

The requirements shown in this chapter also apply to mobile health-care units and doctors' and dentists' offices. This chapter is to be used in conjunction with *Article 517* whose scope applies to *electrical construction and installation criteria in health-care facilities that provide services to human beings.*

GENERAL

The following requirements are applicable to patient care areas of all health-care facilities. These requirements do not apply to business offices, corridors, and waiting rooms, in clinics, medical and dental offices, and outpatient facilities. **Patient care areas** may be in general care areas or they may be in critical care areas, either of which can be classified as a wet location. A *wet location* is one that is subject to wet conditions while patients are present. These conditions include standing fluids on the floor or drenching of the work area, either of which condition is intimate with the patient or staff. Routine housekeeping procedures and incidental spillage of fluids do not define a wet location (Figure 17-2). Another type of bath facility is shown in Figure 17-3.

Wiring and Protection

All branch-circuits serving patient care areas shall be provided with a ground path for fault-current by installation in a metal raceway system or a cable armor or sheath that qualifies as an equipment grounding return path. Type AC, Type MC, and Type MI cables must have an outer sheath that is identified as an acceptable grounding return path.

In patient care areas, the grounding terminals of all receptacles must be grounded by an insulated copper conductor installed in a metal raceway or by a Type AC, Type MC, or Type MI cable, provided the outer armor or sheath is identified as an acceptable grounding return path.

Figure 17-1 Typical residential custodial care facility

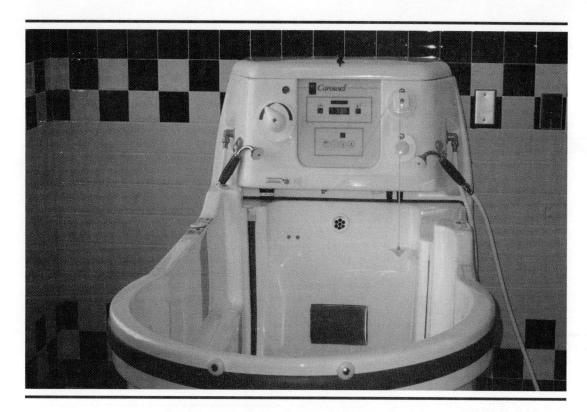

Figure 17-2   Wet location health-care facility

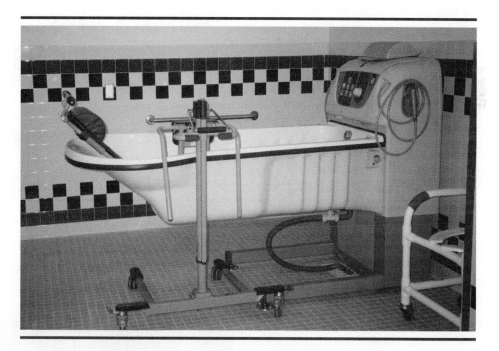

Figure 17-3   Nursing home resident bath area

In patient care areas, a patient vicinity shall be recognized. This is an area in which patients are normally cared for. The patient vicinity is any space with surfaces likely to come into contact with the patient or an attendant who might touch the patient. Typically, this is an area extending 6 ft (1.8 m) beyond the perimeter of the bed, and extending not less than 7½ ft (2.5 m) above the floor (Figure 17-4).

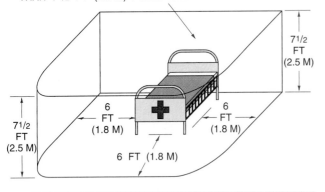

Figure 17-4 Patient vicinity area

Figure 17-5 Listed hospital grade receptacle. Note: Identifying green dot

General Care Areas

Patient Bed Branch-Circuits. In **general care areas**, each patient bed location shall be supplied by at least two branch-circuits, one from the normal system and one from the emergency system. This requirement does not apply to patient bed locations in clinics, medical and dental offices, psychiatric and substance abuse centers, or sleeping rooms in nursing homes and limited-care facilities.

Patient Bed Receptacles. Each patient bed location must be provided with a minimum of four receptacles. They are permitted to be of the single or duplex types or a combination of both. All receptacles must be listed as *hospital grade* and be so identified. Hospital grade receptacles are marked *hospital grade* and have a green dot on the face of the receptacle. Each receptacle must be grounded by means of an insulated copper conductor (Figure 17-5). Figures 17-6 and 17-7 illustrate receptacles required in patient bed locations.

Critical Care Areas

Patient Bed Branch-Circuits. In **critical care areas**, each patient bed location shall be fed from at least two branch-circuits, one from the normal system and one from the emergency system. At least one branch-circuit must supply a receptacle outlet at the bed location.

Figure 17-6 Typical patient bed area with required receptacles

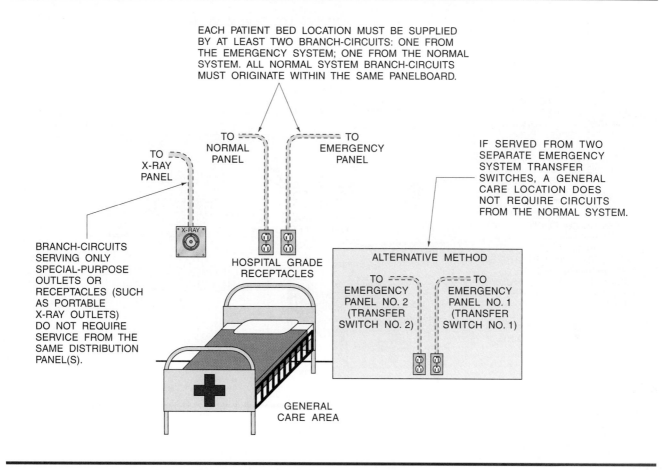

Figure 17-7 Branch-circuits (general care area)

Patient Bed Receptacles. Each patient bed location must be provided with at least six receptacles. At least one of the receptacles must be connected to a normal system branch-circuit or to an emergency system branch-circuit supplied by a different transfer switch from the other receptacles at the same location. All receptacles must be listed *hospital grade* and be so identified. The receptacles are permitted to be single, duplex, or a combination of both. Each receptacle shall be grounded to the reference grounding point by an insulated copper grounding conductor (Figure 17-8).

Patient Equipment Grounding Point. Each patient care area must have a reference grounding point. This reference grounding point is the ground bus of the panelboard or isolated power system supplying the patient care area. As an option, a patient vicinity is permitted to have a patient equipment grounding point. The patient vicinity grounding point acts as a collection point for redundant grounding for the purpose of maintaining equipotential grounding of electric appliances serving the patient area or for grounding other items in order to eliminate electromagnetic interference problems.

Equipotential grounding is a term used to describe a condition where all electrically conductive materials are kept at an equal potential (voltage) to ground and thereby to each other. Each patient bed vicinity can be served by only one reference grounding point, but more than one patient vicinity can be served by one reference grounding point. The reference grounding point is an extension of the grounding terminal bus in either of the two panels supplying that patient vicinity. The reference grounding point is connected to the panel ground bus by an insulated equipment grounding conductor not smaller than 10 AWG copper. An equipment bonding jumper, not smaller than 10 AWG copper, shall be used to connect the grounding terminal to the patient equipment grounding point (Figure 17-9).

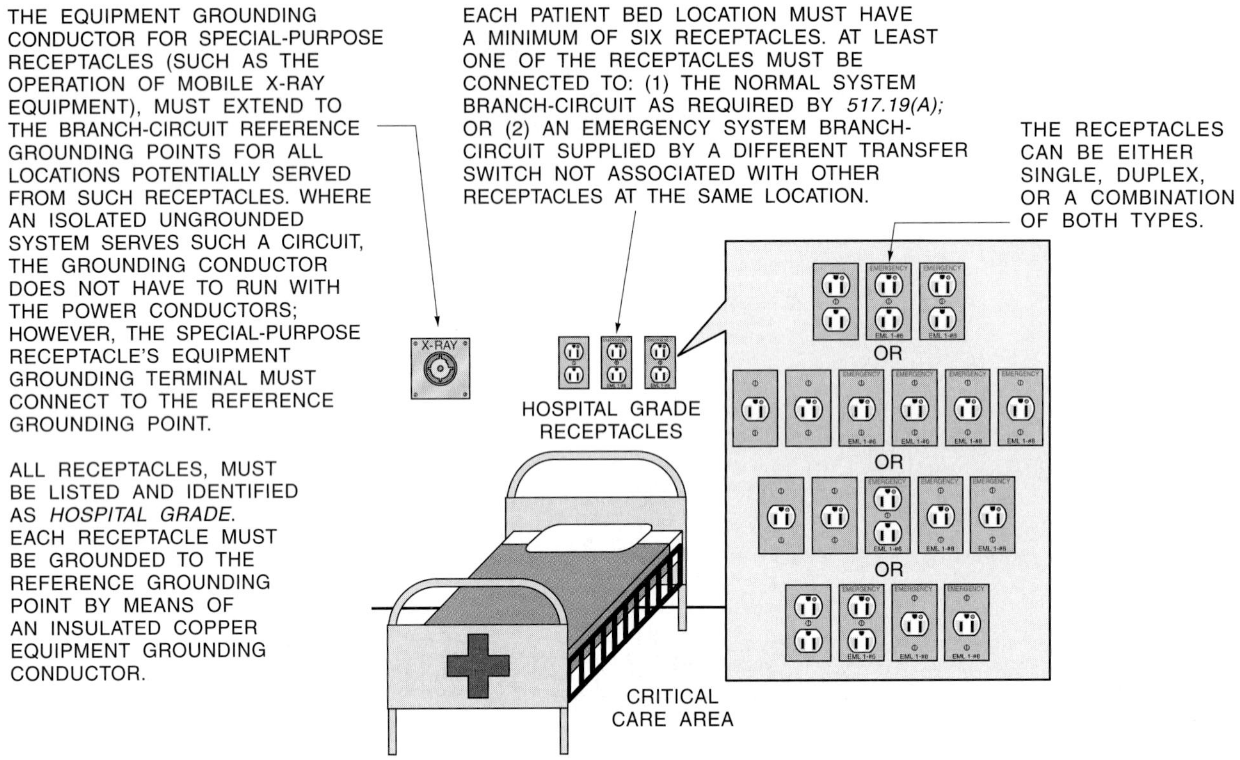

CHAPTER 17 Health-Care Facilities

THE EQUIPMENT GROUNDING CONDUCTOR FOR SPECIAL-PURPOSE RECEPTACLES (SUCH AS THE OPERATION OF MOBILE X-RAY EQUIPMENT), MUST EXTEND TO THE BRANCH-CIRCUIT REFERENCE GROUNDING POINTS FOR ALL LOCATIONS POTENTIALLY SERVED FROM SUCH RECEPTACLES. WHERE AN ISOLATED UNGROUNDED SYSTEM SERVES SUCH A CIRCUIT, THE GROUNDING CONDUCTOR DOES NOT HAVE TO RUN WITH THE POWER CONDUCTORS; HOWEVER, THE SPECIAL-PURPOSE RECEPTACLE'S EQUIPMENT GROUNDING TERMINAL MUST CONNECT TO THE REFERENCE GROUNDING POINT.

ALL RECEPTACLES, MUST BE LISTED AND IDENTIFIED AS *HOSPITAL GRADE*. EACH RECEPTACLE MUST BE GROUNDED TO THE REFERENCE GROUNDING POINT BY MEANS OF AN INSULATED COPPER EQUIPMENT GROUNDING CONDUCTOR.

EACH PATIENT BED LOCATION MUST HAVE A MINIMUM OF SIX RECEPTACLES. AT LEAST ONE OF THE RECEPTACLES MUST BE CONNECTED TO: (1) THE NORMAL SYSTEM BRANCH-CIRCUIT AS REQUIRED BY *517.19(A)*; OR (2) AN EMERGENCY SYSTEM BRANCH-CIRCUIT SUPPLIED BY A DIFFERENT TRANSFER SWITCH NOT ASSOCIATED WITH OTHER RECEPTACLES AT THE SAME LOCATION.

THE RECEPTACLES CAN BE EITHER SINGLE, DUPLEX, OR A COMBINATION OF BOTH TYPES.

Figure 17-8 Patient bed area requirements for hospital grade receptacles and combinations permitted

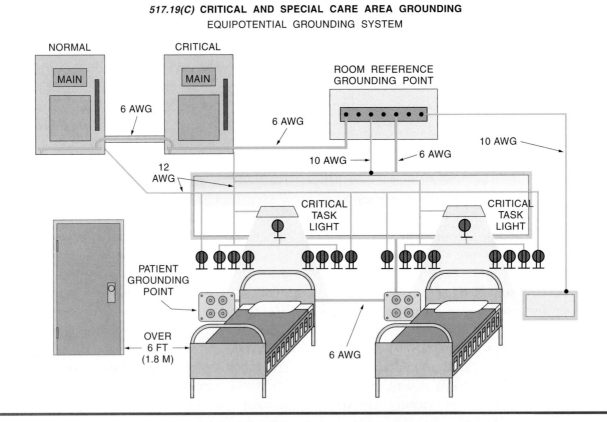

Figure 17-9 Equipotential grounding point

ESSENTIAL ELECTRICAL SYSTEMS

General

The **essential electrical system** is a system that is capable of supplying a limited amount of lighting and power service, which is considered essential for life safety and orderly continuation of procedures during the time the normal electrical service is interrupted for any reason. The essential electrical system varies for hospitals, nursing homes and limited-care facilities, and other health-care facilities.

Essential Electrical Systems for Hospitals

Essential electrical systems for hospitals require two separate systems. These two systems shall be the emergency system and the equipment system. The emergency system is limited to life safety and critical care. These are designated the life safety branch and the **critical branch**. The equipment system shall supply major electrical equipment necessary for patient care and basic hospital operation.

Transfer Switches. Each branch of the emergency system and each equipment system shall have at least one automatic transfer switch (Figure 17-10).

One transfer switch shall be permitted to serve one or more branches or systems in a facility with a maximum demand of 150 kVA (Figure 17-11).

Wiring Requirements. The life safety branch and the critical branch of the emergency system shall be kept independent of all other wiring and equipment and shall not enter the same raceways, boxes, or cabinets with each other or other wiring. The wiring of the emergency system of a hospital shall be mechanically protected by installation in Nonflexible Metal Raceways or shall be wired with Type MI cable.

The branches of the emergency system shall be installed and connected so that functions of the emergency system shall be automatically restored to operation within 10 seconds after interruption of the normal source.

Life Safety Branch. The life safety branch shall supply power for the following lighting, receptacles, and equipment:

1. Illumination of means of egress and exit signs, such as lighting required for corridors, stairways, and all necessary ways of approach to exits.

2. Alarms and alerting systems, such as fire alarms and those required for the piping of nonflammable medical gases.

3. Communications systems where used for issuing instructions during emergency conditions.

4. Generator set location selected receptacles and task illumination battery charger for emergency battery-powered lighting units.

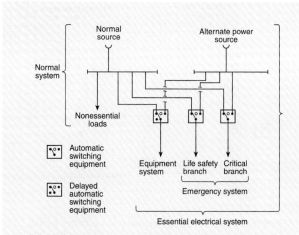

FPN Figure 517.30, No. 1 Hospital — minimum requirement for transfer switch arrangement.

Figure 17-10 Essential electrical system—hospital multiple transfer switches (Reprinted with permission from NFPA 70-2002)

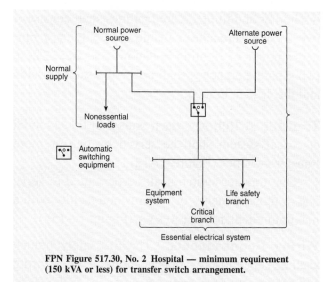

FPN Figure 517.30, No. 2 Hospital — minimum requirement (150 kVA or less) for transfer switch arrangement.

Figure 17-11 Essential electrical system—hospital single transfer switch (Reprinted with permission from NFPA 70-2002)

5. Elevator cab lighting, control, communications, and signal systems for elevators.
6. Automatically operated doors used for building egress.

Critical Branch. The critical branch of the emergency system shall supply power for task illumination, fixed equipment, selected receptacles, and special power circuits serving the following areas:

1. Critical care areas that utilize anesthetizing gases
2. Isolated power systems
3. Patient care areas
4. Nurse call systems
5. Blood, bone, and tissue banks
6. Telephone equipment
7. General care beds (at least one receptacle)

Sources of Power. Essential electrical systems shall have a minimum of two independent sources of power: (1) a normal source generally supplying the entire electrical system, and (2) one or more alternate sources when normal power is interrupted. These sources may be generators or an external utility service when the normal source consists of a generating unit located on the premises.

Essential Electrical Systems for Nursing Homes and Limited Care

Facilities. Essential electrical systems for nursing homes and limited-care facilities shall consist of two separate branches capable of supplying a limited amount of lighting and power service. These two separate branches shall be the life safety branch and the critical branch.

Transfer Switches. Each branch of the essential electrical system shall be served by one or more transfer switches (Figure 17-12).

One transfer switch shall be permitted to serve one or more branches or systems in a facility with a maximum demand of 150 kVA (Figure 17-13).

Wiring Requirements. The life safety branch shall be kept entirely independent of all other wiring and equipment and shall not enter the same raceways, boxes, or cabinets of other circuits that are not a part of the life safety branch. The life safety branch shall be installed and connected to the alternate source of power so that all functions shall be automatically restored within 10 seconds after interruption of the normal source.

Life Safety Branch. The life safety branch shall supply power for the following lighting, receptacles, and equipment:

1. Illumination of means of egress and exit signs, such as lighting required for corridors, stairways, landings, and exit doors and all ways of approach to exits.

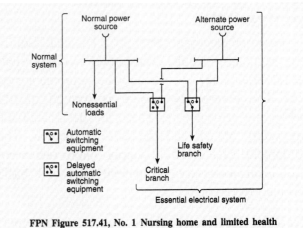

FPN Figure 517.41, No. 1 Nursing home and limited health care facilities — minimum requirement for transfer switch arrangement.

Figure 17-12 Essential electrical system—nursing homes and limited-care facilities (Reprinted with permission from NFPA 70-2002)

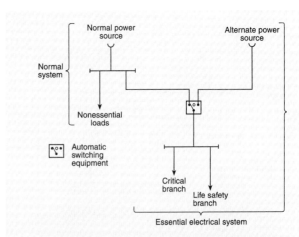

FPN Figure 517.41(B), No. 2 Nursing home and limited health care facilities — minimum requirement (150 kVa or less) for transfer switch arrangement.

Figure 17-13 Essential electrical system—nursing home and limited-care facilities single transfer switch (Reprinted with permission from NFPA 70-2002)

2. Alarms and alerting systems, such as fire alarms and those required for the piping of nonflammable medical gases.
3. Communications systems where used for issuing instructions during emergency conditions.
4. Generator set location selected receptacles and task illumination.
5. Elevator cab lighting, control, communications, and signal systems for elevators.

Critical Branch. The critical branch shall be installed and connected to the alternate source of power so that the following equipment shall be automatically restored to operation at appropriate time-lag intervals following the restoration of the life safety branch to operation.

Delayed Automatic Connection:

1. Patient care areas
2. Sump pumps
3. Smoke control and stair pressurization systems
4. Kitchen hood supply and exhaust systems
5. Ventilating systems for airborne infectious isolation rooms

Delayed Automatic or Manual Connection:

1. Heating equipment for patient rooms
2. Elevator services to allow temporary use of elevators to release trapped passengers
3. Additional illumination, receptacles, and equipment shall be permitted to be connected only to the critical branch

Sources of Power. Essential systems must have a minimum of two independent sources of power: (1) a normal source generally supplying the entire electrical system, and (2) one or more alternate sources for use when the normal source is interrupted.

Alternate Source of Power. The alternate source of power shall be a generator driven by some form of prime mover and located on the premises. Unless the normal source consists of generating units, the alternate source can be either another generator or an external utility service. Nursing homes are permitted to use a battery system.

Isolated Power Systems. An **isolated power system** is comprised of an isolating transformer, a line isolation monitor, and its ungrounded circuit-conductors. Each isolated power circuit shall be controlled by a switch that has a disconnecting pole in each isolated circuit-conductor. Isolation shall be accomplished by having no electrical connection between primary and secondary windings, by means of motor generator sets, or by means of suitably isolated batteries.

Circuits supplying primaries of isolating transformers shall operate at not more than 600 volts between conductors of each circuit. All circuits supplied from such secondaries shall be ungrounded and shall have an approved overcurrent device in each conductor. Circuits supplied directly from batteries or from motor generator sets shall be ungrounded and protected by overcurrent in each conductor.

An isolation transformer shall not supply more than one operating room, except where an induction room serves more than one operating room. The isolated circuits of the induction room shall be permitted to be supplied from the isolation transformer of any one of the operating rooms.

Conductor Identification. The isolated circuit-conductors shall be identified as follows:

1. Isolated conductor No. 1—Orange
2. Isolated conductor No. 2—Brown

For 3-phase systems, the third conductor shall be identified as yellow. Where isolated circuit-conductors supply 125-volt, single-phase, 15- and 20-ampere receptacles, the orange conductor shall be connected to the terminal on the receptacle that is identified for connection of the grounded circuit-conductor.

Line Isolation Monitor. Each isolated power system shall be provided with a continually operating line isolation monitor that indicates total hazard current. The monitor shall be designed so that a green signal lamp remains lighted when the system is adequately isolated from ground. An adjacent red signal lamp and an audible warning signal shall be energized when the total hazard current from either isolated conductor to ground reaches a value of 5 mA.

X-ray Installations. A disconnecting means of adequate capacity for at least 50% of the input required for the momentary rating or 100% of the input required for the long-time rating of the **X-ray** equipment, whichever is greater shall be provided in the supply circuit.

The disconnecting means shall be operable from a location readily accessible from the X-ray location. Transformers that are a part of X-ray equipment shall not be required to comply with *Article 450*.

SUMMARY

The installation criteria and wiring methods used in health-care facilities is intended to minimize the electrical hazard risks by maintaining low-potential differences between exposed conductive surfaces that are likely to become energized and could be contacted by a patient.

Health-care facilities covers a broad range of areas or locations designed to facilitate or ease the burden on persons who are unable to care for themselves or need assistance to maintain adequate living standards.

Hospitals, nursing homes, residential custodial care and other medical assistance areas form the basis for the requirements of *Article 517*. It is an extremely difficult task to separate the areas where these requirements are necessary and the areas where the care does not justify the stringent requirements outlined here. While common sense may seem to be the rule, it is much more justified to have specific requirements that dictate the best manner in which to make electrical installations that satisfy the needs of the patients.

Article 517 begins with pointing out the areas of concern and attempts to eliminate those areas where these stringent requirements are not necessary. To prevent overzealous enforcement of unnecessary precautions, *Part II* qualifies areas where the requirements of *Article 517* do not apply.

Generally, power to health-care facilities is provided by two systems: (1) the normal power, and (2) the essential electrical system that is energized upon interruption of the normal source of power. The essential electrical systems require two separate systems. These are the emergency and the equipment systems. The emergency system is divided into two systems: (1) the life safety, and (2) the critical branch. As shown in Figure 17-9, three transfer switches are fed from the normal source and the alternate source. When the normal source of power is interrupted, the transfer switches *transfer* the load to the alternate source of power. As shown in Figure 17-10, transfer can be done with one transfer switch if the maximum demand on the essential system does not exceed 150 kVA.

Proper grounding techniques are essential in health-care facilities to prevent noncurrent-carrying conductive materials from becoming energized where within reach of patients. Sensitive areas, such as operating rooms and induction rooms, require isolated wiring systems. These systems are ungrounded and must be monitored to assure the integrity of the isolation system. Isolation systems are generally provided through the use of isolating transformers.

REVIEW QUESTIONS

MULTIPLE CHOICE

1. The wiring method for patient care areas permits the use of:
   ____ A. Type MI cable.
   ____ B. Type AC cable.
   ____ C. Type MC cable.
   ____ D. all of the above.

2. A patient vicinity is an area extending _____ above the floor.
   _____ A. 6 ft (1.8 m)
   _____ B. 7½ ft (2.5 m)
   _____ C. 12 ft (3.7 m)
   _____ D. none of the above

3. In general care areas, each patient location bed must be supplied by at least _____ branch-circuit(s).
   _____ A. three
   _____ B. four
   _____ C. two
   _____ D. one

4. In critical care areas, each patient bed location must be provided with at least _____ receptacles.
   _____ A. two
   _____ B. three
   _____ C. four
   _____ D. six

5. The reference grounding point is connected to the panel ground bus by an insulated equipment grounding conductor not smaller than _____ AWG.
   _____ A. 8
   _____ B. 6
   _____ C. 12
   _____ D. 10

6. Essential electrical systems for hospitals require _____ separate systems.
   _____ A. two
   _____ B. three
   _____ C. six
   _____ D. none of the above

7. On a 3-phase system in an isolated system, the third conductor shall be identified as _____.
   _____ A. yellow
   _____ B. green
   _____ C. white
   _____ D. black

8. Patient care areas of all health-care facilities do not apply to :
   _____ A. business offices.
   _____ B. corridors.
   _____ C. waiting rooms.
   _____ D. all of the above.

9. A reference grounding point is the:
   _____ A. ground bus in the panel supplying the area.
   _____ B. collection point for redundant grounding.
   _____ C. grounding electrode.
   _____ D. neutral bus in the lighting panel.

304 CHAPTER 17 Health-Care Facilities

10. All receptacles at patient bed locations must be marked:
    ____ A. hospital grade.
    ____ B. 20-ampere rated.
    ____ C. at least 18 in. (450 mm) above the floor.
    ____ D. none of the above.

TRUE OR FALSE

1. A wet location is one that is subject to wet conditions while patients are present.
   ____ A. True
   ____ B. False
   Code reference _____.

2. In critical areas, each patient bed location must be fed from at least three branch-circuits.
   ____ A. True
   ____ B. False
   Code reference _____.

3. In general care areas, each patient bed location must be provided with a minimum of two receptacles.
   ____ A. True
   ____ B. False
   Code reference _____.

4. In general care areas, each patient bed location must be supplied by at least four branch-circuits.
   ____ A. True
   ____ B. False
   Code reference _____.

5. Hospital grade receptacles are marked *hospital grade* and have a green dot on the face of the receptacle.
   ____ A. True
   ____ B. False
   Code reference _____.

6. Each patient care area must have a reference grounding point.
   ____ A. True
   ____ B. False
   Code reference _____.

7. In hospitals, there can only be one transfer switch from the normal power source to the essential electrical system.
   ____ A. True
   ____ B. False
   Code reference _____.

8. One transfer switch shall be permitted to serve one or more branches or systems in a facility with a maximum demand of 150 kVA.
   ____ A. True
   ____ B. False
   Code reference _____.

9. The life safety branch and the critical branch wiring may be run in the same raceway.
   ____ A. True
   ____ B. False
   Code reference _____ .

10. An isolated power system must be provided with an equipment grounding conductor.
    ____ A. True
    ____ B. False
    Code reference _____ .

FILL IN THE BLANKS

1. A patient care area is any portion of a _____ where patients are intended to be examined or treated.

2. The patient vicinity is the space with surfaces likely to be _____ or an attendant who can touch the patient.

3. All receptacles must be listed as _____ grade and be so _____ .

4. In critical care areas, each patient bed location shall be fed from _____ branch-circuits.

5. At least one of the receptacles must be connected to a _____ system branch-circuit or to an _____ system branch-circuit supplied by a different transfer switch.

FREQUENTLY ASKED QUESTIONS

Question 1: Are all of the receptacles at patient bed locations in general care areas required to be on the emergency system?

Answer: In health-care facilities, the auxiliary or alternate power sources that provide power to designated areas and functions are called the essential electrical system. This essential electrical system is divided into two branches designated as the emergency system and the normal equipment system. The emergency system is further divided into two branches called the life safety branch and the critical branch (see Figure 17-10). *NEC® 517.18* requires at least two branch-circuits, one from the normal and one from the emergency. It does not specify which branch of the emergency, that is, the life safety or the critical branch. *NEC® 517.33(A)(8)(a)* requires that the critical branch shall supply at least one duplex receptacle for each general care bed. Based on the previous text, only one duplex receptacle in a general care patient bed location need be on the critical branch of the emergency system.

Question 2: What is the difference between the two branches of the emergency system of an essential electrical system?

Answer: The life safety branch of the emergency system of the essential electrical system supplies power for the functions listed in *517.32*. No function other than those listed in that section shall be connected to the life safety branch. Briefly these functions are: (a) illumination of means of egress, (b) exit signs, (c) alarm and alerting systems, (d) communications systems, (e) generator set location, (f) elevators, and (g) automatic doors.

The critical branch of the emergency system of the essential electrical system supplies power to the functions listed in *517.33(A)*. Briefly these functions are: (1) critical areas that utilize anesthetizing gases; (2) isolated power systems; (3) patient care areas selected receptacles; (4) patient care task illumination; (5) nurse call systems; (6) blood, bone, and tissue banks; (7) telephone equipment rooms; (8) task illumination, selected receptacles, and selected power circuits for general care beds, etc.; and (9) additional illumination, receptacles, and power circuits needed for effective hospital operation.

Question 3: How many transfer switches are required for an essential electrical system in a nursing home?

Answer: *NEC® 517.41 No. 2* requires that each branch of the essential electrical system be served by one or more transfer switches. This would mean at least two—one for the critical branch and one for the life safety branch. However, if a facility has a maximum demand on the essential electrical system of 150 kVA, one transfer switch is permitted (see Figure 17-13).

Question 4: What is an isolated power system used for in a hospital health-care facility?

Answer: An isolated power system is used to derive ungrounded circuit-conductors. In operating rooms and their anesthetizing induction rooms, it is considered to be less dangerous to use ungrounded systems where sparking or an arc due to ground-faults is eliminated. The problem with the ungrounded system is if one phase conductor becomes accidentally grounded, there is no problem or indication and if another ground should occur on an opposite phase, we have a short-circuit between phases at phase voltage. This could result in severe arcing or burning and cause severe damage. If this were to occur in an operating room or induction room where oxygen or other flammable anesthetics are used, serious damage or death could result from explosion or fire.

For this reason, a line isolation monitor is employed to detect if the system is adequately isolated from ground. A green signal lamp remains lighted when the system is adequately isolated from ground. A red signal lamp and an audible warning signal are energized if the leakage to ground from any isolated conductor reaches a value of 5 mA. Further information on line isolation monitoring can be found in *517.160(B)*.

Question 5: Are isolated circuit-conductors required to be color-coded?

Answer: Yes. *NEC® 517.160(A)(5)* requires that isolated circuit-conductors be identified with isolated conductor No. 1—orange, isolated conductor No. 2—brown, and for 3-phase systems, the third conductor—yellow.

If the isolated conductors feed 125-volt, single-phase, 15- or 20-ampere receptacles, the orange conductor shall be connected to the terminal on the receptacle identified for connection to the grounded conductor.

Question 6: Where are the audible and visual indicators of the line isolation monitor required to be located?

Answer: According to *517.160(B)(1)*, the audible and visual indicators may be located remote from the isolated area. *NEC® 517.19(E)*—the audible and visual indictors can be located at the nurses' station. This appears to be a logical place for them.

Question 7: What is the purpose in all the grounding safeguards required for health-care facilities if they do away with grounding in areas such as operating rooms?

Answer: The non-grounding aspect has to do with areas where flammable or explosive gases or compounds are used to reduce the possibility of a damaging explosion. Grounding techniques required in other areas pertain to electric shock or electrocution. In a health-care facility, conductive paths from the patient's body to a grounded object is difficult to prevent because the path may be established accidentally or by instrumentation connected directly to the patient's body. To control the electric shock hazard, we must limit the amount of electric current that may flow in an electric circuit in which the patient's body is a part. Control of the electric current that may flow in a circuit that involves the patient's body is done by raising the resistance of the electric circuit by insulating exposed surfaces that could become energized, and by reducing the potential differences between exposed conductive surfaces.

Question 8: What method is used to ground receptacles and fixed electric equipment in patient care areas?

Answer: *NEC® 517.13(A) and (B)* require that all branch-circuits serving patient care areas must be provided with a ground path for fault-current by installation in a metal raceway or a cable armor or sheath assembly that qualifies as an equipment grounding conductor return path. Type AC, Type MC, and Type MI cables shall have an outer metal armor or sheath that is identified as an acceptable grounding return path. In patient care areas, the grounding terminals of all receptacles shall be grounded by an insulated grounding conductor. The grounding conductor must be installed in metal raceways with the branch-circuit conductors supplying these receptacles. This means that redundant grounding is required for patient care area circuits supplying receptacles or fixed electrical equipment. The metal raceway or cable armor must be supplemented with an insulated equipment grounding conductor.

Question 9: What is the difference between a reference grounding point and a patient equipment grounding point in a health-care facility?

Answer: The *reference grounding point* is the ground bus of the panelboard supplying the patient care area. If more than one panelboard is serving the same patient vicinity, they must be bonded together with an insulated copper conductor not smaller than 10 AWG. The *patient equipment grounding point* is a terminal that collects the redundant grounding conductors of electrical appliances. This patient equipment grounding point is required to be connected to the reference grounding point by a bonding jumper not smaller than 10 AWG.

The patient equipment grounding point is generally a copper bus bar or strip located in the patient vicinity area to which all exposed conductive surfaces would be bonded by means of a 10 AWG grounding conductor.

Question 10: Where are GFCIs required in a health-care facility?

Answer: The use of GFCIs is restricted for usage only where interruption of power under fault conditions can be tolerated. Wet locations for patient care areas are required to have GFCI protection if interruption of power will not cause more severe problems—*517.20(A)*. GFCI protection for personnel is not required for receptacles installed in critical areas where the basin and toilet are installed within the patient room.

CHAPTER 18

Mobile Homes, Manufactured Homes, and Mobile Home Parks

OBJECTIVES

After completing this chapter, the student should understand:
- requirements for a mobile home electrical service.
- mobile home disconnecting means.
- branch-circuit requirements.
- receptacle locations.
- grounding requirements.
- how to calculate the feeder assembly electrical load.

KEY TERMS

Arc-Fault Circuit-Interrupter (AFCI): An electrical device designed to open the circuit and stop the arc fault and its accompanying high intensity heat before a fire is likely to ignite

Bonding: The joining of metallic parts to form an electrically conductive path that will ensure electrical continuity, and the capacity to conduct safely any current likely to be imposed

Ground-Fault Circuit-Interrupter (GFCI): A device intended to protect personnel from electrical shock by de-energizing the circuit within a given time when the current to ground exceeds a given value

Leg: A portion of a circuit, such as the conductor on either side of neutral in a 3-wire circuit

Manufactured Home: A structure that is transportable in one or more sections, 8 ft (2.5 m) or more in width and 40 ft (12 m) or more in length in the traveling mode, and when erected on-site 320 sq ft (30 sq m) or more. It is built on a chassis and designed to be a dwelling with or without a permanent foundation where connected to utilities and electric systems contained therein

Mast Weatherhead: A conduit raised above the ground to a specified height with a conduit fitting used to allow a conductor entry while preventing weather entry

Mobile Home: A factory-assembled structure or structures (transportable in one or more sections) that is built on a permanent chassis and designed to be used as a dwelling without permanent foundation where it is connected to utilities

Panelboard: A panel which includes buses, main disconnect, and branch-circuit overcurrent protection

Pipe Heating Cable: High-resistance cable that supplies heat to prevent water pipes from freezing

Power-Supply Cord: An electrical cord listed for mobile home use, designed for the purpose of delivery of energy from the electrical supply to the distribution panelboard in the mobile home

CHAPTER 18: Mobile Homes, Manufactured Homes, and Mobile Home Parks

INTRODUCTION

This chapter will cover the electrical conductors and equipment installed in mobile or manufactured homes. The requirements for **mobile homes**, manufactured homes, and mobile home parks are contained in *Article 550*. A mobile home is built on a permanent chassis and transported to the mobile park site and set on a foundation that is not considered a permanent foundation. A **manufactured home** is built on a chassis and transported to the building site and set in place with or without a permanent foundation. The following text will explain the different requirements for each type of unit relating to the service feeder and internal wiring. Figure 18-1 shows a typical mobile home park.

GENERAL

Some mobile homes are not intended as dwelling units; for example, contractor's on-site offices, construction site dormitories, banks, clinics, and mobile stores. Mobile homes used for these purposes are not required to meet all the provisions of *Article 550* pertaining to number or capacity of circuits required. It shall, however, meet all the requirements of *Article 550* with respect to a 120-volt or 120/240-volt system.

Mobile homes installed in other than mobile home parks are subject to the requirements of *Article 550*. All mobile homes intended for connection to wiring systems rated 120/240-volts, 3-wire ac, with a grounded neutral must comply with the requirements for mobile homes.

Electrical Power Supply

NEC® 550.10 requires that the power supply to a mobile home must be a feeder assembly consisting of a listed 50-ampere **power-supply cord** with a securely attached plug cap. The ampere rating of the cord may be reduced to 40-amperes, providing the mobile home was manufactured with gas or oil central-heating equipment and gas-cooking appliances.

The power cord must be attached permanently to the distribution **panelboard** while the free end terminates in an attachment plug cap. The cord must be a 4-wire cord, one conductor of which must be identified by a continuous green color or a continuous green color with one or more yellow stripes for use as the grounding conductor.

The plug cap must be a 3-pole, 4-wire grounding type, rated 50 amperes, 125/250 volts. Receptacle and plug configuration is shown in Figure 18-2.

The overall length of the power cord is not to

Figure 18-1 Typical mobile home park

CHAPTER 18 Mobile Homes, Manufactured Homes, and Mobile Home Parks

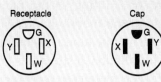

Figure 550.10(C) 50-ampere, 125/250-volt receptacle and attachment plug cap configurations, 3-pole, 4-wire, grounding-types, used for mobile home supply cords and mobile home parks.

Figure 18-2 Receptacle configuration, NEC® Figure 550.10(C), (Reprinted with permission from NFPA 70-2002)

exceed 36.5 ft (11 m) or be less than 21 ft (6.4 m) as measured from the end, including bare leads to the face of the attachment plug face. Further, the length from the face of the attachment plug to the point of entry in the mobile home is at least 20 ft (6 m). The power cable must be labeled to show its use and ampacity. The cable requires the following markings: *For use with mobile homes—40 amperes* or *For use with mobile homes—50 amperes* (Figure 18-3). Power cables that pass through walls or floors are required to be protected by a continuous raceway having a maximum size of 1¼ in. (32 mm) from the

THE ATTACHMENT PLUG CAP MUST NOT ONLY BE A 3-POLE, 4-WIRE, GROUNDING TYPE, RATED 50 AMPERES, 125/250 VOLTS WITH A CONFIGURATION AS SHOWN IN 550.10(C), BUT ALSO MUST BE INTENDED FOR USE WITH A 50-AMPERE, 125/250-VOLT RECEPTACLE CONFIGURATION AS SHOWN IN FIGURE 550.10(C). IT MUST BE LISTED, INDIVIDUALLY OR AS PART OF A POWER-SUPPLY CORD ASSEMBLY, FOR SUCH PURPOSE, AND MUST BE MOLDED TO (OR INSTALLED ON) THE FLEXIBLE CORD AT THE POINT WHERE THE CORD ENTERS THE ATTACHMENT PLUG CAP 550.10(C).

IN A RIGHT-ANGLE CAP CONFIGURATION THE GROUNDING MEMBER MUST BE FARTHEST FROM THE CORD 550.10(C).

THE MOBILE HOME POWER SUPPLY MUST BE A FEEDER ASSEMBLY CONSISTING OF A SINGLE LISTED 50-AMPERE MOBILE HOME POWER-SUPPLY CORD HAVING AN INTEGRALLY MOLDED (OR SECURELY ATTACHED) PLUG CAP, OR A PERMANENTLY INSTALLED FEEDER 550.10(A).

THE POWER-SUPPLY CORD MUST BEAR THE MARKING: *FOR USE WITH MOBILE HOMES — 40 AMPERES* OR *FOR USE WITH MOBILE HOMES — 50 AMPERES* 550.10(E).

THE CORD MUST BE A 4-CONDUCTOR LISTED TYPE. ONE CONDUCTOR MUST BE IDENTIFIED BY A CONTINUOUS GREEN COLOR, OR A CONTINUOUS GREEN COLOR WITH ONE OR MORE YELLOW STRIPES, FOR USE AS THE GROUNDING CONDUCTOR 550.10(B).

A SUITABLE CLAMP (OR EQUIVALENT) MUST BE PROVIDED AT THE DISTRIBUTION PANEL-BOARD KNOCKOUT TO AFFORD CORD STRAIN RELIEF AND EFFECTIVELY PREVENT STRAIN TRANSMISSION TO THE TERMINALS WHEN THE POWER-SUPPLY CORD IS HANDLED AS INTENDED 550.10(B).

A MOBILE HOME POWER-SUPPLY CORD, IF PRESENT, MUST BE PERMANENTLY ATTACHED TO THE DISTRIBUTION PANELBOARD EITHER DIRECTLY OR VIA A JUNCTION BOX ALSO PERMANENTLY CONNECTED THERETO, WITH THE FREE END TERMINATING IN AN ATTACHMENT PLUG CAP 550.10(B).

FROM THE END OF THE POWER-SUPPLY CORD (INCLUDING BARED LEADS) TO THE FACE OF THE ATTACHMENT PLUG CAP, THE CORD MUST NOT BE LESS THAN 21 FT (6.4 M) NOR MORE THAN 36½ FT (11 M) IN LENGTH. FROM THE FACE OF THE ATTACHMENT PLUG CAP TO THE MOBILE HOME ENTRY POINT, THE CORD MUST BE AT LEAST 20 FT (6 M) LONG 550.10(D).

Figure 18-3 Mobile home power-supply cord

panelboard to the underside of the mobile home floor (Figure 18-4). If the load exceeds 50 amperes, or where a permanent feeder is used, the supply must be as follows:

1. A **mast weatherhead** installation according to *Article 230*, containing four insulated color-coded feeder conductors. One of which is the grounding conductor.

2. A metal raceway or RMC from the disconnecting means in the mobile home to a junction box, or a fitting to the raceway on the underside of the mobile home. The manufacturer of the mobile home is required to provide written installation instructions that state the proper feeder conductor sizes for the raceway and the size of the junction box.

Disconnecting Means

NEC® 550.11 permits the power feeder disconnecting means and the branch-circuit equipment to be combined as a single assembly and designed as a

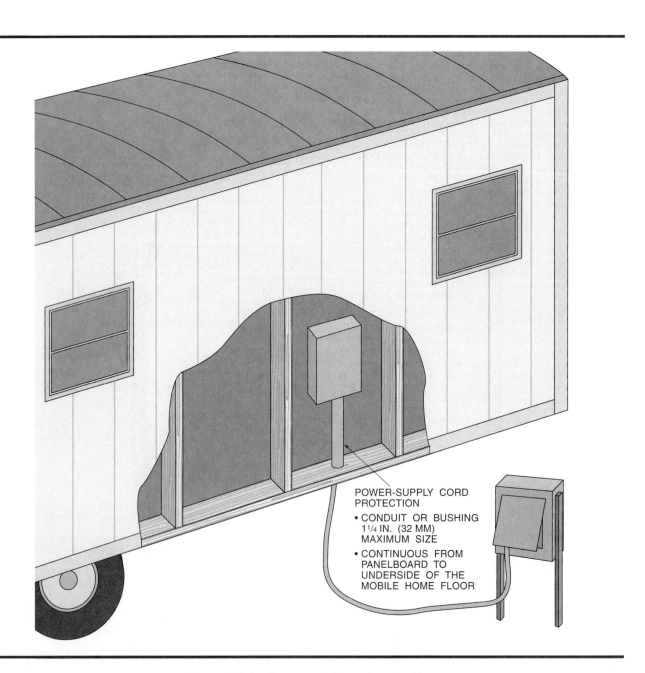

Figure 18-4 **Power-supply cord protection**

CHAPTER 18 Mobile Homes, Manufactured Homes, and Mobile Home Parks

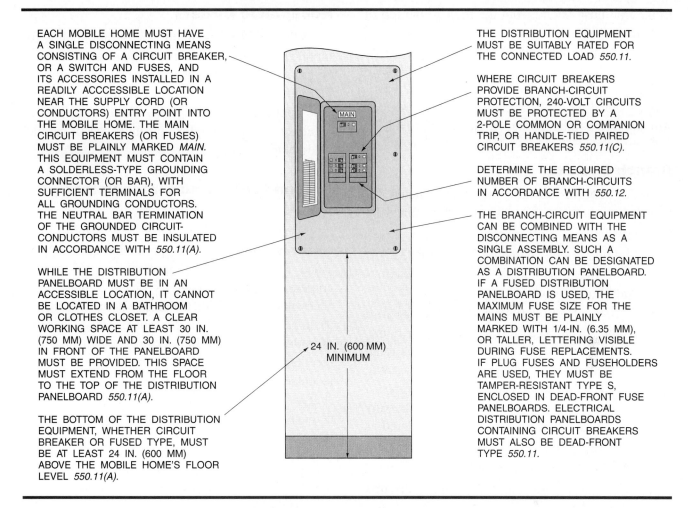

Figure 18-5 Mobile home panelboard

distribution panelboard. If the panelboard uses fuses for the main disconnect, the fuse size must be marked with ¼ in. (6 mm) high letters and visible when the fuses are changed. Also, the main circuit breakers or fuses are to be clearly marked *main* as shown in Figure 18-5.

The distribution panelboard rating should be at least 50 amperes and consist of a 2-pole circuit breaker rated 40 amperes for a 40-ampere supply cord or 50 amperes for a 50-ampere supply cord. *NEC® 550.11(A)* requires that a panelboard using a disconnect switch and fuses must have the fuseholders rated at 60 amperes. The panelboard location must be accessible, not located in bathrooms or clothes closets, and at least 24 in. (600 mm) above the mobile home floor. A clear working space (Figure 18-6) in front of the panelboard should be at least 30 in. (750 mm) in width, 30 in. (750 mm) in

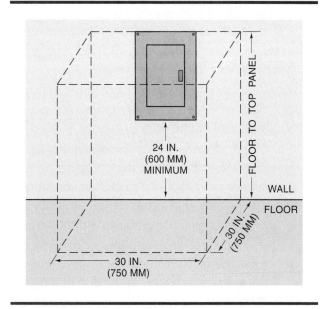

Figure 18-6 Accessible—clear work area

front, and must extend from the floor to the top of the panelboard.

A nameplate is required adjacent to the feeder assembly entrance that has the following information engraved on it: *This Connection for 120/240-volt, 3-pole, 4-wire, 60-Hertz, _____ ampere supply.* Ampere rating to be marked in the space provided.

Branch-Circuits

NEC® 550.12 contains the requirements for branch-circuits. Branch-circuits are determined by the following:

1. Lighting circuits based on 3 VA/sq ft (33 VA/sq m) times the outside dimensions (length × width) of the mobile home.

    Example: Two sections, 10 ft wide × 50 ft long, to form a 20 ft × 50 ft mobile home. 3 VA/sq ft × 1000 sq ft = 3000 VA total; 3000 VA/120 volts = 25 amperes total; number of 15-ampere circuits = 25 amperes/15-ampere per circuit = 1.7 circuits. Any fraction of a circuit must be considered as an additional circuit.

    Two 15-ampere lighting circuits [*550.12(A)*]

2. Two 20-ampere small-appliance circuits [*550.12(B)*]

3. One 20-ampere laundry circuit [*550.12(C)*]

4. One 20-ampere bathroom circuit [*550.12(E)*]

5. General Appliances—furnace, water heater, range, air conditioner, dishwasher, waste disposer, and others [*550.12(D)*]. There may be one or more circuits of adequate rating in accordance with the following:

    a. Ampere rating of fixed appliances not over 50% of circuit rating if receptacles are on the same circuit

    b. Fixed appliances on a circuit without lighting outlets, the sum of rated amperes should be not more than the branch-circuit rating. Continuous duty loads or motor loads are not to exceed 80% of the circuit rating

    c. The rating of cord- and plug-connected appliances on a circuit having no other outlets are restricted to 80% of the circuit rating

    d. Rating of the range circuit depends on range demand shown in *Article 550.18(B)*.

Receptacle Outlets

NEC® 550.13 contains the requirements for mobile home receptacle outlets. All receptacles must be grounding type and be rated at 15- or 20-ampere, 125-volt (Figure 18-7), except for those serving specific appliances (Figure 18-8). The receptacles can be either single or duplex and must accept parallel blade plugs.

Ground-Fault Circuit-Interrupters (GFCI)

In accordance with *550.13(B)*, **ground-fault circuit-interupter (GFCI)** protection must be provided for all 125-volt, single-phase, 15- or 20-ampere receptacles outdoors or in compartments accessible from outside the mobile home (Figure 18-9), in bathrooms including receptacles that are a part of a luminaire (lighting fixture) (Figure 18-10), receptacle outlet(s) serving kitchen countertops, and receptacles within 6 ft (1.8 m) of a wet bar sink (Figure 18-11). Receptacles serving appliances in dedicated spaces, such as a dishwasher, waste disposer, refrigerator, freezer, and laundry equipment do not require GFCI protection (Figure 18-12).

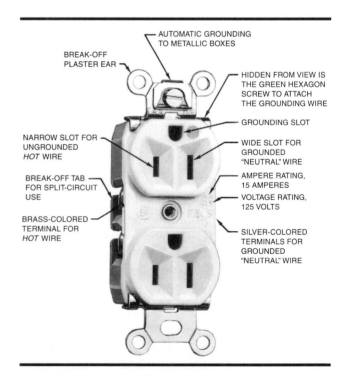

Figure 18-7 15-ampere, 125-volt grounding-type receptacle

CHAPTER 18 Mobile Homes, Manufactured Homes, and Mobile Home Parks 315

Figure 18-8 20-ampere, 250-volt single grounding-type receptacle for specific appliances—air conditioner

Figure 18-9 Receptacle outlet accessible from outside the mobile home

Figure 18-10 Bathroom receptacle ground-fault circuit-interrupter protected

Figure 18-11 Receptacles installed within 6 ft (1.8 m) of wet bar sink must be GFCI protected

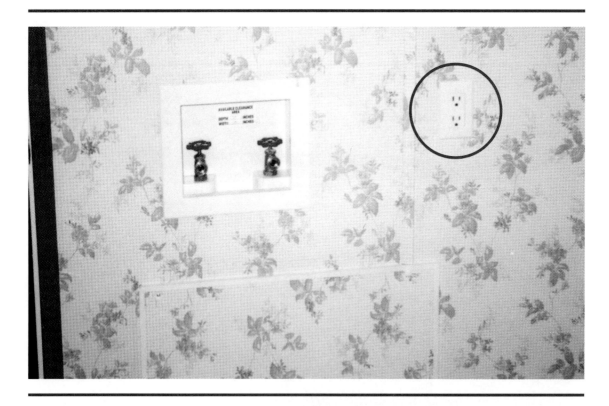

Figure 18-12 Laundry receptacle does not require ground-fault circuit-interrupter protection

Receptacle outlets are to be installed at wall spaces 2 ft (600 mm) wide or more so that no point along the floor is more than 6 ft (1.8 m) measured horizontally from an outlet in that space (Figure 18-13). Receptacle outlets are required in the following locations:

Figure 18-13 Receptacles should not be more than 12 ft (3.7 m) apart (*210.52(A)(1)*)

1. Over or adjacent to countertops in the kitchen, at least one on each side of the sink if countertops are on each side and are 12 in. (300 mm) or more in width (Figure 18-14)
2. Adjacent to the refrigerator and freestanding gas range space. A duplex receptacle can serve both a countertop and refrigerator (Figure 18-15)
3. Countertop spaces for built-in vanities (Figure 18-16)
4. Countertop spaces under wall-mounted cabinets (Figure 18-17)
5. In the wall at the nearest point to where a bar type counter attaches to the wall

CHAPTER 18 Mobile Homes, Manufactured Homes, and Mobile Home Parks

Figure 18-14 Receptacles should not be more than 24 in. (600 mm) from each side of kitchen sink (*210.52(C)(1)*)

Figure 18-15 Refrigerator and countertop receptacles

Figure 18-16 Countertop receptacle at bathroom vanity

Figure 18-17 Kitchen countertop receptacle

320 CHAPTER 18 Mobile Homes, Manufactured Homes, and Mobile Home Parks

6. In the wall at the nearest points where a fixed room divider attaches to the wall

7. In the laundry area with 6 ft (1.8 m) of the intended location of the clothes washer (Figure 18-18)

8. At least one receptacle (outdoor accessible) at grade level and not more than 6½ ft (2.0 m) above grade (Figure 18-19)

9. At least one receptacle outlet installed in bathrooms within 36 in. (900 mm) of the outside edge of each basin. Installation of receptacles in bathroom cabinets is prohibited (Figure 18-20).

The water utility line to a mobile home is protected from freezing during the winter months by a "**pipe heating cable**." A receptacle outlet is required and must be located within 2 ft (600 mm) of the water inlet, and connected to an interior branch-circuit other than the small-appliance circuit. It is acceptable to use the bathroom receptacle circuit or a circuit where all of the outlets are on the load-side of a feedthrough GFCI (Figure 18-21).

Figure 18-18 Laundry area receptacle

Figure 18-19 Outside receptacle less than 6 ft (1.8 m) above grade

Figure 18-20 Receptacle outlet installed in bathroom at 36 in. (900 mm) of the outside edge of each basin (*210.52(D)*)

Figure 18-21 Receptacle for heat trace on water pipe

Figure 18-22 Receptacle outlet not permitted above baseboard heater

in or within reach [30 in. (750 mm)] of a shower or bath tub. *NEC® 550.13(F)(3)* prohibits the installation of receptacles above electric baseboard heater due to the problem of possible deterioration of the insulation on electrical cords resting on top of an electric baseboard (Figure 18-22).

Receptacles are not required in the wall space occupied by built-in kitchen or wardrobe cabinets; in a wall space behind doors that open against the wall surface; on lattice-type room dividers which are less than 8 ft (2.5 m), and within 6 in. (150 mm) of the floor; and wall space afforded by bar type counters.

Because of the possibility of dirt or water intrusion, receptacle outlets are not permitted to be installed in a face-up position in any countertop. Receptacle outlets are not permitted to be installed

Wiring Methods

Wiring methods and materials set forth in *NEC® Chapter 3* must be used for mobile homes unless limited by *Article 550*. *NEC® 550.15* does not permit aluminum conductors, aluminum alloy conductors, and aluminum core conductors such as copper-clad aluminum to be used for branch-circuit wiring.

Nonmetallic boxes can only be used with nonmetallic cable or nonmetallic raceways. Where Nonmetallic and Metal-Sheathed cable pass through 2 × 2 studs or other frames where the 1¼ in. (32 mm) clearance between the cable and the inside or outside of the studs is not maintained, steel plates not less than 0.053 in. (1.35 mm) wall thickness are required on either side of the cable (Figure 18-23).

Connecting a range, clothes dryer, or similar appliance, a length of Metal-Sheathed cable or FMC not less than 3 ft (900 mm) is required to permit movement of the appliance. *NEC® 550.15(F)* requires that where RMC or IMC is terminated at the enclosure with a locknut and bushing connection, two locknuts are required on inside and outside of the enclosure (Figure 18-24).

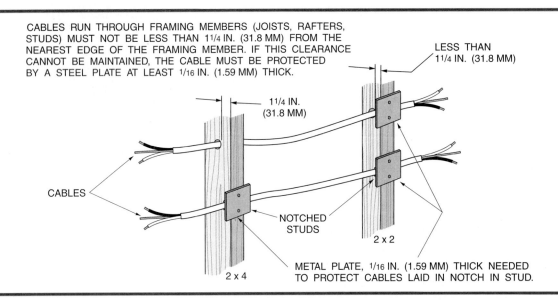

Figure 18-23 Wiring protection when 1¼ in. (32 mm) clearance is not maintained

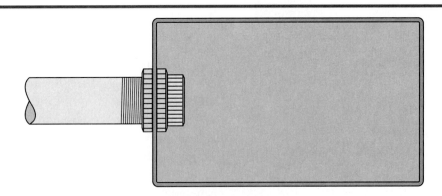

Figure 18-24 Double locknut and bushing connection

Wiring that is exposed to weather and under-chassis is required to be protected by RMC or IMC and the conductors must be suitable for wet locations.

Grounding and Bonding

In accordance with *550.16*, grounding of electrical and nonelectrical metal parts in a mobile home is required through the connection to the grounding bus in the distribution panelboard. The grounding conductor is connected to the grounding bus through the green-colored insulated conductor in the supply cord (or the feeder wiring) to the service ground in the service-entrance equipment that is located adjacent to the mobile home location. The service equipment shall be located in sight not more than 30 ft (9.0 m) from the exterior wall of the mobile home it serves (Figure 18-25).

Figure 18-25 Service location—not on mobile home

The Code does permit service equipment to be located elsewhere on property, provided that a disconnecting means for the service equipment is located in sight or not more than 30 ft (9.0 m) from the exterior wall of the mobile home it serves. The mobile home frame or the appliance frames should not be connected to the grounded (neutral) conductor. When service equipment is installed in or on a manufactured home, as set forth in *550.32(B)*, the neutral (grounded) bus and grounding bus are connected in the distribution panelboard (Figure 18-26).

Noncurrent-carrying metal parts that may become energized must be bonded to the grounding terminal or enclosure of the distribution panelboard. A **bonding** conductor is required between the distribution panelboard and an accessible terminal on the chassis. The bonding conductor must be solid or stranded, insulated or bare, and must be a minimum 8 AWG copper. Metallic piping and ducts, such as water lines, waste pipes, and air-circulating ducts are considered bonded if they are connected to the terminal on the chassis. The metal roofs and side panels are considered bonded if they overlap and are secured to the frame with metallic fasteners, and if the lower metallic exterior covering is secured by metal fasteners at the chassis cross member by two metal straps.

In accordance with *550.16(A)(1)*, the grounded conductor (neutral) must be insulated from the grounding conductors and from equipment enclosures. This requirement is to prevent neutral current to flow through the equipment grounding conductors.

Load Calculation

Based on a 3-wire, 120/240-volt supply with 120-volt loads balanced on each side of neutral, the following method is required for computing the

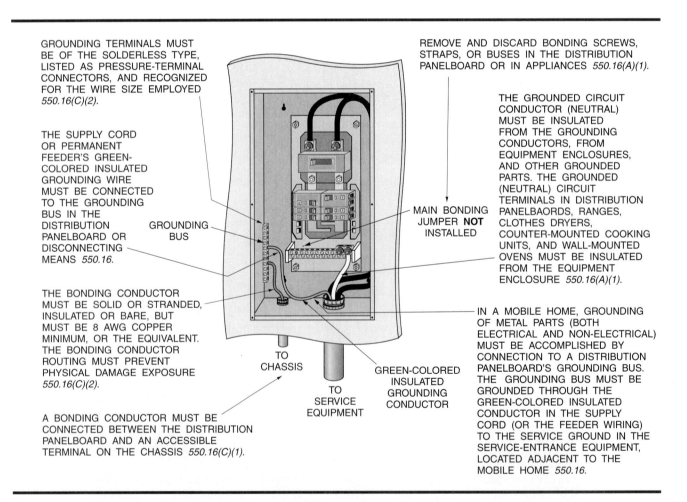

Figure 18-26 Grounding at panelboard

supply-cord and distribution panelboard load for each mobile home.

- Lighting load: Multiply the floor area, outside dimensions (length and width), times 3 VA/sq ft (33 VA/sq m).
- Small appliance load: Number of 20-ampere small-appliance circuits times 1500 volt-amperes.
- Laundry load: 1500 volt-amperes for each circuit.
- Net Load: First 3000 total volt-amperes at 100%, plus remainder at 35%, equals net volt-amperes to be divided by 240 volts to obtain the current for each **leg**.
- Total load for determining power supply: The total load is the sum of the following:
  - Lighting, small-appliance, and laundry loads as calculated previously, current per leg.
  - Nameplate amperes for motors and heater loads: Omit the smaller of heating and cooling loads and add 15 amperes per leg when using a 40-ampere power cord when the air conditioner is not installed.
  - Twenty-five percent of the current of the largest motor listed above.
  - Total of the nameplate amperes for waste disposer, dishwasher, water heater, clothes dryer, wall-mounted oven, and cooking units. Where the number of appliances exceeds three, use 75% of the total.
  - Determine the current for a freestanding range (as distinguished from wall ovens and cooktop units) by dividing the values in *Table 550.18* by 240 volts (Figure 18-27).

Nameplate Rating (watts)	Use (volt-amperes)
0 – 10,000	80 percent of rating
Over 10,000 – 12,500	8,000
Over 12,500 – 13,500	8,400
Over 13,500 – 14,500	8,800
Over 14,500 – 15,500	9,200
Over 15,500 – 16,500	9,600
Over 16,500 – 17,500	10,000

Figure 18-27 *NEC® 550.18(B)(5)*, Range calculations for power-supply load (Reprinted with permission from NFPA 70-2002)

- If outlets or circuits are provided for other than factory-installed appliances, include the anticipated load.

Load Calculation Example

A mobile home outside measurement (coupler excluded) is 12 ft × 50 ft; it has two small-appliance circuits, laundry circuit, ⅓-horsepower blower motor for the gas furnace rated 7.2 amperes at 120 volts, dishwasher rated 420 volt-amperes at 120 volts, and a 6500-watt range.

Find the total electrical load in amperes.

1. Lighting, small-appliance, and laundry load.

   Lighting:
   (12 ft × 50 ft) × 3 VA/sq ft =   1800 VA
   Small appliance:
   2 circuits × 1500 VA =   3000 VA
   Laundry circuit:
   1 circuit × 1500 VA =   1500 VA
                             Total =   6300 VA

   Net volt-amperes =
       3000 VA + (6300 VA − 3000 VA × 0.35)
           = 4155 VA

   Amperes per leg =
       $\frac{4155 \text{ VA}}{240 \text{ volts}}$ = 17.3 amperes

2. Amperes per leg.

Load	Leg A	Leg B	Neutral
Lighting, small appliance, laundry	17.3	17.3	17.3
Blower motor (7.2 × 1.25 = 9 @ 120 volts)	9.0		9.0
Dishwasher (420 VA ÷ 120 volts)		3.5	
Range (6500 w × 0.8 ÷ 240 volts)	21.7	21.7	15.0
Total	48.0 A	42.5 A	41.3 A

A minimum 50-ampere cable would be required.

Arc-Fault Circuit-Interrupter Protection (AFCI)

AFCI protection is required for all branch-circuits that supply 125-volt, single-phase, 15- or 20-ampere outlets installed in bedrooms of mobile homes or manufactured homes. The AFCI is designed to open the circuit and stop the arc-fault and its accompanying high-intensity heat before a fire occurs. Typical conditions where arc-faults may occur are damaged conductor insulation and loose electrical connections.

Mobile Home Park Power Distribution

The electrical distribution system to mobile home lots must be single phase, 120/240 volts. The wiring systems shall be calculated on the larger of the following: (1) 16,000 volt-amperes for each mobile home lot, or (2) The load calculated by *550.18* for the largest typical mobile home suitable for the lot. The feeder or service load can be calculated using the demand factors in *Table 550.31* (Figure 18-28).

Mobile home service equipment must be rated no less than 100 amperes at 120/250 volts. Provisions must be made for connecting a mobile home feeder assembly by a permanent wiring method. Power outlets used as mobile service equipment are permitted to contain receptacles rated 50 amperes with appropriate overcurrent protection.

NEC® 550.32(D) requires that mobile home service equipment must provide a means for connecting a mobile home accessory building (or structure) or additional electrical equipment located outside a mobile home by a fixed wiring method. This means a spare overcurrent device or a space for an additional overcurrent device must be made available in the panel for this purpose.

The outdoor service equipment enclosure should be located at least 2 ft (600 mm) above grade to bottom of enclosure, and the highest position of the operating handle is not to exceed 6.5 ft (2 m) (Figure 18-29).

Marking

Where a 125/250-volt receptacle is used in mobile home service equipment, the service equipment shall be marked as follows: **TURN DISCONNECTING SWITCH OR CIRCUIT BREAKER OFF BEFORE INSERTING OR REMOVING PLUG. PLUG MUST BE FULLY INSERTED OR REMOVED.**

Table 550.31 Demand Factors for Services and Feeders

Number of Mobile Homes	Demand Factor (percent)
1	100
2	55
3	44
4	39
5	33
6	29
7–9	28
10–12	27
13–15	26
16–21	25
22–40	24
41–60	23
61 and over	22

Figure 18-28 *NEC® Table 550.31*, Demand factor for services and feeders (Reprinted with permission from NFPA 70-2002)

SUMMARY

The requirements for wiring mobile homes and manufactured homes are similar, but they are necessarily somewhat different due to mobile homes and manufactured homes being *factory built* and delivered to the homesite in sections or units. *NEC® 550.19* covers the wiring methods required for interconnecting or joining portions of a circuit that must be electrically joined and that are located in adjacent sections. Mobile or manufactured homes are permitted to have disconnecting means with branch-circuit protective equipment in each unit or section. This protective equipment must be located in a manner such that, after joining or assembly of the units, the requirements of *550.5* will be met.

CHAPTER 18 Mobile Homes, Manufactured Homes, and Mobile Home Parks

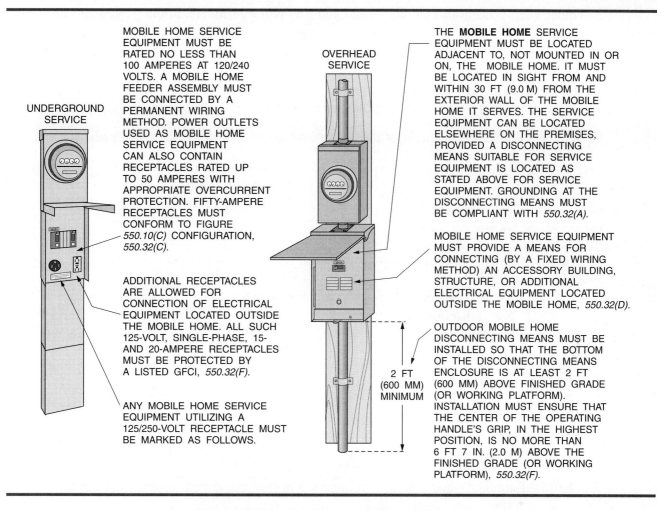

Figure 18-29 Mobile home service equipment

REVIEW QUESTIONS

MULTIPLE CHOICE

1. The maximum rated current for a mobile home power-supply cord is
   - A. 40 amperes.
   - B. 50 amperes.
   - C. 60 amperes.
   - D. 100 amperes.

2. How is the power-supply cord's plug cap specified as to the number of poles and wires?
   - A. Single-pole, 4-wires
   - B. Double-pole, 3-wires
   - C. Three-pole, 4-wire
   - D. Three-pole, 3-wires

3. Under what conditions can the neutral and grounding terminals in the panelboard be connected?
   ___ A. Never
   ___ B. When the service equipment is in or on the manufactured home
   ___ C. When all conductors have an insulated cover
   ___ D. When a 3-conductor cable supplies power to the mobile home

4. What is the maximum length of the power-supply cord from the face of the plug cap to the end including the bare lead?
   ___ A. 22 ft (6.7 m)
   ___ B. 20 ft (6 m)
   ___ C. 50 ft (11 m)
   ___ D. 36.5 ft (11 m)

5. What type of disconnecting means is required in each mobile home?
   ___ A. Snap switch
   ___ B. Circuit breaker rated for maximum load
   ___ C. Fused disconnect switch rated at 50 amperes
   ___ D. Not required

6. The lighting load is determined by calculating the floor area of the mobile home and multiplying the area by how many volt-amperes per sq ft?
   ___ A. 2 VA/sq ft
   ___ B. 3 VA/sq ft
   ___ C. 3.5 VA/sq ft
   ___ D. 4 VA/sq ft

7. The small-appliance and laundry circuits require how many volt-amperes per circuit?
   ___ A. 1500
   ___ B. 1200
   ___ C. 3000
   ___ D. 4500

8. The demand table for mobile home ranges reduces the nameplate rating of 13,000 volt-amperes to what demand value used in determining total load? [use *550.18(B)(5)*]
   ___ A. 80%
   ___ B. 13 kVA
   ___ C. 8 kVA
   ___ D. 8.4 kVA

9. What is the maximum distance that the mobile home service equipment can be located from the exterior wall of the mobile home?
   ___ A. 20 ft (6 m)
   ___ B. 25 ft (7.5 m)
   ___ C. 30 ft (9 m)
   ___ D. 35 ft (11 m)

10. If the mobile home load requirement exceeds 50 amperes, how is the power supplied?
    ___ A. Dual 50-ampere power-supply cords
    ___ B. Use a 100-ampere power-supply cord
    ___ C. By use of a mast weatherhead
    ___ D. None of the above

CHAPTER 18 Mobile Homes, Manufactured Homes, and Mobile Home Parks

TRUE OR FALSE

1. It is not necessary to have a disconnecting means in a mobile home if there is a disconnect at the service equipment, located within sight of and not more than 30 ft (9 m) from the exterior wall of the mobile home.
   ____ A. True
   ____ B. False
   Code reference _____ .

2. The mobile home power-supply cord requires plug cap to be 2-pole, 3-wire, rated 50 amperes, and 125/250 volts.
   ____ A. True
   ____ B. False
   Code reference _____ .

3. All receptacles in the mobile home must be of the grounding type.
   ____ A. True
   ____ B. False
   Code reference _____ .

4. Two outdoor receptacle outlets are required for each mobile home.
   ____ A. True
   ____ B. False
   Code reference _____ .

5. It is a requirement that the neutral and grounding terminals be connected together in the panelboard.
   ____ A. True
   ____ B. False
   Code reference _____ .

6. The mobile home frame is required to be grounded through the water line.
   ____ A. True
   ____ B. False
   Code reference _____ .

7. AFCIs are designed to protect against shock hazard.
   ____ A. True
   ____ B. False
   Code reference _____ .

8. It is not required that the service equipment be located in sight of the mobile home.
   ____ A. True
   ____ B. False
   Code reference _____ .

9. The bathroom receptacle luminaires (lighting fixtures) are required to be GFCI protected.
   ____ A. True
   ____ B. False
   Code reference _____ .

10. The outdoor receptacle for the water-pipe heating tape can be part of the bathroom circuit.
    ____ A. True
    ____ B. False
    Code reference _____ .

CHAPTER 18 Mobile Homes, Manufactured Homes, and Mobile Home Parks

FILL IN THE BLANKS

1. The power supply to the mobile home is a feeder consisting of not more than one _____-ampere mobile home power-supply cord with an integral plug cap.

2. The disconnecting means and branch-circuit protection are combined into a single assembly and designated as _____.

3. The load of _____ volt-amperes is used for each small-appliance and laundry circuit.

4. Bedroom branch-circuits are protected by _____ circuit breakers.

5. A 13-kW, 240-volt range has a total load of _____ amperes for determining power requirements and conductor sizing.

6. The mobile home chassis is bonded to the panelboard using a minimum _____ size copper conductor.

7. Receptacles are _____ permitted in the shower or bathtub space or within _____ in reach.

8. A disconnecting means is required in each mobile home consisting of _____ or _____.

9. The main disconnect must be plainly marked _____.

10. Aluminum conductors, aluminum alloy conductors, and aluminum core conductors are not _____ for use as branch-circuits.

FREQUENTLY ASKED QUESTIONS

Question 1: Why do they have all these requirements for outlet location and spacing when it's already shown in *Article 210*?

Answer: The requirements you find in *Article 210* are for dwelling units of standard construction. The requirements for mobile homes and manufactured homes are in Chapter 5, Special Occupancies, and are different in many respects due to the factory construction or assembly of these units and their delivery to job sites in multiple sections.

Question 2: What is the difference in service installation between a mobile home and a manufactured home?

Answer: Service equipment is generally not installed on or in a mobile home. The mobile home is fed with a 4-wire power-supply cord that is connected with an attachment plug cap to the mobile home park electrical distribution equipment. The bared leads are fed through the mobile home roof, floor, or exterior wall. Generally, a 1¼-in. raceway is run from the disconnecting means in the mobile home to the underside of the mobile home with provisions for attachment to a junction box. The manufacturer must provide written instructions stating the proper feeder conductor size for the raceway and the size of the junction box to be used.

A manufactured home generally has a conventional service, either overhead or underground, run to the building through a utility meter. In accordance with *550.32(B)* the service must include a grounding electrode, main bonding jumper, and a red warning label that states the following: **WARNING—DO NOT PROVIDE ELECTRICAL POWER UNTIL THE GROUNDING ELECTRODE(S) IS INSTALLED AND CONNECTED (SEE INSTALLATION INSTRUCTIONS).**

Question 3: Can a mobile home be fed with a conventional service from a utility company source?

Answer: If the calculated load is more than 50 amperes, a mast weatherhead installation can be made. This installation shall be in accordance with *550.10(I)(1)*.

Question 4: How is the *mobile home park* distribution calculated without knowing the loads on each mobile home that will be connected to it?

Answer: *NEC® 550.31* allows demand factors to be applied to a required calculation method. The calculation is made on the larger two criteria. Either a load of 16,000 volt-amperes for each mobile home lot, or a load calculated in accordance with *550.18* for the largest mobile home that each lot will accommodate. *Table 550.31* is used to apply the demand factor to the calculation.

Question 5: What are the requirements for GFCI protection for mobile homes?

Answer: According to *550.13(B)*, GFCI protection is required for outdoor receptacles; receptacles in compartments accessible from outside the unit; receptacles in bathrooms, including receptacles in luminaires (lighting fixtures); receptacles serving kitchen countertops; receptacles located within 6 ft (1.8 m) of a wet bar; and receptacle outlet(s) for the connection of pipe heating cables.

CHAPTER 19

Recreational Vehicles, Recreational Vehicle Parks, and Park Trailers

OBJECTIVES

After completing this chapter, the student should understand:
- low-voltage systems.
- combination electrical systems.
- grounding and bonding.
- calculation of loads.
- 120-volt vehicle system requirements.
- means for connecting to power source.
- vehicle parks electrical distribution system.

KEY TERMS

Converter: A device that changes electrical energy from one form to another, as from alternating current to direct current

Dead-Front (as applied to switches, circuit breakers, switchboards, and distribution panelboards): Designed, constructed, and installed so that no current-carrying parts are normally exposed on the front

Disconnecting Means: The necessary equipment, usually consisting of a circuit breaker or switch and fuses and their accessories, located near the point of entrance to supply conductors in a recreational vehicle and intended to constitute the means of cutoff for the supply to that recreational vehicle

Frame: Chassis rail and any welded addition thereto of metal thickness of 0.053 in. (1.35 mm) or greater

Low-Voltage: An electromotive force rated 24 volts, nominal, or less, supplied from a transformer, converter, or battery

Park Trailer: A unit that is built on a single chassis, mounted on wheels, and having a gross trailer area not exceeding 400 sq ft (37.2 sq m) in the set-up mode

Power-Supply Assembly: The conductors, including ungrounded, grounded, and equipment grounding conductors, grommets, or devices installed for the purpose of delivering energy from the source of electrical supply to the distribution panel within the recreational vehicle

Recreational Vehicle: A vehicle-type unit primarily designed as temporary living quarters for recreational, camping, or travel use, which either has its own motive power or is mounted on or drawn by another vehicle. The general entities are travel trailer, camping trailer, truck camper, and motor home

Recreational Vehicle Park: A plot of land upon which two or more recreational vehicle sites are located, established, or maintained for occupancy by recreational vehicles of the general public as temporary living quarters for recreation or vacation purposes

Recreational Vehicle Site: A plot of ground in a recreational vehicle park intended for the accommodation of a recreational vehicle, tent, or other individual camping unit on a temporary basis

INTRODUCTION

This chapter presents the requirements for installing electrical conductors and equipment within or on recreational vehicles, the conductors that connect recreational vehicles to a supply of electricity, and the installation of electrical equipment and devices within a **recreational vehicle park**. It also covers the electrical requirements for park trailers (Figure 19-1 and Figure 19-2).

GENERAL

The electrical conductors and equipment installed within and on recreational vehicles are categorized as low-voltage systems, 120- or 120/240-volt systems, other power systems, and combination electrical systems (Figure 19-3).

Articles 551 and *552* of the *NEC*® provide requirements for **recreational vehicles**, recreational parks, and **park trailers**. In addition, laws in many states follow closely the requirements of the NFPA 501C-1996, *Standards on Recreational Vehicles*.

Low-Voltage Systems

The **low-voltage** wiring within the recreational vehicle that is used in place of the 120-volt supply is subject to *551.10* of the *NEC*.® The automotive vehicle circuits, such as the headlights for roadway light-

Figure 19-1 Typical recreational trailer

Figure 19-2 Typical recreation vehicles

ing, brake lights, and turn-signals lighting, are subject to federal and state laws and the *NEC*.

Recreational vehicles can use the **frame** or chassis as the return path to the electrical supply. However, it is not the intent of the *NEC* to allow the sidewalls or the roof to serve as the ground return path. Stranded copper conductors with installation ratings of 194°F (90°C) for interior installations and 257°F (125°C) for all engine compartment wiring or any under-chassis wiring conductors that are located

Figure 19-3 Typical electrical characteristics of recreational vehicles and trailers

less than 18 in. (450 mm) from any component of an internal combustion engine exhaust system must be used. Listed conductors are marked as required by the listing agency. Usually, the marking consists of manufacturer's name or logo, temperature rating, wire size, conductor material, and insulation thickness.

Wiring methods for low-voltage protection against physical damage must be provided. Conductors must be adequately secured to the vehicle structure, and conductor splicing must be mechanically and electrically secure without solder with any bare ends covered with an insulation equivalent to the existing conductor insulation. Battery and dc circuits must be physically separated by at least 0.5 in. (12.7 mm) space from circuits from a different power source, and ground connections to the chassis or frame must be in an accessible location and mechanically secured. The chassis-grounding terminal of the battery must be bonded to the vehicle chassis with a minimum 8 AWG copper conductor (Figure 19-4). If the battery lead to the positive terminal exceeds a No 8, the grounding cable size must be increased to be of equal size.

Storage battery installation should be in a vapor-tight area to the interior and vented directly to exterior of the vehicle. Batteries cannot be installed in a compartment containing spark or flame-producing equipment. However, they may be installed in the engine generator compartment if the only charging source is from the engine generator.

The low-voltage circuit wiring must be protected by overcurrent protective devices that are not rated in excess of the ampacity of the copper conductors (see Figure 19-5).

Circuit breakers or fuses shall be of an approved type. Fuseholders are required to be marked with the maximum fuse size, and the fuses and circuit breakers must be protected against shorting and physical damage by a cover or equivalent means. DC appliances which consume large amounts of currents,

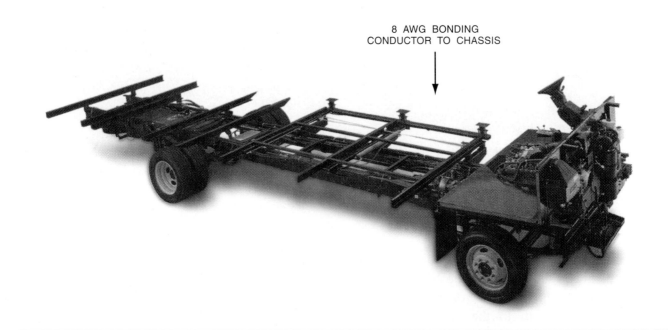

Figure 19-4 Recreational vehicle chassis or frame. Battery bonded to chassis with 8 AWG conductor

Table 551.10(E)(1) Low-Voltage Overcurrent Protection

Wire Size (AWG)	Ampacity	Wire Type
18	6	Stranded only
16	8	Stranded only
14	15	Stranded or solid
12	20	Stranded or solid
10	30	Stranded or solid

Figure 19-5 *NEC® Table 551.10(E)(1)*, Low-voltage overcurrent protection (Reprinted with permission from NFPA 70-2002)

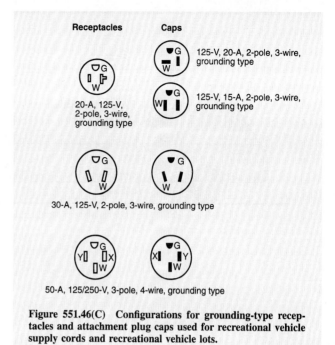

Figure 551.46(C) Configurations for grounding-type receptacles and attachment plug caps used for recreational vehicle supply cords and recreational vehicle lots.

Figure 19-6 Reprinted with permission from NFPA 70-2002

such as heater blowers, compressors, pumps, and other motor-driven appliances, should be installed according to the manufacturer's instructions. Switches and luminaires (lighting fixtures) must be listed.

Nominal 120-Volt or 120/240-Volt Systems

The electrical equipment and material of recreational vehicles indicated for connection to a wiring system rated at 120 volts or 120/240 volts must be listed and installed according to *551.60*. The means to connect the power source to the vehicle is accomplished by a **power-supply assembly** which is part of the vehicle during manufacturing or is supplied by the factory. The power-supply assembly can be a separate cord with a female connector and molded attachment plug cap that attaches to a flanged surface inlet (male, recessed-type, motor-based attached plug) that is wired directly to the distribution panel. A second means is a power-supply assembly permanently connected to the distribution panel terminals and having an attached plug (see Figure 19-6) on the supply end.

Where the power cord passes through the walls or floor, protection is required by either a conduit or bushing. While the vehicle is in transit, the cord assembly requires protection against corrosion and mechanical damage. Labeling is required for the electrical entrance of each recreational vehicle. The label must be permanently affixed to the exterior skin, at (or near) the point of entrance of the power-supply cord. The label 3 in. (76 mm) × 1¾ in. (44.5 mm) minimum size made of etched, metal-stamped, or embossed brass, stainless steel, or anodized or alclad aluminum not less than 0.0020 in. (506 μm) thick.

The point of entrance of the power-supply assembly is specified to be within 15 ft (4.5 m) from the rear, on the left (road) side, or at the rear, left of the longitudinal center of the vehicle, within 18 in. (450 m) of the outside wall (Figure 19-7).

Each recreational vehicle containing a 120-volt electrical system must contain one of the following circuits to supply lights, receptacle outlets, and fixed appliances:

- One 15-ampere circuit consisting of a 15-ampere switch and fuse or one 15-ampere circuit breaker
- One 20-ampere circuit consisting of a 20-ampere switch and fuse or one 20-ampere circuit breaker
- A maximum of five 15- or 20-ampere circuits with a distribution panelboard rated at 120 volts with a 30-ampere rated main power-supply assembly
- For six or more circuits, a 50-ampere, 120/240-volt power supply is required

The branch-circuit overcurrent devices are rated as follows: (1) not more than the circuit conductors,

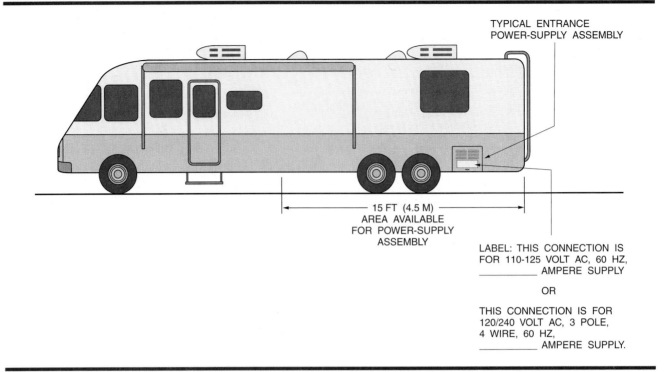

Figure 19-7 Power supply entrance location on recreational vehicle

and (2) not more than 150% of a single appliance rated 13.3 amperes or more and supplied by an individual branch-circuit, but (3) not more than the overcurrent protection size marked on an air conditioner or other motor-operated appliance.

Each recreational vehicle is required to have only one of the following main power-supply assemblies (power cable):

- Fifteen-ampere main power-supply assembly or larger for recreational vehicles with one 15-ampere circuit

- Twenty-ampere main power-supply assembly or larger for recreational vehicles with one 20-ampere circuit

- Thirty-ampere main power-supply assembly or larger for recreational vehicles with 30-ampere rated main power panelboard

- Fifty-ampere, 120/240-volt power-supply assembly or larger for recreational vehicles where six or more circuits are used

The distribution panelboard must be listed and rated for the purpose used. The grounded termination bar must be insulated from the enclosure. While the equipment grounding terminal must be attached inside the metal enclosure, the panel must be located in an accessible location with a working clearance of not less than 24 in. (600 mm) wide and 30 in. (750 mm) deep. Further, the panel must be **deadfront** and consist of one or more circuit breakers or Type S fuseholders. A main **disconnecting means** is required when fuses are used or when more than two circuit breakers are included. For panelboards with two or more branch-circuits, a main overcurrent protective device, not exceeding the panel rating, must be provided.

Receptacle outlets are required on walls 2 ft (600 mm) wide (or more) so that no point along the floor line is more than 6 ft (1.8 m) from an outlet. Exceptions to the spacing are for bath and hall areas, and spaces occupied by kitchen cabinets, wardrobe cabinets, built-in furniture, and behind doors that open fully against a wall surface or similar facilities.

Receptacle outlets are required as follows:

- Adjacent to kitchen countertops that are at least 12 in. (300 mm) wide with at least one on either side of the sink

- Adjacent to the refrigerator and gas range *unless* electrical connection is not needed

- Adjacent to countertop spaces of 12 in. (300 mm) or more in width that cannot be reached from a receptacle without crossing the sink, cooking appliance, or traffic area

GFCIs are required for the following 120-volt 15/20-ampere receptacles:

- Adjacent to bathroom lavatory
- Serving countertop surfaces and are within 6 ft (1.8 m) of a sink or lavatory

A receptacle outlet as part of a listed luminaire (lighting fixture) is permitted. However, these luminaires (fixtures) cannot be installed in tub or tub/shower compartments. Receptacles are *not* permitted to be installed face-up in any countertop or similar surface within the living area.

Other Power Sources

Other sources of ac power, such as a motor generator set, are required to be listed for use in recreational vehicles, be installed according to the listing, and be wired in conformance with *551.30*.

Generator installation requires that it be mounted in such a manner that it is effectively bonded to the recreational vehicle frame. Internal-combustion-driven generators and storage batteries must be secured in place to avoid displacement from vibration and road shock and in a well-ventilated compartment.

Combination Electrical System

Voltage **converters** are used to change the 120-volt alternating current to low-voltage direct current. The converters and transformers must be rated for use in recreation vehicles and designed or equipped to provide over-temperature protection. To determine the converter rating, the following formula shall be applied to the total connected loads of all 12-volt equipment: (1) the first 20 amperes of load at 100%, plus (2) the second 20 amperes of load at 50%, plus (3) all load above 40 amperes at 25%.

The noncurrent-carrying metal enclosure of the voltage converter must be bonded to the vehicle frame with minimum 8 AWG copper conductor. The grounding conductor for the battery and the metal enclosure is permitted to be the same conductor.

In a recreational vehicle equipped with 120- or 120/240-volt ac systems and a low-voltage system, the receptacles and plug caps of the low-voltage system must differ in configuration from those of the ac system. Where a vehicle has a low-voltage system external connection, the connector must have a configuration that will not accept 120 volt.

Recreational Vehicle Parks

Every **recreational vehicle site** with electrical supply is required to be equipped with at least one 20-ampere, 125-volt receptacle. A minimum requirement of 5% of all recreational vehicle sites with electrical supply must be equipped with 50-ampere, 125/250-volt receptacles conforming to *551.71* of the *NEC®* (Figure 19-8).

A minimum of at least 70% of all recreational vehicle sites with electrical supply must be 30-ampere, 125-volt receptacles. All 125-volt, single-phase, 15- and 20-ampere receptacles must have listed GFCI protection.

The distribution system for 50-ampere recreational vehicle sites is derived from a single phase, 120/240-volt, 3-wire system. The 20- and 30-ampere-receptacles' power is derived from any grounded distribution system that supplies 120-volt, single-phase receptacles. The neutral conductors are not to be reduced in size below the size of the ungrounded conductors for site distribution. For permanently connected loads, the 240-volt neutral is permitted to be reduced below the size of the ungrounded conductors.

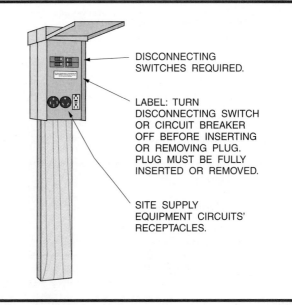

Figure 19-8 Recreational vehicle site supply equipment

Basis of calculations for **recreational vehicle park** site feeders and electric service are:

- Sites equipped with 50-ampere, 120/240-volt service .... 9600 volt-amperes per site

- Sites equipped with 20 and 30-ampere, 120-volt service .... 3600 volt-amperes per site

- Sites equipped with 20-ampere, 120-volt service .... 2400 volt-amperes per site

- Tent sites equipped with 20-ampere, 120-volt service .... 600 volt-amperes per site

Where the site supply has more than one receptacle, the load calculation is based on the highest rated receptacle.

To determine the demand factors for site feeders, see Figure 19-9. The demand applies to all sites. For example, 20 sites calculated at 45% of 3600 volt-amperes results in a permissible demand of 1620 volt-amperes per site or a total of 32,400 volt-amperes for 20 sites.

All electrical equipment in recreational vehicle parks must be grounded according to *Article 250* of the *NEC.* Exposed noncurrent-carrying metal parts of fixed equipment, metal boxes, cabinets, and fittings that are not electrically connected to grounded equipment must be grounded by a continuous equipment grounding conductor run with circuit-conductors from a service transformer of a secondary distribution system. The disconnect or removal of the receptacle or other devices will not interfere with or interrupt the grounding continuity. The neutral conductor must *not* be used as an equipment ground for recreational vehicles or equipment in the recreational vehicle park. The recreational vehicle site supply equipment (see Figure 19-10) is required to be accessible, located not less than 2 ft (600 mm) and not more than 6½ ft (2.0 m) above the ground, and for the 125/250-volt receptacle, the equipment must be marked as follows: *Turn disconnecting switch or circuit breaker off before inserting or removing plug. Plug must be fully inserted or removed.* The marking is required to be placed on the equipment adjacent to the receptacle outlet (Figure 19-10).

Outdoor equipment must be rainproof equipment. Overhead conductors, not over 600 volts, must have a vertical clearance of 18 ft (5.5 m) and a horizontal clearance not less than 3 ft (900 mm) in areas subject to vehicle movement.

Table 551.73 Demand Factors for Site Feeders and Service-Entrance Conductors for Park Sites

Number of Recreational Vehicle Sites	Demand Factor (percent)
1	100
2	90
3	80
4	75
5	65
6	60
7 – 9	55
10 –12	50
13 –15	48
16 –18	47
19 –21	45
22 –24	43
25 –35	42
36 plus	41

Figure 19-9 *NEC® Table 551.73* (Reprinted with permission from NFPA 70-2002)

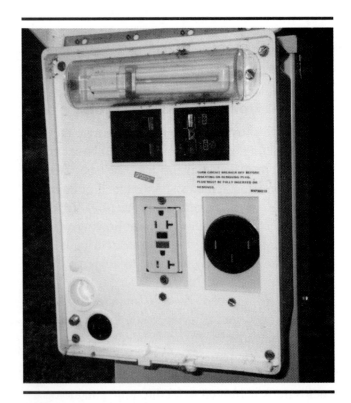

Figure 19-10 Recreational vehicle park site power pedestal

SUMMARY

Recreational vehicles are either the park trailer type or are the motor coach type. They are over the road type vehicles that each night, or for many days and nights, are parked in recreational vehicle parks connected to the park's electrical facilities (Figure 19-11).

The electrical connection pedestals are fed from the recreational park vehicle distribution system (Figure 19-12).

Many parks have bath and shower facilities for use while the recreational vehicles are staying in the park. Figure 19-13 shows a bath and shower facility in the construction stage, plumbing and electrical stub-ups can be seen in the concrete pour.

Recreational vehicle parks have many other amenities, such as picnic tables at each site, potable water connections, and electrical connections (Figure 19-14).

Recreational vehicles come equipped with easily accessible points of connection for both water and electricity (Figure 19-15).

Figure 19-11 Recreational vehicle electrical connection pedestal with various receptacle configurations

Figure 19-12 Recreational vehicle park electrical distribution

Figure 19-13 Bath and shower facility under construction

Figure 19-14 Typical recreational vehicle at park site with electrical and water hook-up

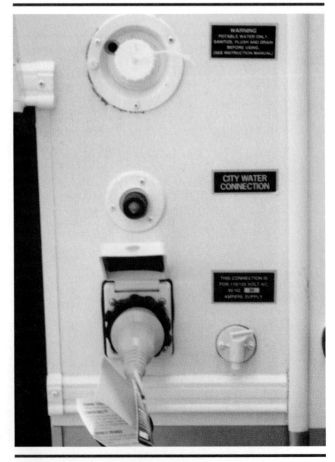

Figure 19-15 Typical connection point on recreational vehicle for potable water and electricity

REVIEW QUESTIONS

MULTIPLE CHOICE

1. A plug cap for 50-ampere, 120/240-volt service has which of the following configurations?
   ___ A. 2-pole, 3-wire, ground type
   ___ B. 2-pole, 4-wire, ground type
   ___ C. 3-pole, 4-wire, ground type
   ___ D. 3-pole, 3-wire, ground type

2. The point of entry of a power-supply assemble into a recreational vehicle must be:
   ___ A. within 15 ft (4.5 m) of the rear.
   ___ B. on the left (road) side of vehicle.
   ___ C. left of longitudinal center within 18 in (450 m) of the outside wall.
   ___ D. all the above.

3. Overhead electrical conductors, not over 600 volts, are required to have a vertical clearance not less than:
   ___ A. 12 ft (3.7 m).
   ___ B. 15 ft (4.5 m).
   ___ C. 18 ft (5.5 m).
   ___ D. 20 ft (6.0 m).

4. If the site supply equipment contains a 120/240-volt receptacle, the equipment is required to be:
   ___ A. rated 600 volts.
   ___ B. rated 100 amperes.
   ___ C. marked regarding the sequence of plug removal.
   ___ D. all the above.

5. What is the load assigned to a site equipped with 50-ampere, 120/240-volt supply?
   ___ A. 3600 volt-amperes
   ___ B. 250 volt-amperes
   ___ C. 9600 volt-amperes
   ___ D. 600 volt-amperes

6. For a recreational vehicle park with 40 sites equipped with 50-ampere, 120/240-volt supply, what is the total load demand for the 40 sites?
   ___ A. 384 kVA
   ___ B. 157 kVA
   ___ C. 480 kVA
   ___ D. 240 kVA

TRUE OR FALSE

1. The battery compartment must be vented to the atmosphere.
   ___ A. True
   ___ B. False
   Code reference _____ .

2. The frame of a recreational vehicle is used as a return path for the low-voltage system.
   ____ A. True
   ____ B. False
   Code reference _____ .

3. A power-supply assembly is used to connect the 120- or 120/240-volt source to the recreational vehicle.
   ____ A. True
   ____ B. False
   Code reference _____ .

4. The recreational vehicle distribution panel is required to separate terminals for grounding and grounded conductors with a jumper connecting the two terminals.
   ____ A. True
   ____ B. False
   Code reference _____ .

5. All receptacles are of the grounding type.
   ____ A. True
   ____ B. False
   Code reference _____ .

6. All recreational vehicles are required to have a permanent label affixed to the exterior skin, at or near the entrance of the power supply, which indicates the voltage, frequency, and ampere supply.
   ____ A. True
   ____ B. False
   Code reference _____ .

FILL IN THE BLANKS

1. Ground connections to the chassis or frame must be in an accessible location, and fuseholders must be _____ with maximum size.

2. The battery bond to the vehicle must be done by _____ AWG and _____ conductor.

3. Aluminum or copper-clad aluminum conductors are not used for bonding if such conductors are exposed to _____ elements.

4. Every recreational vehicle site with electricity must be equipped with at least _____ 125/250 volt-receptacle.

5. The demand factor for 14 sites is _____ %.

6. Recreational vehicles with one 20-ampere circuit would require a _____ main power-supply assembly.

7. Recreational vehicles receptacles are required on walls _____ wide (or more) so no point along the floor line is more than _____ from an outlet.

8. Receptacles serving countertop surfaces and within _____ of the sink need to be _____ protected.

9. Voltage converters are used to change _____ to _____ direct current.

FREQUENTLY ASKED QUESTIONS

Question 1: Why are recreational vehicles required to have internal branch wiring the same as a residence?

Answer: Recreational vehicles are used as dwelling units for long periods of time. The hazards that can arise from the use of electricity are the same in a recreational vehicle as they are in a permanently installed dwelling. *NEC*® *551.41* contains the requirements for receptacle outlets in a recreational vehicle.

Question 2: Are the receptacle outlets in a recreational vehicle required to have GFCI protection?

Answer: *NEC*® *551.42(C)* shows the requirements for GFCI protection for receptacle outlets installed in recreational vehicles. The requirements are basically the same as for dwelling units covering outdoor receptacle outlets, countertop receptacles in the kitchen, and any receptacle outlets in the area of a bathroom or shower.

Question 3: Why does the *NEC*® have so much information about recreational vehicles when they are factory built?

Answer: You probably will not ever wire a recreational vehicle, but in order to provide proper power for the recreational vehicle sites, it is necessary to know and understand the load you are preparing the power pedestal for.

CHAPTER 20

Floating Buildings, Marinas, and Boatyards

This Chapter will be divided into two parts. Part I will cover floating buildings, and Part II will cover marinas and boatyards.

Part I—Floating Buildings

OBJECTIVES

After completing Part I, the student should understand:
- the definition of a floating building.
- service and feeder equipment for floating buildings.
- grounding of floating buildings.

KEY TERMS

Floating Building: A building unit that floats on water, is moored in a permanent location, and has a premises wiring system served through connection by permanent wiring to an electric supply system not located on the premises

Moored: Secured to shore by lines and cables at a particular location

INTRODUCTION

Part I outlines wiring, services, feeders, and grounding of floating buildings (Figure 20-1).

A key requirement for a **floating building** is that it is **moored** in a permanent location. The *NEC®* requirements for installation of wiring and equipment can be found in *Article 553*.

GENERAL

There are many yacht clubs, restaurants, marina office buildings, and commercial stores that cater to the boating public and others that are housed in buildings that generally are boats that are no longer used for water transportation but are instead permanently located on water at a marina or similar docking place. The premises wiring is connected to an electric supply not located on the premises (Figure 20-2).

Services and Feeders

NEC® 553.4 contains the requirements for the location of service equipment for a floating building. The service equipment must be located adjacent to, but not on or in, the building. The intent is that the power supply, whether it is from a utility source or from some other source, must be located on the shore adjacent to the location where the floating building is moored. One set of service conductors or other supply conductors may supply more than one set of service equipment or power supply equipment.

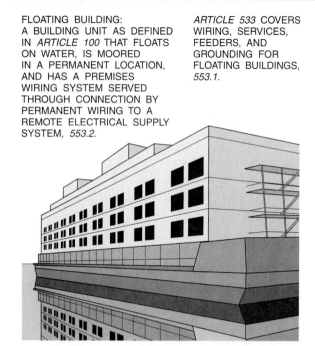

Figure 20-1 Floating building provisions

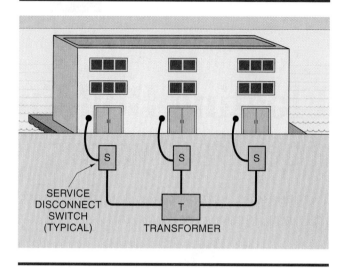

Figure 20-3 Floating office building with multiple feeders

Figure 20-2 Floating building power supply not located on building

NEC® 553.6 requires that each floating building be supplied by a single set of feeder conductors from its service equipment. Where a floating building has multiple occupancies, each occupant is permitted to be supplied by a single set of feeder conductors extended from the occupant's service equipment to the occupant's panelboard (Figure 20-3).

In accordance with *553.7*, the installation of services and feeders must be done in a manner that will maintain flexibility between the floating building and the supply conductors. All wiring must be installed so that motion of the water surfaces and changes in the water level will not result in unsafe conditions.

The use of LFMC or LFNC with approved fittings is permitted for feeders and locations where flexible connections are required for services. Extra-hard usage portable power cables, listed for both wet locations and sunlight resistance, are permitted for a feeder to a floating building where flexibility is required (Figure 20-4).

Grounding

The general requirements for grounding a floating building are found in *553.8*. The grounding must be through connection to a grounding bus in the building panelboard. The grounding bus is connected through a green-colored insulated equipment grounding conductor run with the feeder conductors and connected to a grounding terminal in the service equipment. The grounding terminal in the service equipment must be grounded by connection through

Figure 20-4 Flexible cords to multiple occupancy floating buildings

an insulated grounding electrode conductor to a grounding electrode on shore (Figure 20-5).

The grounded circuit conductor (neutral) shall be an insulated conductor identified in conformance with *200.6*. The neutral conductor must be connected to the equipment grounding terminal in the service equipment, and except for that connection, it must be insulated from the equipment grounding conductors, equipment enclosures, and all other grounded parts. The neutral terminal in the panelboard, and in ranges, clothes dryers, counter-mounted cooking units must be insulated from the enclosures. This requirement is made to ensure that no parallel path is formed between the neutral conductor and any equipment grounding conductor.

Bonding Requirements

NEC® 553.10 contains the requirements for bonding all enclosures and exposed metal parts of electrical systems to the grounding bus. All metal parts in contact with water, all metal piping, and all noncurrent-carrying metal parts that may become energized must be bonded to the grounding bus in the panelboard (Figure 20-6).

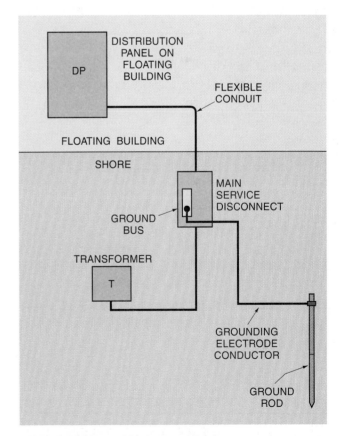

Figure 20-5 Grounding electrode conductor and ground rod for floating building

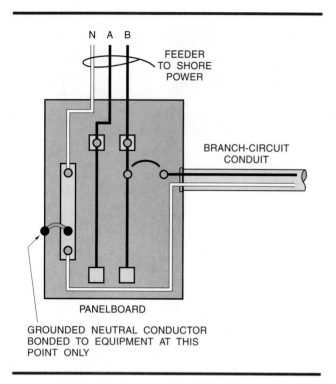

Figure 20-6 Floating building bonding requirements

Where required to be grounded, cord-connected appliances must be grounded by means of an equipment grounding conductor in the cord and a grounding-type attachment plug. All metal parts in contact with the water and all noncurrent-carrying metal parts that may become energized must be bonded to the grounding bus in the panel.

SUMMARY

Floating buildings must be permanently moored. The service equipment for floating buildings must be located adjacent to the building, but cannot be located on or in the building. The feeder from the service equipment to the building must be flexible to compensate for changes in water level or motion. A green equipment grounding conductor must be run with the feeder to the building panelboard grounding bus. Feeder conductors must be sized according to *Article 220*.

Part II—Marinas and Boatyards

OBJECTIVES

After completing Part II, the student should understand:
- electrical datum plane.
- electrical equipment on piers.
- marine power outlets.
- distribution and transformers for yard and piers.
- load calculations for shore power.
- wiring methods for pier wiring.
- grounding for marina equipment.
- disconnecting means for shore power.
- gasoline dispensing stations.

KEY TERMS

Catenary: The curve or angle of a cable suspended from two points

Electrical Datum Plane: The electrical datum plane is defined as follows:

1. In land areas subject to tidal fluctuation, the electrical datum plane is a horizontal plane 2 ft (606 mm) above the highest tide level for the area occurring under normal circumstances, that is, the highest high tide

2. In land areas not subject to tidal fluctuation, the electrical datum plane is a horizontal plane 2 ft (606 mm) above the highest water level for the area occurring under normal circumstances

3. The electrical datum plane for floating piers and landing stages that are (1) installed to permit rise and fall response to water level without lateral movement, and (2) that are so equipped that they can rise to the datum plane established for (1) or (2) above. The electrical datum plane is a horizontal plane 30 in. (762 mm) above the water level at the floating pier or landing stage and a minimum of 12 in. (305 mm) above the level of the deck

Marine Power Outlet: An enclosed assembly that can include receptacles, circuit breakers, fused switches, fuses, watt-hour meters, and monitoring means approved for marine use

INTRODUCTION

Part II covers the installation of wiring and equipment in areas comprising fixed or floating piers, wharfs, docks, and other areas in marinas, boatyards, boat basins, boathouses, yacht clubs, boat condominiums, docking facilities associated with residential condominiums, or similar occupancies. A single, private, noncommercial docking facility for a one-family dwelling is not intended to be covered by this chapter. The requirements for the installation of wiring and equipment can be found in *Article 555* (Figure 20-7).

GENERAL

Yard and pier electrical distribution systems shall not exceed 600 volts phase-to-phase. Transformers and enclosures must be specifically approved for the intended location. The bottoms of transformers must not be located below the **electrical datum plane.** The service equipment for floating docks or marinas

352 CHAPTER 20 Floating Buildings, Marinas, and Boatyards

Figure 20-7 Typical marina and boatyard

shall be located adjacent to, but not on or in, the floating structure. All electrical connections must be located at least 12 in. (305 mm) above the deck of a floating pier and at least 12 in. (305 mm) above the deck of a fixed pier, but not below the electrical datum plane.

NEC® 555.10 requires that electrical equipment enclosures installed on piers above deck level must be securely and substantially supported by structural members independent of any conduit connected to them (Figure 20-8). If enclosures are not attached to mounting surfaces by external means, the internal screw heads must be sealed to prevent seepage of water through the mounting holes.

Circuit Breakers, Switches, Panelboards, and Marine Power Outlets

NEC® 555.11 permits that overcurrent protection for feeders or branch-circuits to be provided by

Figure 20-8 Electrical equipment enclosures supported by structural members

the use of either circuit breakers or fuses. Circuit breakers and switches installed in gasketed enclosures must be arranged to permit required manual operation without exposing the interior of the enclosure. All such enclosures must be arranged with a weep hole to discharge condensation.

Load Calculations for Service and Feeder Conductors

NEC® 555.12 contains the requirements for load calculations. The load for each service and/or feeder circuit-supplying receptacles that provide shore power for boats shall be calculated using the demand factors shown in Figure 20-9.

On many installations, a utility service is run along the marina structure, and taps are made by the utility to separate combination meter(s), disconnect(s), and receptacle(s) enclosures at each berthing facility (Figure 20-10).

Wiring Methods

NEC® 555.13 contains the requirements for the installation and wiring methods required for piers.

Figure 20-10 Utility meter enclosure and receptacle combination fitting

Table 555.12 Demand Factors

Number of Receptacles	Sum of the Rating of the Receptacles (percent)
1 – 4	100
5 – 8	90
9 –14	80
15 –30	70
31 –40	60
41 –50	50
51 –70	40
71-plus	30

Notes:
1. Where shore power accommodations provide two receptacles specifically for an individual boat slip and these receptacles have different voltages (for example, one 30 ampere, 125 volt and one 50 ampere, 125/250 volt), only the receptacle with the larger kilowatt demand shall be required to be calculated.
2. If the facility being installed includes individual kilowatt-hour submeters for each slip and is being calculated using the criteria listed in Table 555.12, the total demand amperes may be multiplied by 0.9 to achieve the final demand amperes.

Figure 20-9 *NEC® Table 555.12* (Reprinted with permission from NFPA 70-2002)

Extra-hard usage portable cable, listed for both wet locations and sunlight resistance, must be used as permanent wiring on the underside of piers (floating or fixed) and where flexibility is necessary for piers composed of floating sections. Temporary wiring to boats may only be used in accordance with *Article 527*.

Installation. Overhead wiring must be carefully installed with consideration to avoiding possible contact with masts or other parts of boats. Conductors and cables must be routed to avoid wiring closer than 20 ft (6.0 m) from the outer edge of the yard that can be used for moving vessels or raising or lowering masts.

Where portable power cables are used, there must be a junction box of approved corrosion-resistant construction with permanently installed terminal blocks on each pier section to which the feeder and feeder extensions are to be connected.

Rigid Metal or Nonmetallic Conduit shall be installed to protect the wiring above decks or piers. The conduit must be connected to the enclosure by full standard threads.

Grounding. An equipment grounding conductor shall be bonded to all metal boxes, metal cabinets, and metal frames of utilization equipment. The equipment grounding conductor must be connected to the grounding terminal of all grounding-type receptacles. The equipment grounding conductor must not be smaller than 12 AWG.

Disconnecting Means for Shore Power Connection

NEC® 555.17 contains the requirements for disconnecting means for shore power connections (Figure 20-11). A disconnecting means must be provided to disconnect each boat from its supply connections. The disconnecting means must be a circuit breaker or switch and must be properly identified as to its use. The disconnecting means must be readily accessible and located not more than 30 in. (762 m) from the receptacle it controls.

Receptacles. In accordance with *555.19*, receptacles must be located not less than 12 in. (300 mm) above the deck surface of the pier and not below the electrical datum on a fixed pier.

Receptacles intended to supply shore power to boats shall be housed on **marine power outlets** listed for the conditions. Means must be provided to reduce strain on the plug and receptacle caused by the weight and **catenary** angle of the shore power cord. The receptacle face shall be at any angle from horizontal to 65° below horizontal (Figure 20-12).

Each single receptacle that provides shore power to boats shall be supplied from a marine power outlet or panelboard by an individual branch-circuit of the voltage, class, and rating corresponding to the rating of the receptacle. Receptacles that provide shore power for boats must not be rated less than 30 amperes and shall be a single outlet type. Receptacles rated not less than 30 amperes and not more than 50 amperes must be of the locking and grounding type. Receptacles rated for 60 amperes or 100 amperes shall be of the pin and sleeve type (Figure 20-13).

Receptacles for use other than shore power, such as in boathouses, storage, or maintenance buildings, must be provided with GFCI protection for personnel.

Where it is necessary to provide power to a mobile crane or hoist in a boatyard, and a trailing cable is used, it must be a listed portable power cable provided with an outer jacket of distinctive color for safety.

Gasoline Dispensing Stations. Electrical wiring and equipment located at or serving gasoline dispensing stations must comply with *Article 514* in addition to the requirements of *Article 555*.

Figure 20-11 Disconnecting means for shore power

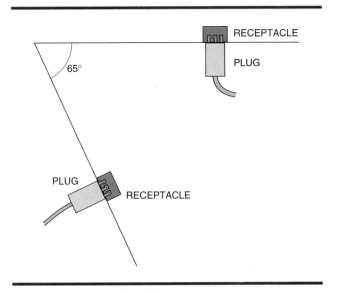

Figure 20-12 Installation method to provide strain relief on the plug and receptacle

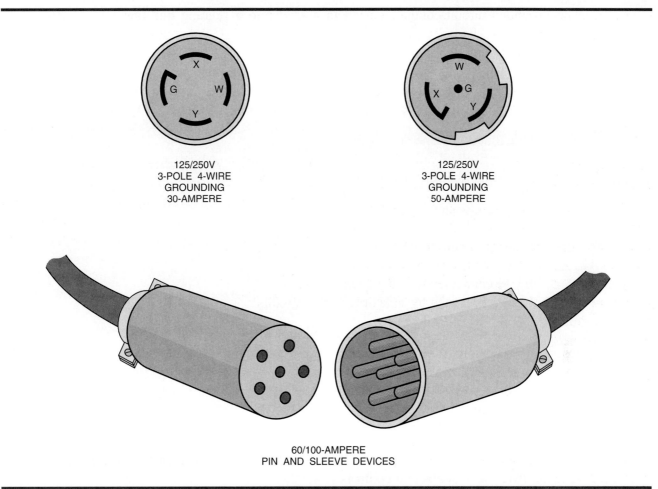

Figure 20-13   Receptacle configurations (NEMA) 30A—50A—60A—100A

SUMMARY

Marinas or boatyards are equipped with shore power located on the shore or on fixed or floating piers. All electrical equipment used for this type of installation must be listed for this particular use. Portable cables must be listed for wet locations and for exposure to sunlight. All enclosures must be of corrosion-resistant material with terminal blocks for conductor terminations. Electrical wiring and equipment must be located where it will not be subjected to submersion in water or where it can be damaged by the normal operation of boat use.

REVIEW QUESTIONS

MULTIPLE CHOICE

1. A feeder to a floating building may be installed by which of the following wiring methods?
   ____ A. LFMC
   ____ B. LFNC
   ____ C. extra-hard usage flexible cable
   ____ D. all of the above

2. The grounded circuit-conductor used in a feeder to a floating building is required to be:
   ____ A. not smaller than 10 AWG.
   ____ B. insulated and identified in conformance with *200.6*.
   ____ C. green with yellow stripes.
   ____ D. none of the above.

3. The electrical datum plane describes:
   ____ A. the rise and fall of the water level in a marina or boatyard.
   ____ B. the number of boats on one service feeder.
   ____ C. the maximum permissible height of a mast.
   ____ D. none of the above.

4. The service equipment for floating docks or marinas shall always:
   ____ A. be located on or in the structure.
   ____ B. be connected by a rigid wiring method.
   ____ C. be connected by a flexible wiring method.
   ____ D. none of the above.

5. Receptacles that provide shore power for boats shall not be rated less than:
   ____ A. 50 amperes.
   ____ B. 60 amperes.
   ____ C. 30 amperes.
   ____ D. 20 amperes.

6. The disconnecting means for shore power connections must be located not more than _____ from the receptacle it controls.
   ____ A. 30 in. (762 mm)
   ____ B. 6 ft (1.8 m)
   ____ C. 10 ft (3.0 m)
   ____ D. 15 ft (4.5 m)

7. The equipment grounding conductor used for marinas and boatyards shall not be smaller than:
   ____ A. 8 AWG copper.
   ____ B. 12 AWG copper.
   ____ C. 10 AWG copper.
   ____ D. none of the above.

8. Receptacles intended to supply shore power to boats shall be housed in:
   ____ A. marine power outlets.
   ____ B. listed enclosures protected from the weather.
   ____ C. listed weatherproof enclosures.
   ____ D. all of the above.

9. A trailing cable supplying power to a mobile crane or hoist in a boatyard must be:
   ____ A. listed portable power cable.
   ____ B. provided with an outer jacket of distinctive color for safety.
   ____ C. both A and B.
   ____ D. all of the above.

10. The demand factor for 20 shore-power receptacles as determined by the sum of the rating of receptacles (percent) is:
    ___ A. 90%.
    ___ B. 80%.
    ___ C. 70%.
    ___ D. 60%.

TRUE OR FALSE

1. Floating buildings are moored in a permanent location.
   ___ A. True
   ___ B. False
   Code reference _____.

2. Floating buildings must always be supplied by a single set of feeders.
   ___ A. True
   ___ B. False
   Code reference _____.

3. The grounded circuit-conductor can be insulated, bare, or covered.
   ___ A. True
   ___ B. False
   Code reference _____.

4. Cord-connected appliances in a floating building shall be grounded by means of an equipment grounding conductor in the cord- and grounding-type attachment plug.
   ___ A. True
   ___ B. False
   Code reference _____.

5. Boatyard and pier distribution systems shall not exceed 600 volts phase-to-phase.
   ___ A. True
   ___ B. False
   Code reference _____.

6. Service equipment for floating docks shall be located on the floating dock.
   ___ A. True
   ___ B. False
   Code reference _____.

7. Overcurrent protection for feeders and branch-circuits for marine power outlets shall be maintained by the use of circuit breakers.
   ___ A. True
   ___ B. False
   Code reference _____.

8. Receptacles that provide shore power to boats must not be rated less than 30 amperes.
   ___ A. True
   ___ B. False
   Code reference _____.

358 CHAPTER 20 Floating Buildings, Marinas, and Boatyards

9. Receptacles must not be located more than 12 in. (305 mm) above the deck surface of the pier.
   ____ A. True
   ____ B. False
   Code reference _____ .

10. Receptacles for uses other than shore power, such as in boathouses, storage, or maintenance buildings, must be provided with GFCI protection.
    ____ A. True
    ____ B. False
    Code reference _____ .

FILL IN THE BLANKS

1. All wiring must be installed so that _____ of the water surface and changes in the _____ will not result in unsafe conditions.

2. The grounded circuit-conductor (neutral) shall be an _____ identified in conformance with *200.6*.

3. A green equipment grounding conductor must be run _____ to the building panelboard _____ bus.

4. Yard and pier distribution systems shall not _____ phase-to-phase.

5. All electrical connections must be located at least 12 in. (305 mm) _____ of a floating pier.

6. Circuit breakers and switches, installed in gasketed enclosures, must be arranged to permit required _____ without exposing the interior of the enclosure.

7. Overhead wiring must be installed to avoid possible contact _____ or other parts of boats.

8. A disconnecting means _____ to disconnect each boat from its supply connections.

9. Each single receptacle that provides shore power to boats shall be supplied from a marine power outlet or panelboard by an _____ of the voltage, class, and rating corresponding to the rating of the receptacle.

10. Electrical wiring and equipment must be located where it will not be subjected to _____ or where it can be damaged by the normal operation of boat use.

FREQUENTLY ASKED QUESTIONS

Question 1: What serves as the grounding electrode for a floating building?
Answer: The service equipment for a floating building is installed adjacent to, but cannot be on or in, the floating building. The service equipment is installed on the shore, and the grounding electrode is established according to *Article 250, Part III*.

This means that the floating building service equipment is grounded on shore in the same way as would be done at most other buildings. It is the feeder to the floating building that is specific to this type of building. The requirements for feeders are governed by *553.7(A)*, which requires flexibility of the feeder so that motion of the water surface and changes in water level will not result in unsafe conditions.

Question 2: Where is the distribution panel located for a floating building?

Answer: The distribution or circuit panel is located on the floating building. It is fed from the shore-mounted service equipment by a single set of feeders in accordance with *553.6*. If a floating building has more than one occupancy, a single set of feeders is permitted to supply each occupancy.

Question 3: What wiring method is required to provide the necessary flexibility for a feeder to a floating building?

Answer: *NEC® 553.7(B)* permits Liquidtight Metal or Liquidtight Nonmetallic Conduit to provide flexibility, but generally a portable power cable is used. The feeder conductors must be sized for the load to be served and no demand factors are permitted to be used. The grounded (neutral) conductor must be an insulated conductor and must be connected to the equipment grounding terminal in the service equipment. The grounded (neutral) conductor must be insulated from the equipment grounding conductor and all other grounded parts.

Question 4: How is the service equipment for boats moored at piers or marinas installed?

Answer: Generally, a combination meter housing and a panel feeding an outlet or outlets is installed at each berthing location. When boats are moored at these locations, a portable power cord is connected from the boat to the receptacle(s) in the panel. Utility owned service conductors are run through the pier and tapped at each berthing location. An equipment grounding conductor is run with the service conductors for grounding purposes.

Question 5: What are shore-power receptacles?

Answer: Shore-power receptacles are the receptacles that the boats are connected to when moored at a pier berth. Sometimes, the shore-power receptacles are fed directly from a combination meter and panelboard by the utility company, and in some areas, the shore-power receptacles are fed from a distribution panel that feeds several panel locations. When this latter method is used, the feeder for the distribution panel can be calculated using the demand factor table shown in *555.12*.

CHAPTER 21

Electric Signs

OBJECTIVES

After completing this chapter, the student should understand:
- signs and outline lighting systems.
- portable or mobile signs.
- ballasts, transformers, and electronic power supplies.
- field-installed skeleton tubing.
- neon tubing.
- cold cathode lighting.

KEY TERMS

Cold Cathode Lighting: Lighting similar to neon but with larger diameter tubing and transformers with a higher mA (milliampere) rating

Electric-Discharge Lighting: Systems of illumination utilizing fluorescent lamps, high-intensity discharge (HID) lamps, or neon tubing

Electrode Connection: Connections made using a connection device, by twisting the wires together, or by use of an electrode receptacle

Electrode Receptacle: A listed device used for terminating secondary conductors to the sign electrode

GTO Cable: Gas-tube sign and oil-burner ignition systems cable. GTO-5, GTO-10, GTO-15 are the UL kilovolt designations for this cable. Generally, this cable is thermoplastic insulated with an outer thermoplastic jacket

Midpoint Return: The transformer secondary midpoint tap that establishes a dual-voltage secondary 90° out of phase with the ends of the transformer windings

Neon Tubing: Electric-discharge tubing filled with various inert gases and manufactured into shapes that form letters, parts of letters, skeleton tubing, outline lighting, other decorative elements, and art forms

Skeleton Tubing: Neon tubing that is itself the sign, or outline lighting, and not attached to an enclosure or sign body

Sign Body: A portion of a sign that may provide protection from the weather but is not an electrical enclosure

INTRODUCTION

This chapter will cover the installation of conductors and equipment for electric signs and outline lighting. Electric signs and outline lighting includes all products and installations utilizing **neon tubing**, such as signs, decorative elements, **skeleton tubing**, or art forms, and installations utilizing listed luminaires (lighting fixtures). Electric signs and outline lighting, fixed, mobile, or portable, must be listed and installed in conformance with that listing except that field-installed skeleton tubing is not required to be listed where installed in conformance with the *NEC*.® Outline lighting is not required to be listed as a system when it consists of listed luminaires (lighting fixtures) wired in conformance with the *NEC*.®

Although there are many types of signs that are manufactured and are available for inspection and listing by a testing laboratory, there are many signs, such as outline lighting, that are field-installed for a particular application, and listing would require a field type of inspection in order to obtain a listing label. For this reason *600.3* permits these signs to be installed without listing as long as they conform to wiring methods in accordance with *NEC®* Chapter 3. This becomes the responsibility of the AHJ.

GENERAL

There are two types of illuminated signs. One is the electric sign which is a fixed, stationary, or portable self-contained electrically illuminated utilization equipment with words or symbols designed to convey information or attract attention. An example of this would be an exit sign. The other is called outline lighting and is an arrangement of incandescent lamps or **electric-discharge lighting** to outline or call attention to certain features such as the shape of a building or the decoration of a window. High-intensity discharge luminaires (lighting fixtures), fluorescent luminaires (lighting fixtures), and neon tubing are all forms of electric discharge lighting.

Markings. *NEC® 600.4* requires that signs and outline lighting systems with incandescent lampholders be marked to indicate the maximum allowable wattage of lamps. These markings must be located where visible during re-lamping. This requirement will ensure that the internal conductors of the sign will not be overloaded.

Branch-Circuits. Branch-circuit requirements are shown in *600.5*. Every commercial building and each commercial occupancy accessible to pedestrians must have at least one outlet in an accessible location at each entrance to each tenant space for sign or outline lighting system use. These sign outlets must be supplied by a branch-circuit with a rating of at least 20 amperes and must not supply other loads. This requirement is in connection with *220.3(B)(6)* which requires a minimum of 1200 volt-amperes to be computed for each branch-circuit, ensuring that sign outlets will be available for the tenants and that ample capacity has been figured for any sign loads.

Branch-circuits that supply signs and outline lighting containing incandescent and fluorescent lighting cannot be rated to more than 20 amperes. Branch-circuits that supply neon tubing installations shall not be rated in excess of 30 amperes. These requirements are designed to keep the amperage rating of the circuit within the amperage rating of the secondary circuit equipment limitations.

Signs using fluorescent or incandescent lighting are usually box or **sign body** type with a glass or equivalent front panel containing the graphics or symbols representing the meaning of the sign. Exit and other directional signs are examples of this type of sign (Figure 21-1).

Neon tubing is generally the sign itself, or is in itself the outline lighting, and is not attached to any box or sign body. Neon advertising signs hanging in windows of retail establishments are examples of neon sign lighting (Figure 21-2).

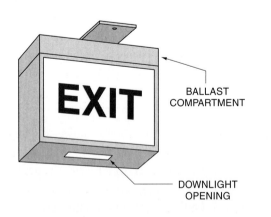

Figure 21-1 Exit sign—electric-discharge lighting

Figure 21-2 Typical neon window advertising sign

Signs and Outline Lighting Systems

Each sign and outline lighting system or feeder circuit or branch-circuit supplying sign or outline lighting system must have an externally operated disconnecting means in accordance with *600.6*. The disconnecting means must be within sight of the sign or outline lighting system and open all ungrounded conductors. Where a sign is out of sight from the disconnecting means, the disconnecting means shall be capable of being locked in the open position. Where the sign or outline lighting system is operated by a controller, the disconnecting means may be located within sight from the controller and shall be designed so that no pole can be operated independently and shall be capable of being locked in the open position. Considerable emphasis is placed on the requirements for a disconnecting means to protect sign maintenance personnel from being subjected to electric shock by some person turning on a sign that is being worked on (Figure 21-3).

NEC® 600.7 requires that all signs and outline lighting systems be grounded. FMC or LFMC that is used to enclose the secondary circuit conductor from a transformer or power supply for use with electric-discharge tubing shall be permitted as a bonding means if the total accumulative length of the conduit in the secondary circuit does not exceed 100 ft (30 m). Small metal parts not exceeding 2 in. (50 mm) in any dimension, not likely to be energized, and spaced at least ¾ in. (19 mm) from neon tubing, does not need bonding (Figure 21-4).

Where Nonmetallic Conduit is used to enclose the secondary circuit conductor from a transformer or power supply and a bonding conductor is required, the bonding conductor must be installed separately from the nonmetallic conduit and be spaced at least 1½ in. (38 mm) from the conduit (Figure 21-5).

Arcing of secondary conductors to ground is a serious problem with electric-discharge sign lighting systems due to the high voltages used, particularly with electric discharge tubing systems using various inert gases.

Portable or Mobile Signs

NEC® 600.10 requires that portable or mobile signs must be adequately supported and readily movable without the use of tools. They must be powered by use of an attachment plug. All cords used for portable or mobile signs must be junior hard service or hard service types as shown in *400.4* and must have an equipment grounding conductor.

Portable or mobile signs must be provided with factory-installed GFCI protection for personnel. The GFCI must be integral with the attachment plug or must be located in the power-supply cord within 12 in. (305 mm) of the attachment plug. Cords used with portable or mobile signs in dry locations shall be SP-2, SPE-2, SPT-2, or heavier, and the cord shall not exceed 15 ft (4.4 m) in length (Figure 21-6).

Ballasts, Transformers, and Electronic Power Supplies

NEC® 600.21 requires that all ballasts, transformers, and electronic power supplies be located where accessible and must be securely fastened in place. Ballasts, transformers, and electronic power supplies must be located as near as possible to the lamps or neon tubing to keep the secondary conductors as short as possible.

Where ballasts, transformers, or electronic power supplies are used in wet locations, they must

Figure 21-3 Disconnecting means for building sign

364 CHAPTER 21 Electric Signs

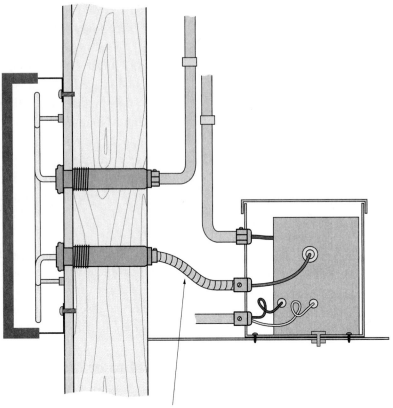

Figure 21-4 Flexible Metal Conduit used as bonding means

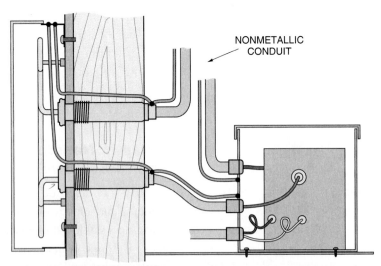

Figure 21-5 Nonmetallic Conduit spacing from bonding conductor

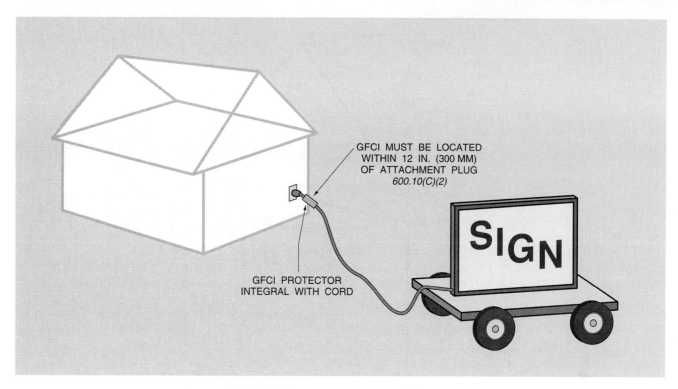

Figure 21-6 Mobile sign with ground-fault protection

be of the outdoor type and must be protected by placement in a sign body or separate enclosure.

Where ballasts, transformers, or electronic power supplies are not installed in a sign body, a working space must be provided at each location 3 ft (900 mm) high, 3 ft (900 mm) wide, by 3 ft (900 mm) deep. Ballasts, transformers, and electronic power supplies are permitted to be located in attic or soffit locations provided there is an access door at least 3 ft (900 mm) by 2 ft (600 mm) and a passageway with a suitable permanent walkway extending from the point of entry to each component (Figure 21-7).

Ballasts, transformers, and electronic power supplies are permitted to be located above suspended ceilings provided their enclosures are securely fastened in place, and they are not dependent upon the suspended ceiling for support. Ballasts, transformers, and electronic power supplies installed in suspended ceilings must not be connected to branch-circuits by flexible cords (Figure 21-8). Ballasts must be listed and shall be thermally protected as required by *600.22*.

Transformers and electronic power supplies shall have secondary circuit ground-fault protection unless: (1) they have isolated ungrounded secondaries and maximum open-circuit voltage of 7500 volts or less, and (2) they have integral porcelain or glass secondary housing for the neon tubing and require no field wiring of the secondary circuit. Secondary circuit voltage shall not exceed 15,000 volts under any load condition. The voltage-to-ground of any output terminals of the secondary circuit shall not exceed 7500 volts under any load conditions. Secondary output circuits shall not be connected in series or parallel. A transformer or power supply must be marked to indicate that it has secondary circuit ground-fault protection. This secondary circuit ground-fault protection is not required to be integral with the transformer.

Field-Installed Skeleton Tubing

NEC® 600.31 requires that neon secondary circuit conductors, 1000 volts or less, must be insulated, listed for the purpose, and not smaller than 18 AWG. The conductors shall be installed using any wiring method included in Chapter 3 of the *NEC®* suitable for the conditions. Bushings must be provided for wires passing through an opening in metal.

NEC® 600.32 permits neon secondary circuit conductors, over 1000 volts to be installed on insu-

366 CHAPTER 21 Electric Signs

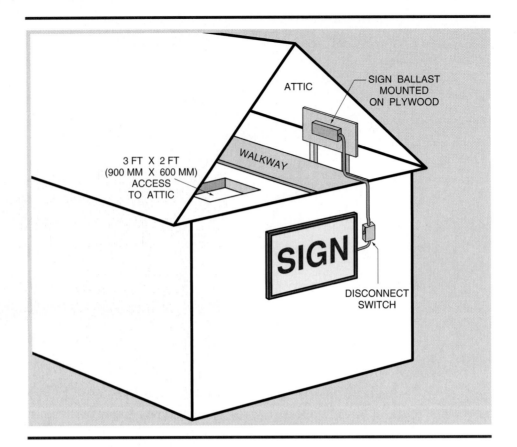

Figure 21-7 Sign ballast located in attic space

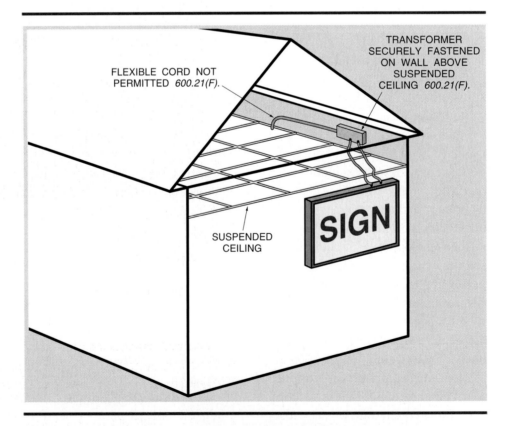

Figure 21-8 Transformers above suspended ceiling

lators, in RMC, IMC, RNC, LFNC, FMC, LFMC, EMT, metal enclosures, or other equipment listed for the purpose and shall be installed in accordance with the requirements of Chapter 3 of the *NEC*.® Conduit or tubing shall be a minimum of trade size ½ (16) and may contain only one conductor. Secondary conductors must be separated from each other and from all objects other than insulators or neon tubing by a spacing not less than 1½ in. (38 mm). **GTO cable** requires no spacing between the cable insulation and the conduit or tubing. Conductors shall be insulated and not smaller than 18 AWG.

Conductors shall be permitted to be run between the ends of neon tubing or to the secondary circuit **midpoint return** of transformers or electronic power supplies. The distance of secondary circuit conductors from the high-voltage terminal of a transformer or electronic power supply to the first neon tube electrode shall not exceed: (1) 20 ft (6 m) when installed in metal conduit or tubing for a transformer or electronic power supply rated 45 mA or less, (2) 50 ft (15 m) when installed in Nonmetallic Conduit. All other sections of secondary circuit conductors shall be as short as practicable (Figure 21-9).

Figure 21-9 Maximum length of secondary conductors, *600.32(J)*

Neon Tubing

Neon is a term that has come to be used to describe a specific type of gas discharge lighting that is created by passing electricity through a gas. Several different gases are used to create neon signs. When electrically stimulated, neon gas creates a red light, argon gas creates a blue light, helium produces a yellow light, and krypton produces a white light with a purple tint. Neon and argon are the most commonly used, although combinations are used to make almost any color. Neon signs operate at high voltages ranging from 2000 to 15,000 volts that are produced by a transformer or electronic power supply.

The secondary circuit voltage required depends on the length of the neon lamp tubing used in the sign. The highest voltage rating for a neon sign transformer is 15,000 volts. If the length of the tubing is greater than can be illuminated with 15,000 volts, two or more transformers are used. Even though the secondary circuit current is in the milliampere range, the voltage levels present a danger for shock or fire if the installation is not properly installed and insulated.

Electrode connections to the neon tubing are accomplished by the requirements of *600.42*. Terminals of the electrode must not be accessible to unqualified persons. Connections are made by the use of a connection device, twisting the wires together, or by the use of an **electrode receptacle**. The neon tubing and conductor must be supported not more than 6 in. (150 mm) from the electrode connection (Figure 21-10).

Cold Cathode Lighting

No direct reference by name is made in *Article 600* to another type of gaseous discharge lighting called **cold cathode lighting**. *Article 410, Part XIV*, titled *Special Provisions for Electric Discharge Lighting Systems of More than 1000 Volts*, contains the requirements for cold cathode lighting systems. Cold cathode lighting refers to a special type of custom manufactured fluorescent lamp. Cold cathode lamps can be bent or shaped to form almost any configuration. Cold cathode tubing can range from around 300 lumens per linear ft up to 700 lumens per linear ft for a single row of cold cathode lamps.

Transformers or ballasts used for cold cathode lighting shall not have a secondary open circuit

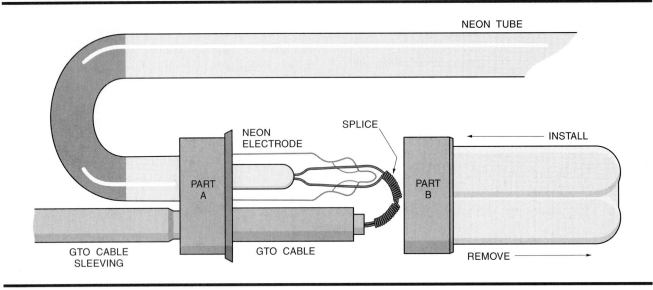

Figure 21-10 Neon electrode connections

voltage exceeding 15,000 volts. The secondary current rating shall not be more than 120 milliamperes if the open circuit voltage is over 7500 volts and not more than 240 milliamperes if the open circuit voltage is 7500 volts or less.

Each cold cathode secondary circuit of tubing having an open circuit voltage exceeding 1000 volts must have a clear legible marking indicating the open circuit voltage.

SUMMARY

Electric sign installation is a specialty trade. The installing electrician and the electrical inspector must clearly understand the operation and installation procedures to ensure a safe installation.

Neon lamps used for signs are of the high-voltage type and are fed from standard branch-circuits through small stepup transformers. The length of the neon tubing determines the voltage required from the transformer. The maximum voltage used on neon lamps is 15,000. When greater lengths of neon tubing than can be handled by 15,000 volts are necessary, two or more transformers may be used as required.

All signs, whether of the incandescent type or the gaseous discharge type, must meet certain criteria. All equipment must be enclosed in metal boxes and, unless fed from ungrounded circuits, must be grounded. Each sign must be controlled in accordance with *600.6(A)*, by an externally operated disconnecting means located within sight from the sign. Control switches, flashers, and similar devices used for controlling transformers and electronic power supplies must be rated for controlling inductive loads or have a current rating not less than twice the current rating of the transformer.

A sign outlet must be provided for each commercial occupancy, in accordance with *600.5(A)*, for each tenant space. The outlet must be supplied by a branch-circuit rated at least 20 amperes and that supplies no other loads. Branch-circuits that supply signs and outline lighting using incandescent or fluorescent forms of illumination must not be rated in excess of 20 amperes. Branch-circuits that supply neon tubing installations must not be rated in excess of 30 amperes.

In accordance with *600.10*, portable or mobile signs must be provided with factory-installed GFCI protection for personnel. The GFCI can be a part of the attachment plug or may be installed in the flexible cord not more than 12 in. (305 mm) of the attachment plug.

Ballasts, transformers, and electronic power supplies are permitted to be installed above suspended ceilings provided they are not dependent upon the ceiling grid for support. Ballasts, transformers, and electronic power supplies installed in suspended ceilings must not be connected to the branch-circuit by flexible cord.

Neon tubing must be spaced not less than ¼ in. (6 mm) from the nearest surface, other than its support. Listed tube supports must be used to support neon tubing. Terminals of neon tubing electrodes must not be accessible to unqualified persons. Connections must be made between the terminals and the tubing by use of a connection device, twisting the wires together, or by use of an electrode receptacle.

In accordance with *600.32(A)*, when neon secondary circuit conductors exceed 1000 volts, the conductors must be installed on insulators or be in RMC, IMC, RNC, LFNC, FMC, LFMC, EMT, metal enclosures, or other equipment listed for the purpose. Conduit or tubing can contain only one conductor and the conduit or tubing must be a minimum trade size ½ (16).

REVIEW QUESTIONS

MULTIPLE CHOICE

1. Cords supplying indoor portable signs must have a maximum length not exceeding:
   - ____ A. 6 ft (1.8 m).
   - ____ B. 15 ft (4.5 m).
   - ____ C. 12 ft (3.7 m).
   - ____ D. 10 ft (3.0 m).

2. A branch-circuit supplying a neon tubing installation shall not be rated in excess of:
   - ____ A. 20 amperes.
   - ____ B. 15 amperes.
   - ____ C. 10 amperes.
   - ____ D. 30 amperes.

3. Neon secondary circuit conductors over 1000 volts shall not be smaller than:
   - ____ A. 18 AWG.
   - ____ B. 16 AWG.
   - ____ C. 14 AWG.
   - ____ D. 22 AWG.

4. The maximum secondary circuit voltage for a neon sign transformer is:
   - ____ A. 7500 volts.
   - ____ B. 5000 volts.
   - ____ C. 10,000 volts.
   - ____ D. 15,000 volts.

5. Sign or outline lighting system equipment shall be at least _____ above areas accessible to vehicles unless protected from physical damage.
   - ____ A. 14 ft (4.3 m)
   - ____ B. 15 ft (4.5 m)
   - ____ C. 10 ft (3.0 m)
   - ____ D. 12 ft (3.7 m)

TRUE OR FALSE

1. Branch-circuits supplying neon tubing installations shall not be rated in excess of 30 amperes.
   - ____ A. True
   - ____ B. False

   Code reference _____ .

2. The disconnecting means for a sign or outline lighting system shall be within sight from the sign or outline lighting system that it controls.
   ____ A. True
   ____ B. False
   Code reference _____ .

3. All cords supplying portable or mobile signs must have an equipment grounding conductor.
   ____ A. True
   ____ B. False
   Code reference _____ .

4. Sign ballasts are not permitted to be located in attics and soffits.
   ____ A. True
   ____ B. False
   Code reference _____ .

5. Neon secondary circuit conductors, 1000 volts or less, shall not be smaller than 14 AWG.
   ____ A. True
   ____ B. False
   Code reference _____ .

6. Equipment having an open-circuit voltage exceeding 1000 volts shall not be installed in dwelling units.
   ____ A. True
   ____ B. False
   Code reference _____ .

7. Sign outlets at commercial occupancies must be supplied by a branch-circuit with a rating of less than 30 amperes.
   ____ A. True
   ____ B. False
   Code reference _____ .

8. Neon secondary conductors, 1000 volts or less, must be insulated, listed for the purpose, and not smaller than 18 AWG.
   ____ A. True
   ____ B. False
   Code reference _____ .

9. The highest voltage rating for a neon transformer is 5000 volts.
   ____ A. True
   ____ B. False
   Code reference _____ .

10. More than one neon secondary conductor over 1000 volts can be run together in one conduit.
    ____ A. True
    ____ B. False
    Code reference _____ .

FILL IN THE BLANKS

1. Ballast, transformers, and electronic power supplies installed in suspended ceilings must not be connected to the branch-circuit by _____ .

2. Conductors are permitted to run between the ends of neon tubing or to the secondary circuit _____ of transformers or electronic power supplies listed for the purpose and provided with terminals at the midpoint.

3. Electrode connections must be made by use of a connection device, or by twisting the wires together, or use of an _____ receptacle.

4. The disconnecting means must be _____ of the sign or outline lighting system that it controls.

5. Metal parts of a building shall not be permitted as a _____ conductor or an equipment grounding conductor.

6. Branch-circuits that supply signs and outline lighting containing incandescent and fluorescent lighting cannot be rated _____ 20 amperes.

7. Portable or mobile signs must be provided with _____ GFCI protection for personnel.

8. Each sign and outline lighting system or feeder-circuit or branch-circuit supplying sign or outline lighting system must have an _____ disconnecting means in accordance with *600.6*.

9. Where ballasts, transformers, or electronic power supplies are used in _____ , they must be of the outdoor type and must be protected by placement in a sign body or separate enclosure.

10. GTO cable requires no spacing between the _____ and the conduit or tubing. Conductors shall be insulated and not smaller than 18 AWG.

FREQUENTLY ASKED QUESTIONS

Question 1: What is the reason for the requirement in mobile or portable signs for a GFCI to be integral with the attachment plug or in the power supply cord within 12 in. (305 mm) of the attachment plug? Why can't I just plug it into a GFCI receptacle?

Answer: So that wherever you put that sign, the protection will be there. There may not be a ground-fault protected receptacle everywhere you put that sign. This way, the protection is part of the sign, and where the sign goes, the ground-fault protection also goes. Requiring the protection in the attachment plug, or within 12 in. (305 mm) of the attachment plug, assures protection of the entire length of flexible cord along with the sign. See *600.10(B)(2)*.

Question 2: Are the secondary circuit conductors from a neon sign tubing transformer required to be in a raceway?

Answer: Not according to *600.32(A)*. This section permits secondary circuit conductors over 1000 volts to be installed on insulators. This is written for field-installed systems and does not apply to GTO cable used in listed systems. GTO cable is required to be installed inside a raceway or sign enclosure for listed signs.

Question 3: What is GTO cable and where is it shown in the *NEC*®?

Answer: Gas tube and oil burner ignition cable is not shown in the *NEC.*® This cable has been evaluated by Underwriters Laboratories and is listed as GTO-5, GTO-10, and GTO-15 for the kilovolt designations for this cable. The cable is generally thermoplastic insulated with an outer thermoplastic jacket. It is a special use conductor and is not listed under *Article 310*, which lists conductors for general wiring.

Question 4: What does *600.6(A)(1)* mean where it says the disconnecting means must be in sight from the sign, but if it's not in sight from the sign it must be capable of being locked open?

Answer: It is possible for a disconnect switch to be within sight from only a part of a sign. A sign may run around two sides of a building, and the disconnect might only be in sight from a portion of the sign. In an instance like this, the disconnecting means must be capable of being locked in the open position. The disconnecting means must always be in sight of the sign or a section thereof, but, where it is out of sight from any section of the sign, it must be capable of being locked in the open position.

Question 5: How does neon lighting work?

Answer: Neon lighting consists of a gas like neon, argon, or krypton at low pressure in a glass tube. At each end of the tubing there are metal electrodes. When you apply high voltage to the electrodes, the gas vaporizes and electrons flow through the gas. The electron flow excites the neon atoms, and they emit light that we can see. Neon emits red light, argon emits blue light, and krypton emits a yellowish light. These gases are combined with mercury and with tinting of the tubing to create approximately 30 different colors.

The high voltage to the metal electrodes is produced by a transformer, which is permitted to have a secondary circuit voltage as high as 15,000 volts. The secondary voltage required depends upon the length of the neon tubing. *NEC® 600.23(C)* permits secondary voltages to be not higher than 15,000 volts. If the length of the tubing is greater than can be illuminated with 15,000 volts, two or more transformers are used. *NEC® 600.23(E)* prohibits secondary circuit outputs from being connected in parallel or series. Tubing charts are readily available showing voltage ratings of transformers required for the various lengths of tubing that may be used.

The voltage is stepped up from 120 volts to 15,000 volts, which means the transformer has a turns ratio of approximately 125 to 1, and, if a 30-ampere circuit is applied, the maximum current that can be applied is approximately 240 milliamperes. This is consistent with *600.22(D)*, which limits the secondary current rating of transformers and electronic power supplies to not more than 300 mA.

Question 6: How is fluorescent lighting different from neon lighting?

Answer: Fluorescent light works on a similar principle, but with fluorescent, a low-pressure mercury vapor is used. When ionized, the mercury vapor emits ultraviolet light. The interior of a fluorescent lamp is coated with phosphor, and the phosphor accepts the energy of the ultraviolet photons (electromagnetic radiation) and emits visible photons. The light we see from a fluorescent tube is the light given off by the phosphor coating inside the tube. This action is caused when the phosphor begins to fluoresce (emits radiation), and this is where the name fluorescent comes from.

Question 7: How is the connection to the neon electrode required to be made?

Answer: When the neon tubing is constructed, an electrode is installed at each end of the tubing. This electrode has a conductor called the lead wire attached to it that may be used to attach an electrode receptacle or may be spliced to the GTO cable. *NEC® 600.42(B)* permits either method to be used. The connections must be mechanically secure and must be in an enclosure listed for the purpose. Generally, this enclosure is a boot that slips over the connection point and provides protection for the electrode connection.

Question 8: What is the electrode shell?

Answer: The electrode shell is the metal cylinder inside the glass envelope of the electrode. It is the source of the electric discharge in the lamp. Lead wires are used to suspend the electrode shell in the glass envelope and supply the electricity from the transformer or electronic power source.

Question 9: How is neon tubing required to be supported?

Answer: *NEC® 600.41* requires neon tubing to be supported by listed tube supports. One manufacturer uses *C* clips which encircle approximately 75% of the tubing diameter. These clips provide a standoff of 1¾ in. (45 mm) to accommodate a *double-back* tube thickness and provide the spacing required between the tubing and the nearest surface other than the support. Light shines through the transparent clips, making them virtually invisible.

Question 10: Is GTO cable sleeving required and what is it?

Answer: GTO sleeving is not required in the *NEC®* but it is not prohibited from use. It is used as an added protection to GTO cable, both when used as open conductors on insulators and when enclosed in a raceway. The sleeving is intended to sleeve no more than one GTO cable. The manufacturers of this sleeving feel that the use of this product over the GTO cable in the conduit will prevent the cable from coming into contact with the grounding plane created by the conduit or tubing. The sleeving will help in reducing the effects of exposure to ozone generated by the high voltages used in this application.

Question 11: Can I locate the ballast for a neon sign above the drywall ceiling of the show window where the sign is located?

Answer: If there is an access door to the area above the ceiling, the access door must be at least 3 ft (900 mm) by 2 ft (600 mm) and there must be a passageway at least 3 ft (900 m) high by 2 ft (600 mm) wide, extending from the point of entry to each component.

CHAPTER 22

Office Furnishings

OBJECTIVES

After completing this chapter, the student should understand:
- partition wireways.
- partition interconnections.
- lighting accessories.
- fixed-type partitions.
- freestanding type partitions.
- freestanding type partitions, cord- and plug-connected.

KEY TERMS

Fixed-Type Partitions: Permanently secured to the building surfaces

Freestanding Partitions: Not fixed or secured to the building surfaces

Lighting Accessories: A lighting unit specifically listed for use with office furnishings

Mechanically Contiguous: Physically connected together

INTRODUCTION

This chapter covers electrical equipment, **lighting accessories**, and wiring systems used to connect, or that are contained within, or installed on relocatable-wired partitions. These office furnishings are portable and consist of panels, workstations, and pedestal-style systems that may be mechanically interconnected to form an office furnishing system. These assemblies shall be installed and used only as provided for in *Article 604*.

GENERAL

NEC® 605.2 requires that wiring systems shall be identified as suitable for providing power for lighting accessories and appliances in wired partitions. The partitions shall not extend from floor to ceiling unless where permitted by the AHJ. Relocatable-wired partitions shall be permitted to extend to the ceiling but shall not penetrate the ceiling (Figure 22-1).

Partition Wireways

According to *605.3*, all conductors and connections shall be contained within wiring channels of metal or other material identified as being suitable for the conditions of use. Wiring channels must be free from projections or other conditions that may damage conductor insulation (Figure 22-2).

Partition Interconnections

NEC® 605.4 requires that the electrical connections between partitions shall be a flexible assembly

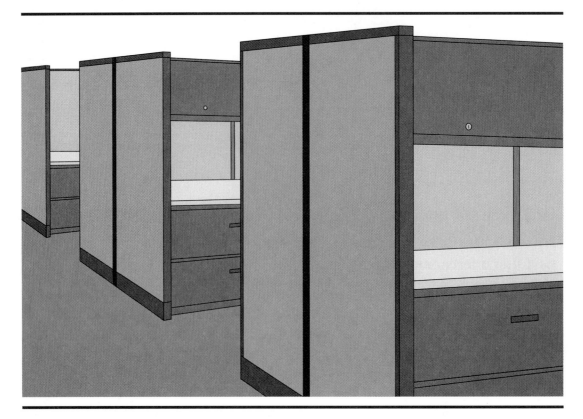

Figure 22-1 Relocatable office partition—not penetrating ceiling

Figure 22-2 Office partition wiring channel

identified for use with wired partitions (Figure 22-3) or shall be permitted to be installed using flexible cord, provided all of the following conditions are met:

1. The cord must be extra-hard usage type with 12 AWG or larger conductors with an insulated equipment grounding conductor.

2. The partitions are **mechanically contiguous**.

3. The cord is not longer than necessary for maximum positioning of the partitions but in no case to exceed 2 ft (600 mm).

4. The cord is terminated at an attachment plug and cord connector with strain relief.

Lighting Accessories

In accordance with *605.5*, lighting accessories must be listed and identified for use with wired partitions and must comply with all of the following (see Figure 22-4):

1. A means of secure attachment or support must be provided.

2. Where cord- and plug-connections are provided, the cord length must be suitable for the intended application but must not exceed 9 ft (2.7 m) in length.

3. The cord must not be smaller than 18 AWG and shall be of the hard-usage type.

4. The flexible cord must contain an equipment grounding conductor.

CHAPTER 22 Office Furnishings   377

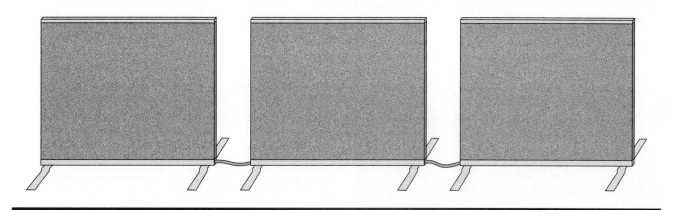

Figure 22-3   Flexible assembly connection between office partitions

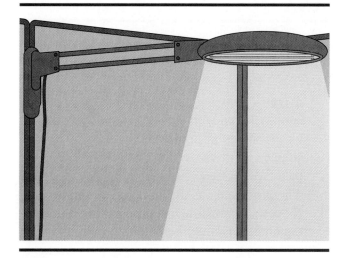

Figure 22-4   Lighting accessories for relocatable office partitions

Figure 22-5   Fixed wired partition connection to premises wiring system

5. Convenience receptacles are not permitted in lighting accessories.

Fixed-Type Partitions

NEC® 605.6 requires that **fixed-type partitions** (wired partitions that are fixed or secured to the building surface) must be permanently connected to the building electrical system by one of the wiring methods of Chapter 3 (Figure 22-5).

Freestanding-Type Partitions

In accordance with *NEC® 605.7*, **freestanding partitions** (not fixed to the building surfaces) are permitted to be permanently connected to the building electrical system by one of the wiring methods of Chapter 3 (Figure 22-6).

Freestanding-Type Partitions, Cord- and Plug-Connected

NEC® 605.8 requires that individual partitions of the freestanding type, or groups of individual partitions that are electrically connected, mechanically contiguous, and do not exceed 30 ft (9.0 m) when assembled, are permitted to be connected to the building's electrical system by a single flexible cord and plug, provided all of the following conditions are met (see Figure 22-7).

1. The flexible power-supply cord must be extra-hard usage type with 12 AWG or larger conductors with an insulated equipment grounding conductor and not exceeding 2 ft (600 mm) in length.

CHAPTER 22 Office Furnishings

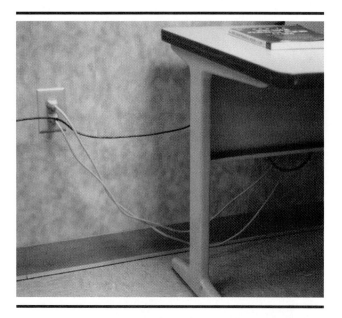

Figure 22-6 Free-standing office partition connection to premises wiring systems

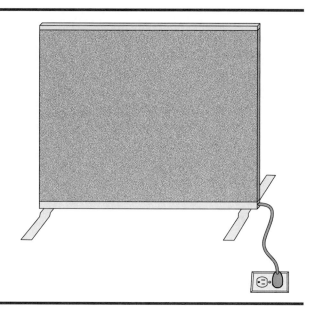

Figure 22-7 Free-standing type office partitions cord- and plug-connected

2. The receptacle supplying power shall be on a separate circuit serving only panels, and no other loads, and shall be located not more than 12 in. (300 mm) from the partition that is connected to it.

3. Individual partitions or groups of interconnected individual partitions shall not contain more than thirteen 15-ampere, 125-volt receptacle outlets.

4. Individual partitions or groups of interconnected individual partitions shall not contain multiwire circuits.

SUMMARY

Relocatable office furnishings are listed in three types:

- Type I—A system that includes all parts and contains *prewired* modular raceways and accessories and only requires a *quick-connect* type of electrical connection. A Type I system may be shipped with the accessories installed in the panel, or the accessories may be field installed where marked for use in the system. A means for permanent wiring connections to the branch-circuit supply are provided.

- Type II—A system that provides raceways and devices for routing and termination of wiring. *All wiring is done in the field.*

- Type III—A system not intended to be wired that has no provision for routing or termination of wiring. Partitions that extend to the ceiling, or that are used to support the building structure, are not listed as office furnishings.

REVIEW QUESTIONS

MULTIPLE CHOICE

1. Freestanding-type partitions, cord- and plug-connected, must not exceed _____ when assembled.
   - ___ A. 9 ft (2.7 m)
   - ___ B. 10 ft (3.0 m)
   - ___ C. 20 ft (6.0 m)
   - ___ D. 30 ft (9.0 m)

2. Where cord- and plug-connections are provided for lighting accessories, the cord must not exceed _____ in length.
   ____ A. 9 ft (2.7 m)
   ____ B. 10 ft (3.0 m)
   ____ C. 12 ft (3.7 m)
   ____ D. 15 ft (4.5 m)

3. The flexible power-supply cord for freestanding partitions must be extra-hard usage with _____ or larger conductors.
   ____ A. 10 AWG
   ____ B. 14 AWG
   ____ C. 12 AWG
   ____ D. 16 AWG

4. Individual partitions or groups of interconnected individual partitions shall not contain more than _____ 15-ampere, 125-volt receptacle outlets.
   ____ A. 10
   ____ B. 13
   ____ C. 12
   ____ D. 15

5. Flexible cord may be used to interconnect between partitions provided:
   ____ A. the cord is 12 AWG or larger.
   ____ B. the partitions are mechanically contiguous.
   ____ C. the flexible cord must have an insulated equipment grounding conductor.
   ____ D. all of the above.

TRUE OR FALSE

1. Relocatable-wired partitions shall be permitted to extend to the ceiling but shall not penetrate the ceiling.
   ____ A. True
   ____ B. False
   Code reference _____ .

2. Flexible cord used for electrical interconnection between partitions may not be longer than 3 ft (900 mm).
   ____ A. True
   ____ B. False
   Code reference _____ .

3. Individual partitions or groups of interconnected individual partitions shall not contain multiwire circuits.
   ____ A. True
   ____ B. False
   Code reference _____ .

4. Individual partitions or groups of interconnected individual partitions shall not contain more than ten 15-ampere, 125-volt receptacle outlets.
   ____ A. True
   ____ B. False
   Code reference _____ .

5. The receptacle supplying power to cord- and plug-connected partitions must be on a separate circuit serving no other loads.
   ____ A. True
   ____ B. False
   Code reference _____ .

6. Fixed-type partitions must be permanently connected to the building electrical system by one of the wiring methods discussed in Chapter 3.
   ____ A. True
   ____ B. False
   Code reference _____ .

7. Convenience receptacles are not permitted in lighting accessories.
   ____ A. True
   ____ B. False
   Code reference _____ .

8. Individual partitions of the free-standing type that are electrically connected and do not exceed 40 ft (12.0 m) when assembled are permitted to be connected to the building electrical system by a single flexible cord.
   ____ A. True
   ____ B. False
   Code reference _____ .

9. All conductors and connections shall be contained within wiring channels of metal or other material identified as suitable for the conditions of use.
   ____ A. True
   ____ B. False
   Code reference _____ .

10. Flexible cord is not permitted to interconnect partitions that are not mechanically contiguous.
    ____ A. True
    ____ B. False
    Code reference _____ .

FILL IN THE BLANKS

1. Wiring channels must be free from _____ or other conditions that may damage the conductor insulation.

2. Relocatable-wired partitions shall be permitted to extend to the ceiling but shall not _____ the ceiling.

3. Where cord and plug connections are provided, the cord length must be suitable for the intended application but must not exceed _____ in length.

4. Lighting accessories must be listed and identified for use with wired partitions, and the cord must not be smaller than _____ and shall be of the hard-usage type.

5. Office furnishings are portable and consist of panels, workstations, and pedestal-style systems that may be _____ to form an office furnishing system.

FREQUENTLY ASKED QUESTIONS

Question 1: Why are multiwire circuits not permitted in office furnishings partitions?

Answer: The receptacles used in office partitions are likely to be used for office equipment that has electronic switching devices that will create nonlinear loads and will enhance the possibilities of harmonic distortion, resulting in overheating of the neutral conductor. *NEC® 605.8(D)*, however, only requires this protection for freestanding-type partitions that are cord- and plug-connected.

Question 2: What is the reasoning behind limiting the number of receptacle outlets in individual or groups of interconnected partitions to not more than thirteen 15-ampere, 125-volt outlets?

Answer: Assuming that the supply receptacle is derived from a 20-ampere rated circuit (20 A × 120 V = 2400 volt-amperes) and using *220.3(B)(9)* which requires computing receptacle outlets at 180 volt-amperes (2400 ÷ 180 = 13) we find 13 receptacles at 180 VA equals 2340 VA.

Question 3: Are office partitions prewired, or, are they supposed to be field wired?

Answer: Office partitions listed as Type I are prewired from the manufacturer and require only a connection to a feed, and those listed as Type II are a system that provides raceways for field wiring.

Question 4: Why are office partitions not permitted to be ceiling height?

Answer: *NEC® 605.2* permits ceiling height partitions contingent upon approval of the AHJ. Office partitions are intended to be portable and relocatable. They are not intended to be used as walls. Where permitted by the AHJ, relocatable-wired partitions may extend to the ceiling but may not penetrate the ceiling.

Question 5: Is there a minimum size branch-circuit rating required for the receptacle outlet supplying power to a freestanding-type, cord- and plug-connected partition?

Answer: The branch-circuit rating of the receptacle supplying power to freestanding-type, cord- and plug-connected partitions can be either 15- or 20-ampere. It must be a separate branch-circuit and can serve no other loads.

CHAPTER 23

Cranes and Hoists

OBJECTIVES

After completing this chapter, the student should understand:
- hazardous location requirements.
- wiring methods.
- contact conductors.
- disconnecting means.
- overcurrent protection.
- control.
- grounding.

KEY TERMS

Brake Coil: A coil winding used to electrically brake and hold a motor

Contact Conductor: Bare copper trolley wires that run parallel to the crane runway

Monorail Hoist: A hoist that is suspended from a single rail or track

Runway Conductor Disconnecting Means: Disconnecting means, either fused or circuit breaker type, used to disconnect the crane feeder circuit from the main collector contact conductors

Runways: The length and width of the area traversed by the crane

Suspended Pushbutton Station: Cord-connected pushbutton control for crane operation suspended by flexible cord for operation from floor level

Trolley Wires: Alternate terminology for contact conductors

INTRODUCTION

This chapter covers the installation of electric equipment and wiring used in connection with cranes, **monorail hoists**, hoists, and all **runways**. Installation requirements can be found in *Article 610*.

GENERAL

Cranes and hoists are very specialized pieces of equipment that are generally installed by the manufacturer of the equipment. There are many types of cranes and hoists that are specifically designed to meet the user's needs. There are cab-operated bridge cranes. There are cranes operated by a suspended control station. Some cranes are top-running on the girders and some are under-running, suspended from the girders (Figure 23-1).

Cranes and hoists are generally operated using three motors. A bridge motor is used to move the crane through the runway. A trolley motor is required to move the trolley between the runway rails and a hoist motor is required to lift the load. Some cranes have a fourth auxiliary motor that is used to hoist smaller loads.

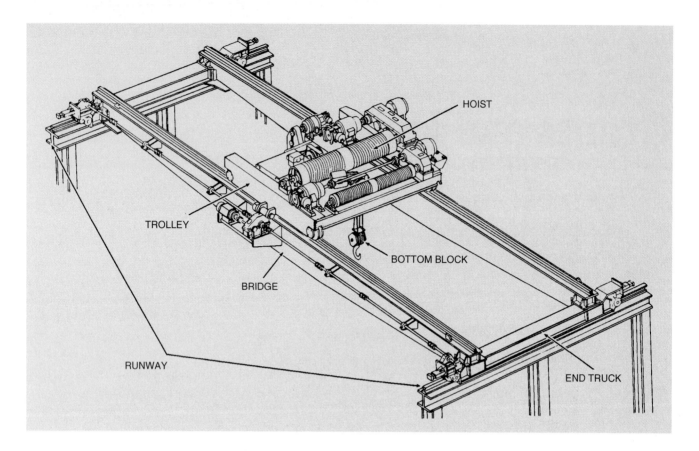

Figure 23-1 Bridge crane with top-running trolley. *Courtesy of* Harnischfeger

Hazardous Location Requirements

All equipment that operates in locations that are hazardous because of the presence of flammable gases or vapors must conform to *Article 501*. All equipment used in locations that are hazardous because of the presence of combustible dust must conform to *Article 502*, and all equipment used in locations that are hazardous because of the presence of easily ignitable fibers or flyings must conform to *Article 503*.

Where a crane or hoist operates over combustible fibers in Class III locations, the electric crane or hoist is not permitted to be grounded by *250.22* and *503.13*. This reduces the chance that sparks from faulted equipment will fall onto the combustible fibers, causing a fire (Figure 23-2).

Where a crane or hoist operates over combustible material, the braking resistors must be located in a cabinet that will not emit flames or molten metal (Figure 23-3), or they must be located in a cage or cab, constructed of noncombustible material that encloses the sides of the cage or cab

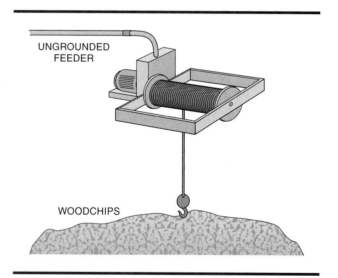

Figure 23-2 Cranes that operate over combustible fibers are not permitted to be grounded

to a point at least 6 in. (150 mm) above the top of the resistors.

Contact conductors must be located or guarded so as to be inaccessible to other than qualified

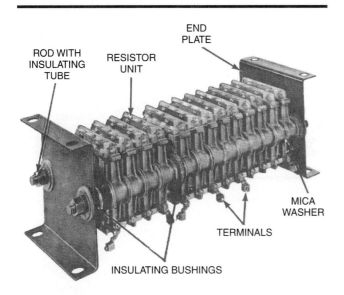

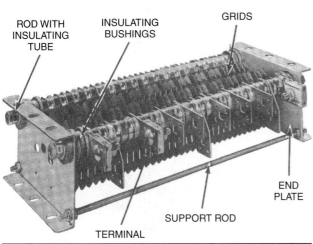

Figure 23-3 Secondary resistors. *Courtesy of Harnischfeger*

persons and must be protected against accidental contact with foreign objects. The current collectors must be arranged and guarded so as to confine normal sparking and prevent escape of sparks or hot particles. In Class III, Division 1 and 2 locations, the power supply to contact conductors must be isolated from other systems and be equipped with a ground detector that will give alarm and automatically de-energize the contact conductors in case of a fault to ground or will give a visual and audible alarm as long as power is supplied to contact conductors and the ground-fault remains.

Cranes and hoists that operate in the area of electrolytic cell line working zones, as shown in *668.32*, are not required to be grounded. The portion of an overhead crane or hoist that contacts the energized electrolytic cell or energized attachments must be insulated from ground.

Wiring Methods

Wiring methods for cranes and hoists are given in *Article 610, Part II*. Conductors used for the wiring of cranes and hoists shall be enclosed in raceways or be Type AC cable with insulated grounding conductor, Type MC cable, or Type MI cable, unless otherwise permitted in *610.11*. Contact conductors, sometimes called **trolley wires**, are not required to be enclosed in raceways. Flexible connections to motors and similar equipment may be flexible stranded conductors in FMC, LFMC, LFNC, Multiconductor cable, or an approved Flexible Nonmetallic enclosure. Where flexibility is required for power or control to moving parts, a cord suitable for the purpose may be used if suitable strain relief and protection from physical damage is provided.

Where Multiconductor cable is used with a **suspended pushbutton station**, the station must be supported in some manner that will relieve the strain on the conductors. Strain relief cable connectors are readily available, such as basket grips, that fasten to the pushbutton enclosure and encircle the cable in a manner that tightens or holds the cable to the enclosure when strain is put on the cable (Figure 23-4).

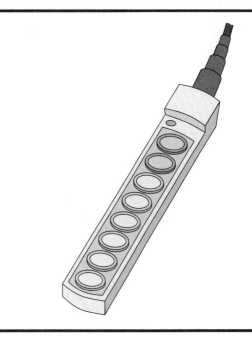

Figure 23-4 Suspended pushbutton station with strain relief connector

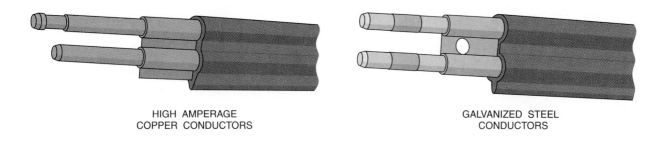

Figure 23-5 Contact conductors

Contact Conductors. Contact conductors along runways, crane bridges, and monorails are permitted to be bare and can be copper, aluminum, steel, or other alloys or combinations in the form of hard-drawn wire, tees, angles, tee rails, or other stiff shapes (Figure 23-5).

NEC® *610.14* shows the rating and size of conductors used in the wiring of cranes and hoists. *Table 610.14(A)* shows the ampacities of insulated copper conductors used with short-term rated crane and hoist motors (Figure 23-6).

Table 610.14(A). Ampacities of Insulated Copper Conductors Used with Short-Time Rated Crane and Hoist Motors. Based on Ambient Temperature of 30°C (86°F). Up to Four Conductors in Raceway or Cable.[1] Up to 3 ac[2] or 4 dc[1] Conductors in Raceway or Cable

Maximum Operating Temperature	75°C (167°F)		90°C (194°F)		125°C (257°F)		Maximum Operating Temperature
Size (AWG or kcmil)	Types MTW, RHW, THW, THWN, XHHW, USE, ZW		Types TA, TBS, SA, SIS, PFA, FEP, FEPB, RHH, THHN, XHHW, Z, ZW		Types FEP, FEPB, PFA, PFAH, SA, TFE, Z, ZW		Size (AWG or kcmil)
	60 Min	30 Min	60 Min	30 Min	60 Min	30 Min	
16	10	12	—	—	—	—	16
14	25	26	31	32	38	40	14
12	30	33	36	40	45	50	12
10	40	43	49	52	60	65	10
8	55	60	63	69	73	80	8
6	76	86	83	94	101	119	6
5	85	95	95	106	115	134	5
4	100	117	111	130	133	157	4
3	120	141	131	153	153	183	3
2	137	160	148	173	178	214	2
1	143	175	158	192	210	253	1
1/0	190	233	211	259	253	304	1/0
2/0	222	267	245	294	303	369	2/0
3/0	280	341	305	372	370	452	3/0
4/0	300	369	319	399	451	555	4/0
250	364	420	400	461	510	635	250
300	455	582	497	636	587	737	300
350	486	646	542	716	663	837	350
400	538	688	593	760	742	941	400
450	600	765	660	836	818	1042	450
500	660	847	726	914	896	1143	500

Figure 23-6 *NEC*® *Table 610.14(A)* (Reprinted with permission from NFPA 70-2002)

Secondary Resistor Conductors. The secondary resistors are connected to the motor windings of wound-rotor motors so that the motor characteristics become similar to a variable speed motor to permit greater control of the movement of the hoist during hoisting operations. Where the secondary resistor is separate from the controller, the minimum size of the conductors between the controller and the resistor shall be calculated by multiplying the motor secondary current by the appropriate factor from *Table 610.14(B)* (Figure 23-7) and selecting the conductor from *Table 610.14(A)* (Figure 23-6).

Contact conductors must have an ampacity not less than that required by *Table 610.14*.

Disconnecting Means. The requirements for disconnecting means for cranes and hoists are shown in *Article 610, Part IV*. A **runway conductor disconnecting means** must be provided between the runway contact conductors and the power supply. The disconnecting means can be a motor-circuit switch, circuit breaker, or molded-case switch. The disconnecting means must be readily accessible and operable from the ground or floor level. It must be capable of being locked in the open position and must open all ungrounded conductors simultaneously. It must be located in view of the runway contact conductors.

The continuous ampere rating of the switch or circuit breaker required shall not be less than 50% of the combined short-time ampere rating of the motors, nor less than 75% of the sum of the short-term ampere rating of the motors required for a single motion.

The runway supply conductors and main contact conductors of a crane or monorail hoist must have overcurrent protection that is not greater than the largest rating or setting of any branch-circuit protective device, plus the sum of the nameplate ratings of all the other loads with the application of the demand factors from *Table 610.14(E)* (Figure 23-8).

Branch-circuits for cranes, hoists, and monorail hoists must have overcurrent protection in accordance with *610.42*. Overcurrent protection shall be provided by fuses or by inverse-time circuit breakers that have a rating in accordance with *Table 430.152* (Figure 23-9). Where two or more motors operate a single motion, the sum of their nameplate current

Table 610.14(B) Secondary Conductor Rating Factors

Time in Seconds		Ampacity of Wire in Percent of Full-Load Secondary Current
On	Off	
5	75	35
10	70	45
15	75	55
15	45	65
15	30	75
15	15	85
Continuous Duty		110

Figure 23-7 NEC® Table 610.14(B) (Reprinted with permission from NFPA 70-2002)

Table 610.14(E) Demand Factors

Number of Cranes or Hoists	Demand Factor
2	0.95
3	0.91
4	0.87
5	0.84
6	0.81
7	0.78

Figure 23-8 NEC® Table 610.14(E), Demand Factors (Reprinted with permission from NFPA 70-2002)

ratings shall be considered as that of a single motor. Where two or more motors are connected to the same branch-circuit, each tap conductor to an individual motor must have an ampacity not less than one-third of that branch-circuit. Each motor must be protected from overload in accordance with *610.43*.

Where taps to motor-control circuits originate on the load side of a branch-circuit protective device, each tap and piece of equipment shall be protected in accordance with *430.72*. Taps to **brake coils** shall be permitted without overcurrent protection. This will prevent failure of a braking or brake holding operation due to the opening of an overcurrent device.

Overload Protection. The requirements for overcurrent protection may be found in *Article 610, Part V*. Motors, motor controllers, and branch-

Table 430.52 Maximum Rating or Setting of Motor Branch-Circuit Short-Circuit and Ground-Fault Protective Devices

	Percentage of Full-Load Current			
Type of Motor	Nontime Delay Fuse[1]	Dual Element (Time-Delay) Fuse[1]	Instantaneous Trip Breaker	Inverse Time Breaker[2]
Single-phase motors	300	175	800	250
AC polyphase motors other than wound-rotor				
Squirrel cage — other than Design E or Design B energy efficient	300	175	800	250
Design E or Design B energy efficient	300	175	1100	250
Synchronous[3]	300	175	800	250
Wound rotor	150	150	800	150
Direct current (constant voltage)	150	150	250	150

Note: For certain exceptions to the values specified, see 430.54.
[1]The values in the Nontime Delay Fuse column apply to Time-Delay Class CC fuses.
[2]The values given in the last column also cover the ratings of nonadjustable inverse time types of circuit breakers that may be modified as in 430.52(C), Exception No. 1 and No. 2.
[3]Synchronous motors of the low-torque, low-speed type (usually 450 rpm or lower), such as are used to drive reciprocating compressors, pumps, and so forth, that start unloaded, do not require a fuse rating or circuit-breaker setting in excess of 200 percent of full-load current.

Figure 23-9 *NEC® Table 430.152, Maximum Rating or Setting of Motor Branch-Circuit Short-Circuit and Ground-Fault Protective Devices* **(Reprinted with permission from NFPA 70-2002)**

circuit conductors must be protected from overload by one of the means shown in *610.43*. These requirements are as follows:

1. A single motor is considered as being protected where the branch-circuit overcurrent device meets the rating requirements of *610.42*.
2. An overload relay element in each ungrounded circuit conductor with all relay elements protected from short-circuit by the branch-circuit protection.
3. Thermal sensing devices, sensitive to motor temperature or to temperature and current, that are thermally in contact with the motor windings. A hoist or trolley motor is considered to be protected if the sensing device is connected in the hoist upper limit switch circuit so as to prevent further hoisting during an overload condition or an over-temperature condition of either motor.

Control. The requirements for controllers can be found in *Article 610, Part VI*. Each motor shall be provided with an individual controller except that a single controller may be used to control two or more motors that drive a single hoist, carriage, truck, or bridge. One controller may be used to be switched between motors where only one motor is operated at a time and if the horsepower rating of the controller is not less than the horsepower rating of the largest motor.

Conductors of control circuits must be protected against overcurrent. Control circuits are considered as protected by overcurrent devices that are set at not more than 300% of the ampacity of the control conductors, except taps to control transformers will be considered as protected where the secondary circuit is protected by a device rated or set at not more than 200% of the rated secondary current of the transformer and not more than 200% of the ampacity of the control circuit conductors.

Where the opening of the control circuit would constitute a hazard, for example, in the control circuit of a hot metal crane, the control circuit conductors are considered as being protected by the branch-circuit overcurrent devices.

A limit switch or other device must be used to prevent the load block from passing the safe upper limit of travel of all hoisting mechanisms.

Where controls are enclosed in cabinets, the doors must open to at least 90° or be removable.

Grounding. The requirements for grounding cranes, monorail hoists, hoists, and accessories can be found in *Article 610, Part VII*. All exposed non-current-carrying metal parts of cranes, monorail hoists, hoists, and accessories, including pendant controls, must be metallically joined together into a continuous electrical conductor so that the entire crane or hoist will be grounded in accordance with *Article 250*. Moving parts that have metal-to-metal bearing surfaces are considered electrically connected to each other through the bearing surfaces for grounding purposes. The trolley frame and bridge frame shall be considered as electrically grounded through the bridge and trolley wheels and its respective tracks, unless paint or other insulating materials prevent reliable metal-to-metal contact. In this case, a separate grounding conductor shall be provided.

SUMMARY

A typical crane is illustrated in Figure 23-10. The unit shown has a single trolley, hoist, and individual bridge drives. An auxiliary hoist may be added to the crane. The auxiliary hoist is a supplemental hoisting unit usually designed to handle lighter loads at higher speeds than the main hoist.

The bridge consists of two girders tied together at the ends by the end trucks. The bridge drive motor(s), speed reducer(s), and control cabinets are mounted on the front platform. The trolley runs on tracks set on top of the girders. On cab-controlled cranes, the cab is attached to the underside of the front girder.

Usually two types of bridge drive systems are used. A center gear case drive is standard for bridge spans up to 60 ft and consists of a single electric motor supplying power to a double-reduction gear case. The gear case is connected by means of shafts to the drive wheels (Figure 23-11).

The individual drive system consists of a single motor, gear case, shaft, and drive wheel on each end of the bridge. The two individual drives are used to propel the bridge along the runway (Figure 23-12).

A typical trolley drive system is shown in Figure 23-13. Motor torque is transmitted through the reduction gears to the trolley drive wheels. The trolley is moved along rails attached to the tops of the bridge girders. The motor speed and direction of

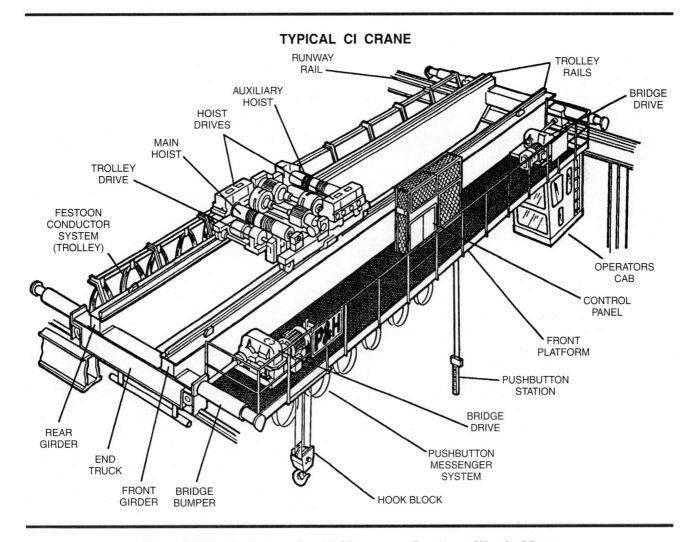

Figure 23-10 Typical overhead bridge crane. *Courtesy of* Harnischfeger

390 CHAPTER 23 Cranes and Hoists

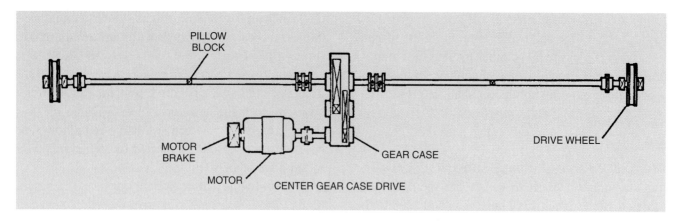

Figure 23-11　Bridge center gear case drive. *Courtesy of* Harnischfeger

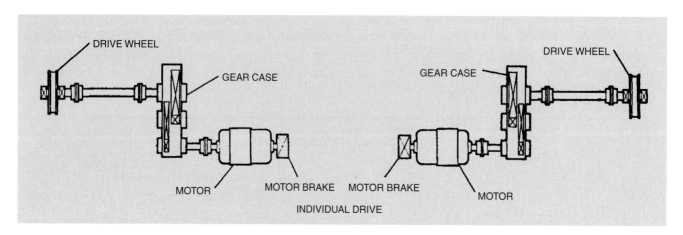

Figure 23-12　Bridge individual drive. *Courtesy of* Harnischfeger

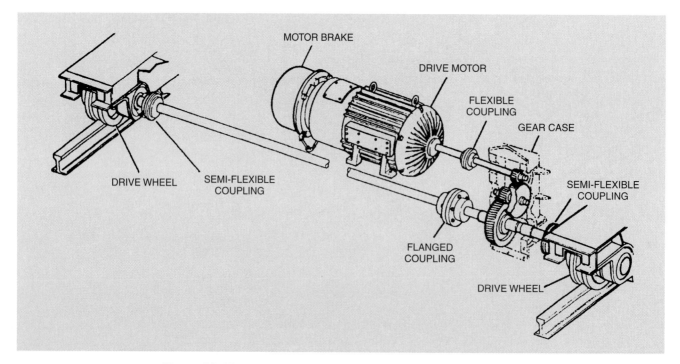

Figure 23-13　Typical trolley drive. *Courtesy of* Harnischfeger

rotation are determined by the position of the master lever or depression of the pendant pushbuttons, as applicable (Figure 23-14).

The function of a hoist system is to raise and lower loads. The loads are raised or lowered by winding up or paying out a wire rope on a hoist drum. All hoist systems consist of a gear case, drum, drum-bearing pedestal, drive motor, and motor brake. A typical hoist set-up is shown in (Figure 23-15).

Drive motors generally are ac wound-rotor motors, but some ac squirrel-cage motors are used with adjustable frequency controls. The drive motor brake is wired into the control system such that its brake coil energizes each time electrical power is applied to the drive motor primary. The motor brake releases automatically when power is applied to the drive motor primary and allows the drive motor to rotate. When power is removed from the drive motor primary, it is also simultaneously removed from the brake circuit, and the brake automatically resets.

Under normal operating conditions, the motor brake is intended for use primarily as a holding brake. In the event of a power loss to the drive motor for any reason while raising or lowering a load, the motor brake will set and hold the load in a suspended state.

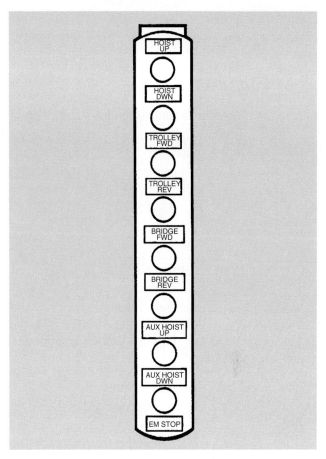

Figure 23-14 Typical suspended pushbutton station. *Courtesy of* Harnischfeger

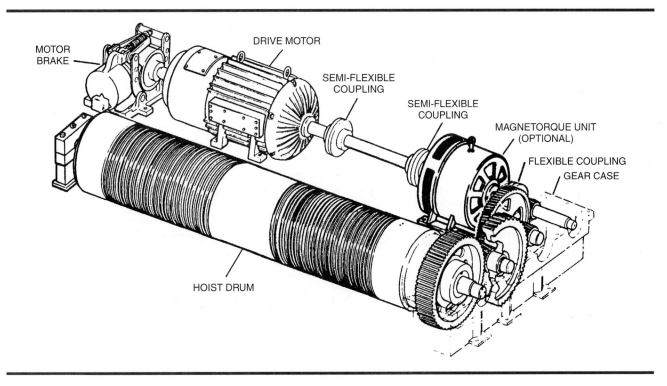

Figure 23-15 Typical hoist system. *Courtesy of* Harnischfeger

REVIEW QUESTIONS

MULTIPLE CHOICE

1. The minimum ampacity of the power-supply conductors for multiple motors on a single crane or hoist shall be the nameplate full-load ampere rating of the largest motor, plus _____% of the nameplate full-load ampere rating of the next largest motor using the column of *Table 610.14(A)* that applies to the longest time-rated motor.
    ____ A. 50
    ____ B. 75
    ____ C. 40
    ____ D. 25

2. If an auxiliary hoist motor is used with a bridge crane, there will be _____ motors.
    ____ A. three
    ____ B. four
    ____ C. two
    ____ D. five

3. A crane or hoist operating over combustible fibers is not permitted to be _____ .
    ____ A. single phase
    ____ B. three phase
    ____ C. grounded
    ____ D. isolated

4. Conductors used for the wiring of cranes and hoists shall be enclosed in raceways or be:
    ____ A. Type MC cable.
    ____ B. Type MI cable.
    ____ C. Type AC cable with an insulated grounding conductor.
    ____ D. all of the above.

5. The disconnecting means for cranes and hoists must be:
    ____ A. located in view of the runway conductors.
    ____ B. readily accessible.
    ____ C. capable of being locked in the open position.
    ____ D. all of the above.

TRUE OR FALSE

1. Where two or more motors operate a single motion, the sum of their nameplate current ratings shall be considered as that of a single motor.
    ____ A. True
    ____ B. False
    Code reference _____ .

2. Taps to brake coils are required to have overcurrent protection based on the ampacity of the tap conductors.
    ____ A. True
    ____ B. False
    Code reference _____ .

3. Control circuits are considered as protected by overcurrent devices that are set at not more than 500% of the ampacity of the control conductors.
   ____ A. True
   ____ B. False
   Code reference _____ .

4. Where opening the control circuit would constitute a hazard, the control circuit conductors are considered as being protected by the branch-circuit conductors.
   ____ A. True
   ____ B. False
   Code reference _____ .

5. Moving parts that have metal-to-metal bearing surfaces are considered to be electrically connected to each other through the bearing surfaces for grounding purposes.
   ____ A. True
   ____ B. False
   Code reference _____ .

FILL IN THE BLANKS

1. A _____ or other device must be used to prevent the load block from passing the safe upper limit of travel of all hoisting mechanisms.

2. Where controls are enclosed in cabinets, the doors must open to at least _____ or be removable.

3. The function of a hoist system is to _____ loads. The loads are raised or lowered by winding up or paying out a _____ on a hoist drum.

4. Under normal operating conditions, the motor brake is intended for use primarily as a _____ brake.

5. Taps to _____ shall be permitted without overcurrent protection. This will prevent failure of a _____ operation due to the opening of an overcurrent device.

FREQUENTLY ASKED QUESTIONS

Question 1: The braking action on cranes and hoists seems complicated. How does this braking work?

Answer: Cranes and hoists generally have three motors. There is one for each of the crane motions. One is the bridge motor for movement through the runway, one is the trolley or carriage motor for traversing the bridge, and one is the hoist motor for raising or lowering loads. There is sometimes a fourth motor that is an auxiliary motor for hoisting operations of lesser loads at faster speeds.

Each motor has a motor brake. The motor electric brake consists of a brake coil and brake shoes or a brake disc. The shoe type is spring set and electrically released. The disc type is spring set and magnetically released. The electric control to either type is identical. Operating the motion device (the electrical control circuit device) associated with the crane motion desired, closes the circuit which applies power to both the motor and the brake coil. Returning the motion control device to the "OFF" position opens the circuit to both the motor and the brake coil, and the spring sets the brake.

Question 2: What is the purpose of the secondary resistors shown in *610.14(B)*?

Answer: The ac motors generally used on cranes and hoists are the wound-rotor type. In a wound-rotor motor, the currents produced in the rotor are fed into slip rings mounted on the end of the rotor. These currents are taken through slip ring brushes into an external control device that is usually a variable resistor (secondary resistor) through which the motor characteristics can be changed. In this manner, we can vary the starting torque and current, the operating speed, and the acceleration of the motor up to full-load speed. There is a big loss in motor efficiency when used in this manner, and for this reason, the wound-rotor motor is used where high starting torque and variable speed are desired, such as in cranes, hoists, or elevators.

Question 3: What is a bridge drive system?

Answer: The function of the bridge drive is to propel the bridge along the bridge runway. Power from an electric motor is transmitted to a gear case and suitable shafts to the drive wheels at the ends of the bridge girder. Bridge drives can be accomplished by a single motor with the output shaft extending from both ends of the motor to two gear cases located near each end of the bridge girder. Bridge drive can also be accomplished by individual wheel drive using two motors, one at each end of the bridge. Bridge spans over 60 ft in length usually use the two-motor drive system.

Question 4: What is a motion controller?

Answer: A motion controller is the electric device that is operated to activate the motion (up–down, forward–reverse) desired. It is used to control whatever movement or motion is required. It may govern the acceleration, speed, retardation, or stopping of a moving equipment.

Question 5: How can the crane runway track be used as a conductor of current for one phase of a 3-phase system?

Answer: This can only be done as shown in *610.21(F)*. It must be done through an isolating transformer with a secondary that does not exceed 300 volts. The track or rail serving as a conductor must be effectively grounded at the transformer and is also permitted to be grounded by the fittings used for the suspension of the track or rail to the building or structure (Figure 23-16).

CHAPTER 23 Cranes and Hoists 395

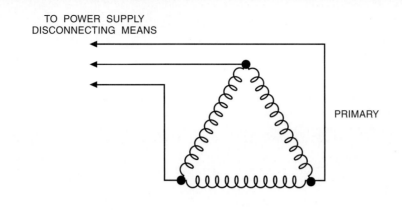

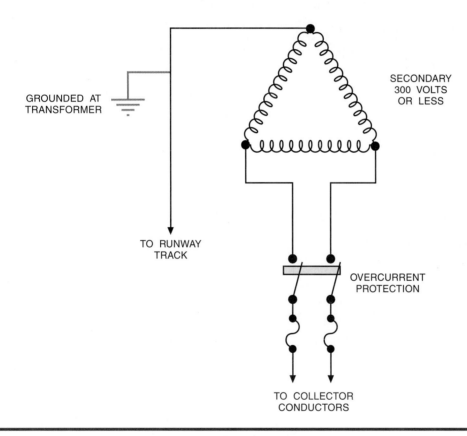

Figure 23-16 Three-phase isolating transformer with secondary phase connected to runway track

CHAPTER 24

Elevators, Dumbwaiters, Escalators, Moving Walks, Wheelchair Lifts, and Stairway Chair Lifts

OBJECTIVES
After completing this chapter, the student should understand:
- voltage limitations.
- conductors.
- wiring methods.
- installation of conductors.
- traveling cables.
- disconnecting means and control.
- overcurrent protection.
- machine rooms.
- grounding.
- emergency and standby power.

KEY TERMS

Control System: The overall system governing the starting, stopping, direction of motion, acceleration, speed, and retardation of the moving member

Hoistway: The cavity or structural opening through which the elevator car travels or is hoisted

Motion Controller: The electric device that governs the acceleration, speed, retardation, and stopping of the moving member

Operating Device: The car switch, pushbuttons, key or toggle switches, or other devices used to activate the operation controller

Operation Controller: The electric device that initiates the starting, stopping, and direction of motion in response to a signal from an operating device

Runway: The structural cavity or opening in the building or structure into which the escalator is inserted

Signal Equipment: Includes audible and visual equipment, such as chimes, gongs, lights, and displays that convey information to the user

Wellway: The cavity or structural opening in the building or structure into which the moving walk is installed

CHAPTER 24

398 Elevators, Dumbwaiters, Escalators, Moving Walks, Wheelchair Lifts, and Stairway Chair Lifts

INTRODUCTION

This chapter covers the installation of electrical equipment and **control system** wiring used in connection with elevators, dumbwaiters, escalators, moving walks, wheelchair lifts, and stairway chairlifts. Installation requirements can be found in *Article 620*.

GENERAL

The installation of power wiring, equipment wiring, and equipment for elevators, dumbwaiters, escalators, moving walks, wheelchair lifts, and stairway chairlifts is generally done by more than one contractor. The electrical contractor supplies the power wiring to the elevator machine room or machinery space and provides branch-circuits for the car lighting and other equipment. The elevator constructors install and wire all controls, door operator wiring, and power wiring to motors and other associated equipment. Elevators are constructed in many different types for different types of buildings or structures (Figure 24-1).

Voltage Limitations

The supply voltage cannot exceed 300 volts between conductors unless otherwise permitted in *620.3(A)* through *(C)*. Branch-circuits to door operator controllers, and door motors and branch-circuits

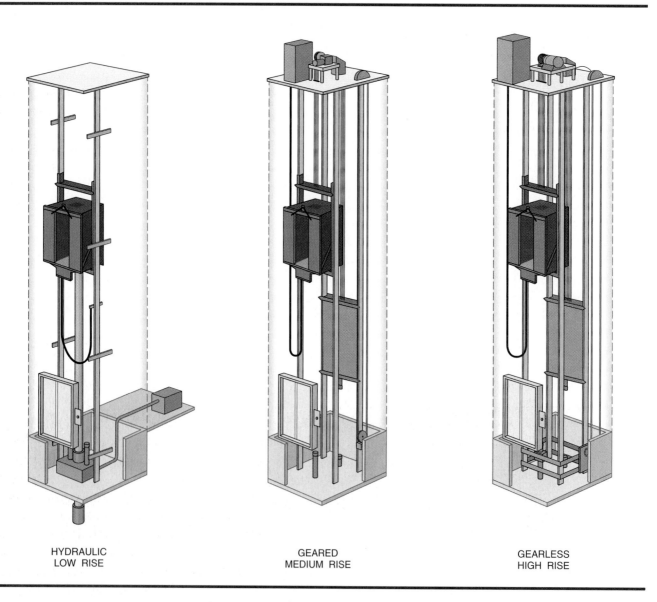

HYDRAULIC LOW RISE   GEARED MEDIUM RISE   GEARLESS HIGH RISE

Figure 24-1 Typical elevator drive types

CHAPTER 24 Elevators, Dumbwaiters, Escalators, Moving Walks, Wheelchair Lifts, and Stairway Chair Lifts

and feeders to motor controllers, driving machine motors, machine brakes, and motor-generator sets must not have a voltage in excess of 600 volts. Power conversion and functionally associated equipment, including interconnecting wiring, are permitted to have higher voltages provided all such equipment and wiring is listed for the higher voltages. Where the voltage exceeds 600 volts, warning labels or signs that read DANGER—HIGH VOLTAGE must be attached to the equipment and must be visible (Figure 24-2). Lighting circuit branch-circuit voltages must not exceed those specified in *Article 410*. Heating and air-conditioning equipment branch-circuits located on an elevator car must not exceed 600 volts.

In general, the working clearances around electrical equipment must follow the requirements shown in *110.26(A)*, except where only qualified persons will work on the equipment, *620.5* permits these requirements to be waived as follows: Flexible connections to equipment are permitted so that this equipment can be repositioned to meet clearance requirements of *110.26(A)* (Figure 24-3).

Motion controllers, **operation controllers**, and disconnecting means for dumbwaiters, escalators, moving walks, wheelchair lifts, and stairway chairlifts

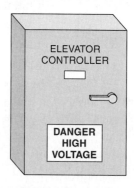

Figure 24-2 Equipment enclosure with high-voltage warning

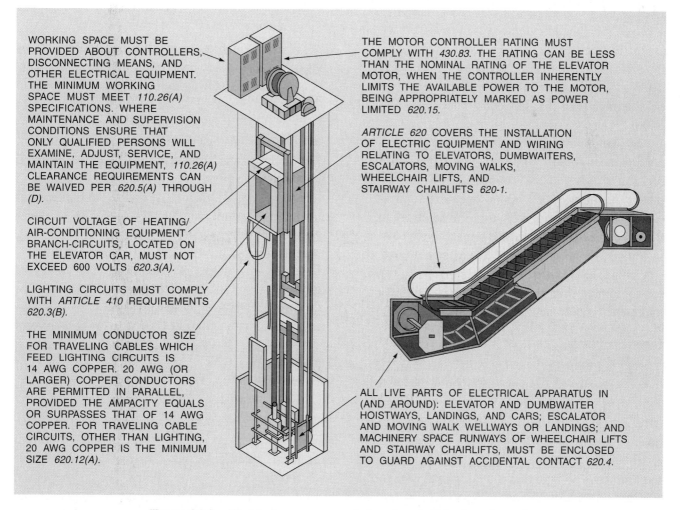

Figure 24-3 Working space around elevator electrical equipment

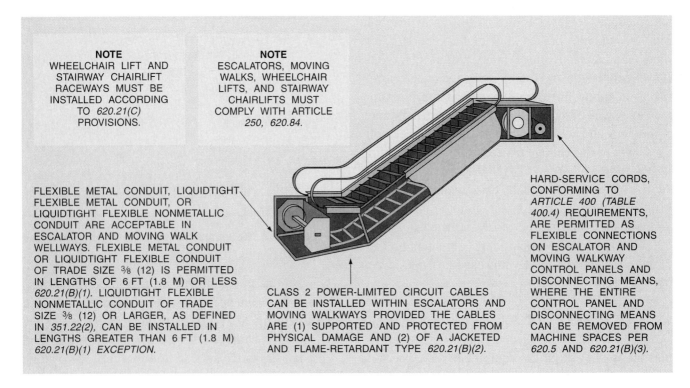

Figure 24-4 Flexible wiring method to reposition escalator equipment for servicing

are many times installed in the same space or pit with the driving machine, and the controllers and disconnecting means are permitted to be connected with a flexible means so that this equipment can be pulled out of the pit or space for servicing (Figure 24-4).

In many elevator types, controllers and disconnecting means are installed in the **hoistway** or on the car and these may be connected with a flexible means. Controllers for door operators and other electrical equipment installed in the hoistway or on the car can also be connected with a flexible means to allow for the equipment to be repositioned for servicing. The clearance requirements are also waived for equipment that is not required to be serviced or maintained while the equipment is energized.

Conductors

The requirements for conductors used in elevators, dumbwaiters, escalators, moving walks, wheelchair lifts, and stairway chairlifts are found in *Article 620, Part II. NEC® 620.11* requires that hoistway door interlock wiring must be suitable for a temperature of not less than 392°F (200°C). This requirement will help to ensure that in the event of a fire, the doors will operate. All conductors used on this equipment must be flame-retardant.

Traveling cables used as flexible connections between the elevator or dumbwaiter car or counterweight and the raceway must be of a type shown in *Table 400.4* (Figure 24-5). Elevator traveling cables for operating and control circuits contain nonmetallic fillers as necessary to maintain concentricity. Cables have steel supporting members as required for suspension by *620.41*. Where steel supporting members are used, they must run through the center of the cable assembly and not be cabled with copper strands of any conductor (see *Note 5, Table 400.4*). Elevator cables in sizes 20 AWG to 14 AWG are rated 300 volts, and sizes 10 through 2 AWG are rated 600 volts (see *Note 10, Table 400.4*).

The minimum size of conductors for lighting used in traveling cables is 14 AWG copper. 20 AWG copper conductors may be used in parallel, provided the ampacity is equivalent to at least that of 14 AWG. *NEC® 620.13* requires that feeder and branch-circuit conductors shall have an ampacity as follows: With generator field control, the conductor ampacity must be based on the nameplate current rating of the driving motor of the motor-generator set that drives the elevator motor. The heating of the conductors is based on the current rating of the motor-generator driving motor, rather than the rating

Table 400.4 Flexible Cords and Cables (See 400.4.)

Trade Name	Type Letter	Voltage	AWG or kcmil	Number of Conductors	Insulation	Nominal Insulation Thickness[1]			Braid on Each Conductor	Outer Covering	Use	
						AWG or kcmil	mm	mils				
Lamp cord	C	300 600	18–16 14–10	2 or more	Thermoset or thermoplastic	18–16 14–10	0.76 1.14	30 45	Cotton	None	Pendant or portable	Dry locations	Not hard usage
Elevator cable	E See Note 5. See Note 9. See Note 10.	300 or 600	20–2	2 or more	Thermoset	20–16 14–12 12–10 8–2	0.51 0.76 1.14 1.52	20 30 45 60	Cotton	Three cotton, Outer one flame-retardant & moisture-resistant. See Note 3.	Elevator lighting and control	Unclassified locations	
						20–16 14–12 12–10 8–2	0.51 0.76 1.14 1.52	20 30 45 60	Flexible nylon jacket				
Elevator cable	EO See Note 5. See Note 10.	300 or 600	20–2	2 or more	Thermoset	20–16 14–12 12–10 8–2	0.51 0.76 1.14 1.52	20 30 45 60	Cotton	Outer one Three cotton, flame-retardant & moisture-resistant. See Note 3.	Elevator lighting and control	Unclassified locations	
										One cotton and a neoprene jacket. See Note 3.		Hazardous (classified) locations	
Elevator cable	ET See Note 5. See Note 10.	300 or 600	20–2	2 or more	Thermoplastic	20–16 14–12 12–10 8–2	0.51 0.76 1.14 1.52	20 30 45 60	Rayon	Three cotton or equivalent. Outer one flame-retardant & moisture-resistant. See Note 3.	Unclassified locations		
	ETLB See Note 5. See Note 10.	300 or 600							None				
	ETP See Note 5. See Note 10.	300 or 600							Rayon	Thermoplastic	Hazardous (classified) locations		
	ETT See Note 5. See Note 10.	300 or 600							None	One cotton or equivalent and a thermoplastic jacket			
Portable power cable	G	2000	12–500	2–6 plus grounding conductor(s)	Thermoset	12–2 1–4/0 250–500	1.52 2.03 2.41	60 80 95		Oil-resistant thermoset	Portable and extra hard usage		

5. Elevator traveling cables for operating control and signal circuits shall contain nonmetallic fillers as necessary to maintain concentricity. Cables shall have steel supporting members as required for suspension by 620.41. In locations subject to excessive moisture or corrosive vapors or gases, supporting members of other materials shall be permitted. Where steel supporting members are used, they shall run straight through the center of the cable assembly and shall not be cabled with the copper strands of any conductor.

In addition to conductors used for control and signaling circuits, Types E, EO, ET, ETLB, ETP, and ETT elevator cables shall be permitted to incorporate in the construction, one or more 20 AWG telephone conductor pairs, one or more coaxial cables, or one or more optical fibers. The 20 AWG conductor pairs shall be permitted to be covered with suitable shielding for telephone, audio, or higher frequency communications circuits; the coaxial cables consist of a center conductor, insulation, and shield for use in video or other radio frequency communications circuits. The optical fiber shall be suitably covered with flame-retardant thermoplastic. The insulation of the conductors shall be rubber or thermoplastic of thickness not less than specified for the other conductors of the particular type of cable. Metallic shields shall have their own protective covering. Where used, these components shall be permitted to be incorporated in any layer of the cable assembly but shall not run straight through the center.

9. Insulations and outer coverings that meet the requirements as flame retardant, limited smoke, and are so listed, shall be permitted to be marked for limited smoke after the code type designation.

10. Elevator cables in sizes 20 AWG through 14 AWG are rated 300 volts, and sizes 10 through 2 are rated 600 volts. 12 AWG is rated 300 volts with a 0.76-mm (30-mil) insulation thickness and 600 volts with a 1.14-mm (45-mil) insulation thickness.

Figure 24-5 Partial *NEC®* Table 400.4 with *Notes 5, 9,* and *10* (Reprinted with permission from NFPA 70-2002)

of the elevator motor, which represents actual (but short time) and intermittent full-load values. Conductors supplying a single motor must have an ampacity no less than the percentage determined in *430.22(A)* for continuous duty motors and *430.22(E)* for other than continuous duty. Elevator motors are inherently intermittent duty, and the conductors are sized for duty cycle service as shown in *Table 430.22(E)* (Figure 24-6).

Conductors supplying a single motor controller must have an ampacity not less than the motor controller nameplate current rating plus all other connected

Table 430.22(E) Duty-Cycle Service

Classification of Service	Nameplate Current Rating Percentages			
	5-Minute Rated Motor	15-Minute Rated Motor	30- & 60-Minute Rated Motor	Continuous Rated Motor
Short-time duty operating valves, raising or lowering rolls, etc.	110	120	150	—
Intermittent duty freight and passenger elevators, tool heads, pumps, drawbridges, turntables, etc. (for arc welders, see 630.11)	85	85	90	140
Periodic duty rolls, ore- and coal-handling machines, etc.	85	90	95	140
Varying duty	110	120	150	200

Figure 24-6 *NEC® Table 430.22(E), Duty Cycle Service* (Reprinted with permission from NFPA 70-2002)

loads. Conductors supplying a single-power transformer must have an ampacity not less than the nameplate current rating of the power transformer plus all other connected loads.

Conductors supplying more than one motor, motor controller, or power transformer must have an ampacity not less than the sum of the nameplate current ratings of the equipment plus all other connected loads. The ampere ratings of motors to be used in the total are to be determined by *Table 430.22(E)*, *430.24*, and *430.24, Exception No. 1*.

Feeder conductors of less ampacity than required by *620.13* are permitted subject to the requirements of *Table 620.14* (Figure 24-7).

Wiring Methods

Conductors in hoistways, escalators and moving walk **wellways**, wheelchair lifts, stairway chairlift **runways**, and machinery spaces, in or on cars, and in machine and control rooms, not including the traveling cables connecting the car to counterweight and hoistway wiring are required by *620.21* to be installed in RMC, IMC, EMT, RNC, or wireways or must be Type MC, MI, or AC cable unless otherwise shown in *620.21(A)*, *(B)*, and *(C)*. Wiring methods are shown in *620.37* (Figure 24-8).

Wellways and runways are terms used to designate the structural openings in a building or structure where the escalator or moving walk is placed. The wiring methods described are used to install the raceway or cable used to make the necessary connections to the controls or driving motors of this equipment.

For elevators, FMC, LFMC, or LFNC is permitted in hoistways. Flexible conduits of trade size 3/8 (12), not exceeding 6 ft (1.8 m) in length are permitted on elevator cars if securely fastened in place. In machine rooms and machinery spaces, FMC, LFMC, or LFNC of trade size 3/8 (12) or larger, but not exceeding 6 ft (1.8 m), is permitted between

Table 620.14 Feeder Demand Factors for Elevators

Number of Elevators on a Single Feeder	Demand Factor
1	1.00
2	0.95
3	0.90
4	0.85
5	0.82
6	0.79
7	0.77
8	0.75
9	0.73
10 or more	0.72

Figure 24-7 *NEC® Table 620.14, Feeder Demand Factors for Elevators* (Reprinted with permission from NFPA 70-2002)

CHAPTER 24 Elevators, Dumbwaiters, Escalators, Moving Walks, Wheelchair Lifts, and Stairway Chair Lifts

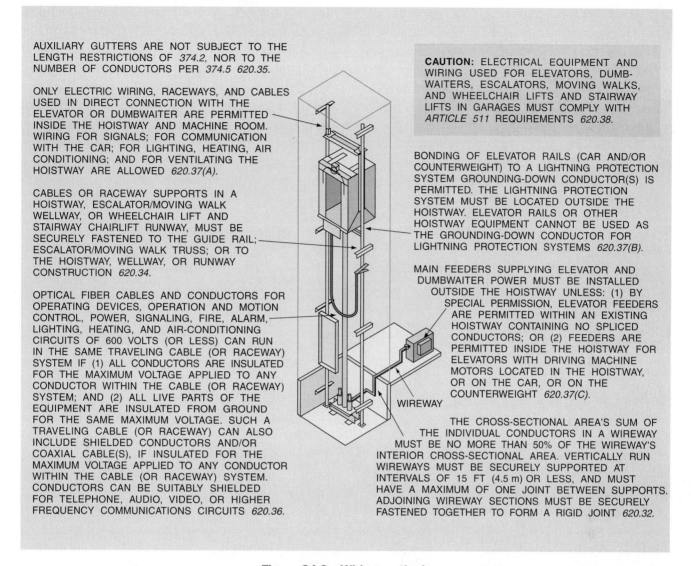

Figure 24-8 Wiring methods

control panels, machine motors, machine brakes, motor-generator sets, disconnecting means, and pumping unit motors and valves.

For escalators, trade size ⅜ (12) FMC, LFMC, or LFNC in lengths not exceeding 6 ft (1.8 m) is permitted in escalator and moving walk wellways.

For wheelchair lifts and stairway chairlifts, FMC or LFMC is permitted to be used in trade size ⅜ (12) in lengths not exceeding 6 ft (1.8 m).

In escalators, wheelchair lifts, and stairway chairlifts, trade size ⅜ (12) or larger, LFNC, as defined in *356.2(2)* is permitted to be used in lengths in excess of 6 ft (1.8 m).

The requirements for branch-circuits for car lighting, receptacles, ventilation, heating, and air-conditioning can be found in *620.22*. The overcurrent devices for this equipment must be located in the machine room/machinery space. A separate branch-circuit shall supply the car lights, receptacles, auxiliary lighting power source, and ventilation on each elevator car. The auxiliary lighting power is a battery-powered unit, and the branch-circuit holds the unit equipment relay open, unless the branch-circuit is opened and the unit battery equipment relay contacts close the battery circuit to auxiliary power lamps.

A dedicated branch-circuit is required by *620.22(B)* to supply the air-conditioning and heating units on each elevator car. The overcurrent device protecting the branch-circuit must be located in the elevator machine room/machinery space.

A separate branch-circuit is required by *620.23(A)* to supply the machine room/machinery space lighting and receptacles. The required lighting cannot be connected to the load side of a GFCI. This requirement will assure that there will not be a loss of necessary lighting due to a ground-fault on other components of the circuit. A light switch for the machine room or machinery space must be located at the point of entry to the machine room or machinery space.

At least one 125-volt, single-phase, duplex receptacle shall be provided in each machine room and machinery space. All 125-volt, single-phase, 15- or 20-ampere receptacles must have GFCI protection for personnel (see *620.85*) (Figure 24-9).

A separate branch-circuit is required by *620.24* to supply the hoistway pit lighting and receptacles. The lighting switch must be readily accessible from the pit access door, and the lighting outlet shall not be connected to the load of a GFCI. Hoistway pit receptacles are required to be ground-fault protected, except for receptacles that supply power to a sump pump. Any emergency or safety device, such

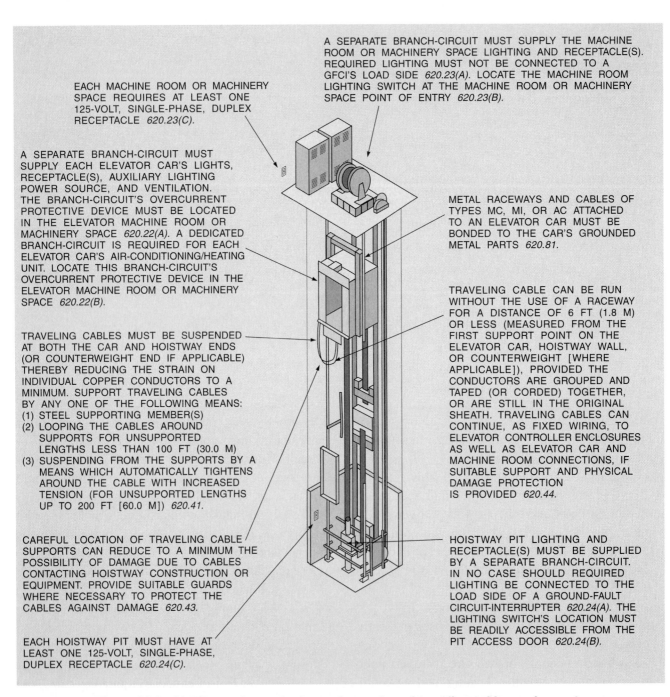

Figure 24-9 Lighting and receptacle requirements and traveling cable requirements

as a sump pump, should not be on a ground-fault protective device which may trip and put the pump out of commission when it is needed.

Installation of Conductors

The installation of conductors in wireways must be done according to *620.32*. These requirements are different from those found in *Article 376, Metal Wireways* and *Article 378, Nonmetallic Wireways*. For elevators and escalators, the sum of the cross-sectional area of the individual conductors in a wireway shall not be more than 50% of the interior cross-sectional area of the wireway. This is a major change from the 20% permitted by *Articles 376* and *378*. *NEC® 620.35* permits the length of auxiliary gutters and the number of conductors not to be subject to the restrictions of *374.2* or *374.5*.

In accordance with *620.37*, only electric wiring used directly in connection with the elevator or dumbwaiter, including wiring for signals; for communication with the car; for lighting, heating, air conditioning, and ventilating the car; for fire-detecting systems; for pit sump pumps; and for heating, lighting, and ventilating the hoistway, shall be permitted inside the hoistway and the machine room. The elevator constructors have always been very careful about the wiring permitted in these areas. In the interest of safety, it is better that no one other than qualified elevator personnel be permitted to work in elevator machinery rooms and that wiring not directly associated with the elevator performance not be permitted in elevator hoistways.

Bonding of elevator rails (car and counterweight) to a lightning protection system, grounding down conductor shall be permitted, but the lightning protection down conductor cannot be located within the hoistway and elevator rails cannot be used as the grounding down conductor.

Main feeders supplying power to elevators and dumbwaiters must be installed outside the hoistway, except feeders are permitted inside the hoistway for elevators with driving machine motors located in the hoistway or on the car or counterweight. Increasingly new innovations in elevator design have taken the driving motors out of the machine room and put them either on the counterweight, on the car, or in the hoistway. Machine rooms have given way to machinery spaces, and new terminology relating to control rooms and control spaces is being used.

By special permission, feeders for elevators shall be permitted within an existing hoistway if no conductors are spliced within the hoistway.

Traveling Cables

NEC® 620.41 requires traveling cables to be supported properly at both ends to reduce the strain on the individual copper conductors. This is accomplished by the steel-supporting member of the cable or by looping the cable around supports for unsupported lengths less than 100 ft (30 m). Supports that automatically tighten around the cable must be used when tension is increased for unsupported lengths up to 200 ft (60 m) (Figure 24-10).

Traveling cable conductors are permitted by *620.44* to be run without the use of a raceway for a distance not exceeding 6 ft (1.8 m) from the first point of support on the elevator car or hoistway wall, provided the conductors are grouped together, taped or corded, or in the original sheath. This means that where the traveling cable suspension is terminated on either end, instead of terminating in a junction box and using a raceway system to complete the run, it is permitted to tape the traveling cable conductors together and run to the termination point, not exceeding 6 ft (1.8 m).

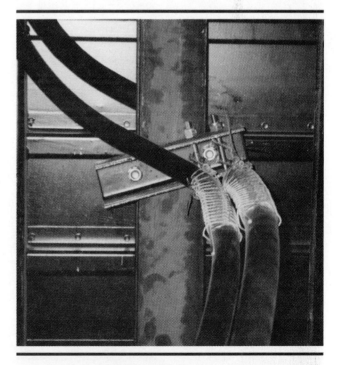

Figure 24-10 View inside hoistway of traveling cable secured below elevator car

Traveling cables are permitted to be continued to elevator controller enclosures and to the elevator car and machine room connections as fixed wiring, provided they are supported and protected from physical damage

Disconnecting Means

NEC® 620.51 requires that a single means for disconnecting all ungrounded main power-supply conductors be provided for each unit. The disconnecting means for the main power-supply conductors shall not disconnect the branch-circuit required in *620.22, 620.23,* and *620.24* (Figure 24-11).

No provision is permitted to be made to open or close this disconnecting means from any other part of the premises. Opening this disconnecting means must not disconnect the car lighting, receptacles, ventilation, heating or air-conditioning, the machinery room or machinery space lighting and receptacles, or the hoistway pit lighting and receptacles.

In accordance with *620.51,* the disconnecting means for escalators and moving walks must be installed in the space where the controller is located. The disconnecting means for wheelchair lifts and stairway chairlifts must be located within sight of the motor controller.

NEC® 620.53 requires that elevators have a single means of disconnect for all ungrounded car light, receptacles, and ventilation power supply conductors for that elevator car. The disconnecting means shall be capable of being locked in the open position and shall be located in the machine room or machinery space for that elevator car.

NEC® 620.54 requires that elevators have a single means for disconnecting all ungrounded car heating and air-conditioning power-supply conductors for that car. The disconnecting means shall be capable of being locked in the open position and shall be located in the machine room/machinery space for that elevator car.

Overcurrent Protection

NEC® 620.61 contains the requirements for overcurrent protection for elevators, escalators, dumbwaiters, moving walks, wheelchair lifts, and stairway chairlifts. **Operating devices** and **signal equipment** must be protected against overcurrent in accordance with *725.23* and *725.24* of the *NEC.®* duty on elevator and dumbwaiter driving machine motors and driving motors of motor generators used with generator field control shall be rated as intermittent and shall be protected against overload in accordance with *430.33*. This overload protection may be provided by the branch-circuit short-circuit and ground-fault protective device (Figure 24-12). The duty rating on escalator and moving walk driving machine motors shall be rated as continuous, and these motors shall be protected against overload in accordance with *430.32,* which has the requirements for continuous duty motors. Escalator and moving walk driving machine motors and driving motors of motor-generator sets shall be protected against running overload as provided in *Table 430.37* (Figure 24-13).

Figure 24-11 Typical machine room disconnect panel

CHAPTER 24 Elevators, Dumbwaiters, Escalators, Moving Walks, Wheelchair Lifts, and Stairway Chair Lifts

Figure 24-12 Relay control and overcurrent protection

Table 430.37 Overload Units

Kind of Motor	Supply System	Number and Location of Overload Units, Such as Trip Coils or Relays
1-phase ac or dc	2-wire, 1-phase ac or dc ungrounded	1 in either conductor
1-phase ac or dc	2-wire, 1-phase ac or dc, one conductor grounded	1 in ungrounded conductor
1-phase ac or dc	3-wire, 1-phase ac or dc, grounded neutral	1 in either ungrounded conductor
1-phase ac	Any 3-phase	1 in ungrounded conductor
2-phase ac	3-wire, 2-phase ac, ungrounded	2, one in each phase
2-phase ac	3-wire, 2-phase ac, one conductor grounded	2 in ungrounded conductors
2-phase ac	4-wire, 2-phase ac, grounded or ungrounded	2, one per phase in ungrounded conductors
2-phase ac	Grounded neutral or 5-wire, 2-phase ac, ungrounded	2, one per phase in any ungrounded phase wire
3-phase ac	Any 3-phase	3, one in each phase*

*Exception: An overload unit in each phase shall not be required where overload protection is provided by other approved means.

Figure 24-13 *NEC® Table 430.37*, Overload units required for elevator motors (Reprinted with permission from NFPA 70-2002)

The duty on wheelchair lift and stairway chair-lift driving machine motors shall be rated as intermittent. These motors shall be protected against overload in accordance with *430.33*.

Motor feeder short-circuit and ground-fault protection and motor branch-circuit short-circuit and ground-fault protection shall be in accordance with *Table 430.52* (Figure 24-14).

Where more than one driving machine disconnecting means is supplied by a single feeder, the overcurrent protective devices in each disconnecting means shall be selectively coordinated with any supply side overcurrent protective devices. This requirement is required so that a short-circuit or ground-fault in one of the driving machine circuits will not open the supply side overcurrent device and shutdown all of the units on that feeder or circuit (Figure 24-15).

Machine Room

NEC® 620.71 contains the requirements for machine rooms (Figure 24-16).

Elevator, dumbwaiter, and moving walk driving machines; motor-generator sets; motor controllers; and disconnecting means shall be installed in a room or enclosure set aside for that purpose unless otherwise authorized as follows:

- Motor controllers shall be permitted outside these spaces, provided they are in enclosures with doors or removable panels that are capable of being locked in the closed position, and the disconnecting means is located adjacent to or is a part of the motor controller.

Table 430.52 Maximum Rating or Setting of Motor Branch-Circuit Short-Circuit and Ground-Fault Protective Devices

Type of Motor	Percentage of Full-Load Current			
	Nontime Delay Fuse[1]	Dual Element (Time-Delay) Fuse[1]	Instantaneous Trip Breaker	Inverse Time Breaker[2]
Single-phase motors	300	175	800	250
AC polyphase motors other than wound-rotor				
Squirrel cage — other than Design E or Design B energy efficient	300	175	800	250
Design E or Design B energy efficient	300	175	1100	250
Synchronous[3]	300	175	800	250
Wound rotor	150	150	800	150
Direct current (constant voltage)	150	150	250	150

Note: For certain exceptions to the values specified, see 430.54.
[1] The values in the Nontime Delay Fuse column apply to Time-Delay Class CC fuses.
[2] The values given in the last column also cover the ratings of nonadjustable inverse time types of circuit breakers that may be modified as in 430.52(C), Exception No. 1 and No. 2.
[3] Synchronous motors of the low-torque, low-speed type (usually 450 rpm or lower), such as are used to drive reciprocating compressors, pumps, and so forth, that start unloaded, do not require a fuse rating or circuit-breaker setting in excess of 200 percent of full-load current.

Figure 24-14 Maximum Rating or Setting of Motor Branch-Circuit Short-Circuit and Ground-Fault Protective Devices, *NEC® Table 430.52* (Reprinted with permission from NFPA 70-2002)

- Motor-controller enclosures for escalators or moving walks are permitted to be in the balustrade on the side away from the moving steps or moving treadway. A balustrade is the railing and handrail that encloses the moving steps or moving treadway. If the disconnecting means is an integral part of the motor controller, it must be operable without opening the enclosure.
- Elevators with driving machines located on the car, counterweight, or in the hoistway and driving machines for dumbwaiters, wheelchair lifts, and stairway chairlifts shall be permitted outside the spaces specified in *620.71*.

Figure 24-15 More than one driving machine on a single feeder

Grounding

Electric elevators are required by *620.82* to have all metal enclosures for electrical equipment, the frames of all motors, elevator machines, and controllers grounded.

Escalators, moving walks, wheelchair lifts, and stairway chairlifts are required by *620.84* to have all metal parts grounded.

The requirements for GFCI protection for personnel are shown in *620.85*. Each 125-volt, single-phase, 15- and 20-ampere receptacle installed in pits, in hoistways, on elevator cartops, and in escalator and moving walk wellways shall be of the GFCI type. This requirement makes it mandatory for each receptacle to be of the GFCI type and that protection by being downstream is not acceptable.

All 125-volt, single-phase, 15- or 20-ampere receptacles installed in machine rooms and machinery spaces must have GFCI protection but may be protected by upstream receptacles.

A single receptacle supplying a permanently installed sump pump does not require GFCI protection. It is not good practice to put a sump pump motor on a circuit that may become de-energized by nuisance tripping or by a ground-fault.

CHAPTER 24 Elevators, Dumbwaiters, Escalators, Moving Walks, Wheelchair Lifts, and Stairway Chair Lifts

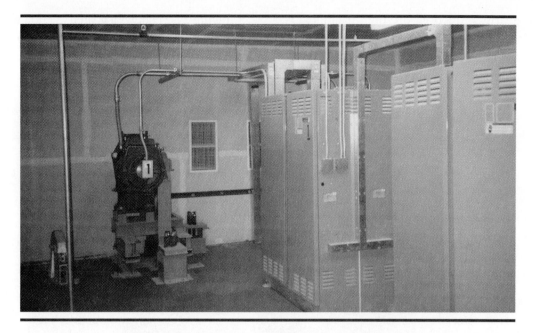

Figure 24-16 Typical modern elevator machine room

Emergency and Standby Power Systems

The requirements for elevator emergency and standby power systems are shown in *620.91*. An elevator is permitted to be powered either by an emergency or a standby system.

The disconnecting means required by *620.51* shall disconnect the elevator from both the emergency or standby power system and the normal power system.

SUMMARY

The elevator industry is changing rapidly with new technological advances being introduced everyday. The standard machine room for elevator driving machines is giving way to driving machines being placed on the elevator cartop, on the counterweight, or in the hoistway. Motor-generator sets are being replaced with power transformers and adjustable speed drives (Figure 24-17).

A typical machine room with circa 1930–1950

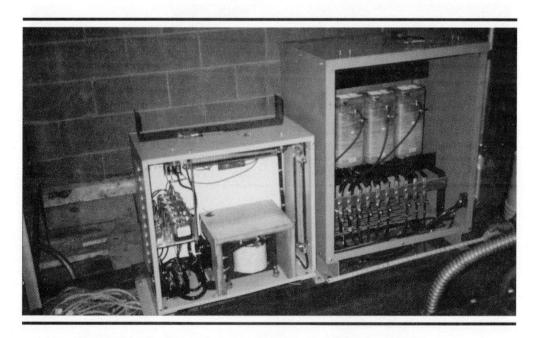

Figure 24-17 Power input transformer and filter for DC-SCR drive

high-speed gearless traction driving machines is shown in Figure 24-18.

A present day typical machine room with modern solid-state control (DC-SCR drive) is shown in Figure 24-19.

Figure 24-20 shows a gearless traction machine with modern DC-SCR drive control.

Figure 24-21 shows a geared traction machine circa 1990 with ACVF (variable frequency drive) control.

The *NEC*® 2002 edition contains many new advances and methods of installing elevator equipment in areas not permitted before. These new requirements correlate the *NEC*® with the elevator safety standard ANSI 17.1.

Figure 24-18　Typical machine room with circa 1930–1950 high-speed gearless traction machines

Figure 24-19　Typical modern machine room with solid-state control drive

CHAPTER 24 Elevators, Dumbwaiters, Escalators, Moving Walks, Wheelchair Lifts, and Stairway Chair Lifts

Figure 24-20 Modernization of 1930–1950 high-speed gearless traction machine to modern DC-SCR drive control

Figure 24-21 Geared traction machine with ACVF control

REVIEW QUESTIONS

MULTIPLE CHOICE

1. Branch-circuits to door operator controllers are limited to _____ volts.
   ____ A. 125
   ____ B. 240
   ____ C. 300
   ____ D. 600

2. Heating and air-conditioning circuits located on an elevator car must not exceed _____ volts.
   ____ A. 600
   ____ B. 300
   ____ C. 240
   ____ D. 125

3. GFCI type receptacles are required to be installed:
   ____ A. in hoistway pits.
   ____ B. on elevator cartops.
   ____ C. in escalator wellways.
   ____ D. all of the above.

4. Conductors installed in wireways used in elevator machine rooms shall not exceed _____% of the cross-sectional area of the wireway.
   ____ A. 20
   ____ B. 30
   ____ C. 40
   ____ D. 50

5. The disconnecting means for the main power-supply conductors shall not disconnect the branch-circuit required in:
   ____ A. 620.22.
   ____ B. 620.23.
   ____ C. 620.24.
   ____ D. all of the above.

TRUE OR FALSE

1. Elevator door motor branch-circuits are limited to 600 volts.
   ____ A. True
   ____ B. False
   Code reference _____ .

2. For lighting circuits used in traveling cables, 20 AWG copper conductors may be used in parallel.
   ____ A. True
   ____ B. False
   Code reference _____ .

3. Type AC cable used in elevator hoistways must have an insulated copper equipment grounding conductor as part of the cable assembly.
   ____ A. True
   ____ B. False
   Code reference _____ .

4. Auxiliary lighting power is required to be on a separate branch-circuit.
   ____ A. True
   ____ B. False
   Code reference _____ .

5. A separate branch-circuit must supply the machine room or machinery space lighting and receptacle circuits, and the circuit must be GFCI protected.
   ____ A. True
   ____ B. False
   Code reference _____ .

6. Receptacles installed in machine rooms or machinery spaces must be of the GFCI type.
   ____ A. True
   ____ B. False
   Code reference _____ .

7. Receptacle outlets for hoistway pit sump pumps must be GFCI protected.
   ____ A. True
   ____ B. False
   Code reference _____ .

8. For escalators, the sum of the cross-sectional area of the individual conductors shall not exceed 30% of the interior cross-sectional area of the wireway.
   ____ A. True
   ____ B. False
   Code reference _____.

9. A lightning down conductor of a lightning protection system is permitted to be bonded to the elevator rails.
   ____ A. True
   ____ B. False
   Code reference _____.

10. The disconnecting means for escalators and moving walks must be located in the space where the controller is located.
    ____ A. True
    ____ B. False
    Code reference _____.

FILL IN THE BLANKS

1. Escalator and moving walkway driving machine motors of motor-generator sets shall be protected against _____ as provided in *Table 430.37*.

2. The _____ on elevator and dumbwaiter driving machine motors and driving motors of motor generators used with generator field control shall be rated as _____ and shall be protected against overload in accordance with *430.33*.

3. Motor-controller enclosures for escalators or moving walks are permitted to be in the _____ from the moving steps or moving treadway.

4. A _____ supplying a permanently installed _____ does not require GFCI protection.

5. The disconnecting means required by *NEC®* _____ shall disconnect the elevator from both the _____ power system and the normal power system.

6. The _____ is the car switch, pushbuttons, key or toggle switches, or other devices used to activate the operation controller.

7. Flexible connections to equipment are permitted so that this equipment can be _____ to meet clearance requirements of *NEC®* _____.

8. Elevator traveling cables for _____ circuits contain nonmetallic fillers as necessary to maintain concentricity.

9. For wheelchair lifts and stairway chairlifts, FMC or LFMC is permitted to be used in _____ in lengths not exceeding _____.

10. A separate branch-circuit shall supply the car lights, _____, auxiliary lighting power source, and _____ on each elevator car.

FREQUENTLY ASKED QUESTIONS

Question 1: I'm doing a remodeling job in a multi-story building. Can I run new feeders to new feeder panels on each floor in the elevator hoistway?

Answer: *NEC® 620.37(A)* permits only such electric wiring, raceways, and cables used directly in connection with the elevator to be inside the hoistway. There is a danger that the raceways or cables may come loose and interfere with the travel of the car and cause damage (or worse) to the elevator equipment associated with the hoistway. The feeders you want to run in the hoistway would be difficult to maintain in that they would be virtually inaccessible without using the elevator car for access.

Question 2: Can I tap off the main power-supply disconnecting means to a panel for lighting, receptacles, and HVAC equipment in the elevator car?

Answer: *NEC® 620.51* requires that the disconnecting means for the main power-supply conductors for each unit shall not disconnect the branch-circuits for lighting, receptacles, ventilation, heating, and air-conditioning on the elevator car, in the machine room/machinery space or hoistway pit.

Question 3: How do I size the feeder conductors for a wheelchair lift?

Answer: Wheelchair lift driving machine motor duty is classified as intermittent. *NEC® 620.61(B)(4)* permits these driving machine motors to be protected against overload in accordance with *430.33*, and *430.33* permits the branch-circuit short-circuit and ground-fault protection to be in accordance with *Table 430.152*. The feeder conductors must be sized according to *620.13(A)*.

Question 4: I have three driving machines for an elevator. Should I install a single means of disconnect for all the driving machines?

Answer: *NEC® 620.51* requires that a single means of disconnect be provided for each unit, and where there are multiple units, one disconnecting means must be provided to disconnect all the motors and control valve operating magnets. This provides a main control for the elevator and individual control for each unit.

Question 5: Is the traveling cable required to be terminated in a junction box on the elevator car at its first point of contact with the car?

Answer: Traveling cable conductors are permitted by *620.44* to be run open on the car for a distance not to exceed 6 ft (1.8 m) to a termination point, provided the conductors are grouped together, taped, or in the original sheath.

CHAPTER 25

Electric Vehicle Charging System

OBJECTIVES

After completing this chapter, the student should understand:
- charging equipment wiring methods.
- charging equipment construction.
- charging equipment control and protection.
- charging equipment locations.
- charging equipment ventilation requirements.

KEY TERMS

Electric Vehicle Connector: A device that by insertion into an electric vehicle inlet establishes an electrical connection to the electric vehicle for the purpose of charging and information exchange. This is part of the electric vehicle coupler

Electric Vehicle Coupler: A mating electric vehicle inlet and electric vehicle connector set

Electric Vehicle Inlet: The device on the electric vehicle into which the electric vehicle connector is inserted for charging and information exchange. This is part of the electric vehicle coupler. The electric vehicle inlet is considered to be part of the electric vehicle and not a part of the electric vehicle supply equipment

Electric Vehicle Interlock: A device that de-energizes the electric vehicle connector and its cable whenever the electric connector is uncoupled from the electric vehicle

Electric Vehicle Supply Equipment: The conductors, including the ungrounded, grounded, and equipment grounding conductors, and the electric vehicle connectors, attachment plugs, and all other fittings, devices, power outlets, or apparatus installed specifically for the purpose of delivering energy from the premises wiring to the electric vehicle

Non-Vented Storage Battery: A hermetically sealed battery that has no provision for release of excessive gas pressure, or for the addition of water or electrolyte, or for external measurement of electrolyte specific gravity

Personnel Protection System: A system of personnel protection devices and constructional features that, when used together, provide protection against electric shock. These devices are other than the GFCI

INTRODUCTION

Article 625 covers the electrical conductors and equipment external to an electric vehicle that connect an electric vehicle to a supply of electricity by conductive or inductive means, and the installation of equipment and devices related to the charging of electric vehicle **non-vented storage batteries**. For the purpose of this chapter, electric motorcycles, and off-road self-propelled electric vehicles, such as golf carts, industrial trucks, hoists, and lifts are not included.

GENERAL

Electric vehicles and electric vehicle charging systems are coupled in two different manners. One is the conductive method where connection is accomplished by direct connection between the **electric vehicle inlet** and the **electric vehicle connector**. The other method is the inductive means where, for the purpose of charging and information exchange, a plate or disk is inserted into the electric vehicle inlet, and where, without direct contact, the coupling is accomplished through inductive means similar to transformer action (Figure 25-1).

Unless other voltages are specified, the nominal ac system voltages of 120, 120/240, 208Y/120, 240, 480Y/277, 480, 600Y/347, and 600 volts shall be used.

Charging Equipment Wiring Methods

The **electric vehicle coupler** (*625.9*) must have a configuration that is noninterchangeable with wiring devices in other electrical systems. This will prevent accidental energizing from an unintended source. The electric vehicle coupler must be provided with a means to prevent unintentional disconnection. The electric vehicle coupler must be provided with a grounding pole, unless listed as suitable for the purpose in accordance with *Article 250*. If a grounding pole is provided, the electric vehicle coupler shall be designed so that the grounding pole connection is the first to make and the last to break contact (Figure 25-2).

Figure 25-1 Electric vehicles—inductive (top), conductive (bottom)

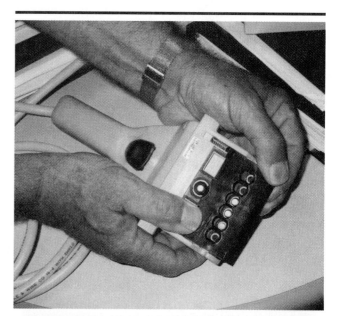

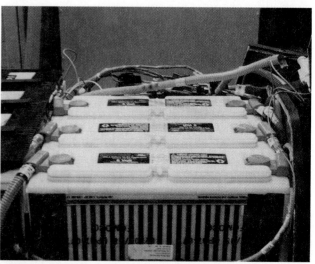

Figure 25-2 EV connector (top), Non-venting battery (bottom)

Charging Equipment Construction

NEC® 625.13 permits **electric vehicle supply equipment** rated at 125-volts, single-phase, 15- or 20-amperes, that meet the requirements of *625.18, 625.19,* and *625.29,* to be cord- and plug-connected. All other electric vehicle supply equipment shall be permanently connected and fastened in place.

In accordance with *625.14,* electric vehicle charging loads are considered to be continuous loads. Electric vehicle supply equipment must be provided with an **electric vehicle interlock** (*625.18*) to de-energize the electric vehicle connector and its cable whenever the electric vehicle connector is uncoupled from the electric vehicle. An interlock is not required for portable cord- and plug-connected equipment intended for connection to receptacle outlets rated 125-volts, single-phase, 15- or 20-amperes. Electric vehicle supply equipment receptacles rate 125-volts, single-phase, 15- or 20-amperes must be marked to indicate if ventilation is required. If ventilation is required, the charging equipment receptacle shall be switched, and the mechanical ventilation shall be interlocked through the switch supply power to the receptacle (Figure 25-3).

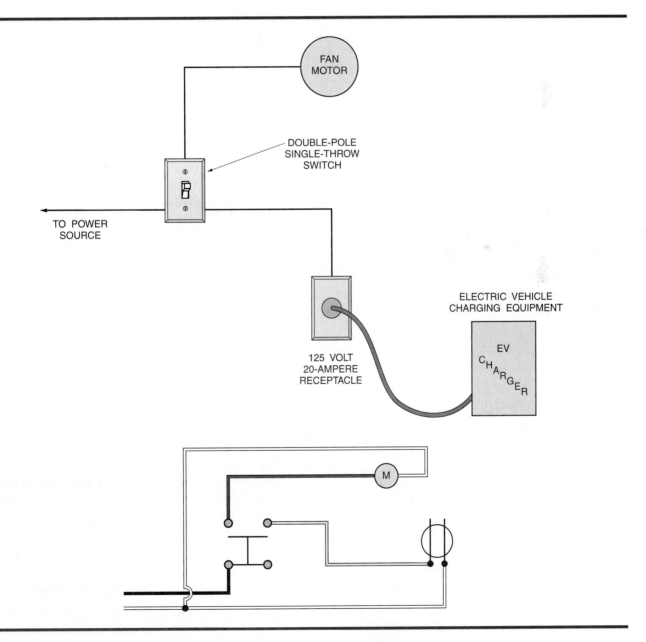

Figure 25-3 Fan switch interlock wiring diagram

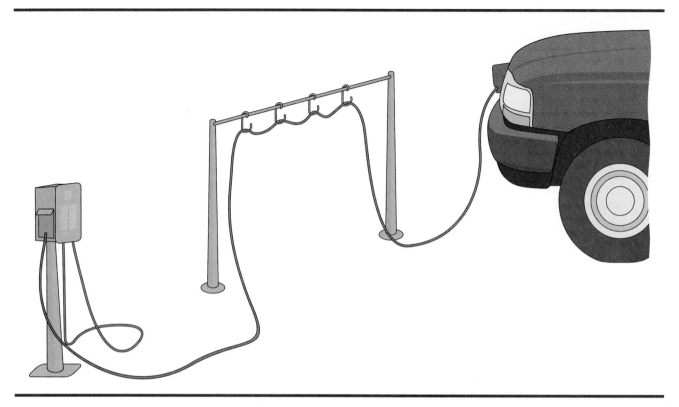

Figure 25-4 Charging cable management system for cables over 25 ft (7.5 m) in length

The Electric Vehicle Supply Equipment Cable (*625.17*) shall be a type identified by the letters EV in *Article 400* and *Table 400.4*. The overall length of the cable shall not exceed 25 ft (7.5 m) unless the Electric Vehicle Supply Equipment Cable is suspended from overhead or is intended for portable use in the facility and is equipped with a cable management system (Figure 25-4).

NEC® 625.19 requires that the electric vehicle supply equipment must be provided with an automatic means to de-energize the cable conductors and electric vehicle connector upon exposure to strain that could result in cable rupture or separation of the cable from the electric connector and exposure of live parts.

Automatic means of de-energization is not required for portable cord- and plug-connected equipment intended for connection to receptacles rated at 125-volts, single-phase, 15- or 20-amperes.

Charging Equipment Control and Protection

Overcurrent protection for feeders and branch-circuits supplying electric vehicle supply equipment shall be sized for continuous duty and must have a rating of not less than 125% of the maximum load of the electric vehicle supply equipment. In accordance with *625.23*, for electric vehicle supply equipment rated more than 60 amperes or more than 150 volts to ground, a disconnecting means must be provided in a readily accessible location. The disconnecting means must be capable of being locked in the open position (Figure 25-5).

A **personnel protection system** must be provided in accordance with *625.22* for protection of personnel from shock. The protection system must be a listed system of personnel protection devices and constructional features. If cord- and plug-connected electric vehicle supply equipment is used, the interrupting device of a listed personnel protection system shall be provided and shall be an integral part of the attachment plug or shall be located in the power supply cable not more than 12 in. (300 mm) from the attachment plug (Figure 25-6).

A means must be provided so that upon loss of power from the utility company or other electric system, energy cannot be back-fed through the electric vehicle supply equipment to the premises wiring system. *NEC® 625.25* prohibits the electric vehicle from serving as a standby power supply.

CHAPTER 25 Electric Vehicle Charging System 419

Figure 25-5 Disconnecting means for electric vehicle charging equipment

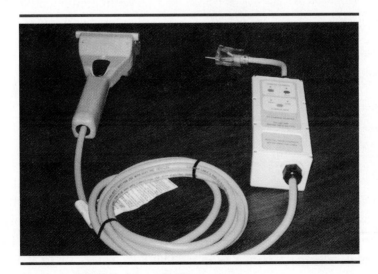

Figure 25-6 Protection device in cable

Charging Equipment Locations

Indoor sites as shown in *625.29* include, but are not limited to, integral, attached, and detached residential garages; enclosed and underground parking and underground parking structures; repair and non-repair commercial garages; and agricultural buildings.

The electric vehicle supply equipment shall be located to permit direct connection to the electric vehicle. The coupling means of the electric vehicle supply equipment must be stored or located at a height of not less than 18 in. (450 mm) and not more than 4 ft (1.2 m) above floor level (Figure 25-7).

Where electric vehicle non-vented batteries are used or where the electric vehicle supply equipment is listed or labeled as suitable for charging electric vehicles without ventilation and marked in accordance with *625.15(B)*, mechanical ventilation shall not be required.

Figure 25-7 Electric vehicle charger installed in a residential garage

Charging Equipment Ventilation

Where the electric vehicle supply equipment is listed or labeled as suitable for charging electric vehicles that require ventilation for indoor charging and marked in accordance with *625.15(C)*, mechanical ventilation such as a fan, must be provided. The ventilation must include both supply and exhaust equipment and must be permanently installed and located to intake from and vent directly to the outdoors. Positive pressure ventilation systems shall only be permitted in buildings or areas that have been specifically designed and approved for that application. Mechanical ventilation requirements shall be determined by one of the following methods:

1. **Table Values:** For supply voltages and currents, specified in *Table 625.29(D)*, the minimum requirements shall be as specified in *Table 625.29(D)* for each of the number of vehicles that can be charged at one time (Figure 25-8).

2. **Other Values:** For supply voltages and currents other than specified in *Table 625.29(D)*, the minimum ventilation requirements shall be calculated by means of the following general formulas as applicable.

   Single Phase:

   $$\frac{\text{Ventilation in cubic}}{\text{meters per minute}} = \frac{\text{volt-amperes}}{1718}$$

   $$\frac{\text{Ventilation in cubic}}{\text{feet per minute}} = \frac{\text{volt-amperes}}{48.7}$$

   Three Phase:

   $$\frac{\text{Ventilation in cubic}}{\text{meters per minute}} = \frac{\text{volt-amperes} \times 1.732}{1718}$$

   $$\frac{\text{Ventilation in cubic}}{\text{feet per minute}} = \frac{\text{volt-amperes} \times 1.732}{48.7}$$

   Ventilation requirements shall be permitted to be determined per calculations specified in an engineering study.

Supply Circuits

The supply circuit to the mechanical ventilation equipment shall be electrically interlocked with the electric vehicle supply equipment and must remain energized during the entire electric vehicle charging cycle. Electric vehicle supply equipment shall be marked in accordance with *625.15*. Elec-

Table 625.29(D)(1) Minimum Ventilation Required in Cubic Meters per Minute (m³/min) for Each of the Total Number of Electric Vehicles That Can Be Charged at One Time

Branch-Circuit Ampere Rating	Branch-Circuit Voltage						
	Single Phase			3 Phase			
	120 V	208 V	240 V or 120/240 V	208 V or 208 Y/120 V	240V	480 V or 480 Y/277 V	600 V or 600 Y/347 V
15	1.1	1.8	2.1	—	—	—	—
20	1.4	2.4	2.8	4.2	4.8	9.7	12
30	2.1	3.6	4.2	6.3	7.2	15	18
40	2.8	4.8	5.6	8.4	9.7	19	24
50	3.5	6.1	7.0	10	12	24	30
60	4.2	7.3	8.4	13	15	29	36
100	7.0	12	14	21	24	48	60
150	—	—	—	31	36	73	91
200	—	—	—	42	48	97	120
250	—	—	—	52	60	120	150
300	—	—	—	63	73	145	180
350	—	—	—	73	85	170	210
400	—	—	—	84	97	195	240

Figure 25-8 *NEC® Table 625.29(D)(1), Ventilation Requirements* (Reprinted with permission from NFPA 70-2002)

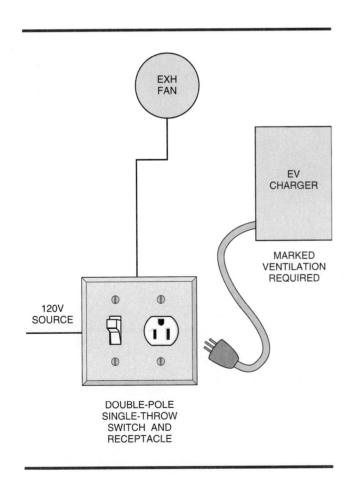

Figure 25-9 Interlock device

tric vehicle supply equipment receptacles rated at 125 volts, single-phase, 15- and 20-amperes shall be marked in accordance with *625.15(C)* and shall be switched, and the mechanical ventilation system shall be electrically interlocked through the switch supply power to the receptacle (Figure 25-9).

Outdoor sites shall include, but not be limited to, residential carports, driveways, curbside, open parking structures, parking lots, and commercial charging facilities (Figure 25-10).

The electric vehicle supply equipment shall be located to permit direct connection to the electric vehicle. Unless specifically listed for the purpose and location, the coupling means shall be stored or located at a height of not less than 24 in. (600 mm) and not more than 4 ft (1.2 m) above the parking surface.

SUMMARY

Article 625 covers electric vehicle charging equipment and does not cover the electric vehicle itself. Every means has been taken to provide safety for the users of this equipment, and it is recognized that users of this equipment may be standing in damp or wet locations while handling electrical equipment.

Figure 25-10 Electric vehicle charger in commercial area

NEC® 625.22 requires a listed system of personnel protection that has been specifically designed to protect the user of this equipment. *NEC® 625.18* requires an interlock that de-energizes the electric vehicle connector and its cable whenever the electric connector is uncoupled from the electric vehicle. Provisions are made in *625.19* for the automatic de-energization of cable conductors and the electric vehicle connector upon exposure to strain that could result in cable rupture or separation of the cable from the electric connector and exposure of live parts.

Means have been provided in accordance with *625.25*, such that upon loss of voltage from the utility or other source of supply, energy cannot be backfed through the electric vehicle supply equipment to the premises wiring system. Electric vehicles are not permitted to serve as a standby power supply.

REVIEW QUESTIONS

MULTIPLE CHOICE

1. Electric vehicles not covered by *Article 625* include:
   ____ A. golf carts.
   ____ B. tractors.
   ____ C. electric motorcycles.
   ____ D. all of the above.

2. The electric vehicle that is a part of the electric vehicle coupler is a(n):
   ____ A. inlet.
   ____ B. connector.
   ____ C. grounding pole.
   ____ D. none of the above.

3. The electric vehicle supply equipment cable shall be:
   ____ A. Type EV.
   ____ B. hard service SO cord.
   ____ C. not smaller than 10 AWG.
   ____ D. not larger than 10 AWG.

4. The supply voltage for electric vehicle charging equipment shall not be greater than:
   ____ A. 120 volts.
   ____ B. 240 volts.
   ____ C. 480 volts.
   ____ D. 600 volts.

5. Electric vehicles may be connected to a supply of electricity by _____ means.
   ___ A. conductive
   ___ B. inductive
   ___ C. all of the above
   ___ D. none of the above

TRUE OR FALSE

1. Electric vehicle charging loads shall be considered to be continuous loads.
   ___ A. True
   ___ B. False
   Code reference _____ .

2. The overall length of an electric vehicle charging cable shall not exceed 25 ft (7.5 m) unless the electric vehicle supply equipment cable is suspended from overhead.
   ___ A. True
   ___ B. False
   Code reference _____ .

3. The grounding pole connection must be the last to make and the first to break.
   ___ A. True
   ___ B. False
   Code reference _____ .

4. If an electric vehicle supply equipment is rated more than 30 amperes or more than 150 volts to ground, a disconnecting means shall be provided in an accessible location.
   ___ A. True
   ___ B. False
   Code reference _____ .

5. The electric vehicle can be used as a standby power supply if listed for the purpose.
   ___ A. True
   ___ B. False
   Code reference _____ .

FILL IN THE BLANKS

1. The electric vehicle supply equipment shall be located to permit _____ to the electric vehicle.

2. The coupling means of the electric vehicle supply equipment must be stored or located at a height of not less than _____ and not more than _____ above floor level.

3. Overcurrent protection for feeders and branch-circuits supplying electric vehicle supply equipment shall be sized for _____ and must have a rating of not less than _____ of the maximum load of the electric vehicle supply equipment.

4. The electric vehicle coupler must be provided with a _____ unintentional disconnection.

5. The supply circuit to the mechanical ventilation equipment shall be _____ interlocked with the electric vehicle supply equipment and must remain _____ through the entire electric vehicle charging cycle.

FREQUENTLY ASKED QUESTIONS:

Question 1: If there are two ways to connect the charging equipment to the electric vehicle, how do they determine which type to install at commercial places where facilities are provided?

Answer: Commercial electric vehicle chargers have a charging cable on each side, each with a different type of vehicle connector. At the present time, when an electric vehicle is purchased by a homeowner, the matching electric vehicle charging supply equipment for that vehicle is a part of the package.

Question 2: How are electric vehicle owners protected from the dangers of electric shock when hooking up their vehicles to a commercial electric vehicle charger?

Answer: *NEC® 625.22* requires a personnel protection system for this purpose. A ground-fault protection system may consist of one or more components that provide protection against electric shock for different portions of the electric vehicle supply equipment circuitry, which may be operating at frequencies other than 50/60 hertz, at direct current potentials and/or voltages above 150 volts to ground. The protection system will include a GFCI with additional circuitry to detect and respond to a fault in portions of the electric vehicle supply equipment operating on direct current or at high frequency. For systems operating at voltages above 150 volts to ground, the protective system includes monitoring systems to ensure that proper grounding is provided and maintained during charging.

Question 3: How does being in a hazardous classified area affect the electric vehicle charging equipment?

Answer: *NEC® 625.28* requires that where electric vehicle charging equipment or wiring is installed in a hazardous classified location, the requirements of *Articles 500* through *516* shall apply. This provision may be required in commercial garages and agricultural buildings. The height requirements for locating or storing electric vehicle supply charging equipment are designed to keep this equipment out of the hazardous area.

Question 4: What will happen if someone forgets to unplug the charging cable before moving his electric vehicle?

Answer: *NEC® 625.19* requires that the electric vehicle supply equipment be provided with an automatic means to de-energize the cable conductors and electric vehicle connector upon exposure to strain that could result in rupture or separation of the cable from the electric connector and exposure of live parts. This requirement is not required for portable cord- and plug-connected electric vehicle supply equipment intended for connection to receptacle outlets rated at 125-volts, single-phase, 15- or 20-amperes.

Question 5: Is there such a thing as portable electric vehicle charging equipment that can be carried to a dead electric vehicle and be used to charge the vehicle if a suitable power receptacle is available?

Answer: Provisions for that kind of charging equipment is covered in *625.19*.

CHAPTER 26

Information Technology Rooms

OBJECTIVES

After completing this chapter, the student should understand:
- special requirements for ITE rooms.
- supply circuits and interconnecting cables.
- cables not in ITE rooms.
- disconnecting means in ITE rooms.
- uninterruptible power supplies.
- grounding in ITE rooms.

KEY TERMS

Fire/Smoke Dampers: A movable plate located in a duct to control the movement of air through the duct. The damper may be positioned from a fully opened to a fully closed position. This positioning may be done by automatic or manual means

Heating/Ventilating/Air-Conditioning (HVAC): A multipurpose system usually packaged as one unit to perform all of the heating, ventilating, or air-conditioning functions required

Interconnecting Cables: Data processing cables or assemblies used to interconnect data processing units

Raised Floors: A listed floor assembly consisting of a support structure and cover that allows unlimited access to the underfloor area by removal of one or more of the covers

INTRODUCTION

Article 645 covers equipment, power-supply wiring, equipment interconnecting wiring, and grounding of information technology equipment and systems, including terminal units, in an information technology equipment room. The intent of this article is to allow for some modification or relaxing of the wiring methods required in the Code if the ITE room complies with the requirements of *Article 645.2*.

GENERAL

Information technology equipment is considered to be special equipment by the *NEC.*® In the 1990 edition of the *NEC,*® the concept of information technology equipment rooms was introduced into the Code. At that time, the terminology "Electronic Computer Data Processing Equipment" was used. The thought was to modify the requirements for wiring methods as long as the requirements for the construction of the room were adhered to. While the underfloor area was still a plenum, if it was self-contained, it was permitted to deviate from the more stringent requirements of *300.22* regarding plenum areas. Cables with a DP designation and other cables with certain fire-resistant qualities able to pass a vertical flame test can be used under the **raised floor**.

Special Requirements for ITE Rooms

The following requirements are set forth in *645.2* for computer room construction to be classified as an "Information Technology Equipment" (ITE) room.

1. A disconnecting means in the room complying with *645.10*.
2. A separate **heating/ventilating/air-conditioning** system or a **fire/smoke damper** controlled shared HVAC system.
3. Listed ITE is installed.
4. Occupancy limited to trained personnel.
5. Separation of the room from other occupancies by fire-resistant walls, floors, and ceilings with protected openings.

A requirement that the construction of the room be in compliance with all applicable local building ordinances was removed for the *NEC®* 2002 Code because the requirement is not electrical in nature.

Supply Circuits and Interconnecting Cables

The requirements for supply circuits and interconnecting cables are shown in *645.5*. Branch-circuit conductors, supplying one or more units of a data processing system, must have an ampacity not less than 125% of the total connected load.

The data processing system is permitted to be connected to a branch-circuit either by a cord and attachment plug or hard-wired. Power-supply cords are limited to 15 ft (4.5 m) not only to allow for an efficient interchange of equipment but also to discourage leaving cords or placing cords in a manner that will subject them to physical damage. The placement of power-supply receptacles properly will prevent cords from being damaged after installation.

Interconnecting cables may be used to interconnect separate data processing units. These cables and cable assemblies must be listed for the purpose, and if run on the floor, they must be protected from damage. For above floor applications, any listed cable can be used according to its listing data. If an interconnect cable supplied with ITE is not listed, it is generally restricted to a length of 10 ft (3 m) and must meet certain flame test requirements. It is required to have a tag or marking associating it with the listed ITE. Most interconnecting cables are not supplied with the equipment and are selected at the time of installation to conform to the room dimensions. If these cables selected are not listed, then it is the responsibility of the AHJ to determine the approval for their use. Listed cables are readily available and should always be used.

Under Raised Floors

The more commonly used method of installing interconnect cables is under the raised floor. Power cables, communications cables, interconnecting cables, and receptacles associated with the ITE can be used under the raised floor. Interconnecting cables run under the raised floor are not required to be installed in raceways, if they are a type specified in *645.5(D)(5)(c)*, such as listed Type DP cable and others having adequate fire-resistant characteristics.

NEC® 645.5(D) permits power cables, communication cables, interconnecting cable, and receptacles associated with the ITE to be under a raised floor provided (1) the raised floor is of suitable construction and the area under the floor is accessible (Figure 26-1), (2) the branch-circuit supply conductors to receptacles or field-wired equipment are in RMC, RNC, IMC, EMT, ENT, metal wireway, nonmetallic wireway, surface metal raceway with cover, nonmetallic surface raceway, FMC, LFMC, or LFNC, Type MI cable, or Type AC cable, (3) the ventilation system is so arranged with approved smoke detection devices that upon detection of fire

Figure 26-1 Raised floor in ITE room showing access method using suction grip

or products of combustion in the underfloor space, the circulation of air will cease, (4) openings in raised floors for cables protect the cables against abrasions and minimize the entrance of debris beneath the floor, (5) cables other than Type MI, MC, or AC and those complying with a, b, and c following shall be listed as Type DP cable having adequate fire-resistant characteristics suitable for use under raised floors of an ITE room.

a. Interconnecting cables enclosed in a raceway.

b. Interconnecting cables listed with equipment manufacturers prior to July 1, 1994, being installed with that equipment.

c. Cable type designations Type TC (*Article 336*); Types CL2, CL3, and PLTC (*Article 725*); Types NPLF and FPL (*Article 760*); Types OFC and OFN (*Article 770*); Types CM and MP (*Article 800*); Type CATV (*Article 820*). These designations shall be permitted to have an additional letter P or R or G. Green or green/yellow insulated single-conductor cables, 4 AWG and larger, marked *for use in cable trays* or *for CT use* shall be permitted for equipment grounding.

and (6) abandoned cables shall not be permitted to remain.

All of the cables shown in *645.5(D)(5)(c)* have been listed as having adequate fire-resistant characteristics which make them acceptable for use under a raised floor in an ITE room. The plenum wiring method requirements found in *330.22* have been relaxed for installation under a raised floor in an ITE room. Cables extending beyond the ITE room are required by *645.6* to follow the applicable requirements of the *NEC*.

NEC® 645.10 requires a means of disconnect to be provided to disconnect power to all electronic equipment in the ITE room. There shall also be a similar means to disconnect the power to all dedicated HVAC systems serving the room and cause all required fire/smoke dampers to close. The controls for these disconnecting means shall be grouped and identified and must be readily accessible at the principal exit doors. A single means to control both the electronic equipment and the HVAC systems is permitted. Where a pushbutton is used as a means to disconnect power, pushing the button in shall disconnect the power (Figure 26-2).

This disconnect is usually accomplished by using a switch to activate a shunt trip circuit breaker

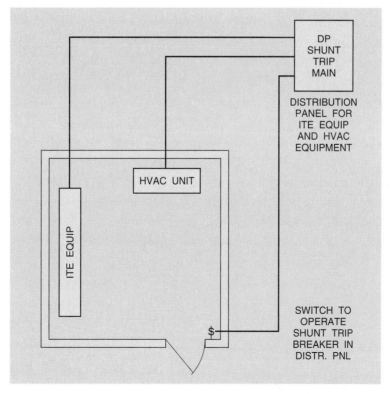

Figure 26-2 Disconnect at principal door to ITE room shunt trip opens main breaker

controlling the distribution panel that feeds the ITE room and HVAC system. The purpose of the disconnect is to be used as an emergency shut-off to protect personnel and equipment in the case of a fire, flood, failure of structural building components, or hazardous contamination, any of which may be the result of a problem in other parts of the building.

Uninterruptible Power Supplies

NEC® 645.11 requires that unless otherwise permitted in (1) or (2), UPS systems within the ITE room and their supply and output circuits shall also comply with *645.10*. The disconnecting means shall also disconnect the battery from its load. This requirement assures that when the disconnecting means required in *645.10* is opened, the backup UPS system does not come in to replace the lost power (Figure 26-3).

Grounding

All exposed noncurrent-carrying metal parts of an information technology system shall be grounded in accordance with *Article 250* or shall be double insulated. Where signal reference structures are installed, they shall be bonded to the equipment grounding system provided for the ITE.

SUMMARY

The use of *Article 645* is contingent upon the ITE room meeting all of the requirements outlined in *645.2*. If all of these requirements are *not* met, then the other requirements of the *NEC®* are applicable. The plenum rules that were relaxed in *Article 645* for an ITE room cannot be used, and the more restrictive requirements of *300.22* must be adhered to.

Grounding of the equipment in an ITE room is essential, and particular attention should be paid to the proper recommended bonding procedures for the metal supporting members of the raised floor.

Abandoned cables should be removed promptly so as not to obstruct the airflow under the raised floor and to eliminate the increased fuel load they represent.

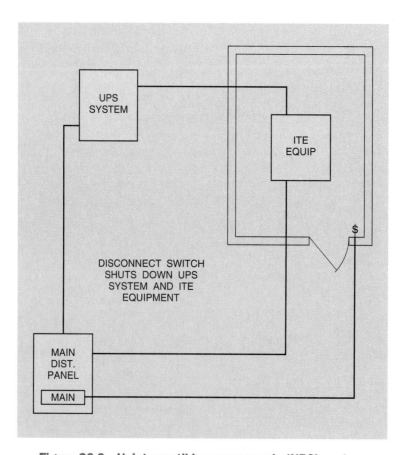

Figure 26-3 Uninterruptible power supply (UPS) system

REVIEW QUESTIONS

MULTIPLE CHOICE

1. The disconnecting means in an ITE room must disconnect:
   ____ A. lighting.
   ____ B. receptacle outlets.
   ____ C. electronic equipment.
   ____ D. none of the above.

2. Power-supply cords to electronic equipment are limited to _____ in length.
   ____ A. 6 ft (1.8 m)
   ____ B. 15 ft (4.5 m)
   ____ C. 10 ft (3.0 m)
   ____ D. none of the above

3. The following cables are permitted to be used under raised floors:
   ____ A. Type MI.
   ____ B. Type MC.
   ____ C. Type AC.
   ____ D. all of the above.

4. Branch-circuits supplying one or more units of a data processing system must have an ampacity not less than _____ of the total connected load.
   ____ A. 100%
   ____ B. 125%
   ____ C. 115%
   ____ D. none of the above

5. Green or green/yellow insulated single-conductor cables, _____ and larger, marked *for use in cable trays* or *for CT use* shall be permitted for equipment grounding.
   ____ A. 8 AWG
   ____ B. 6 AWG
   ____ C. 4 AWG
   ____ D. none of the above

TRUE OR FALSE

1. The underfloor area in an ITE room is not considered a plenum.
   ____ A. True
   ____ B. False
   Code reference _____ .

2. Cables with a DP designation and other cables with fire-resistant qualities can be used under the raised floor.
   ____ A. True
   ____ B. False
   Code reference _____ .

3. Occupancy in an ITE room is limited to personnel needed for the maintenance and operation of the equipment in the ITE room.
   ____ A. True
   ____ B. False
   Code reference _____ .

4. Power-supply cords are limited to 20 ft (6.0 m) to allow for efficient interchange of equipment.
   ____ A. True
   ____ B. False
   Code reference _____ .

5. Where a pushbutton is used as a means to disconnect power, pushing the button in shall disconnect the power.
   ____ A. True
   ____ B. False
   Code reference _____ .

FILL IN THE BLANKS

1. Branch-circuit conductors supplying one or more units of a data processing system must have an ampacity _____ of the total connected load.

2. These cables and cable assemblies must be listed for the purpose, and if _____ , they must be protected from damage.

3. All of the cables shown in *645.5(D)(5)(c)* have been listed as having adequate _____ which make them acceptable for use under a raised floor.

4. A _____ to control both the electronic equipment and the HVAC systems is permitted.

5. Where _____ structures are installed, they shall be bonded to the equipment grounding system provided for the ITE.

FREQUENTLY ASKED QUESTIONS

Question 1: What is the reason for having a separate ITE room instead of the old computer rooms?

Answer: The restrictions on computer rooms caused a lot of unnecessary expense without much benefit derived from the expense. The ITE room, as shown in *645.2*, defines the requirements for this room. It is a self-contained room that relaxes the restrictions on the use of cables and other wiring under the raised floor and permits cables to be used without a raceway.

Question 2: Can the ITE room utilize HVAC systems from other parts of the building?

Answer: *NEC® 645.2* permits any HVAC system that serves other occupancies to be used if fire or smoke dampers are provided at the point of penetration of the room boundary. These dampers must operate on the activation of smoke detectors and also by operation of the disconnecting means required by *645.10*.

Question 3: What is the purpose of the disconnecting means required in *645.10* for ITE rooms?

Answer: *NEC® 645.10* requires two disconnecting means that are readily accessible at the principal exit doors of the ITE room. One disconnect shall disconnect power to all electronic equipment. The second disconnecting means must disconnect power to all dedicated HVAC systems serving the room and cause all required smoke or fire dampers to close. A single means to control both the electronic equipment and HVAC systems is permitted.

Generally, a switch that operates a shunt-trip breaker is employed to disconnect power to the distribution panel that supplies power to the ITE room.

The purpose of the disconnecting means requirement is *emergency* shutdown of all electronic equipment and dedicated HVAC systems in the case of fire, flood, failure of structural building components, personnel safety, or hazardous contamination. ITE rooms may be above, below, or alongside areas that are equipped with sprinklers, areas that may be classified as hazardous, areas that have water, drainage, steam, or other contaminants. *NEC® 645.1(5)* requires that the ITE room be separated from other occupancies by fire-resistant walls, floors, and ceilings with protected openings. Sprinkler systems are generally required for ITE rooms, and activation by any means could be disastrous to electronic equipment. The intent of *dedicated* HVAC systems is in conformance with *645.2(2)*. HVAC systems dedicated to the ITE room must be completely shutdown by means of the disconnecting means required in *645.10*.

CHAPTER 27

Fire Pumps

OBJECTIVES

After completing this chapter, the student should understand:

- power sources.
- continuity of power.
- transformers for fire pumps.
- power wiring.
- voltage drop.
- equipment location.
- control wiring.

KEY TERMS

Engine Driven Fire Pump Motors: Mechanical engines, usually diesel operated, which drive fire pumps

Jockey Pump: A pressure maintenance pump installed to keep a predetermined pressure in the sprinkler piping system

On-Site Power Production Facility: The normal on-site supply of electric power for the site that is expected to be constantly producing power

On-Site Standby Generator: Alternate supply of electric power that is automatically activated upon loss of normal power

Supervised Connection: Fire pump connection supervised by a central station or a remote station signal device

INTRODUCTION

Article 695 covers the installation of electric power sources and interconnecting circuits and switching and control equipment dedicated to fire pump drivers (Figure 27-1).

"It is the intent that the fire-pump motor will attempt to run under any conditions of loading and not be automatically disconnected by an overcurrent protection device."

The Standard for the Installation of Stationary Pumps for Fire Protection, NFPA 20–1999, covers the performance, maintenance, and acceptance testing of the fire pump system and the internal wiring of the components of the system (Figure 27-2).

GENERAL

Fire pumps are a requirement of the fire ordinances and building department ordinances of individual municipalities. The National Fire Protection Association (NFPA) requirements are found in NFPA 20 and extracts from this can be found in NFPA 70, *NEC.® Article 695* was introduced into the Code for the 1996 edition of the *NEC.®* Only the requirements

434 CHAPTER 27 Fire Pumps

Figure 27-1 Picture of fire pump, fire pump controller, jockey pump, and jockey pump controller

Figure 27-2 Fire inspector conducting inspection

for premises electrical wiring and the supply source are in the *NEC.*®

Power Sources

NEC® *695.3* requires that power sources for fire pumps must be reliable and capable of carrying indefinitely the sum of the locked-rotor current of the fire pump motor and the pressure maintenance pump motor (**jockey pump**) and the full-load current of the associated fire pump accessory equipment. The power source for an electric motor-driven fire pump shall be one or more of the following:

- **Electric Utility Service Connection:** Fire pump power service from a serving utility may be supplied as a separate service or by a tap located ahead of, and not within, the same cabinet, enclosure, or vertical switchboard section as the service disconnecting means (Figure 27-3). A tap ahead of the service disconnecting means must be installed in accordance with the requirements for service-entrance conductors. A tap ahead of the service disconnecting means must be marked for identification, as shown in *230.2*, and must be sufficiently remote from

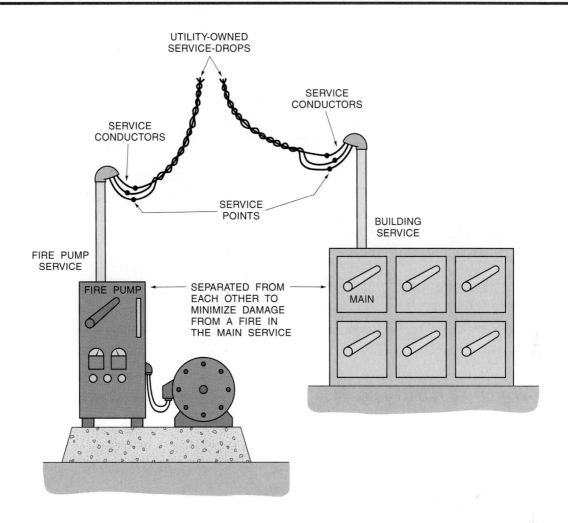

Figure 27-3 Fire pump connection to separate service-drop

the normal service disconnecting means, as required by *230.72(B)*, to minimize the possibility of simultaneous interruption of service (Figure 27-4).

- **On-Site Power Production Facility:** A fire pump is permitted to be supplied by an on-site power production facility. This source facility must be located and protected to minimize the possibility of damage by fire.

  Where reliable power cannot be obtained from a utility company source, the power must be supplied from an approved combination of two or more of either of such sources, or from an approved combination of feeders constituting two or more power sources from separate utility services (see Figure 27-5), or one or more of such power sources in combination with an **on-site standby generator**. An on-site standby generator shall be of sufficient capacity to allow normal starting and running of the motors driving the fire pumps while supplying all other simultaneously operated load. A tap ahead of the on-site generator disconnecting means is not be required (Figure 27-6).

Continuity of Power

The continuity of power to fire pump motors is critical, and *695.4* requires direct connection of the supply conductors to the fire pump controller or that they be connected through no more than one disconnecting means, which must be supervised. The disconnecting means must be suitable for use as service

436 CHAPTER 27 Fire Pumps

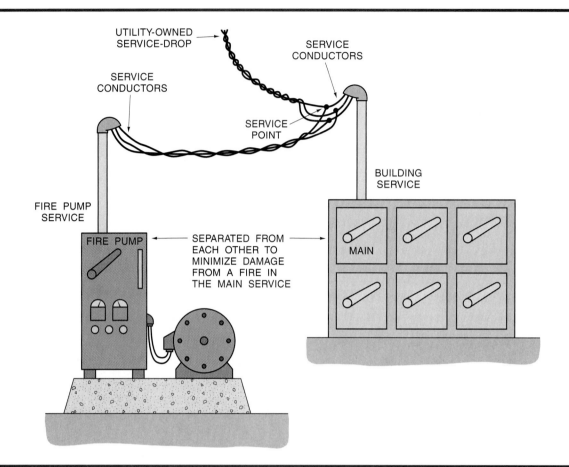

Figure 27-4 Fire pump connection to utility drop ahead of main disconnect

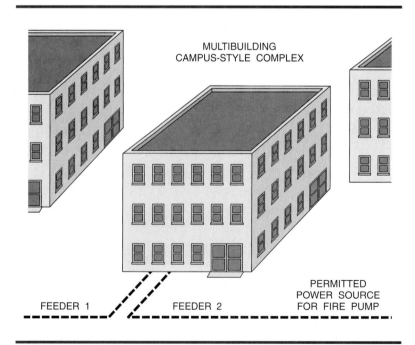

Figure 27-5 Multiple feeder source from two separate utility services

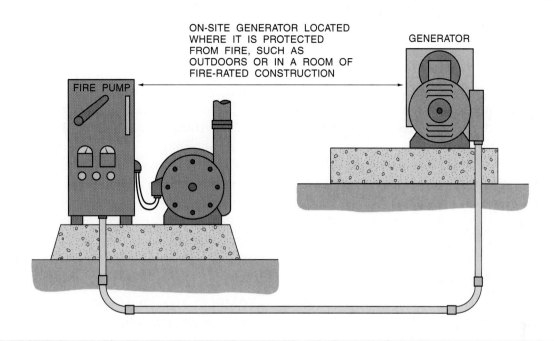

Figure 27-6 On-site generator power source for fire pump

equipment, lockable in the closed position, and located sufficiently remote from other building or other fire pump source disconnecting means that inadvertent simultaneous operation would be unlikely.

The disconnecting means shall be **supervised connection** in the closed position by one of the following methods: (1) A central station or remote station signal device; (2) local signaling that will cause the sounding of an audible alarm; or (3) locking the disconnect in the closed position or sealing of the disconnect with weekly recorded inspections where the disconnecting means are located within fenced-in enclosures or in buildings under control of the owner.

Let's review the permitted power sources for a fire pump installation. Fire pump motors are not required to have alternate sources of power by NFPA 20, *Standard for the Installation of Centrifugal Fire Pumps* or by NFPA 70, *NEC*,® but often more than one source is required by the AHJ in areas where the utility source may not be considered as a reliable source.

A *fire pump motor power source* may be a utility service supplying normal power to the site. The fire pump motor feed would be required to be tapped ahead of the main disconnecting means but not within the same cabinet, enclosure, or vertical switchboard section as the disconnecting means (Figure 27-4).

A *fire pump motor power source* may be a separate utility service provided solely for the purpose of supplying power for the fire pump motor (see Figure 27-3). Utility supplied power sources may be supplemented, either by user design criteria or by local municipal ordinances, by an on-site standby generator.

A *fire pump motor power source* may be an on-site production facility (separately derived system) supplying the normal supply of electric power for the site. While it is not clearly stated in the *NEC*,® this source should be a tap ahead of the main disconnect, supplying all other simultaneously operated normal loads. The **on-site power production facility** (separately derived system) may be supplemented by a utility provided power source, either in a co-generation mode or as a standby source.

A *fire pump motor power source* may be an on-site standby generator provided solely for the purpose of supplying power to the fire pump motor. A tap ahead of the on-site generator disconnecting means is not required. An on-site standby generator may be supplemented by an alternate power source, such as a utility source or an on-site production facility.

A *fire pump motor power source* may be for multibuilding campus-style complexes, two or more feeder sources as one power source, or as more than one power source where such feeders are derived from separate utility services (Figure 27-5).

The purpose of all the suggested means for supplying power to fire pump motors is to obtain a reliable power source preferably with an alternate source. The intent is to keep the fire pump running under any circumstances under fire conditions.

Transformers for Fire Pumps

NEC 695.5 contains the requirements for the use of transformers in fire pump installations. Where the system voltage is different from the utilization voltage of the fire pump motor, a transformer is permitted to be installed in between the system supply and the fire pump controller.

Where a transformer supplies an electric motor-driven fire pump, it must be rated at 125% of the fire pump motor and *pressure maintenance pump motor* load and 100% of the associated fire pump accessory equipment supplied by the transformer.

The primary overcurrent protective device must be set or selected to carry indefinitely the sum of the locked-rotor current of the fire pump motor and the pressure maintenance pump motor and the full-load current of the associated fire pump accessory equipment. Secondary overcurrent protection is not permitted (Figure 27-7).

EXAMPLE: Using a 4160/480-volt, 3-phase transformer dedicated as a fire pump motor power source

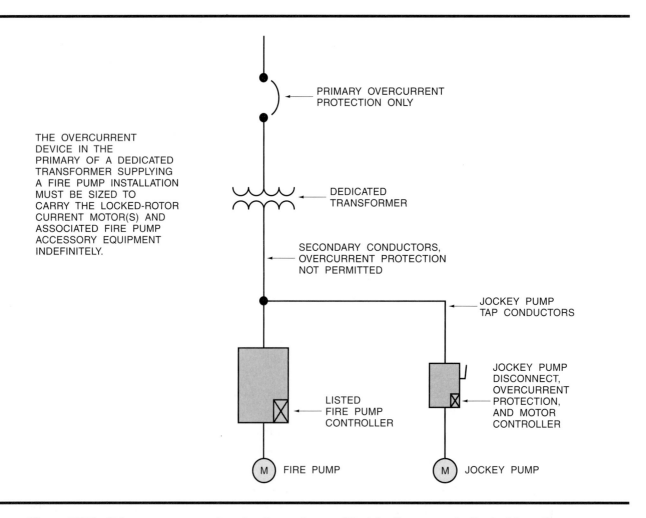

Figure 27-7 Primary overcurrent protection only permitted for fire pump dedicated transformer

to supply power to a 50-horsepower, 460-volt, 3-phase fire pump motor with a code letter F and a ¾ horsepower, 460-volt, 3-phase jockey pump motor with a code letter D, calculate the minimum size transformer required and the minimum size primary overcurrent protective device required.

Step #1: Calculate minimum size transformer required:

Table 430.150
50-hp, 3-phase FLA =	65.0 amperes
¾-hp, 3-phase FLA =	1.6 amperes
Total FLA =	66.6 amperes

In accordance with *695.5(A)*:

$$66.6 \text{ A} \times 1.25 = 83.25 \text{ amperes}$$

$$\text{Transformer kVA} = \frac{V \times A \times 1.732}{1000} =$$

$$\frac{480 \times 83.25 \times 1.732}{1000} = 69.21$$

The minimum size transformer permitted is 69.21 kVA. The next larger size transformer is 75 kVA.

Step #2: Calculate the minimum size primary overcurrent protection permitted:

The minimum size primary overcurrent protection must allow the transformer to carry the locked-rotor currents of the fire pump motor and the jockey pump.

Table 430.7(B)

$$5.59 \text{ kVA/hp} \times 50 \text{ hp} = 279.5 \text{ kVA}$$

$$\text{LRA (50 hp)} = \frac{279.5 \text{ kVA}}{480 \times 1.732} = 336.2 \text{ amperes}$$

$$4.49 \text{ kVA/hp} \times \text{¾ hp} = 3.37 \text{ kVA}$$

$$\text{LRA (¾ hp)} = \frac{3.37 \text{ kVA}}{480 \times 1.732} = 4.1 \text{ amperes}$$

Total Secondary LRA = 340.3 amperes

The equivalent locked-rotor amperes on the primary side can be found as follows:

$$\text{LRA}_{primary} =$$

$$\frac{\text{secondary voltage}}{\text{primary voltage}} \times \text{locked-rotor amperes}_{secondary}$$

$$\frac{480 \text{ volts}}{4160} \times 340.3 = 39.3 \text{ amperes}$$

The value of 39.3 amperes represents the minimum size overcurrent device permitted on the primary of the dedicated transformer. The next higher size overcurrent device is 40 amperes.

Power Wiring

NEC® 695.6 contains the requirements for power wiring for fire pump motors. Service conductors must be physically routed outside the building. Where the service conductors cannot be physically routed outside the building, they are permitted to be run through buildings if they are installed in accordance with *230.6*. This *Section* considers conductors to be out of the building if installed under not less than 2 in. (50 mm) of concrete beneath a building or where installed within a building in a raceway that is encased in concrete or brick that is not less than 2 in. (50 mm) thick. This requirement applies to all supply conductors on the load side of the service disconnecting means that constitute the normal source of supply to that fire pump motor.

Fire pump supply conductors on the load side of the final disconnecting means and overcurrent devices must be kept entirely independent of all other wiring. Fire pump supply conductors may only supply loads that are directly associated with the fire pump system. Fire pump conductors may run through a building if they are encased in 2 in. (50 mm) of concrete, or are within an enclosed construction dedicated to the fire pump circuits and having a minimum rating of 1-hour fire resistive, or if they are a listed cable or listed electrical circuit protection system with a minimum 1-hour fire rating. The supply conductors located in the electrical equipment room, where they originate, and in the fire pump room are not required to have the minimum fire separation or fire-resistance rating.

Conductors supplying a fire pump motor, pressure maintenance pump, and associated fire pump accessory equipment shall have a rating of not less than 125% of the fire pump motor and the pressure maintenance pump plus 100% of the associated accessory equipment.

Power circuits must not have automatic protection against overloads. Branch-circuit and feeder conductors must be protected against short-circuit only. All wiring from the controllers to the pump motors shall be in RMC, IMC, LFMC, or LFNC Type LFNC-B or Type MI cable.

Voltage Drop

The voltage at the controller line terminals must not drop more than 15% below normal (controller rated voltage) under motor starting conditions. The voltage at the motor terminals shall not drop more than 5% below the voltage rating of the motor when the motor is operating at 115% of the full-load current rating of the motor.

Equipment Location

Controllers and transfer switches for motor-driven fire pumps must be located as close as practicable to the motors and must be within sight from the motors. **Engine-driven fire pump motor** controllers must be located as close as is practical to the engines they control and must be within sight from the engines. Storage batteries for diesel engine drives must be rack-supported above the floor, secured against displacement, and located where they will not be subject to physical damage or flooding with water.

Control Wiring

External control circuits that extend outside the fire pump room must be arranged so that failure of any external circuit (open or short-circuit) shall not prevent the operation of a pump from all other internal or external means. Breakage, disconnecting, shorting of the control wires, or loss of power to these circuits may cause continual running of the pump motor but must not prevent the controller from starting the fire pump due to causes other than these external control circuits.

No other under-voltage, phase-loss, frequency-sensitive, or other sensors shall be installed that automatically or manually prohibit actuation of the motor contactor. No remote device shall be installed that will prevent automatic operation of the transfer switch.

All wiring between the controller and the diesel engine shall be stranded and sized to carry continuously the charging or control circuits.

All electric motor-driven fire pump control wiring shall be in RMC, IMC, LFMC, LFNC Type B (LFNC-B), or Type MI cable.

A control conductor installed between the fire pump power transfer switch and the standby generator supplying the fire pump during loss of normal power shall be kept entirely independent of all other wiring.

SUMMARY

The important thing to remember concerning fire pump installation is that when a fire pump is called upon to operate, it must operate continuously without interruption until manually disconnected. The requirements for premises wiring and supply sources are shown in *Article 695*.

Power sources must be chosen based on available sources that meet the requirements of *Article 695*. Generally, a fire pump power service is supplied by a separate utility service or a tap ahead of the main disconnect of the service supplying a premises. Power may also be obtained from an on-site standby generator. Generally, two sources of power are used as a means to provide power to a fire pump. One source is used as a backup for the other. The intent is to keep the fire pump in operation under any circumstances under fire conditions.

No more than one disconnecting means may be used for the fire pump controller and the disconnecting means must be supervised in the closed position. The fire pump supply conductors must be kept separate from all other wiring so that any problem with other wiring will not jeopardize the fire pump wiring.

Article 695 is one of the few places in the *NEC*® where voltage drop consideration is required. These requirements may be found in *695.7*. It is critical that every means be used to ensure the proper operation of a fire pump.

All external controls must be arranged so that failure of any control will not prevent the operation of the fire pump. All control wiring must be run in RMC, IMC, LFMC, LFNC-B or Type MI cable. No remote control device may be installed that will prevent automatic operation of the transfer switch.

REVIEW QUESTIONS

MULTIPLE CHOICE

1. Power sources for motor-driven fire pumps must be capable of carrying indefinitely:
   _____ A. the locked-rotor current of the fire pump.
   _____ B. the locked-rotor current of the pressure maintenance pump.
   _____ C. the full-load current of associated fire pump accessories.
   _____ D. all of the above.

2. The main disconnecting means for a fire pump must be supervised by:
   _____ A. a central station signal device.
   _____ B. a local audible signal alarm.
   _____ C. locking or sealing in the closed position.
   _____ D. all of the above.

3. Where a transformer supplies an electric motor-driven fire pump, it must be rated at _____% of the fire pump motor and pressure maintenance pump motor load.
   _____ A. 200
   _____ B. 150
   _____ C. 125
   _____ D. 100

4. Where fire pump conductors are installed within a building, they are considered as being out of the building if installed in a raceway that is encased in concrete or brick that is not less than _____ thick.
   _____ A. 1 in. (25 mm)
   _____ B. 2 in. (50 mm)
   _____ C. 1¼ in. (32 mm)
   _____ D. none of the above

5. Voltage at the controller terminal must not drop more than _____% below normal (controller rated voltage) under motor starting conditions.
   _____ A. 3
   _____ B. 5
   _____ C. 10
   _____ D. 15

TRUE OR FALSE

1. Power sources for fire pumps must be reliable and capable of carrying indefinitely the sum of the locked-rotor current of the fire pump and the pressure maintenance pump and the full-load current of the associated fire pump accessory equipment.
   _____ A. True
   _____ B. False
   Code reference _____.

2. All fire pump motor power sources must have an alternate source of power that is automatically energized when the normal power source fails.
   _____ A. True
   _____ B. False
   Code reference _____.

442 CHAPTER 27 Fire Pumps

3. The requirements for fire pump installations are based on the need for a reliable power supply that will allow the fire pump to run indefinitely during a fire.
   ____ A. True
   ____ B. False
   Code reference _____.

4. Where a transformer supplies an electric motor-driven fire pump, it must be rated at 125% of the fire pump motor and pressure maintenance pump load and 125% of the associated fire pump accessory equipment supplied by the transformer.
   ____ A. True
   ____ B. False
   Code reference _____.

5. Fire pump supply conductors on the load side of the final disconnecting means and overcurrent devices must be kept entirely independent of all other wiring.
   ____ A. True
   ____ B. False
   Code reference _____.

FILL IN THE BLANKS

1. A fire pump motor power source may be a _____ supplying normal power to the site.

2. Where the system voltage is different from the _____ of the fire pump motor, transformers are permitted to be installed in between the system supply and the fire pump controller.

3. Where a transformer supplies an electric motor-driven fire pump, it must be rated at _____ of the fire pump motor and pressure maintenance load.

4. The voltage at the _____ terminals must not drop more than 15% below normal (controller rated voltage) under motor starting conditions.

5. All wiring between the controller and the diesel engine shall be stranded and sized to carry continuously the _____ circuits.

FREQUENTLY ASKED QUESTIONS

Question 1: I have a utility service that feeds directly from the meter fitting, through the wall of the fire pump room, to the fire pump combination main disconnect and controller. The supply conductors are in a rigid metal raceway and run 10 feet before they connect to the main disconnect. Does this raceway have to be enclosed with concrete or brick?

Answer: *NEC® 695.6(A)* requires supply conductors to be run outside the building and shall be run as service-entrance conductors. These service conductors can be installed in accordance with *230.6* by encasing the raceway in 2 in. (50 mm) of concrete.

Question 2: Is the raceway enclosing the fire pump motor-circuit conductors from the controller to the fire pump required to be encased in concrete?

Answer: *NEC® 695.12(A)* requires electric motor-driven fire pump controllers to be located as close as is practicable to the motors they control and to be within sight from the motors. *NEC® 695.6(E)* requires that all wiring from controllers to the pump motors be in RMC, IMC, LFMC, or LFNC. There is no requirement for encasing the raceway in concrete.

Question 3: What is an on-site power production facility? Even with the definition shown, it isn't clear.

Answer: It's a facility that produces power on site. It also is the normal supply of electric power for the site and is expected to be constantly producing power. That makes it a separately derived system to my way of thinking. It can be used as a source of supply for an electric motor-driven fire pump, either alone or backed-up by another source, as shown in *695.3*.

Question 4: *Article 695* doesn't say anything about utility metering if a utility source is used for the fire pump motor source. How can a meter withstand the combined locked-rotor currents of the fire pump and pressure maintenance pump and the current of all the associated equipment?

Answer: Yes. A 50-horsepower fire pump and all the associated accessories, amounting to about 80 amperes on a 3-phase, 480-volt service, could be handled easily by a 200-ampere meter. This meter probably could not handle the locked-rotor currents involved that may run as high as 400 amperes. The internal meter wiring is considerably smaller than the 200-ampere rating of the meter and, in a worst case situation, may not handle the fire pump load as required. Using current transformer metering is the safest way to handle this situation.

Question 5: What protects fire pump motors from overloads?

Answer: Fire pump motors are not protected against overloads. It is the intent that the fire pump motor will attempt to run under any conditions of loading and not be automatically disconnected by an overcurrent device. When the pump is needed, it is required that it run itself to destruction if necessary.

CHAPTER 28

Class 1, Class 2, Class 3 Circuits

OBJECTIVES

After completing this chapter, the student should understand

Class 1 circuits:
- power source requirements.
- overcurrent protection.
- overcurrent device locations.
- wiring methods.
- conductors of different circuits.
- circuit-conductors.

Class 2 and Class 3 circuits:
- power sources.
- circuit marking.
- wiring methods—supply side.
- wiring methods—load side.
- installation of conductors and equipment.
- application of listed Class 2, Class 3, and PTLC cables.
- listing and marking of Class 2, Class 3, and PTLC cables.

KEY TERMS

Class 1 Circuit: The portion of the wiring system between the load side of the overcurrent device or power-limited supply and the connected equipment. The voltage and power limitations are in accordance with *725.21*

Class 2 Circuit: The portion of the wiring system between the load side of a Class 2 power source and the connected equipment. Due to its power limitations, a Class 2 circuit considers safety from a fire initiation standpoint and provides acceptable protection from electric shock

Class 3 Circuit: The portion of the wiring system between the load side of a Class 3 power source and the connected equipment. Due to its power limitations, a Class 3 circuit considers safety from a fire initiation standpoint. Since higher levels of voltage and current than Class 2 are permitted, additional safeguards are specified to provide protection from an electric shock hazard that could be encountered

INTRODUCTION

Article 725 covers remote-control, signaling, and power-limited circuits that are not an integral part of a device or appliance. The circuits described herein are characterized by usage and electrical power limitations that differentiate them from electric light and power circuits. Alternative requirements are given with regard to minimum wire sizes, derating factors, overcurrent protection, insulation requirements, wiring methods, and materials.

The rules or requirements for Class 1, 2, and 3 circuits are less restrictive with regard to overcurrent protection, raceway fill, and derating. Careful thought must be given to determine the power-limiting requirements for these circuits.

GENERAL

Class 1, **Class 2**, and **Class 3 circuits** must be installed in a neat and workmanlike manner. Cables and conductors, exposed on the surfaces of ceilings and sidewalls, shall be supported by structural components of the building in such a manner that cable and conductors shall not be damaged by normal building use. Class 1, Class 2, and Class 3 conductors are generally run in cable systems and do not have the protection of a raceway. This makes it *extremely important* that the installation be done in a workmanlike manner. A typical Class 1 circuit is shown in Figure 28-1.

In Figure 28-1, the remote-control circuit supply is provided by a circuit breaker other than the motor

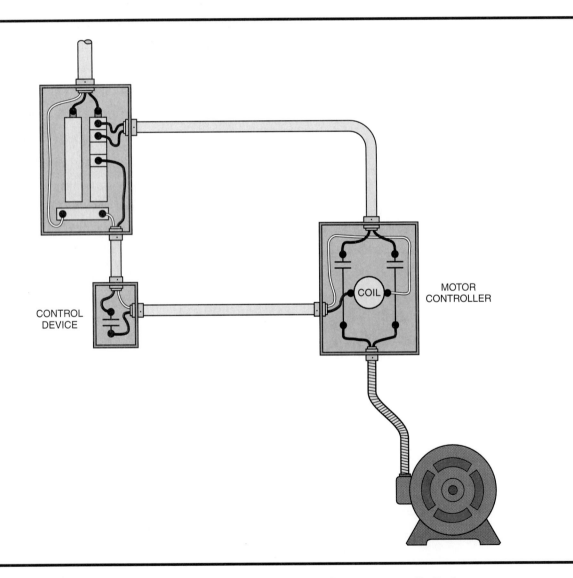

Figure 28-1 Class 1 remote-control circuit—non-power limited

circuit supply. It is from a non-power limited supply, and the output voltage must not be more than 600 volts. It must comply with all the requirements of *NEC*, Chapter 3.

A typical power-limited Class 1 circuit is shown in Figure 28-2. The power supply is from a power-limiting source of 1000 volt-amperes or less, and the output voltage is more than 30 volts but less than 600 volts.

In Figure 28-2, the power supply is provided by a transformer that is rated less than 1000 volt-amperes. This power-limited circuit is fed from a 120/24-volt, 750-volt-ampere transformer. This is a common application for providing feeders to damper motors used in HVAC systems.

NEC® 725.8 contains the requirements for safety-control equipment. Remote-control circuits for safety-control equipment shall be classified as Class 1 if the failure of the equipment to operate introduces a direct fire or life hazard. Figure 28-3 shows a steam boiler with a safety control (high-pressure switch) to de-energize the power supply if the boiler pressure reaches a set pressure. This is a Class 1 non-power limited circuit.

Figure 28-3 also shows a Class 2 power-limited circuit. The 120/24-volt transformer is rated at 50 volt-amperes which is less than the 100 volt-ampere maximum permitted for Class 2 circuits rated 30 volts or less.

Figure 28-4 is an example of another Class 2 circuit.

Room thermostats, water temperature regulating devices, and similar controls used in conjunction with electrically controlled household heating and air-conditioning shall not be considered safety-control equipment. These circuits originate in an inherently power-limited transformer source and are generally considered to be Class 2 circuits.

Where damage to remote-control circuits of safety-control equipment would introduce a hazard,

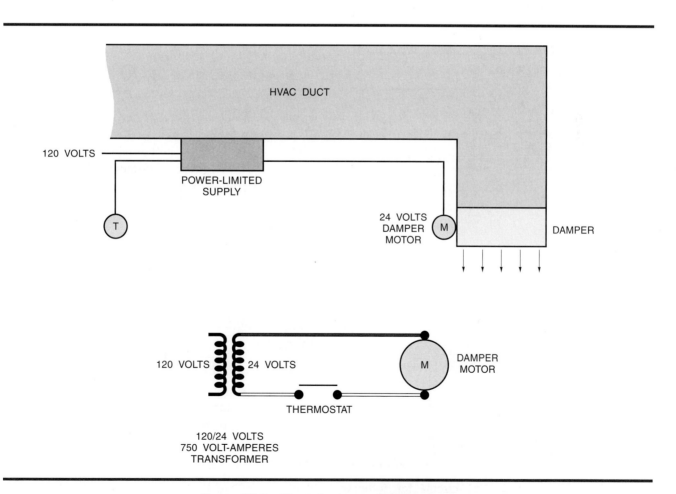

Figure 28-2 Class 1 power—limited circuit

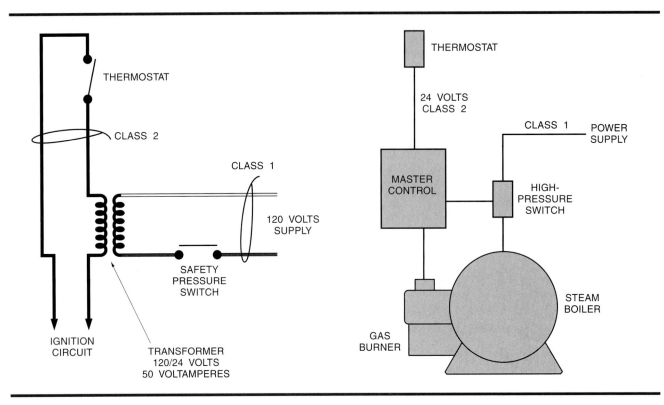

Figure 28-3   Class 1 safety control circuit; Class 2 power-limited thermostat circuit

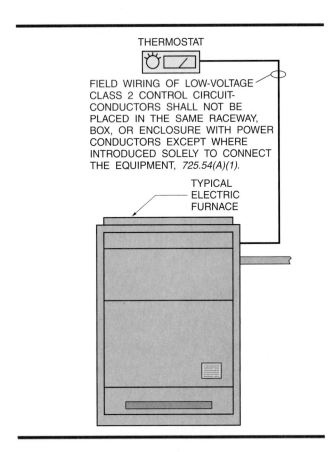

Figure 28-4   Class 2 control circuit for furnace thermostat

all conductors of such remote-control circuits shall be installed in RMC, IMC, RNC, EMT, Type MI cable, Type MC cable, or be otherwise suitably protected from physical damage.

Class 1, Class 2, and Class 3 circuits shall be identified at terminal and junction locations in a manner that will prevent unintentional interference with other circuits during testing and servicing.

Class 1 Circuits

Power Source Requirements. *NEC® 725.21* contains the power source requirements. The power source of Class 1 remote-control or signaling circuits shall not be more than 600 volts and must comply with the requirements of *NEC,®* Chapters 1 through 4.

Overcurrent Protection. In accordance with *725.23,* overcurrent protection for conductors 14 AWG and larger shall be provided in accordance with the conductor ampacity without applying the derating factors of *310.15,* relating to more than three current-carrying conductors in a raceway or a cable. Overcurrent protection shall not exceed 7 amperes for 18 AWG conductors and 10 amperes for 16 AWG conductors. These ampacities are based on those found in *Table 400.5(A)*.

Overcurrent Device Location. *NEC® 725.24* contains the requirements for Class 1 circuit overcurrent device location. Overcurrent devices shall be located where the conductor to be protected receives its supply.

Class 1 feeder-circuit conductors are permitted to be tapped, without overcurrent protection at the tap, where the circuit overcurrent device is sized to protect the tap.

Class 1 circuit-conductors (14 AWG and larger) that are tapped from the load-side of the overcurrent protective device of a controlled light and power circuit shall require only short-circuit and ground-fault protection and shall be permitted to be protected by the branch-circuit protective device where the rating is not more than 300% of the ampacity of the Class 1 circuit conductor. Figure 28-5 illustrates Class 1 conductors tapped from a lighting power source to control a lighting contactor by means of a photo-cell control. No overcurrent devices are required for this tap in accordance with *725.24(C)*. The 20-ampere overcurrent provided for the lighting contactor is not more than 300% of the ampacity of the Class 1 conductors.

Class 1 circuit-conductors supplied by the secondary of a single-phase transformer, having only a 2-wire (single-voltage) secondary, shall be permitted to be protected by the branch-circuit overcurrent protection provided on the primary of the transformer, provided this protection is in accordance with *450.3* and does not exceed the value determined by multiplying the secondary conductor ampacity by the secondary to primary transformer voltage ratio. Transformer secondary conductors, other than 2-wire, are not considered protected by the primary overcurrent protection (Figure 28-6).

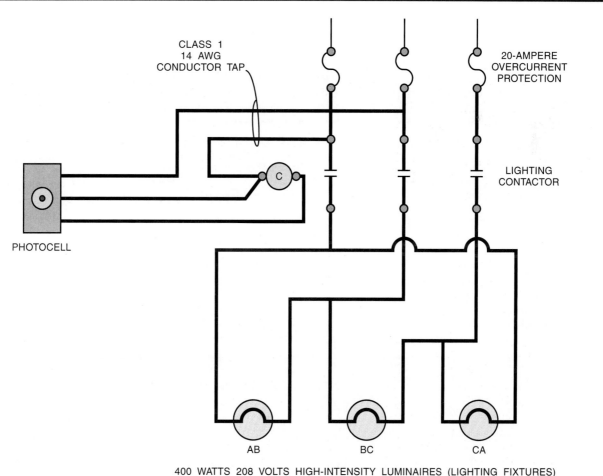

Figure 28-5 Class 1 conductors tapped from power source

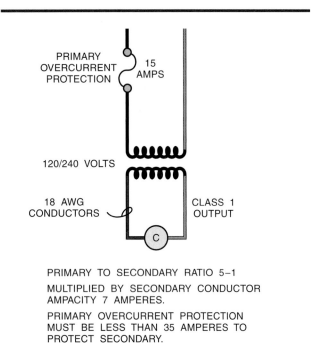

PRIMARY TO SECONDARY RATIO 5–1
MULTIPLIED BY SECONDARY CONDUCTOR
AMPACITY 7 AMPERES.

PRIMARY OVERCURRENT PROTECTION
MUST BE LESS THAN 35 AMPERES TO
PROTECT SECONDARY.

Figure 28-6 Two-wire secondary protected by primary overcurrent protection

Wiring Methods. *NEC® 725.25* requires that installations of Class 1 circuits be in accordance with appropriate articles in *NEC,® * Chapter 3. While this section does not enumerate which sections of Chapter 3 are appropriate, it is understood that splices and taps must be made in accordance with *300.15* and that the installation and use of boxes must conform to *Article 370*. Requirements regarding the use and installation of raceways and cable are found in Chapter 3 of the *NEC.®* Figure 28-7 shows a lower voltage Class 1 signaling system for a nurse's call system used in health-care facilities.

Conductors of Different Circuits. *NEC® 725.26* covers the requirements for conductors of different circuits in the same cable, cable tray, enclosure, or raceway. Two or more Class 1 circuits shall be permitted to occupy the same cable, cable tray, enclosure, or raceway without regard as to whether the individual circuits are ac or dc, provided all conductors are insulated for the maximum voltage of any conductor in the cable, cable tray, enclosure, or raceway.

Class 1 circuits are permitted to be installed with power conductors in a cable, enclosure, or raceway only where the equipment powered is functionally associated. Figure 28-8 shows Class 1 conductors in a motor-controller enclosure with power conductors as permitted by *725.26* where the equipment powered is functionally associated. In this illustration, the Class 1 conductors are used to feed the coil

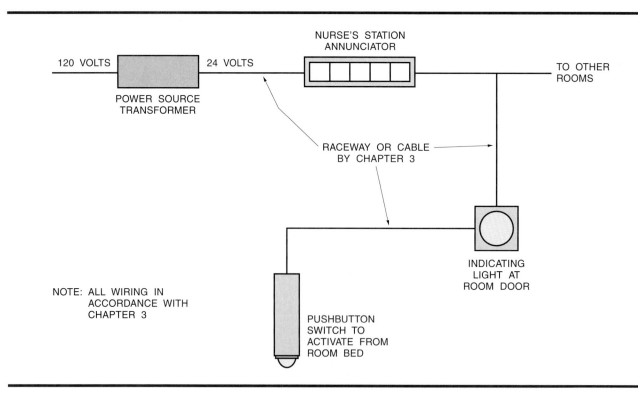

Figure 28-7 Nurse call system—Class 1 circuit operating at 24 volts, *725.25*

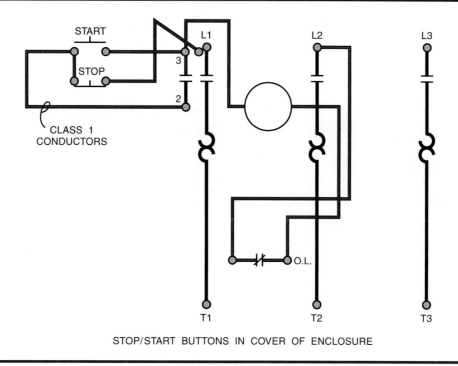

STOP/START BUTTONS IN COVER OF ENCLOSURE

Figure 28-8 Class 1 conductors functionally associated with power conductors permitted in same enclosure, 725.26

control circuit through the stop/start buttons. Class 1 circuits are permitted to be installed in factory or field-assembled control centers.

Class 1 circuits and power-supply circuits are permitted as underground conductors in a manhole if the Class 1 or power-supply conductors are in a metal-enclosed cable or Type UF cable; the conductors are permanently separated from the power-supply conductors by a continuous firmly fixed nonconductor, such as flexible tubing, in addition to the insulation on the wire. The conductors are permanently and effectively separated from the power-supply conductors and securely fastened to racks, insulators, or other approved supports.

Class 1 circuit-conductors may be run in cable trays where the Class 1 circuit-conductors and power-supply conductors, not functionally associated with the Class 1 circuit-conductors, are separated by a solid fixed barrier of a material compatible with the cable tray or where the power-supply or Class 1 circuit-conductors are in a metal-enclosed cable (Figure 28-9).

Class 1 Circuit Conductors. *NEC® 725.27* contains the requirements for conductors to be used in Class 1 circuits. Conductors of sizes 18 AWG and

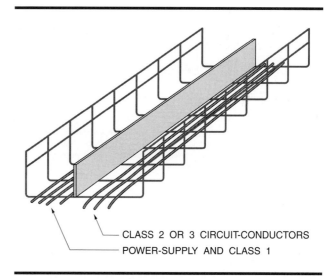

Figure 28-9 Solid fixed barrier separating power and Class 2 and 3 circuits in a cable tray

16 AWG are permitted to be used, provided they supply loads that do not exceed the ampacities shown in *400.5* and are installed in a raceway, an approved enclosure, or a listed cable. Conductors larger than 16 AWG shall not supply loads greater than the ampacities shown in *310.15*. Flexible cords shall comply with *Article 400* (Figure 28-10).

Table 400.5(A) Allowable Ampacity for Flexible Cords and Cables [Based on Ambient Temperature of 30°C (86°F). See 400.13 and Table 400.4.]

Size (AWG)	Thermoplastic Types TPT, TST	Thermoset Types C, E, EO, PD, S, SJ, SJO, SJOW, SJOO, SJOOW, SO, SOW, SOO, SOOW, SP-1, SP-2, SP-3, SRD, SV, SVO, SVOO / Thermoplastic Types ET, ETLB, ETP, ETT, SE, SEW, SEO, SEOW, SEOOW, SJE, SJEW, SJEO, SJEOW, SJEOOW, SJT, SJTW, SJTO, SJTOW, SJTOO, SJTOOW, SPE-1, SPE-2, SPE-3, SPT-1, SPT-1W, SPT-2, SPT-2W, SPT-3, ST, SRDE, SRDT, STO, STOW, STOO, STOOW, SVE, SVEO, SVT, SVTO, SVTOO		Types HPD, HPN, HSJ, HSJO, HSJOO
		A+	B+	
27*	0.5	—	—	—
20	—	5**	***	—
18	—	7	10	10
17	—	—	12	13
16	—	10	13	15
15	—	—	—	17
14	—	15	18	20
12	—	20	25	30
10	—	25	30	35
8	—	35	40	—
6	—	45	55	—
4	—	60	70	—
2	—	80	95	—

*Tinsel cord.
**Elevator cables only.
***7 amperes for elevator cables only; 2 amperes for other types.
+The allowable currents under subheading A apply to 3-conductor cords and other multiconductor cords connected to utilization equipment so that only 3 conductors are current-carrying. The allowable currents under subheading B apply to 2-conductor cords and other multiconductor cords connected to utilization equipment so that only 2 conductors are current carrying.

Figure 28-10 NEC® Table 400.5(A) (Reprinted with permission from NFPA 70-2002)

Number of Conductors in Cable Trays and Raceways, and Derating.
NEC® 725.28 gives the requirements for Class 1 circuit-conductors in cable trays and raceways. Where only Class 1 conductors are in a raceway, the number of conductors are determined in accordance with 300.17, which gives the basic requirements for the number and size of conductors in a raceway. The derating factors given in 310.15(B)(2) shall only apply if the conductors carry continuous loads in excess of 10% of the ampacity of the conductor (Figure 28-11).

Where power-supply and Class 1 circuit-conductors are permitted in a raceway, the number of conductors permitted shall be in accordance with 300.17. The derating factors given in 310.15(B)(2)(a) shall apply to all conductors where the Class 1 circuit-conductors carry continuous loads in excess of 10% of the ampacity of each conductor and where the total number of conductors is more than three. Derating factors shall apply to the power-supply conductors only where the Class 1 circuit-conductors do not carry continuous loads in excess of 10% of the ampacity of each conductor and where the number of power-supply conductors is more than three.

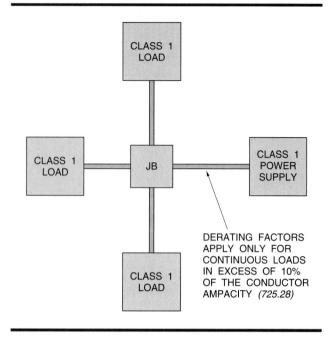

Figure 28-11 Derating factors not applicable

Class 2 and Class 3 Circuits

Power Sources. Power sources for Class 2 or a Class 3 circuit shall be a listed Class 2 or Class 3 transformer; a listed Class 2 or Class 3 power supply; listed information technology equipment limited power circuits; a dry-cell battery shall be considered an inherently limited Class 2 power source, provided the voltage is 30 volts or less, and the capacity is equal to or less than that available from series connected No. 6 carbon zinc cells. Class

2 or Class 3 power sources shall not have the output connections paralleled or otherwise interconnected unless listed for that purpose.

Circuit Marking. Class 2 or Class 3 power sources shall be durably marked where plainly visible to indicate each circuit that is a Class 2 or Class 3 circuit (Figure 28-12).

Wiring Methods on the Supply Side. In accordance with *725.51*, conductors and equipment on the supply side of the power source for Class 2 or Class 3 circuits shall be installed in accordance with the appropriate requirements of *NEC,®* Chapters 1 through 4 (Figure 28-13).

An overcurrent device, rated at not more than 20 amperes, must be used to protect transformers or other devices supplied from electric light or power circuits.

Wiring Methods on the Load Side. In accordance with *725.52*, Class 2 and Class 3 circuits on the load side of the power source are permitted to be installed using wiring methods in accordance with *725.25*. Conductors on the load side of the power source shall be insulated in accordance with the

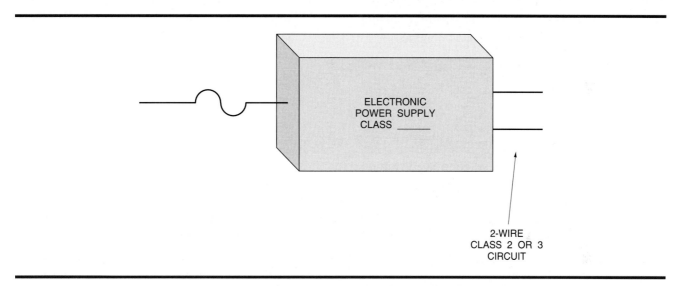

Figure 28-12 Class 2 or Class 3 power supplies marked

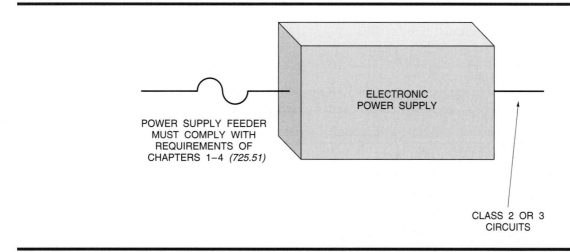

Figure 28-13 Supply wiring for power source for Class 2 or Class 3 circuits.

requirements of *725.71* and shall be installed in accordance with *725.54* and *725.61*.

Installation of Conductors and Equipment. Class 2 and Class 3 circuit-conductors shall be separated from electric light, power, Class 1, non-power-limited fire alarm circuit-conductors, and medium network-powered broadband communications cables in accordance with *725.54*.

Class 2 or Class 3 circuit-conductors must be installed in RMC, RNC, intermediate grade conduit, LFNC, or EMT in hoistways.

Application and Listing and Marking of Class 2 and Class 3 Conductors

The requirements for the application, listing, and marking of Class 1 and Class 2 conductors are located in *725.61* and *725.71*.

SUMMARY

Class 1 Circuit

Class 1 circuits are divided into two types: (1) power-limited or (2) remote-control and signaling circuits.

Power-Limited Circuits. Power-limited circuits are circuits that are used for functions other than remote-control or signaling but in which the power supply is limited to specified limits. The power-limited circuit is that portion of the circuit between the load side of the power supply and the equipment powered. An example of a Class 1 power-limited circuit is shown in Figure 28-2. This drawing demonstrates the use of a Class 1 power-limited circuit derived from a 120/24-volt transformer rated at 750 volt-amperes to supply power to damper motors rated at 24 volts.

Another application of a Class 1 power-limited circuit is for lighting systems operating at 30 volts or less. Figure 28-14 illustrates the use of landscape lighting rated at 12 volts fed from a 120/12-volt transformer rated at 1000 volt-amperes or less. Following are six general rules relating to Class 1 power-limited circuits:

1. Limited to 30 volts or less.
2. 1000 volt-amperes or less.
3. Power-supply transformer protected per *Article 450*.
4. Power sources other than transformers limited to 2500 volt-amperes.
5. Power sources other than transformers, fault-current limited to 10,000 volt-ampere for one minute or less.
6. Power sources other than transformers—overcurrent protection limited to 167% of volt-amperes of source divided by volts.

Remote Control and Signaling Circuits

1. 600 volts or less.
2. No limit on power source volt-amperes.
3. Transformers per *Article 450*.

Class 2 Circuits

0 to 20 volts	100 volt-amperes	5 amperes
21 to 30 volts	100 volt-amperes	3.3 amperes
31 to 150 volts	0.5 volt-amperes	5 milliamperes

Class 3 Circuits

31 to 100 volts	100 volt-amperes (inherently-limited)
31 to 150 volts	100 volt-amperes (not inherently-limited)

CHAPTER 28 Class 1, Class 2, Class 3 Circuits 455

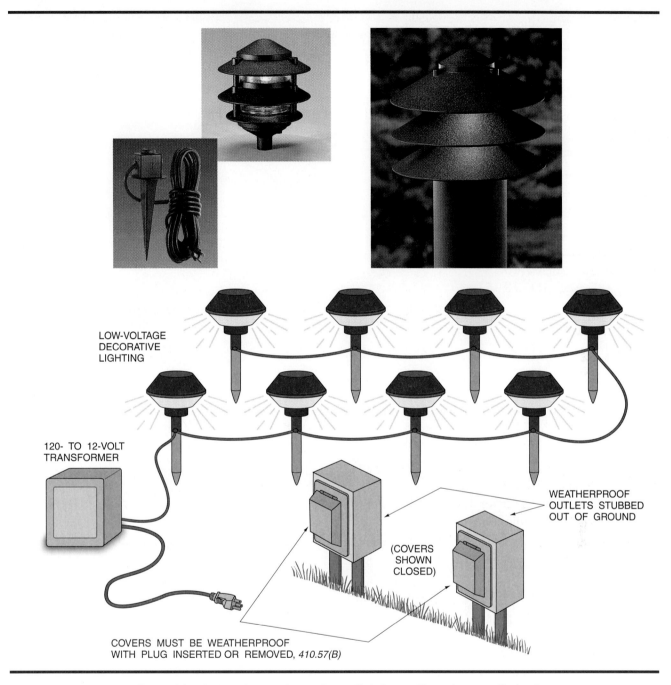

Figure 28-14 Landscape low-voltage lighting—Class 1 power-limited

REVIEW QUESTIONS

MULTIPLE CHOICE

1. Class 1 remote-control circuits shall not exceed _____ volts.
   - ____ A. 24
   - ____ B. 50
   - ____ C. 120
   - ____ D. 600

2. Class 1 remote-control circuit-conductors shall not be smaller than _____ AWG.
   - ____ A. 12
   - ____ B. 14
   - ____ C. 16
   - ____ D. 18

3. The current-carrying capacity for 16 AWG conductors shall not exceed _____ amperes.
   - ____ A. 10
   - ____ B. 7
   - ____ C. 12
   - ____ D. 14

4. The derating factors given in *310.15(B)(2)* shall only apply if the conductors carry continuous loads in excess of _____ % of the ampacity of the conductor.
   - ____ A. 14
   - ____ B. 10
   - ____ C. 125
   - ____ D. none of the above

5. In hoistways, Class 2 and Class 3 circuit-conductors shall be installed in:
   - ____ A. RMC.
   - ____ B. RNC.
   - ____ C. Intermediate Grade Conduit.
   - ____ D. all of the above.

TRUE OR FALSE

1. The power source of Class 1 remote-control or signaling circuits shall not be more than 300 volts.
   - ____ A. True
   - ____ B. False

   Code reference _____ .

2. Class 1 circuits are permitted to be installed with power conductors in a raceway, only where the equipment powered is functionally associated.
   - ____ A. True
   - ____ B. False

   Code reference _____ .

3. Remote-control circuits for safety-control equipment shall be classified as Class 2 if the failure of the equipment to operate introduces a direct fire or life hazard.
   ____ A. True
   ____ B. False
   Code reference _____ .

4. Class 1 feeder circuit-conductors are permitted to be tapped, without overcurrent protection at the tap, where the circuit overcurrent device is sized to protect the tap.
   ____ A. True
   ____ B. False
   Code reference _____ .

5. The derating factors given in *310.15(B)(2)(a)* shall apply to all conductors where the Class 1 conductors carry continuous loads in excess of 15% of the ampacity of each conductor and where the total number of conductors is more than three.
   ____ A. True
   ____ B. False
   Code reference _____ .

FILL IN THE BLANKS

1. Room thermostats, water temperature regulating devices, and similar controls used in conjunction with electrically controlled household heating and air-conditioning shall not be considered _____ equipment.

2. Class 1, Class 2, and Class 3 circuits _____ at terminal and junction locations in a manner that will prevent unintentional interference with other circuits during testing and servicing.

3. Where power-supply and Class 1 circuit-conductors are _____ , the number of conductors permitted shall be in accordance with *300.17*.

4. Class 2 and Class 3 circuits _____ of the power source shall be permitted to be installed using wiring methods in accordance with *725.25*.

5. Class 2 and Class 3 circuit-conductors _____ from electric light, power, Class 1, non-power-limited fire alarm circuit-conductors, and medium network powered broadband communications cables.

FREQUENTLY ASKED QUESTIONS

Question 1: What is the difference between a signal circuit and a remote-control circuit?
Answer: A signal circuit powers a device that gives an audible or visual alarm indication. A doorbell or enunciator are signaling devices. A remote-control circuit exercises control over a device to control other circuits. A motor control or relay are remote-control devices.

Question 2: Are we required to follow the requirements of *NEC®* Chapter 3 when Class 2 and Class 3 circuits are installed in raceway systems that are covered in Chapter 3?
Answer: Yes. There are no requirements in *Article 725* that would exempt Class 2 and Class 3 wiring in raceways from the requirements of Chapter 3.

Question 3: What is a low-voltage circuit? The *NEC*® does not seem to give a clear answer to this.

Answer: The *NEC*® does not define low voltage. Generally speaking, the *NEC*® requirements classify circuits as being *600 volts or less* or *over 600 volts*. In *110.26(A)(1)(b)*, low voltage is shown as 30 volts root-mean-square ac or 60 volts dc. *Article 720* has requirements for circuits and equipment operating at less than 50 volts, and *Article 411* has requirements for lighting systems operating at 30 volts or less. *Article 725* generally uses 30 volts maximum for power-limited Class 1 circuits but permits non-power-limited Class 1 circuits of 600 volts or less.

Question 4: Where can Class 3 circuits be used?

Answer: Class 3 circuits are used where higher power capacity is required than can be obtained when using a Class 2 circuit. For example, a nurses' call system could use a 100-volt-ampere transformer power source over 30-volts as opposed to using a Class 2 power source limited to 0.5 volt-amperes over 30 volts.

Question 5: What is meant by safety-control equipment in relation to remote-control circuits?

Answer: *NEC*® 725.8 contains the requirements for safety-control equipment. Safety-control equipment could be a pressure switch on a steam boiler or similar control. Remote-control circuits for safety-control equipment are classified as Class 1 circuits if the failure of the control equipment introduces a direct fire or life hazard.

Where damage to remote-control circuits of safety-control equipment, all conductors of such remote-control circuits must be installed in RMC, IMC, RNC, EMT, Type MI cable, Type MC cable, or be otherwise suitably protected from physical damage.

APPENDIX A

Inductive Reactance

When current flows through a coil, a magnetic field is set up around the coil. If the current is alternating, the magnetic field will vary (buildup and collapse). The buildup and collapse of the magnetic field will, in itself, produce another current in the coil. This self-induced current is in opposition to the initial current. This opposition is known as *inductive reactance* and is expressed in ohms. Its symbol is X_L.

Capacitive Reactance

Alternating current will pass around a capacitor, but the process of alternately charging and discharging the capacitor produces some opposition. This opposition is termed capacitive reactance and is expressed in ohms. Its symbol is X_C.

Impedance

Unlike dc circuits that contain purely resistive opposition, ac circuits contain inductive reactance and capacitive reactance as well as resistance opposition to current flow. AC circuits may contain only inductive reactance, only capacitive reactance, or only resistance, but this is highly unlikely and would more probably contain a combination of all of them. This opposition in ac circuits is called *impedance*. Its symbol is Z.

Phase Angle

In alternating circuits, the voltage and current need not be in step. The voltage may lead the current or the current may lead the voltage. The amount of lead or lag is termed *phase angle* and is expressed by the Greek letter theta (θ). In a purely resistive circuit, there will be no lead or lag and the voltage and current will be in phase. The phase angle is θ.

In a pure inductive circuit, the voltage leads the current by 90°. The phase angle is +90°. In practical use, most ac circuits are a combination of all three factors, and the ratio of reactance to resistance determines the phase angle. This may be calculated vectorially or by means of trigonometry.

Power Factor

The power factor of a circuit is directly related to resistance, reactance, impedance, and phase angle. In any reactive circuit, the current cannot be in phase with the voltage because of the reactance. The amount of angular displacement between the current and the voltage depends on the ratio of resistance to reactance.

Because the voltage and current do not reach maximum at the same instant, the real power of the circuit must be $E \times I$ multiplied by some factor less than 1. The factor is called the *power factor* of the load and is equal to the cosine θ of the angle between the voltage and the current. Therefore:

$$\text{Power Factor} - \text{cosine } \theta = \frac{R}{Z}$$

The term *apparent power* is applied when the reactive power factor of an ac circuit is disregarded. To obtain true power in an ac circuit, it is necessary to multiply the apparent power by the cosine of the phase angle. Since the phase angle is always less than 90° and the corresponding cosine is a fraction of 1, the true power will always be less than the apparent power. The ratio between true power and apparent power is known as power factor and can be expressed as:

$$\text{Power Factor} = \text{cosine phase angle} = \frac{R}{Z}$$

APPENDIX B

Transformer Basics

- The voltage of a transformer is stepped up when the secondary has more turns than the primary.
- The voltage of a transformer is stepped down when the primary has more turns than the secondary.
- The current across the terminals of a transformer is stepped up when the primary has more turns than the secondary.
- The current across the terminals of a transformer is stepped down when the secondary has more turns than the primary.

- **Example:** A transformer has a primary voltage of 480 volts and 100 turns. How many turns must the primary have for a 240-volt secondary? The voltage and turns ratio are directly proportional. The voltage ratio is 480 to 240 or 2 to 1. 1 divided by 2 equals .50. 100 turns times .50 equals 50 turns on the secondary.
- **Example:** A transformer has a primary voltage of 240 volts and 300 turns. How many turns must the secondary have for a 480-volt secondary? The voltage and turns ratio are directly proportional. The voltage ratio is 240 to 480 or 1 to 2. 2 divided by 1 equals 2. 300 turns times 2 equals 600 turns on the secondary.

- The current is inversely proportional to the voltage. When the voltage is increased, the current is decreased.
  - **Example:** A 75-kVA, 3-phase, 4-wire, 480/208-120 V transformer has a full-load primary current of:

  $$I = \frac{W}{E \times 1.73} \quad \text{or} \quad I = \frac{75{,}000}{830} = 90.3 \text{ A}$$

  The secondary current is:

  $$I = \frac{W}{E \times 1.73} \quad \text{or} \quad I = \frac{75{,}000}{360} = 208.3 \text{ A}$$

  The voltage ratio is 480 to 208 or 480 divided by 208 or 2.33 to 1.

CODE INDEX

Numbers printed in **bold** represent NEC references included in figures

A
ANSI 17.1, 410

ARTICLES
Article 100, **186**, 220, **221**, **348**
    Definitions, 189
Article 210, 25
Article 220, 28, 149, 155, 198, 350
    Part II, 155, 157
Article 230, 270, 312
Article 240, 215
Article 250, 288, 340, 388, **400**, 416
    Part E, **110**, 428
Article 334, 105
Article 336, 427
Article 342, 107, 120
Article 342, Table 4, Intermediate metal conduit, 200, 202
Article 344, 107, 118
Article 348, 124
Article 350, 126
Article 352, 107, 126
Article 356, 128
Article 358, 107, 122
Article 360, 122
Article 362, 107, 127
Article 370, 450
Article 376, Metal Wireways, 405
Article 378, Nonmetallic Wireways, 405
Article 400, **400**, 418, 451
Article 410, 399, **399**
Article 410, Part XIV, 367
Article 422, 136
Article 422, Part III, 136
Article 422.61, 136
Article 430, 183
Article 430, Part J, **186**
Article 450, 164, 454
    Transformers, 198
Article 500, 233
Article 501, 233, 234, 384
Article 502, 233, 384
Article 503, 233, 384
Article 504, 233, 241
Article 505, 233, 242
Article 506, 233
Article 507, 233
Article 508, 233
Article 509, 233
Article 510, 233
Article 511, 233, 242
Article 512, 233
Article 513, 233
Article 514, 233, 242, 354
Article 515, 233
Article 516, 233, 244
Article 517, 294, 302
Article 517, Part II, 302
Article 525, 282, 283, 290
Article 527, 282, 288, 290, 353
Article 533, **348**
Article 550, 310, 322
Article 550.18(B), 314
Article 551, 334
Article 552, 334
Article 553, 347
Article 555, 351, 354
Article 600, 367
Article 604, 375
Article 610, 383
Article 610, Part II, 385
Article 610, Part IV, 387
Article 610, Part V, 387
Article 610, Part VI, 388
Article 610, Part VII, 388
Article 620, 398, **399**
Article 620, Part II, 400
Article 625, 415, 421
Article 645, 425, 428
Article 645.2, 425
Article 680, 250
Article 690, 273, 275, **275**
Article 690, Part C, **276**
Article 690, Part E, **275**, **276**
Article 692, 276
Article 695, 433, 440
Article 700, 268
Article 701, 268
Article 702, 268
Article 705, 268
Article 725, 427, 446
Article 760, 427
Article 770, 427
Article 800, 427
Article 820, 427

CHAPTERS
Chapter 3, 252, 260, 322, 447, 450
Chapter 9, Table 1, 83, **110**
Chapter 9, Table 4, **110**
Chapter 9, Table 5, **110**
Chapter 9, Table 8, **110**

FIGURES
Figure 410.8, 101
Figure 517.30, No. 1, Fine Point Notes, Hospital multiple transfer switches, **299**
Figure 517.30, No. 2, Fine Point Notes, Hospital single transfer switch, **299**
Figure 517.41, No. 1, Fine Point Notes, Nursing homes and limited-care facilities, **300**
Figure 517.41(B), No. 2, Fine Point Notes, Nursing homes and limited-care facilities, **300**
Figure 680.8, **251**

NEMA
NEMA 30A, **355**
NEMA 50A, **355**
NEMA 60A, **355**
NEMA 100A, **355**

NFPA
NFPA 20, 433
    Standard for the Installation of Centrifugal Fire Pumps, 437
NFPA 20-1999, 433
NFPA 70, 433, 437
NFPA 101-2000, Life Safety Code, 268
NFPA 501C-1996, Standards on Recreational Vehicles, 334

SECTIONS
Section 110.4, 218
Section 110.9, **110**, 215, **273**
Section 110.10, **110**, 215, 221, 226
Section 110.14, 40
Section 110.26(A), 399, **399**
Section 110.27, 179
Section 110.34, 179
Section 200.6, 349
Section 210.7(A), 29
Section 210.7(B), 29
Section 210.8, 30
Section 210.8(A)(1), 53
Section 210.8(A)(2), 29, 56, **57**
Section 210.8(A)(3), 29, 54, 58, **104**
Section 210.8(A)(5), 29, 49
Section 210.8(A)(6), 50, 51, **51**
Section 210.11(C)(1), 31, 49
Section 210.11(C)(2), 31, 54, **56**

CODE INDEX

Section 210.11(C)(3), 28, 32
Section 210.19(C), 34
Section 210.19(C), Exception No. 2, 34
Section 210.20(A), 157
Section 210.23(A), 32
Section 210.52, 46
Section 210.52(A), 96
Section 210.52(A)(1), 49
Section 210.52(B), 31, 98
Section 210.52(B), Exception No. 2, 31
Section 210.52(B)(1), Exception No. 2, 49
Section 210.52(C), 98, 100
Section 210.52(C)(3), **52**
Section 210.52(D), 100
Section 210.52(E), 29, 104, **104**
Section 210.52(F), 100
Section 210.52(G), 29, **57**, **60**, 103
Section 210.52(H), **59**
Section 210.63, **59**
Section 210.70, 96, 99, 100, 104
Section 210.70(A), **60**
Section 210.70(A)(1), 59, 96
Section 210.70(A)(2), 59, **60**, 104
Section 210.70(A)(2)(c), **61**, 61, 96
Section 215.2(A), 201
Section 220.3(A), 28, 157, 201, 209, 210
Section 220.3(B)(1), 38
Section 220.3(B)(6), 198, 362
Section 220.3(B)(7), 198
Section 220.3(B)(9), 198, 209, 210
Section 220.3(B)(10), 28, **28**
Section 220.3(C), 38
Section 220.4, 32
Section 220.11, 34, 38
Section 220.12(A), 198
Section 220.12(B), 198
Section 220.15, 38, 201
Section 220.16(A), 31, 34, 37
Section 220.16(B), 31, 34, 37
Section 220.17, 38
Section 220.18, 37, 38
Section 220.19, 35, 38
Section 220.20, 156, 201
Section 220.21, 38
Section 220.22, 37, 38, 39, 157, 208
Section 220.30, 150, 151
Section 220.31, 152
Section 220.41(B), 206
Section 225.18, **111**, 285
Section 230.2, 434
Section 230.6, 439
Section 230.24, **110**
Section 230.24(B), **111**
Section 230.26, **110**
Section 230.28, **110**

Section 230.30, **110**
Section 230.41, **110**
Section 230.42, **110**
Section 230.42(B), 37
Section 230.54, **110**
Section 230.70, **110**
Section 230.70(A), **110**
Section 230.71, 83, **110**
Section 230.72, **110**
Section 230.72(B), 435
Section 230.73, **110**
Section 230.74, **110**
Section 230.75, **110**
Section 230.76, **110**
Section 230.77, **110**
Section 230.78, **110**
Section 230.79, **110**, 152
Section 230.79(C), 37, 111
Section 230.80, **110**
Section 230.81, **110**
Section 230.82, **110**, **273**
Section 230.83, **110**
Section 230.84, **110**
Section 230.85, **110**
Section 230.86, **110**
Section 230.87, **110**
Section 230.88, **110**
Section 230.89, **110**
Section 230.90, **110**
Section 230.91, **110**
Section 230.92, **110**
Section 230.93, **110**
Section 230.94, **110**
Section 230.95, **110**
Section 240.3(B), 83
Section 240.4(B), 178, 202, 220
Section 240.6, 83, 178, 185, 202, 220
Section 240.21(C), 170
Section 240.21(C)(2), 178
Section 240.24(E), **110**
Section 240.81, **221**
Section 240.83(A), **221**
Section 240.83(B), **221**
Section 240.83(C), **221**
Section 240.83(E), **221**
Section 240.86, 225
Section 250.2(B), 73
Section 250.2(C), 73
Section 250.20(B), 70
Section 250.22, 70, 384
Section 250.24(B), 73
Section 250.26, 72
Section 250.28, **110**
Section 250.28(D), 80
Section 250.30, 179
Section 250.30(A)(1), 179
Section 250.34(B), 290

Section 250.50, 76, 81, **110**, 179
Section 250.50(A), **110**
Section 250.50(A)(2), 82
Section 250.52, **110**, 179
Section 250.54, **75**
Section 250.56, 78, **80**
Section 250.64, 80
Section 250.64(B), 80, 84, **110**
Section 250.64(C), **110**
Section 250.68, **110**
Section 250.70, **110**
Section 250.92, **110**
Section 250.93, **110**
Section 250.94, **110**
Section 250.95, **110**
Section 250.96, **110**
Section 250.97, **110**
Section 250.98, **110**
Section 250.99, **110**
Section 250.100, **110**
Section 250.101, **110**
Section 250.102, **110**
Section 250.103, **110**
Section 250.104, **110**
Section 250.118, 289
Section 250.118(5), 124
Section 250.118(6), 124
Section 250.122, 289
Section 250.142, **110**
Section 300.4, 105
Section 300.5, 208, 289
Section 300.17, 452
Section 300.18(A), **107**, **120**
Section 300.22, 425, 428
Section 305.5, 285
Section 310.15, 448, 451
Section 310.15(B)(2), 452
Section 310.15(B)(2)(a), 452
Section 310.15(B)(6), 37
Section 312.6(C), **110**
Section 314.16, 106
Section 314.17, 105
Section 314.17(C), **105**
Section 330.22, 427
Section 342.30(A), **121**
Section 344.30(A), **121**
Section 350.60, 124, **125**
Section 358.30(A), **121**
Section 374.2, **403**, 405
Section 374.5, **403**, 405
Section 400.5, 451
Section 400.13, **452**
Section 404.3, **221**
Section 404.7, **221**
Section 406.4(E), **51**
Section 408.3(C), **110**
Section 408.16(A), 172

CODE INDEX 463

Section 408.16(B), 172
Section 408.16(C), 172
Section 408.16(D), 172, 173, 178
Section 408.20, **110**
Section 408.81, **221**
Section 410.8, 61
Section 410.8(A), **63, 102**
Section 410.8(B), **63, 102**
Section 410.8(C), **63, 102**
Section 410.8(D)(1), **63, 102**
Section 410.8(D)(2), **63, 102**
Section 410.8(D)(3), **63, 102**
Section 410.8(D)(4), **102**
Section 410.8(D)(5), **63**
Section 410.57(B), 455
Section 410.67(C), 124
Section 422.11(C), 140
Section 422.12, 31, 32, 141, **142**
Section 422.13, 139
Section 422.14, 141
Section 422.16(B)(1), 137, **137**
Section 422.16(B)(2), 137
Section 422.16(B)(3), 138, **139**
Section 422.31(B), **139**
Section 422.32(B), 139
Section 422.33(A), 136
Section 422.34(C), 139
Section 422.34(D), 139
Section 422.45, 141
Section 430.6, 184, 188
Section 430.6(A)(1), 187
Section 430.7(B), **188**
Section 430.22, 38, 185
Section 430.22(A), 187, 401
Section 430.22(E), 401
Section 430.24, 189
Section 430.32, **186**, 189, 220, 406
Section 430.32(A)(1), 186, 187
Section 430.33, 406, 407
Section 430.36, Fuses, 190
Section 430.37, Devices other than Fuses, 190
Section 430.52, 185, 188
Section 430.52(C), Exception No.1, **185**
Section 430.52(C), Exception No.2, **185**
Section 430.52(C), Exception No. 1, **218**
Section 430.52(C), Exception No. 2, **218**
Section 430.52, Exception No. 1, **408**
Section 430.52, Exception No. 2, **408**
Section 430.54, **218**, **388**, **408**
Section 430.55, **186**
Section 430.72, 189, 387
Section 430.72(B), 187

Section 430.72(B), Exception No. 2, 186
Section 430.83, **399**
Section 430.101, **186**
Section 430.102(A), 185
Section 430.102(B), 185, **186**
Section 430.110, 185
Section 430.110(A), **186**, 187
Section 440.4, 363
Section 450.3(B)(1), 178
Section 450.8(A), 179
Section 450.8(B), 179
Section 450.8(C), 179
Section 450.8(D), 179
Section 450.21(A), 179
Section 450.26, Exceptions, 173
Section 450.52(C), Exception No. 1, **388**
Section 450.52(C), Exception No. 2, **388**
Section 501.4(A), 235
Section 501.4(B), 238
Section 501.5(A), 236
Section 501.5(B), 238
Section 502.4(A), 239
Section 502.4(B), 239
Section 502.5, Fine Point Notes, 242
Section 503.3(A), 241
Section 503.10, **241**
Section 503.13, 70, 384
Section 516.3(B)(1), **245**
Section 516.3(B)(5), **245**
Section 517.19(A), **298**
Section 517.19(C), **298**
Section 517.60, 71
Section 517.160, 176
Section 525.5(B), 285, **288**
Section 525.11, 285
Section 525.20(G), 289
Section 525.21, 285, **287**, 289
Section 525.22, 283, 284
Section 525.23, 287
Section 525.31, 289
Section 527.3(A), 288
Section 527.3(B), **287**, 288, 289
Section 527.3(C), 289
Section 527.4(C), **287**
Section 527.4(F), **287**, 289
Section 550.5, 326
Section 550.10, 310
Section 550.10(A), **311**
Section 550.10(B), **311**
Section 550.10(C), **311**, **327**
Section 550.10(D), **311**
Section 550.10(E), **311**
Section 550.11, 312, **313**
Section 550.11(A), 313, **313**

Section 550.11(C), **313**
Section 550.12, **313**, 314
Section 550.12(A), 314
Section 550.12(B), 314
Section 550.12(C), 314
Section 550.12(D), 314
Section 550.12(E), 314
Section 550.12(F)(3), 321
Section 550.13, 314
Section 550.13(B), 314
Section 550.15, 322
Section 550.15(F), 322
Section 550.16, 323, **324**
Section 550.16(A)(1), **324**, 324
Section 550.16(C)(1), **324**
Section 550.16(C)(2), **324**
Section 550.18, 326
Section 550.18(B)(5), **325**
Section 550.19, 326
Section 550.32(A), **327**
Section 550.32(B), 324
Section 550.32(C), **327**
Section 550.32(D), 326, **327**
Section 550.32(F), **327**
Section 551.10, 334
Section 551.30, 339
Section 551.60, 337
Section 551.71, 205, 339
Section 551.72, 205
Section 551.73, 205
Section 551.73(A), 205
Section 551.77, 205
Section 553.1, **348**
Section 553.2, **348**
Section 553.4, 347
Section 553.6, 348
Section 553.7, 348
Section 553.8, 348
Section 553.10, 349
Section 555.10, 352
Section 555.11, 352
Section 555.12, 204, 353
Section 555.13, 353
Section 555.17, 354
Section 555.19, 354
Section 555.19(A)(4), 204
Section 555.19(B), 204
Section 600.3, 362
Section 600.4, 362
Section 600.5, 362
Section 600.5(A), 198, 368
Section 600.6, 363
Section 600.6(A), 368
Section 600.7, 363
Section 600.10, 363, 368
Section 600.21, 363
Section 600.22, 365

Section 600.31, 365
Section 600.32, 365
Section 600.32(A), 369
Section 600.32(J), 367
Section 600.42, 367
Section 605.2, 375
Section 605.3, 375
Section 605.4, 375
Section 605.5, 376
Section 605.6, 377
Section 605.7, 377
Section 605.8, 377
Section 610.11, 385
Section 610.14, 386
Section 610.42, 387, 388
Section 610.43, 387, 388
Section 620.1, **399**
Section 620.3(A), 398, **399**
Section 620.3(B), 398, **399**
Section 620.3(C), 398
Section 620.4, **399**
Section 620.5, 399, **400**
Section 620.5(A), **399**
Section 620.11, 400
Section 620.12(A), **399**
Section 620.13, 400
Section 620.15, **399**
Section 620.21, 402
Section 620.21(A), 402
Section 620.21(B), 402
Section 620.21(B)(1), **400**
Section 620.21(B)(1), Exception, **400**
Section 620.21(B)(2), **400**
Section 620.21(B)(3), **400**
Section 620.21(C), **400**, 402
Section 620.22, 403, 406
Section 620.22(A), **404**
Section 620.22(B), 403, **404**
Section 620.23, 406
Section 620.23(A), 403, **404**
Section 620.23(B), **404**
Section 620.23(C), **404**
Section 620.24, 404, 406
Section 620.24(A), **404**
Section 620.24(B), **404**
Section 620.24(C), **404**
Section 620.32, **403**, 405
Section 620.34, **403**
Section 620.35, **403**, 405
Section 620.36, **403**
Section 620.37, 402, 405
Section 620.37(A), **403**
Section 620.37(B), **403**
Section 620.37(C), **403**
Section 620.38, **403**
Section 620.41, 400, **404**, 405

Section 620.43, **404**
Section 620.44, **404**, 405
Section 620.51, 406, 409
Section 620.53, 406
Section 620.54, 406
Section 620.61, 406
Section 620.71, 407, 408
Section 620.81, **404**
Section 620.82, 408
Section 620.84, **400**, 408
Section 620.85, 404, 408
Section 620.91, 409
Section 625.9, 416
Section 625.13, 417
Section 625.14, 417
Section 625.15, 420
Section 625.15(B), 420
Section 625.15(C), 420
Section 625.18, 417, 422
Section 625.19, 417, 418, 422
Section 625.22, 418, 422
Section 625.23, 418
Section 625.25, 418, 422
Section 625.29, 417, 420
Section 627.17, 418
Section 645.2, 426, 428
Section 645.5, 426
Section 645.5(D), 426
Section 645.5(D)(5)(c), 426, 427
Section 645.6, 427
Section 645.10, 426, 427, 428
Section 645.11, 428
Section 668.3(C)(3), 71
Section 668.32, 385
Section 680.5(A), **257**
Section 680.6, 256
Section 680.7, 250
Section 680.10, 251
Section 680.20(B)(1), **257**
Section 680.21, 252
Section 680.21(A)(1), 252
Section 680.22, 250, 252
Section 680.22(B), 252
Section 680.23, 254
Section 680.23(A)(2), 176
Section 680.25(B)(1), **257**
Section 680.25(B)(4), **257**
Section 680.25(D), **257**
Section 680.26, 256
Section 680.27, 257
Section 695.3, 434
Section 695.4, 435
Section 695.5, 438
Section 695.6, 439
Section 695.7, 440
Section 705.40, 275

Section 725.8, 447
Section 725.23, 406, 448
Section 725.24, 406, 449
Section 725.24(C), 449
Section 725.25, 450, **450**, 453
Section 725.26, 450
Section 725.27, 451
Section 725.28, 452
Section 725.51, 453, **453**
Section 725.52
    Class 2, 453
    Class 23, 453
Section 725.54, 454
Section 725.54(A)(1), **448**
Section 725.61, 454
Section 725.71, 454

TABLES

Table 4, **200**
Table 4, Article 342, Intermediate metal
    conduit, 200, 202
Table 5, **200**
Table 220.3(A), 28, **28**, 29, 31, 35, 37,
    156, 198, 201
Table 220.11, **31**, 31, 34, 37, 209
Table 220.13, 201
Table 220.19, 34, **34**, **35**, 37, 38, 138
    Column C, 138
Table 220.19, Note 1, 138
Table 220.19, Note 4, **139**
Table 220.20, 156, **156**, 157
Table 220.32, 154, **154**
Table 220.34, **156**
Table 220.36, 157, **158**
Table 220.40, **208**
Table 220.41, **208**
Table 230.34, 155
Table 250.62, **110**
Table 250.64, **110**
Table 250.66, 79, 80, 82, 83, **110**
Table 250.122, 250, 252, **257**
Table 310.13, **166**, **167**, 168
Table 310.15(B)(6), 37, 38, 79, 80, **110**
Table 310.16, 34, **36**, 37, 38, 40, 79, 80,
    82, 83, 167, **168**, 185, **199**, 202
Table 310.17, 166, **166**, 167
Table 314.16(A), 94, **95**
Table 314.16(B), 96, **96**
Table 344.24, 107, 108, **108**, 119, **120**,
    125, 128
Table 344.30(B)(2), 107, **108**, 120, **121**
Table 348.22, 124, **125**
Table 352.30, 108, 127
Table 352.30(B), **108**
Table 352.44(A), 108, 127
Table 352.44(B), 108, 127

Table 360.24(A), **124**, 124
Table 360.24(B), **124**, 124
Table 400.4, 400, **400**, **401**, 418, **452**
Table 400.4, Note 5, 400, **401**
Table 400.4, Note 9, **401**
Table 400.4, Note 10, 400, **401**
Table 400.5(A), 449, **452**
Table 430.7(B), 186, 220, 439
Table 430.22(E), **402**, 402
Table 430.24, 402
Table 430.33(E), 401
Table 430.37, **190**, 190, 406, **407**
Table 430.52, 185, **185**, 187, 188, **188**, 217, **218**, 220, 407, **408**
Table 430.72(B), 189
Table 430.72, Column C, 186
Table 430.147, 184, 188
Table 430.148, 38, 151, 184, 188

Table 430.149, 184, 188
Table 430.150, **184**, 184, 187, 188, 220, 439
Table 430.152, **220**, **388**
Table 450.3(A), 169, **170**
Table 450.3(B), 170, 173, **174**, 202, 203
Table 450.3(B), Note 2, 173
Table 470.72(B), **187**
Table 514.3(B)(1), 242
Table 550.31, 326, **326**
Table 551.10(E)(1), **337**
Table 551.73, **206**, **340**
Table 555.12, 204, **204**, **353**
Table 610.14, 387
Table 610.14(A), 386, **386**, 387
Table 610.14(B), 387, **387**
Table 610.14(E), 387, **387**

Table 620.14, **402**
Table 625.29(D), 420
Table 625.29(D)(1), **421**
Table 680.8, 251, **251**
Table 680.10, 252, **252**
Table C2, **110**
Table C3, **110**
Table C4, **110**
Table C5, **110**
Table C6, **110**
Table C7, **110**
Table C8, **110**
Table C9, **110**
Table C10, **110**
Table C11, **110**
Table C12, **110**
Table C12A, Appendix C, **110**
Table 430.24, Exception No. 1, 402

INDEX

Note: Page numbers in **bold** reference figures.

A

AFCI. *See* Arc-fault circuit-interrupt (AFCI)
AHJ. *See* Authority Having Jurisdiction (AHJ)
Air conditioning, receptacles for, 58
Alternating-current
   circuits, 12–18
      formula to determine quantities in, 14–15
      defined, 12
      rules for, 18
      sine wave, **12**
Ammeter, defined, 1
Ampacity, defined, 23
Ampere, defined, 1
Apparent power, defined, 12
Appliances, modern, **24**
Arc-fault circuit-interrupt (AFCI)
   defined, 309
   mobile home, 325–326
Attached garage
   defined, 45
   receptacles for, 56
Authority Having Jurisdiction (AHJ)
   appliances and, 136
   concrete slabs and, 108–109
   defined, 93, 135
Autoshop, infrared lamp heating appliances, **142**
Autotransformer
   defined, 163
   protection of, 173–175
   voltage connections, **175**

B

Ballasts, electric signs, 363, 365
Baseboard heaters
   electrical, **151**
   receptacle outlets and, **46**
Basements
   receptacle outlets, rough-in inspection procedures, 103–104, 103–104
   receptacles for, 58, **60**
Bathroom
   branch circuits, 32, **33**
   hydromassage, 260–261
   receptacle outlets, rough-in inspection procedures, 100
   receptacles for, 53

Bathtub, hydromassage, defined, 249, 260–261
Battery
   defined, 4
   storage-type
      defined, 267
      emergency system, 269
      non-vented, 415
Benders, **123**
Boat yards
   commercial load calculations, 203–204
   general information on, 351–352
Bonding
   defined, 69, 249, 309
   floating buildings, 349–350
   metal parts, **258**
   mobile home, 323–324
   swimming pools, 256–257
Bonding grid, defined, 249
Bonding jumper
   defined, 69
   fundamentals of, 70–76
   main, 72, 111
      defined, 69
Box fill
   defined, 93
   requirements, **95**
Branch-circuit
   bathroom, 32, **33**
   defined, 23
   electrical signs, 361
   furnace, 31–32
   hallways/stairways/closets, rough-in inspection procedures, 100
   laundry, 31
   lighting, 31
   mobile home, 314
   multiwire, defined, 23
   overcurrent protection, application, 136
      defined, 135
   patient bed
      critical care areas, 296–298
      general care areas, 296
   rating
      Authority Having Jurisdiction (AHJ) and, 136
      defined, 135
   residential electrical, 25–34
   temporary installations, 289

Built-in dishwasher, 137
Bus bar, defined, 1

C

Cable
   nonmetallic sheathed, 105–107
   traveling,
      elevators/dumbwaiters/escalators/moving walks/wheelchair lifts/stairway chair lifts, 405–406
   trays, number of conductors in, 452
Capacitive reactance, 459
Carnivals
   general information on, 282–283
   power sources for, 283–285
   wiring methods, 285
Central electric space-heating, defined, 149
Central heating equipment, 141
Central vacuum outlet assembly, 139
   defined, 135
Charging equipment
   construction, 417–418
   control/protection, 418
   locations, 420
   supply circuit, 420–421
   ventilation, 420
   wiring methods, 416–417
Charging systems, electric vehicle, 416–421
Circuit
   multiwire, 25
   three-wire, for disposal/dishwasher, **39**
Circuit breakers, 4
   defined, 1
   interrupting rating, **221**
   marinas/boat yards, 352–353
   maximum rating of, for transformers over 600 volts, **171**
   voltage rating, 218–219
Circuit components, **5**
Circuit marking, class1/class2/class3 circuits, 452–453
Circuits
   with components, **5**
   finding current flow in, 17
   not permitted to be grounded, **72**
   one-family dwelling, 28–34

467

three-wire
    circuit showing cancelling effects of opposite current flow, **16**
    with parallel loads, **17**
    with parallel loads and one lamp filament, **17**
    showing intentional ground, **16**
Circular raceways, 107
Circuses
    general information on, 282–283
    power sources for, 283–285
    wiring methods, 285
Class 1 circuit, 446–448
    conductors, 451
    defined, 445
Class 2 circuit, 446–448
    defined, 445
Class 3 circuit, 446–448
    defined, 445
Class I locations, 234–236
    defined, 233
Class II locations, 239
    defined, 233
Class III locations, 239
    defined, 233
Closed loop, 2–3
Closets
    dedicated clothes space in, **63, 102**
    fixtures in, **102**
    lighting outlets, rough-in inspection procedures, 100–101
    receptacle outlets, rough-in inspection procedures, 100
Code letter, defined, 183
Cold cathode lighting, 367–368
    defined, 361
Commercial load calculations
    boat yards, 203–204
    farms, 206–208, 206–208
    introduction to, 197
    marinas/boat yards, 203–204
    motels, 209–210, 209–210
    office building, 201–203
    recreational vehicle parks, 205–206
    shore power, 204
    store building, 198–201
Concessions, defined, 281
Concrete slabs, 108–109
    on grade, 109
Conductor, current flow, 4
Conductors
    in ac systems, **74**
    class1/class2/class3 circuits, 450–451
        installation of, 454
    connecting, system to earth, **73**
    elevators/dumbwaiters/escalators/
        moving walks/wheelchair lifts/stairway chair lifts, 400–402
        installation of, 405
    grounding-electrode, 111
    identification, 301
    insulated
        allowable ampacities of, **36, 166, 168, 199**
        dimensions of, **200**
    service-entrance, 111
    service size, residential feeder and service calculations, 37–40
    temperature rating of, **167**
Conduits
    flexible metal, 124–125
        liquidtight, 126
        used as bonding means, 365
    intermediate metal, 120–122
    liquidtight flexible nonmetallic, 128–129
    rigid metal, 118–120
    seals, **236**
        class III locations, **243**
        class I location, 236, 238
        defined, 233
Connected load, defined, 149
Constant-potential transformer, defined, 163
Contact conductors, 386
Continuous-current rating, 219–220
    defined, 215
Continuous loads, defined, 197
Control, of load, 4
Control circuits, 388
Controller, motor, 189–190
    defined, 183
Control system, defined, 397
Control wiring, fire pumps, 440
Converter, defined, 333
Copper grounding-electrode conductor, raceway size, **84**
Cord and plug equipment, swimming pools/fountains, 250–251
Corner grounded 480-volt Delta, **72**
Counter-mounted cooking units, 138
Counter space, receptacles for, 49–52
Cranes, 383–389
    hazardous location requirements, 384–385
    wiring methods, 385–389
Critical branch
    defined, 293
    hospital, 300
    nursing homes, 301
Critical care areas, 296–298
    defined, 293
Current
    full-load, defined, 183
    locked-rotor, 187
        code letters, **188**
        defined, 183
    voltage compared to, **10**
Current flow
    finding in a circuit, 17
    in resistor, 10–12
    through two loads, 15
Current path, **5**
    ground-fault, **75**
Current rating
    continuous, 219–220
    short circuit, 221–223
Current transformer
    defined, 163
    protection of, 175–176
C values, table of, **224**

D

Dead-front, defined, 333
Deck area, electric heat for, 257, **259**
Dedicated space
    clothes closet, **63**
    defined, 45
    receptacles for, 58
Demand factors, 197
    defined, 197
Demand loads, for electric range, **35**
Derating, number of conductors in, 452
Detached garage
    defined, 45
    receptacles for, 56
Dining rooms, receptacle outlets, rough-in inspection procedures, 98–99
Direct-current
    electrical circuit, 3–9
        parallel, 9–12
    parallel circuits, 9–12
    rules for series circuits, 9
Disconnecting means
    cranes/hoists and, 387
    defined, 333
    elevators/dumbwaiters/escalators/moving walks/wheelchair lifts/stairway chair lifts, 406
    location, 111
    mobile home, 312–314
    shore power, 354
Disconnect switch, motor, defined, 183
Dishwasher, **138**
    built-in, 137
    three-wire circuit for, **39**
Disposal, three-wire circuit for, **39**

Dry-niche luminaire (fixure), 254
  defined, 249
Dry-type transformer, defined, 163
Dumbwaiters, 398–409
  conductors, 400–402
    installation of, 405
  disconnecting means, 406
  emergency/standby power systems, 409
  grounding, 408
  installation of conductors, 405
  machine room, 407–408
  overcurrent protection, 406–407
  traveling cables, 405–406
  voltage limitations, 398–400
  wiring methods, 402–405, 402–405
Dusttight, defined, 233
Dwelling
  multi-family, defined, 45
  one-family, defined, 45
  two-family, defined, 45
  unit, 140–141
    defined, 45, 149
    optional load calculations, 150–152, 150–155

E

Edison 3-wire Systems, 14–15
Egress lighting, 284
  defined, 281
Electrical circuit, 3
  alternating-current, 12–18
  direct-current, 3–9
    parallel, 9–12
Electrical cooking top, **34**
Electrical current, defined, 2–3
Electrical equipment, grounding, 72–76
Electrical formulas, 3
Electrical metallic tubing, 122
  defined, 117
  support requirements, **121**
Electrical nonmetallic tubing, 127–128, **276**
  defined, 117
Electrical power supply, mobile home, 310–312
Electrical service, integrated, defined, 267
Electrical shock, effect of, **250**
Electrical systems
  essential, hospital, 299–300
  grounding, 70–72
Electric baseboard heaters, **151**
Electric-discharge lighting, defined, 361
Electricity, defined, 2
Electric power equipment trailers,
  defined, 281
Electric power production sources, interconnected, emergency system, 273–276
Electric range, 32, 34
  demand loads for, **35**
Electric service utility connection, fire pumps, 434–435
Electric signs, 361–369
  branch-circuit, 362
  outline lettering systems, 363
  portable/mobile, 363
Electric space-heating equipment, defined, 149
Electric vehicle
  charging systems, 416–421
  connector, defined, 415
  coupler, defined, 415
  inlet, defined, 415
  interlock, defined, 415
  supply equipment, defined, 415
Electrode connection, defined, 361
Electrode receptacle, defined, 361
Electromotive force, ammeter, 1
Elevators, 398–409
  conductors, 400–402
    installation of, 405
  disconnecting means, 406
  emergency/standby power systems, 409
  grounding, 408
  installation of conductors, 405
  machine room, 407–408
  overcurrent protection, 406–407
  traveling cables, 405–406
  voltage limitations, 398–400
  wiring methods, 402–405
Emergency lighting control, key-operated, **272**
Emergency power systems
  elevators/dumbwaiters/escalators/moving walks/wheelchair lifts/stairway chair lifts, 409
  source of, 3
Emergency systems
  emergency wiring, 272
  generator sets and, 269–270
  hospital, 299–300
  nursing homes, 300–302, 300–302
  separate service, 270–271
  storage battery, 269
  uninterruptible power supply (UPS), 270
  unit equipment, 271–272
Emergency wiring, emergency system, 272
EMT. See Electrical metallic tubing
Engine driven fire pump motors, defined, 433
Equipment-grounding conductor, 72, **75**
Equipment location, fire pumps, 440
Equipotential grounding point, health care facilities, **298**
Escalators, 398–409
  conductors, 400–402
  disconnecting means, 406
  emergency/standby power systems, 409
  grounding, 408
  installation of conductors, 405
  machine room, 407–408
  overcurrent protection, 406–407
  traveling cables, 405–406
  voltage limitations, 398–400
  wiring methods, 402–405
Essential electrical system, defined, 293

F

F values, conversion to M values, **225**
Fairs
  general information on, 282–283
  power sources for, 283–285
  wiring methods, 285
Farms, commercial load calculations, 206–208
Fault-current, calculations, 223–224
Feeder
  conductors, marinas/boat yards, 353
  defined, 23, 149
  floating buildings, 347–348
  optional load calculations, dwelling unit, 150–152
Ferrous, defined, 117
Field-installed skeleton tubing, 365–367
Fire pumps, 433–439
Fire/smoke dampers, defined, 425
Fire triangle, **234**
  defined, 233
Fittings, sealing, **236**, **237**
Fixed-type partitions
  defined, 375
  office furnishings, 377
Fixtures, not permitted in closets, **102**
Flash point, defined, 233
Flat ground joint, **235**
  defined, 233
Flatirons, 141
Flexible metal conduit, 124–125
  defined, 117
  liquidtight, 126
  used as bonding means, 365

Flexible metallic tubing, 122–124
  defined, 117
Floating buildings
  defined, 347
  general information on, 347–350
Forming shell, defined, 249
Formulas, electrical, 3
Fountains
  cord and plug equipment, 250–251
  general information on, 250
  grounding, 250
  overhead conductor clearances, 251
  potential hazards, 250
  underground wiring location, 251
Frame, defined, 333
Freestanding-type partitions, 377
  cord- and plug-connection, 377–378
  defined, 375
Fuel cells, 275–276
Full-load current, defined, 183
Furnace branch-circuit, 31–32
Fuses, 4
  defined, 1
  dual-element, time current characteristic chart of, **222**
  maximum rating of, for transformers over 600 volts, **171**
  selective coordination with, **225**
  voltage rating, 218

G

Garage door opener, receptacles for, **57**
Garages, receptacles for, 56, 58, **60**
Gas furnace, **142**
Gasoline dispensing pumps, hazardous (classified) locations, **244**
Gasoline storage, with concrete retaining wall, **237**
General care access, defined, 293
General care areas, 296
General lighting load
  by occupancy, **28**
  one-family dwelling, 28–31
  residential feeder and service calculations, 34–35
Generator sets
  defined, 267
  portable, 267
    optional standby, **274**
  storage-type, 269–270
GFCI. *See* Ground-fault circuit-interrupter
Ground, defined, 69
Grounded
  conductor, defined, 69
  defined, 16, 69
Ground-fault, 4

current path, **75**
defined, 1
Ground-fault circuit-interrupter (GFCI), **30**, 85
  defined, 23, 69, 249, 309
  grounding, 85
  mobile home, 314–321
  temporary installations, 287
Grounding
  conductor
    continuity assurance, temporary installations, 287
    equipment, defined, 69
  cranes/hoists and, 388–389
  electrical equipment, 72–76
  electrical systems, 70–72
  electrode conductor, 111
    copper raceway size, **84**
    defined, 69
    size according to service-entrance conductor size, **84**
  elevators/dumbwaiters/escalators/moving walks/wheelchair lifts/stairway chair lifts, 408
  floating buildings, 348–349
  fundamentals of, 70–76
  ground-fault circuit-interrupter, 85
  information technology rooms, 428
  inspection procedures, 109–112
  mobile home, 323–324
  multifamily dwelling, 83–85
  one-family dwelling, 76–80
  at panelboard, **324**
  patient equipment, critical care areas, 297, **298**
  swimming pools, 250, 256
  transformer, 179
  two-family dwelling, 80–83
  type transformer, protection of, 176
GTO cable, defined, 361

H

Habitable room, defined, 45
Hallways
  lighting outlets, rough-in inspection procedures, 100–101
  receptacle outlets, rough-in inspection procedures, 100
  receptacles for, 58
Hand benders, **123**
Hard-wired, defined, 135
Hazard (classified) locations
  cranes/hoists and, 384–385
  defined, 233
  general information on, 234
  locations
    class I, 234–236

    class II, 236
    class III locations, 239–240
    class II locations, 239
    class I location, 234–236
  luminaires for, **243**
  near gasoline dispensing pumps, **244**
Health care facilities, 293–302
  electrical systems, essential, 299–302
  general care areas, 296
  wiring/protection of, 294–295
Heater nameplates, information on, **142**
Heating
  central equipment, 141
    gas furnace, **142**
  receptacles for, 58
  surface elements, 138–139
Heating/ventilating/air conditioning (HVAC), defined, 425
Hoists, 383–389
  hazardous location requirements, 384–385
  wiring methods, 385–389
Hoistway, defined, 397
Hospital
  life safety branch, emergency system, 299–300
  transfer switches, emergency system, 299
  wiring, emergency system, 299
Hospital grade, 296
Hot tub, 259–260
  defined, 249
HVAC *See* Heating/ventilating/air conditioning
Hydraulic bender, power jack, **123**
Hydromassage bathtub, 260–261
  defined, 249

I

IMC. *See* Intermediate metal conduit
Impedance, 459
Individual branch-circuit, appliance, defined, 135
Indoor pool area, 253
Inductive action, **14**
Inductive reactance, 12, 459
  defined, 1
Information technology rooms, 425–428
Infrared lamp heating appliances, 140–141
  autoshop, **142**
  defined, 135
Insulated conductors
  allowable ampacities of, **36**
  rated 0 through 2000 volts, **166**, **168**, 199

dimensions of, **200**
Integrated electrical service, defined, 267
Intensity of current, defined, 1
Interconnected electric power production sources, 268
Interconnecting cables, defined, 425
Intermediate metal conduit, 120–122, **200**
  defined, 117
  support requirements, **121**
Interrupting rating, 220
  defined, 215
Intrinsically safe system, 241–242
  defined, 233
Island counter space, receptacles for, 50, **51**, **52**
Isolated power systems
  defined, 293
  nursing homes, 301
Isolation transformer
  defined, 163
  protection of, 176

J

Jockey pump, defined, 433

K

Kitchen
  receptacles
    layout, **98**
    outlets, rough-in inspection procedures, 98–99
    waste disposers, 136, 136–137
    defined, 135

L

Labyrinth joint, **235**
  defined, 233
Lampholders, **6**
Lamps, **6**
Laundry
  branch-circuit, 31
  receptacle, **48**, 54, **55**
    outlets, rough-in inspection procedures, 100
Leg, defined, 309
Legally required standby system, 268
  defined, 309
LFMC. *See* Liquidtight Flexible Metal Conduit
Life safety branch
  emergency system, hospital, 299–300
  nursing homes, 300–301
Lighting, stairway, switch controlled, **61**
Lighting accessories

defined, 375
  office furnishings, 376–377
Lighting branch-circuits, required, 31
Lighting control, emergency, key-operated, **272**
Lighting fixtures. *See* Luminaires (lighting fixtures)
Lighting load
  by occupancy, **28**
  one-family dwelling, 28–31
Lighting outlets
  3-way/4-way switch controlled, **62**
  defined, 45
  hallways/stairways/closets, rough-in inspection procedures, 100–101
  required, kitchens/dining rooms/bathroom, **60**
  residential, required, 59–63
  rough-in inspection procedures, 96–98
    basements/garages, 104
    kitchens/dining rooms, 99
    outdoors, 104
Lights, in a series, **7**
Line isolation monitor, 301–302
Liquidtight flexible metal conduit, 126
  defined, 117
Liquidtight flexible nonmetallic conduit, 128–129
Live parts
  appliances and, 136
  defined, 135
Load calculations
  marinas/boat yards, 353
  mobile home, 324–325
  optional. *See* Optional load calculations
Loads, 4, **5**
  continuous, defined, 197
Locked-rotor current, 187
  code letters, **188**
  defined, 183
Low-voltage
  defined, 333
  recreational parks/vehicles/trailers and, 334–337
Luminaires (lighting fixtures)
  flexible connection at, **236**
  swimming pools, permanently installed, 252–254
  underwater, 254, **255**

M

Machine room, elevators/dumbwaiters/escalators/moving walks/wheelchair lifts/stairway chair lifts, 407–408

Magnetic field, defined, 1
Main bonding jumper, 72, 111
  defined, 69
Manufactured home, defined, 309
Marinas
  commercial load calculations, 203–204
  general information on, 351–352
Marine power outlets, marinas/boat yards, 352–353
Markings, electrical signs, 361
Mast weatherhead, defined, 309
Mechanical contiguous, defined, 375
Metal boxes, **95**
Metallic, defined, 117
Metallic tubing
  electrical, 122
  flexible, 122–124
Metal raceways, 107, 118–126
Midpoint return, defined, 361
Mobile home, 310–326
  arc-fault circuit-interrupt (AFCI), 325–326
  branch-circuit, 314
  defined, 309
  disconnecting means, 312–314
  electrical power supply, 310–312
  ground-fault circuit-interrupter (GFCI), 314–321
  grounding/bonding, 323–324
  load calculations, 324–325
  panelboard, **313**
  power-supply cord, **311**
  receptacle outlets, 314
  service equipment, **327**
  wiring methods, 322–323
Mobile home park, power distribution, 326
Mobile signs, 363
Moored, defined, 347
Motels, commercial load calculations, 209–210
Motion controller, defined, 397
Motor
  alternating-current, full-load current, **184**
  branch-circuit
    determining size of connector, 185
    maximum rating of, **218**
    overload protection, **188**
  branch-circuit overload protection
    determining, 185–186
    determining requirements for, 186–189
  controller, 189–190
  defined, 183
  disconnect, defined, 183

472 INDEX

disconnect switch, determining rating of, 185
installation
    one-line diagram, **188**
    six steps to, 184–189
    protection of, 190
Moving walks, 398–409
    conductors, 400–402
        installation of, 405
    disconnecting means, 406
    emergency/standby power systems, 409
    grounding, 408
    installation of conductors, 405
    machine room, 407–408
    overcurrent protection, 406–407
    traveling cables, 405–406
    voltage limitations, 398–400
    wiring methods, 402–405
Multifamily dwelling
    defined, 45
    grounding, 83–85
    optional load calculations, 153–155
Multiwire circuit, 25
M values, F values converted to, **225**

N

Negative, defined, 1
Neon tubing, 367
    defined, 361
Neutral feeder
    conductor, residential feeder and service calculations, computations, 35–40
    office building load, 202
Nonferrous, defined, 117
No-niche luminary (fixture), 254
    defined, 249
Nonmetallic, defined, 117
Nonmetallic conduit, spacing from bonding conductor, 365
Nonmetallic raceways, 107–108, 126–128
    rigid metal, 126–127
Nonmetallic sheathed cable, 105–107
Nonmetallic tubing
    electrical, 127–128
    defined, 117
Non-vented storage battery, defined, 415
Not all-electric restaurant, 158
Nursing homes, emergency systems, 300–302

O

Office building, commercial load calculations, 201–203

Office furnishings, 375–378
Ohm, defined, 1
Ohm's law, 10–12
    circle, **9**
    defined, 6
    with resistance in series, **9**
    solving equations, 8–9
Oil-insulated transformer
    defined, 163
    protection of, 173
One-family dwelling
    circuits, 28–34
    defined, 45
On-site power
    production facility, 435
        defined, 433
    standby generator, defined, 433
Open processes, electrical area classification for, **245**
Open spray areas, electrical area classification for, **245**
Operating device, defined, 397
Operation controller, defined, 397
Optional load calculations
    dwelling unit, 150–152
        existing, 152–153
    general, 149–150
    multifamily dwelling, 153–155
    restaurants, new, 157–158
    schools, 155–157
    two-family dwelling, 155
Optional standby system, 268
    emergency system, 273
Outdoors
    receptacle outlets, 54, **55**
    rough-in inspection procedures
        lighting outlets for, 104
        receptacle outlets for, 104
Outlets
    box, defined, 23
    defined, 45
    lighting, residential. *See* Lighting outlets
    receptacle, 98–99
        defined, 45
    outlet, 46–58
Outline lettering systems, electric signs, 363
Oven, wall-mounted, **34**, 138
Overcurrent protection devices, 4, **5**
    defined, 1, 197
    elevators/dumbwaiters/escalators/ moving walks/wheelchair lifts/stairway chair lifts, 406–407
    general information on, 215–216
    locations, class1/class2/class3

    circuits, 449
    protection by, 170–172
    purpose of, 216–218
    selection of, 218–226
Overhead clearances, temporary installations, 285, **288**
Overhead conductor clearances, swimming pools/fountains, 251
Overload protection
    cranes/hoists and, 387–388
    maximum rating of, **187**
    relay, defined, 183

P

Panelboard
    defined, 309
    grounding at, **324**
    marinas/boat yards, 352–353
    mobile home, **313**
Parallel circuits
    defined, 9
    direct-current, 9–12
    string of lamps in, **10**
    with two loads, **10**
Parallel loads, three-wire circuit with, **17**
Park trailer, defined, 333
Partitions
    fixed-type, 377
        defined, 375
    freestanding, 375, 377
        cord- and plug-connection, 377–378
    interconnections of, 375–376
Path, current flow, 4, **5**
Patient bed, 296
    branch-circuit
        critical care areas, 296–298
        general care areas, 296
    receptacles for,
        critical care areas, 296–298
        general care areas, 296
Patient care area, defined, 293
Patient equipment, grounding point, critical care areas, 297
Peninsular counter space, receptacles for, 50–52
Personal protection system, 418
    defined, 415
Phase angle, 459
Photovoltaic systems, solar, 275, **276**
Pipe heating cable, defined, 309
Point of attachment, service/grounding inspection procedures, 109
Poly-vinyl chloride, 126
    defined, 117
Pool covers, 257

Portable distribution, defined, 281
Portable generator equipment
    defined, 267
    optional standby, **274**
Portable signs, 363
Portable wiring/equipment, defined, 281
Positive, defined, 1
Potential difference, 26–27
Power
    continuity of, 435–438
    defined, 1
Power distribution, mobile home park, 326
Power factor, 459
    defined, 1, 12
    loss in an alternating-current, **13**
Power jack, hydraulic bender, **123**
Power sources, **5**
    class1/class2/class3 circuits, 452–453
    direct-current, 3
    fire pumps, 434–435
    hospital, 300
    nursing homes, 301
    recreational parks/vehicles/trailers and, 337–339
Power supply
    assembly, defined, 333
    cord, defined, 309
    electric signs, 363, 365
Power wiring, fire pumps, 439
Primary winding, defined, 163
Prime mover, defined, 267
Protection, health care facilities, 294–295

R

Raceways
    circular, 107
    fill, defined, 93
    metal, 107, 118–126
    nonmetallic, 107–108, 126–128
        rigid, 126–127
    number of conductors in, 452
Raised floors
    defined, 425
    information technology rooms, 426–428
Reamer, defined, 93
Rebars, defined, 93
Receptacle, temporary installations, 289
Receptacle outlets
    mobile home, 314
    rough-in inspection procedures, 96
        basements/garages, 103–104
        kitchens/dining rooms, 98–99

Receptacles
    for air conditioning, 58
    arrangement of, **47**
    for basements, 58, **60**
    for bathroom, 53
    counter space, 49–52
        peninsular, 50–52
    for dedicated space, 58
    defined, 23, 45
    floor, **48**
    for garage door opener, 57
    for garages, 56–58, **60**
    hallway, 58
    for heating, 58
    laundry, **48**
    for laundry, 54, **55**
    outdoor, 54, **55**
    outlet
        defined, 45
        residential, 46–58
    for patient bed, critical care areas, 296–298
    patient bed, general care areas, 296
    small-appliance, 49
    types, **47**
Recreational vehicle
    defined, 333
    general information on, 334–341
    low-voltage system, 334–337
    nominal 120-volt or 120/240 volt systems, 337–339
Recreational vehicle parks, 339–340
    commercial load calculations, 205–206
    general information on, 334–341
    low-voltage system, 334–337
    nominal 120-volt or 120/240 volt systems, 337–339
Recreational vehicle trailers
    general information on, 334–341
    low-voltage system, 334–337
    nominal 120-volt or 120/240 volt systems, 337–339
Refrigerator, receptacle, **30**
Relay, overload protection, defined, 183
Required, standby systems, emergency system, 273
Residential
    electrical, branch-circuit, 25–34
    lighting outlets, 59–63
        required, 59–63
    receptacle outlet, 46–58
    rough-in inspection procedures, 94, 96
Residential feeder and service calculations

    general lighting, 34–35
    neutral feeder conductor computation, 35–40
Resistance, defined, 1
Resistance electric heating elements, defined, 135
Resistors, **7**
    conductors, secondary, 387
    current flow in, 10–12
Restaurants
    optional load calculations, new, 157–158
    service, **158**
Rides
    clearance to, 285–288
    defined, 281
    disconnecting means, **286**
Rigid metal conduit, 118–120
    defined, 117
    support requirements, **121**
Rigid nonmetallic conduit, defined, 117
RMC. *See* Rigid metal conduit
Rough-in
    defined, 93
    inspection procedures, 94–109
Runs, minimum bends in one, **107**
Runway, defined, 397

S

Schools, optional load calculations, 155–157
Sealing fittings, **236**, **237**
Seals, conduit, **236**
Secondary resistor conductors, 387
Secondary winding, defined, 163
Selective coordination, 225
    with fuses, **225**
Separate service
    defined, 267
    emergency system, 270–271
Series, parallel circuit, **11**
Series-rated system, 224–225
Service
    conductor size, residential feeder and service calculations, 37–40
    defined, 23, 149
    disconnect means, connection ahead of, **273**
    floating buildings, 347–348
    inspection procedures, 109–112
    optional load calculations, dwelling unit, 150–152
Service-drops, clearances of, 111
Service-entrance
    conductors, 111
        demand factors for site feeders, **206**

Service factor, defined, 183
Service feeder load, defined, 197
Service load, defined, 197
Shades, lamp, **6**
Shock, electrical, effect of, **250**
Shore power, 204
   disconnecting means, 354
Short-circuit, 4
   current rating, 221–223
      defined, 215
   defined, 1
Signal equipment, defined, 397
Sign body, defined, 361
Signs, electrical. *See* Electric signs
Skeleton tubing
   defined, 361
   field-installed, 365–367
Slabs
   concrete, 108–109
   concrete slabs and, on grade, 109
Small-appliance
   load, 31
   receptacles for, 49
Solar photovoltaic systems, 275, **276**
Spas, 259–260
   defined, 249
Speed of response, 220–221
   defined, 215
Splices, temporary installations, 289
Stairway
   lighting outlets, rough-in inspection procedures, 100–101
   lighting switch-controlled, **61**
   receptacle outlets, rough-in inspection procedures, 100
Stairway chair lifts, 398–409
   conductors, 400–402
   disconnecting means, 406
   emergency/standby power systems, 409
   grounding, 408
   installation of conductors, 405
   machine room, 407–408
   overcurrent protection, 406–407
   traveling cables, 405–406
   voltage limitations, 398–400
   wiring methods, 402–405
Standby power systems, elevators/dumbwaiters/escalators/moving walks/wheelchair lifts/stairway chair lifts, 409
Standby systems, 268
   required, emergency system, 273
Storage battery
   defined, 267
   emergency system, 269
Storage-type water heaters, 139–140

Store building, commercial load calculations, 198–201
Stranded conductors, defined, 249
Supervised connection, defined, 433
Supply circuit
   charging equipment, 420–421
   information technology rooms, 426
Surface heating elements, 138–139
Swimming pools, 459–460
   bonding, 256–261
   cord and plug equipment, 250–251
   general information on, 250
   grounding, 250, 256–261
   indoor area, 253
   overhead conductor clearances, 251
   permanently installed
      area receptacles/equipment, 252
      luminaires (lighting fixtures), 252–254
      motors, 252
   potential hazards, 250
   storable, 258–259
   pump wiring, **260**
   switching devices, 253
   underground wiring location, 251–252
Switches
   marinas/boat yards, 352–353
   motor disconnect, defined, 183
   transfer
      hospital, 299
      nursing homes, 300
Switching devices, swimming pools, 253

T

Temporary installations, 288–289
   general information on, 282–283
   power sources for, 283–285
   wiring methods, 285
Tents, defined, 281
Termination boxes, defined, 281
Three-wire circuit
   for disposal/dishwasher, **39**
   with parallel loads, **17**
   and one lamp filament, **17**
   showing cancelling effects of opposite current flow, **16**
   showing intentional ground, **16**
Time current characteristic chart, dual-element fuse, **222**
Toasters, 136
Transfer equipment, defined, 267
Transfer switches
   emergency system, hospital, 299
   nursing homes, 300
Transformer
   2-wire, single phase, **171**

   3-wire, single phase, **171**
   600 volts or less, 169
   action, **14**
      defined, 2
   basics of, 459–460
   connections, **165**
   constant-potential, defined, 163
   construction of, **177**
   current
      defined, 163
      protection of, 175–176
   delta-delta, **171**
   determining
      primary secondary overcurrent protection, 178
      secondary overcurrent protection, 178–179
   dry-type, defined, 163
   electric signs, 363, 365
   fire pumps, 438–439
   general information, 164–170
   grounding, 179
   grounding-type, protection of, 176
   installation, five steps to, 177–179
   isolation
      defined, 163, 249
      protection of, 176
   oil-insulated, 163
      protection of, 173
   over 600 volts, 169
   overcurrent protection device for, 170–172
   protection of, 172–179
   secondary, 15
      grounded, **71**
   secondary of, 15
   swimming pools, permanently installed, 253–254
   voltage
      defined, 163
      protection of, 175
   zigzag wiring of, **177**
Trash compactor, 137
   cord- and plug- type, 136
   defined, 135
Traveling cables
   elevators/dumbwaiters/escalators/moving walks/wheelchair lifts/stairway chair lifts, 405–406
Tubing
   metallic
      electrical, 122
      flexible, 122–124
   nonmetallic, electrical, 127–128
Two-family dwelling
   defined, 45

grounding, 80–83
optional load calculations, 155

U

Underground wiring location, swimming pools/fountains, 251–252
Under raised floors, information technology rooms, 426–428
Underwater luminaires, 254, **255**
Uninterruptible power supply (UPS)
　defined, 267
　emergency system, 270
　information technology rooms, 428
Unit disconnect means, 136
　defined, 135
Unit equipment
　defined, 267
　emergency system, 271–272
UPS. *See* Uninterruptible power supply (UPS)
Utility meter, location, 111
Utility supplied services, 284
　defined, 281
Utilization device, defined, 23
Utilization equipment, 4
　defined, 23

V

Volt, defined, 2
Voltage, current compared to, **10**
Voltage drop
　calculations, 197
　defined, 197
　fire pumps, 440
Voltage limitations, elevators/dumbwaiters/escalators/moving walks/wheelchair lifts/stairway chair lifts, 398–400
Voltage potential, 26–27
Voltage rating
　circuit breakers, 218–219
　defined, 215
　fuses, 218
Voltage transformer
　defined, 163
　protection of, 175
Voltmeter, defined, 2
Volume allowance, defined, 93

W

Wall counter space, receptacles for, 49
Wall-mounted oven, **34**, 138
Water heaters
　defined, 135
　storage-type, 139–140
Watt, defined, 2
Wellway, defined, 397
Wet-niche luminaires (fixture), 254
　defined, 249
Wheelchair lifts, 398–409
　conductors, 400–402
　　installation of, 405
　disconnecting means, 406
　emergency/standby power systems, 409
　grounding, 408
　installation of conductors, 405
　machine room, 407–408
　overcurrent protection, 406–407
　traveling cables, 405–406
　voltage limitations, 398–400
　wiring methods, 402–405
Winding
　primary, defined, 163
　secondary, defined, 163
Wiring
　emergency system, hospital, 299
　health care facilities, 294–295
　nursing homes, 300
　zigzag wiring of, transformer, **177**
Wiring methods
　charging equipment, 416–417
　class1/class2/class3 circuits, 450
　　load side, 453–454
　　supply side, 453
　class III locations, 240–241
　class II locations, 239
　class I locations, 235–236
　control, fire pumps, 440
　cranes/hoists and, 385
　elevators/dumbwaiters/escalators/moving walks/wheelchair lifts/stairway chair lifts, 402–405
　marinas/boat yards, 353–354
　mobile home, 322–323
　power sources for, fire pumps, 439
　temporary installations, 285
　underground, swimming/fountains, 251–252
Withstand capabilities, 226
Withstand rating, defined, 215

X

X-ray
　defined, 293
　installation, 302

Z

Zigzag transformer wiring, **177**